Current Topics in Behavioral Neurosciences

Volume 37

Current Topics in Behavioral Neurosciences provides critical and comprehensive discussions of the most significant areas of behavioral neuroscience research, written by leading international authorities. Each volume offers an informative and contemporary account of its subject, making it an unrivalled reference source. Titles in this series are available in both print and electronic formats.

With the development of new methodologies for brain imaging, genetic and genomic analyses, molecular engineering of mutant animals, novel routes for drug delivery, and sophisticated cross-species behavioral assessments, it is now possible to study behavior relevant to psychiatric and neurological diseases and disorders on the physiological level. The *Behavioral Neurosciences* series focuses on "translational medicine" and cutting-edge technologies. Preclinical and clinical trials for the development of new diagnostics and therapeutics as well as prevention efforts are covered whenever possible.

More information about this series at http://www.springer.com/series/7854

Robert E. Clark · Stephen J. Martin
Editors

Behavioral Neuroscience of Learning and Memory

Springer

Editors
Robert E. Clark
University of California San Diego
School of Medicine
La Jolla
USA

and

VA Medical Center
San Diego
USA

Stephen J. Martin
Medical Research Institute
Ninewells Hospital and Medical School
Dundee
UK

ISSN 1866-3370 ISSN 1866-3389 (electronic)
Current Topics in Behavioral Neurosciences
ISBN 978-3-030-08778-4 ISBN 978-3-319-78757-2 (eBook)
https://doi.org/978-3-319-78757-2

Cover illustration: Artistic representation of oscillatory synchrony and timing of neurons in networks by Gyorgy Buzsaki

Printed on acid-free paper

This Springer imprint is published by Springer Nature
The registered company is Springer International Publishing AG
The registered company address is: Gewerbestrasse 11, 6330 Cham, Switzerland

Preface

Fundamentally, learning can be understood as an experience-based process that acquires or modifies behaviors, knowledge structures, preferences or values. Memory is a closely related concept whereby information that has been learned or "encoded," is maintained and available to be utilized.

Memory, in the colloquial sense, is the type of memory that we are consciously aware of having. These types of memories adorn our mental life. They are the psychological accoutrements of a lifetime of experience. They allow us to reconstruct our past, appreciate our present, and to an extent, predict and control our future. Our memories provide us with a sense of self. They are our connection to others and to our environment. Memories are stored representations of the world and of our self and, as such, they greatly influence how we perceive and interpret our inner and external environments. Yet in actuality, these types of memory make up only a fraction of the stored representations acquired and maintained by the brain. The multiple neuroanatomical regions of the brain support an impressive variety of distinct learning and memory processes; many of these processes are detailed within the chapters of this volume.

Learning and memory have been systematically and extensively studied since the pioneering work of Hermann Ebbinghaus (1850–1909) which began in the late 1870s. Presently, an enormous amount of experimental research is directed toward elucidating the neurobiological substrates that support the many forms of learning and memory as well as those cognitive processes that utilize memory. This work includes the evaluation of neural and behavioral plasticity at all levels of analysis from molecular biology to complex forms of human behavior.

The volume *Behavioral Neuroscience of Learning and Memory* represents a broad collection of articles that examine the neurobiological underpinnings of processes that support memory encoding (learning), storage, organization (and reorganization), retrieval, and the utilization of memory for decision-making. The opening chapter provides a brief history of the behavioral neuroscience of learning and memory and a summary of each subsequent chapter.

La Jolla, USA — Robert E. Clark, Ph.D.
Dundee, UK — Stephen J. Martin, Ph.D.

Contents

A History and Overview of the Behavioral Neuroscience of Learning and Memory 1
Robert E. Clark

Current Topics Regarding the Function of the Medial Temporal Lobe Memory System 13
Robert E. Clark

Neural Activity Patterns Underlying Spatial Coding in the Hippocampus 43
Marielena Sosa, Anna K. Gillespie and Loren M. Frank

What Versus Where: Non-spatial Aspects of Memory Representation by the Hippocampus 101
Howard Eichenbaum

The Functional and Structural Neuroanatomy of Systems Consolidation for Autobiographical and Semantic Memory 119
Adam J.O. Dede and Christine N. Smith

Memory Reconsolidation 151
Josue Haubrich and Karim Nader

The Sensory Neocortex and Associative Memory 177
Dominik Aschauer and Simon Rumpel

The Representational Basis of Working Memory 213
Derek Evan Nee and Mark D'Esposito

The Neuroscience of Human Decision-Making Through the Lens of Learning and Memory 231
Lesley K. Fellows

Basal Forebrain Cholinergic System and Memory 253
M.G. Blake and M.M. Boccia

Habit Formation and the Striatum 275
Barbara J. Knowlton and Tara K. Patterson

The Anatomy and Physiology of Eyeblink Classical Conditioning 297
Kaori Takehara-Nishiuchi

Exploration of the Neurobiological Basis for a Three-System, Multiattribute Model of Memory 325
Raymond P. Kesner

A History and Overview of the Behavioral Neuroscience of Learning and Memory

Robert E. Clark

Abstract Here, I provide a basic history of important milestones in the development of theories for how the brain accomplishes the phenomenon of learning and memory. Included are the ideas of Plato, René Descartes, Théodule Ribot, William James, Ivan Pavlov, John Watson, Karl Lashley, and others. The modern era of learning and memory research begins with the description of H.M. by Brenda Milner and the gradual discovery that the brain contains multiple learning and memory systems that are supported by anatomically discrete brain structures. Finally, a brief overview is provided for the chapters that are included in current topics in *Behavioral Neuroscience—Learning and Memory*.

Keywords Plato · Descartes · Ribot · William James · Ivan Pavlov John Watson · Karl Lashley · H.M · Brenda Milner

Contents

1 Multiple Learning and Memory Systems 6
2 General Overview of the Volume 8

> The only impressions that can be made upon [the brain] are... through the sensory nerve-roots... The currents, once in, must find a way out. In getting out they leave their traces in the paths which they take. The only thing they can do, in short, is to deepen old paths or to make new ones; and the whole plasticity of the brain sums itself up in two words when we call it an organ in which currents pouring in from the sense-organs make with extreme facility paths which do not easily disappear. For, of course, a simple habit,

---

R. E. Clark (✉)
Veterans Affairs San Diego Healthcare System, San Diego, CA 92161, USA
e-mail: reclark@ucsd.edu

R. E. Clark
Department of Psychiatry, University of California, La Jolla, San Diego, CA 92093, USA

Curr Topics Behav Neurosci (2018) 37: 1–11
DOI 10.1007/7854_2017_482
Published Online: 05 January 2018

> like every other nervous event—the habit of snuffling, for example, or of putting one's hands into one's pockets, or of biting one's nails—is, mechanically, nothing but a reflex discharge; and its anatomical substratum must be a path in the system.
>
> William James (1890, p. 112).

The great questions of philosophy, the mind–body problem, and the nature of knowledge were the same questions that laid the foundations for modern psychology and then ultimately for the behavioral neuroscience of learning and memory. William James, a major figure in the development of modern psychology, devoted a full third of his seminal treatise *Principles of Psychology* (1890) to the brain and nervous system. Both James and Wilhelm Wundt, who founded experimental psychology, studied medicine and philosophy, and both men considered themselves physiologists. Their objectives were not to reduce psychology to physiology, but rather to apply the scientific methods of physiology to the study of the mind and then the brain and nervous system. Thus, philosophy gave rise to psychology and a branch of psychology gave rise to biological or physiological psychology, which then grew into behavioral neuroscience. The major topics in the behavioral neurosciences are sensory and motor processes, motivation, emotion, cognition and, as this volume describes, learning and memory.

In the epigraph, William James was writing specifically of learned "habits." James considered these habits to be learned forms of complex reflexive behaviors. The passage is interesting because it clearly identifies and addresses several important issues with respect to the neuroscience of learning and memory. Firstly, in order to analyze learning and memory, it must be verifiable that it has indeed occurred. This has traditionally been accomplished by observing some change in behavior that indexes the learning. This requires that the memory has some neural "pathway" by which it can influence an output (i.e., "find a way out") that can be measured, usually as an observable behavior. Secondly, in order for learning to take place, external stimuli capable of inducing the learning and memory must have access to a "pathway" into the nervous system. Finally, some change or modification must take place within the nervous system that serves as the neural substrate of the learning and the resulting memory. To understand this simple scenario is to understand the nature of the physical processes and changes in the nervous system that form the foundation of learning and memory. This fundamental problem has challenged and perplexed initially philosophers, more recently psychologists, and most recently neuroscientists. The philosopher Plato stated in his *Theaetetus* on the nature of knowledge that "there exists in the mind of man a block of wax, which is of different sizes in different men; harder, moister, and having more or less purity in one than another… Let us say that this tablet is a gift of Memory…" (Plato, circa 396 B.C., lines 191a–196c; reviewed in Campbell 1883). Plato believed that in the minds of men there existed some substance that was modifiable or pliant and served as the substrate for memory.

René Descartes (1596–1650) is primarily remembered as a philosopher and mathematician, but also was substantially interested in anatomy and physiology and could be considered a founding father of objective animal psychology. He viewed and explicitly defined animals as a collection of reflexive automatons. He suggested that memory was representations subserved by physical traces in the brain. In the opinion of Descartes, memory was an activity-dependent facilitation of connections within these traces:

> Thus, when the soul wants to remember something… volition makes the gland lean first to one side and then to another, thus driving the spirits towards different regions of the brain until they come upon the one containing traces left by the object we want to remember. These traces consist simply of the fact that the pores of the brain through which the spirits previously made their way, owing to the presence of this object, have thereby become more apt than the others to be opened in the same way when the spirits again flow towards them. And so the spirits enter into these pores more easily when they come upon them, thereby producing in the gland that special movement which represents the same object to the soul, and makes it recognize the object as the one it wishes to remember (Descartes 1649).

James described this attribute as plasticity. "Plasticity, then, in the wide sense of the word, means the possession of a structure weak enough to yield to an influence, but strong enough not to yield all at once" (James 1890, p. 110). In 1825, as part of his theory of phrenology, Franz Joseph Gall proposed that specific memories were located in specific places in the brain (reviewed by Squire 1987). These thoughts were echoed by James when he described this localized region as an area in the brain "more liable to be abraded than neighboring parts", a "locus minoris resistentia."

It has been understood for more than 125 years that memory changes over time. The French psychologist Théodule Armand Ribot was the first to suggest that memories are not immediately acquired in a form that will persist unchanged, but rather are slowly reorganized. Ribot was also the first to provide a general theory of memory based on the observation of pathological conditions. This theory was described in his monograph, *Diseases of Memory*, published first in French in 1881 (Ribot 1881) and translated into English in 1882 (Ribot 1882). Ribot noted "The destructive process is a regression from the new to the old, from the complex to the simple, from the voluntary to the automatic, from the least organized to the best organized" (Ribot 1882, p. 203). The final sentence of his book reads: "Finally, our pathological study has led us to this general conclusion: Memory consists of a process of organization of variable stages between two extreme limits—the new state, [and] the organic registration." Concerning "organic registration," Ribot wrote: "Modifications established for years in the nervous elements until they have become organic-dynamical associations and groups of associations called into activity hundreds and thousands of times—these remain; they have a great power of resisting destructive agencies. In this manner we explain a paradox of the memory: the new perishes and the old endures." (p. 119). He described this phenomenon as "The Law of Regression, or Reversion." This law now is often referred to as Ribot's Law and can be used to describe the phenomenon of temporally graded retrograde amnesia.

In 1904, Ivan Petrovich Pavlov was awarded the Nobel Prize in medicine for research in which he used dogs to study the physiology of digestion. This research on digestion sets the stage for observing the phenomenon of classical conditioning. As early as 1880, Pavlov and his associates noticed that sham feedings, in which food was eaten but failed to reach the stomach (being lost through a surgically implanted esophageal fistula), produced gastric secretions, just like real food (Clark 2004). Pavlov and his associates not only discovered the learning process of classical conditioning, they then went on to identify almost all of the major components of classical conditioning (i.e., conditioned and unconditioned stimuli and responses), developed the primary paradigms of delay, trace, simultaneous and backward conditioning, and identified and described many of the parameters that influence the process of classical conditioning. They also described acquisition, extinction, spontaneous recovery, and disinhibition, as well as higher order conditioning, second signaling systems, conditioned discriminations, and even experimental neurosis.

Pavlov's work on classical conditioning was essentially unknown in the USA until 1906 when his lecture "The scientific investigation of the psychical faculties or processes in the higher animals" was published in the journal *Science* (Pavlov 1906). Pavlov's book *Conditioned Reflexes* was translated into English by a former student, G.V. Anrep, and published in 1927 (Pavlov 1927). This made all of Pavlov's conditioning work available in English for the first time. The availability of 25 years' worth of Pavlov's research, in vivid detail, led to increased interest in the experimental examination of classical conditioning, an interest that has continued to this day (for review see Clark 2004).

John B. Watson championed the use of classical conditioning as a research tool for psychological investigations. During 1915, his student Karl Lashley conducted several exploratory conditioning experiments in Watson's laboratory. Watson's presidential address, delivered in 1915 to the American Psychological Association, was entitled "The place of the conditioned reflex in psychology" (Watson 1916). Watson was highly influential in the rapid incorporation of classical conditioning into American psychology.

Karl Lashley was the most important early figure in the development of the behavioral neuroscience of learning and memory in the USA. He earned his Ph.D. at Johns Hopkins University working with Watson and was greatly influenced by Watson's developing concepts of "behaviorism," which was the purely objective experimental branch of psychology with a theoretical goal of predicting and controlling behavior (Watson 1913). The physical processes that underlie learning and the storage of memory were termed the "engram" by Richard Semon in 1904 (Semon 1904) and given widespread attention by Lashley (1950) who was perhaps the first person to clearly conceptualize the issue in a framework that would lend itself to experimental analysis. Lashley focused on analyzing the brain mechanisms of learning and memory for the better part of the 1920s. During this period, Lashley's theoretical views of learning and memory were influenced by two synergistic paradigms: first, the localization of function in neurology and second, behaviorism in psychology.

At this point in time, localization of function in the cerebrum was the dominant view of brain organization. Consequently, Lashley believed it should be possible to localize the place in the cerebral cortex where learning changed the brain's organization to store the memory (i.e., the engram). Thus, localization of function and behaviorism were ideally suited schools of thought for merging. Lashley sets about to systematically identify these learning locations—the engrams—in a series of studies culminating in his seminal monograph, *Brain Mechanisms of Intelligence* Lashley (1929). In this long series of experiments, he made lesions of varying sizes and locations in the rat cerebral cortex and tested the rats on mazes differing in difficulty. The conclusions of this series of experiments profoundly changed Lashley's view of the organization of learning and memory and had a remarkable impact on the young field—a field that was then termed "physiological psychology," but would now be referred to as "behavioral neuroscience." Lashley concluded that the locus of the lesions was unimportant; the size, however, was the critical variable, especially for the more difficult mazes. These findings led to the theoretical concepts of "mass action" and "equipotentiality." Mass action stated that what was critical to disrupting memory was the amount of cerebral tissue removed (larger lesion = larger impairment) and equipotentiality stated that all areas of the cerebral cortex were equally important (at least insofar as maze learning was concerned). These concepts of mass action and equipotentiality created vigorous debates in the field. An influential alternative perspective argued that rats were using a variety of sensory cues to solve the mazes; as more of the sensory regions of the cortex were destroyed, fewer and fewer cues remained available to guide performance. Lashley and his associates, however, countered by demonstrating that removing the eyes produced less effect on maze learning than removing the visual area of the cortex. Out of this debate came a long series of lesion-behavior experiments in the 1940s analyzing behavioral "functions" of the cerebral cortex. This work for all intents and purposes failed to settle the matter. This led to Lashley's often quoted and particularly pessimistic observation: "This series of experiments has yielded a good bit of information about what and where the memory trace is not. It has discovered nothing directly of the real nature of the memory trace. I sometimes feel, in reviewing the evidence of the localization of the memory trace, that the necessary conclusion is that learning is just not possible. It is difficult to conceive of a mechanism that can satisfy the conditions set for it. Nevertheless, in spite of such evidence against it, learning sometimes does occur" (Lashley 1950, pp. 477–78). Clearer insights into the organization of learning and memory would have to await the dawning of the modern era of the behavioral neuroscience of learning and memory; this began with the description of patient H. M. by Brenda Milner (Scoville and Milner 1957).

H.M. had a history of minor and major seizures that were unresponsive to antiepileptic medications that were reaching toxic levels. In an effort to control the seizures, a decision was made to utilize an experimental surgical procedure. On September 1, 1953, William Scoville removed H.M.'s medial temporal lobes (MTL) bilaterally. The surgery succeeded in that it reduced the frequency and severity of the seizures. However, H.M. was left with profound amnesia that was readily apparent immediately after surgery (Scoville 1954). In the decades that followed, the evaluation of H.M. and others with similar brain damage revealed several fundamental properties concerning the biological organization of learning and memory.

Importantly, this work revealed that memory was indeed dependent upon the cerebrum and was distinct from other abilities and dispositions like personality, motivation, perception, and intelligence. Damage to the MTL only impaired long-term memory and did not impair working memory. That is, a limited amount of information could still be held for a brief period. This allows, for example, a posed question to be remembered long enough to provide an appropriate answer. It was discovered that there were many other forms of memory that were not disrupted by MTL damage such as habits (see below; multiple memory systems). In fact, the MTL only supports one form of memory termed declarative memory. This is memory in the ordinary or colloquial use of the term. It is the type of memory that we are consciously aware of having. Finally, the MTL is not the ultimate repository of long-term memory. Memories learned shortly before MTL damage are lost, whereas more remote memories are spared and independent of the MTL.

It is notable that these observations were met initially with great skepticism. This was primarily due to the early failures to develop an animal model of H.M.'s memory impairments. These attempts began almost immediately when the same surgeon who performed the resection on H.M., William Scoville, made the same lesions in monkeys (e.g., Correll and Scoville 1965). The problem was that it was not yet understood that humans and experimental animals often approach ostensibly similar tasks using different strategies—strategies that are independent of the MTL. This situation was true for work in the monkey and in the rodent. Experimental work during the 1960s made it clear that the observed impairments in rats with, for example, hippocampal lesions, did not adequately model the human condition (Clark and Squire 2010). Thus, researchers were more likely to interpret their findings within the framework of response inhibition as first outlined by Ivan Pavlov (1927) and less likely to relate their work to human studies of memory-impaired patients. Tellingly, in a review of the literature on the hippocampus and behavior, published a decade after the initial description of H.M., Robert Douglas lamented:

> Hippocampal lesions obviously do not impair learning in general, even when the learning involves retention for long periods of time. Thus, the animal and human data would appear to be in contradiction. This contradiction could be 'resolved' by postulating that the hippocampus has a different basic function in man and beast. Such a solution to this problem is generally unacceptable to physiological psychologists, however. Another possible resolution of this paradox is that the recent memory loss in man is a secondary effect of a different type of primary disorder. The author has chosen the latter course, and suggests that the recent memory loss in man is a genuine phenomenon, but that it is a byproduct of interference during storage and not due to a lack of ability to store, per se (Douglas 1967, p. 424).

1 Multiple Learning and Memory Systems

It is now clear that the tasks given to animals with hippocampal lesions were tasks that animals would learn as a skill even if humans tended to learn the task by consciously memorizing the material. It was suggested that establishing an animal model would require developing tasks that assess the type of memory impaired in human amnesia and not just whatever task was convenient to use (Gaffan 1974). In 1978, Mortimer Mishkin (1978) tested monkeys with lesions designed to mimic those of H.M. and tested them on the delayed nonmatching to sample (DNMS) task. This is a one-trial test of recognition memory. The monkeys displayed an impairment profile that was strikingly similar to H.M.'s profile in that working memory appeared to be spared, while long-term memory was robustly impaired. This groundbreaking study, along with others that followed, finally established an animal model for human MTL amnesia in the monkey (e.g., Zola-Morgan and Squire 1993).

It is now understood that the MTL lobe exclusively supports declarative memory. This is the type of memory that we are aware of having such as memory for facts (semantic) and events (episodic). Declarative memory can be contrasted with nondeclarative memory, which is a collection of different memory abilities that depend on different and discrete brain structures. We do not have conscious access to the content of nondeclarative memory. Rather, these forms of memory are expressed through experience-based changes in performance.

Work in humans identified the phenomenon of priming, which is the improved ability to classify, detect, or produce an item because of a recent experience with the same or related item (Tulving and Schacter 1990). The striatum was discovered to be important for gradual, feedback-guided learning that forms the basis of habit memory (Mishkin et al. 1984; Packard et al. 1989). Simple forms of classical conditioning, like delay eyeblink classical conditioning, were found to be dependent on the cerebellum and associated brainstem circuitry (Christian and Thompson 2003; Clark 2011). Classical conditioning of fear responses is critically dependent on the amygdala, which is thought to be the structure that permanently encodes and stores the hedonic value of the aversive stimulus, used to classically condition the fear response (Fanselow and Gale 2003). Finally, phylogenetically early forms of nonassociative learning like habituation (a decrease or cessation of a response to a stimulus after repeated presentations) and sensitization (a learning process in which repeated administration of a stimulus results in the progressive amplification of a response) are also forms of nondeclarative memory. Figure 1 shows the categorical organization of the mammalian long-term memory systems and the brain structures that support these systems. It is important to note that other abilities that are related to learning and memory are the province of the neocortex, such as working memory, consolidation and reconsolidation of memory, memory-guided decision-making, and neocortical associative memory; all of these forms of learning and memory can be modulated by transmitter systems including the cholinergic innervation of the neocortex.

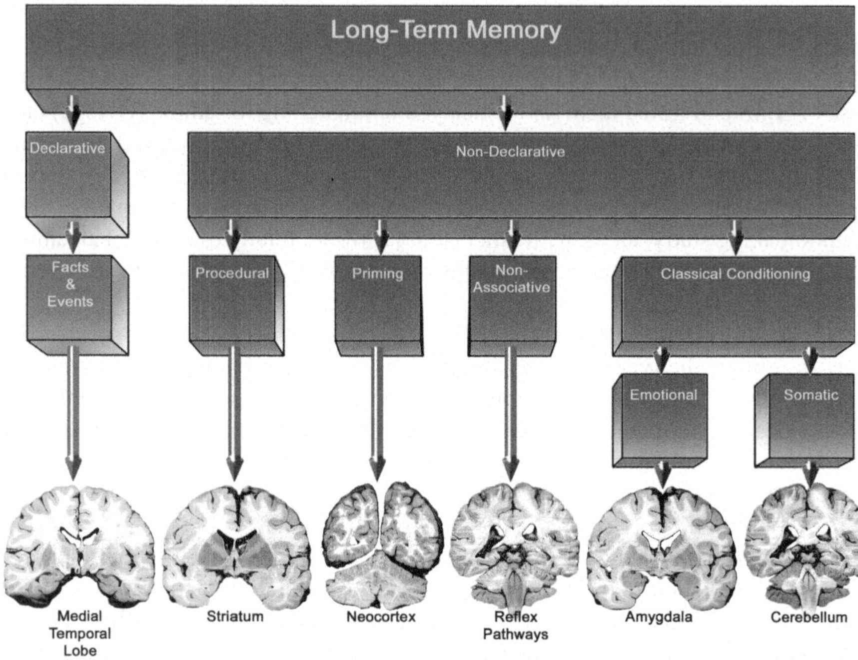

Fig. 1 Memory taxonomy. Long-term memory (purple box) is made up of declarative memory, which includes memory for facts and events (red boxes) and depends on the function of medial temporal lobe (MTL) regions. Nondeclarative memory (blue boxes) is an umbrella term encompassing a collection of learning and memory abilities that are independent of the MTL and are principally supported by anatomically distinct and functionally independent brain regions. The critical brain region for each memory system is highlighted in red in coronal sections through the human brain (adapted from Squire 1992)

2 General Overview of the Volume

In the chapter *Current Topics Regarding the Function of the Medial Temporal Lobe*, Robert Clark describes the discovery of a memory system located in the medial temporal lobe. The current topics include a discussion of whether the perirhinal cortex exclusively supports memory, or if it supports both memory and higher order visual perception. Second, the debate concerning anatomical separation between recollection and familiarity is considered. And finally, a discussion is provided regarding the discord between the human and rodent work on spatial memory. Is spatial memory organized differently between the two species, or does the difference perhaps reflect differences in the working memory capacities of the two species?

Adam Dede and Christine Smith discuss *The Functional and Structural Neuroanatomy of Systems Consolidation for Autobiographical and Semantic Memory*. Systems consolidation theory grew out of the phenomenon whereby remote memories are more resistant to disruption than are more recent memories. This chapter focuses on two types of memory: semantic and episodic. It presents findings from patients with brain damage and neuroimaging studies and evaluates the structural and functional neuroanatomies of past semantic and episodic memory.

Josue Haubrich and Karim Nader describe *Reconsolidation and the Dynamic Nature of Memory*. Reconsolidation is the phenomenon whereby previously consolidated memories are made labile again through reactivation of the memory trace. The evidence for reconsolidation includes all levels of analysis, from molecular and physiological to behavioral, suggesting that it is a fundamental property of memory. Reconsolidation is a topic of active research, and there is growing interest in applying the blockade of reconsolidation and its updating functions as a possible therapeutic tool in several clinical conditions, such as PTSD and drug addiction.

Marielena Sosa, Anna Gillespie, and Loren Frank present the chapter *Neural Activity Patterns Underlying Spatial Coding in the Hippocampus*. Identifying the spatial coding aspects of memory is key to understanding a primary function of the hippocampus. This work is beginning to disambiguate the particular functions of each subfield of the hippocampus including dentate gyrus, CA3, CA2, CA1, and additionally the medial entorhinal cortex. The chapter highlights two classes of activity patterns: local field theta waves and sharp-wave ripples. These two activity patterns both coordinate the activation and reactivation of neuronal ensembles with a precise degree of temporal organization. The goal of this work and other related studies is decoding hippocampal network activity and furthering our understanding of how hippocampal processing relates to memory function.

The late Howard Eichenbaum (1947–2017) descried the Non-spatial Aspects of Memory Representation by the Hippocampus, and advanced the idea that hippocampal network activity goes far beyond simply signaling where an animal is in space. Recently, many experiments using hippocampal lesions and recordings have characterized neural activity in animals and humans, and have revealed a clear and specific role of the hippocampus in the processing of nonspatial information. Evidence is reviewed in support of the idea that the hippocampus organizes the contents of memory in space, in time, and in networks of related memories, with a particular focus on recently discovered hippocampal "time" cells.

In the chapter Exploration of the *Neurobiological Basis for a Three-System, Multi-Attribute Model of Memory*, Raymond Kesner develops his attribute theory of memory systems. Here, memory systems are defined in terms of the type of information that needs to be represented, along with the processes required for the operation of each system. Further, the neurobiological substrates and structural mechanisms of each system and subsystem are considered.

The Representational Basis of Working Memory, by Derek Nee and Mark D'Esposito, describes working memory, where online maintenance allows the manipulation of information even in the absence of external inputs. This is a critical feature for higher-level cognition. The case is made that the relative contributions of sensory cortices and the prefrontal cortex are best understood in terms of the particular processing demands of the memory. If the demands are minimal, then the sensory cortices are sufficient for working memory maintenance. However, when computational demands are increased by requiring representational abstraction or transformational processing, the prefrontal cortex becomes critical.

The chapter *Basal Forebrain Cholinergic System and Memory*, by Mariano Blake and Mariano Boccia, describes how cholinergic neurons modulate cortical arousal and activation. The chapter focuses on several neurotransmitter systems, but particularly the cholinergic system and its influence on basal forebrain neurons. Work is described concerning how these systems are involved in modulating the higher-level cognitive processes important for learning and memory.

The Sensory Neocortex and Associative Memory, by Dominik Aschauer and Simon Rumpel, reviews the latest evidence that some forms of associative memory are critically mediated by sensory cortex. This view can be contrasted with the more traditional perspective that the sensory cortex is simply involved in extracting particular features of the environment. The argument is made that for understanding more complex discriminations, analyzing the activity of single neurons is inadequate. Instead, it appears to be essential to understand the combined activity of larger populations of neurons functioning in cell assemblies.

Lesley Fellows describes *The Neuroscience of Human Decision-Making through the Lens of Learning and Memory*. Decision-making has only recently become a research topic in cognitive neuroscience, and this work is providing a new perspective on the functioning of the orbitofrontal cortex and ventral striatum. It now becoming clearer that decision-making is critically linked with learning and memory systems and that a better way to understand decision-making is by viewing it as a link between memory of the past and the planning of future actions.

The chapter *Habit Formation and the Striatum* by Barbara Knowlton describes work in experimental animals and humans, which indicates that the striatum is crucial for the development of habitual behavior. Habits represent a type of learning that occurs gradually, is relatively inflexible, and independent of the medial temporal lobe structures that underlie declarative memory. Future directions of work in this area are considered, and they include the developmental and long-lasting effects of chronic stress. The chapter also evaluates the potential mechanisms that facilitate the shift in control from a goal directed behavioral system to a habit system.

In the final chapter, Kaori Takehara-Nishiuchi reviews *The Anatomy and Physiology of Eyeblink Classical Conditioning*. The anatomy of this form of learning, at least for delay conditioning, is the most fully described learning circuit in the brain. All of the input and output pathways have been identified as well the brain areas that store the critical physical changes that represent the memory (i.e., the conditioned response). The neurobiology of trace eyeblink conditioning is also considered. Acquisition in trace eyeblink conditioning depends on several forebrain regions, including the medial prefrontal cortex and hippocampus, as well as the cerebellar-brainstem circuit. Details of the computations taking place in these regions remain elusive; nonetheless, recent evidence suggests a view that the forebrain encodes a temporal sequence of the CS, trace interval, and US in a specific environmental context and signals the cerebellar-brainstem circuit to execute the CR when the US occurs.

References

Campbell L (1883) The theaetetus of plato. Oxford University Press
Clark RE (2004) The classical origins of pavlov's conditioning. Integr Physiol Behav Sci 39(4): 279–294
Clark RE (2011) Eyeblink Conditioning and Systems Consolidation: An Ironic Yet Powerful Pairing. Learn Mem 95:118–124
Clark RE, Squire LR (2010) An animal model of recognition memory and medial temporal lobe amnesia: History and current issues. Neuropsychologia 48:2234–2244
Christian KM, Thompson RF (2003) Neural substrates of eyeblink conditioning: acquisition and retention. Learn Mem 11:427–455
Correll RE, Scoville WB (1965) Effects of medial temporal lesions on visual discrimination performance. J comp physiol Psychol. 60:175–181
Descartes, R. (1649). Les passions de l'âme (trans: Stoothoff R). In: The philosophical writings of Descartes, Vol. I. Cambridge University Press, Cambridge (1985)
Douglas RJ (1967) The hippocampus and behavior. Psychol Bull 67(6):416–422
Gaffan D (1974) Recognition impaired and association intact in the memory of monkeys after transaction of the fornix. J Comp Physiol Psychol 88(6):1100–1109
Fanselow MS, Gale GD (2003) The amygdala, fear, and memory. Ann NY Acad Sci 985:125–134
James W (1890) The principles of psychology. Holt, New York
Lashley KS (1929) Brain mechanisms and intelligence. University of Chicago Press, Chicago
Lashley K (1950) In search of the engram. Soc Exp Biol Symp 4:454–482
Mishkin M (1978) Memory in monkeys severely impaired by combined but not separate removal of the amygdala and hippocampus. Nature 273:297–298
Mishkin M, Malamut B, Bachevalier J (1984) Memories and habits: two neural systems. In: Lynch G, McGaugh JL, Weinberger NM (eds) Neurobiology of human learning and memory. Guilford, New York, pp 65–77
Packard MG, Hirsh R, White NM (1989) Differential effects of fornix and caudate nucleus lesions on two radial maze tasks: Evidence for multiple memory systems. J Neurosci 9:1465–1472
Pavlov IP (1906) The scientific investigation of the psychical faculties or processes in the higher animals. Science 24:613–619
Pavlov IP (1927) Conditioned reflexes (trans: Anrep GV). Oxford University Press, London
Plato (circa 396 B.C.). Theaetetus
Ribot T (1881) Les maladies de la mémoire. Baillière, Paris
Ribot T (1882) Diseases of the Memory: an Essay in the Positive Psychology. D. Appleton and Company, New York
Semon R (1904) die Mneme als erhaltendes Prinzip im Wechsel des organischen Geschehens. Wilhelm Engelmann, Leipzig
Scoville WB (1954) The limbic lobe in man. J Neurosurg 11:64–66
Scoville WB, Milner B (1957) Loss of recent memory after bilateral hippocampal lesions. J Neurology, Neurosurg Psychol 20:11–21
Squire LR (1987) Memory and the brain. Oxford University Press, New York
Squire LR (1992) Memory and the hippocampus, A synthesis from findings with rats, monkeys, and humans. Psychol Rev 99:195–231
Tulving E, Schacter DL (1990) Priming and human memory systems. Science 247:301–306
Watson JB (1913) Psychology as the behaviorist views it. Psychol. Rev 20:158–177
Watson JB (1916) The place of the conditioned-reflex in psychology. Psychological Review 23:89–116
Zola-Morgan S, Squire LR (1993) Neuroanatomy of memory. Annual Rev Neurosci 16:547–563

Current Topics Regarding the Function of the Medial Temporal Lobe Memory System

Robert E. Clark

Abstract The first clear insight that the medial temporal lobe of the human brain was in fact a system of anatomically connected structures that were organized into a memory system came in 1957 from the observations by Brenda Milner of the noted amnesic patient H.M. Subsequent work in humans, monkeys, and rodents has identified all of the components of the medial temporal lobe (MTL) that formed the memory system. Currently, work is ongoing to identify the specific contributions each structure in the medial temporal lobe makes towards the formation and storage of long-term declarative memory. The historical background of this work is described including what insights the study of noted neurologic patients H.M. and E.P. provided for understanding the function of the medial temporal lobe. The development of an animal model of medial temporal lobe function is described. Additionally, the insights that lead to the understanding that the brain contains multiple, anatomically discrete, memory systems are described. Finally, three current topics of debate are addressed: First, does the perirhinal cortex exclusively support memory, or does it support both memory and higher order visual perception? Second, is there an anatomical separation between recollection and familiarity? Third, is the organization of spatial memory different between humans and rats, or perhaps the difference is between the working memory capacities of the two species?

Keywords Working memory · Perirhinal · Recollection · Familiarity
Spatial memory · H.M. · E.P.

R. E. Clark (✉)
Veterans Affairs San Diego Healthcare System, San Diego, CA 92161, USA
e-mail: reclark@ucsd.edu

R. E. Clark
Department of Psychiatry, University of California, San Diego, La Jolla, CA 92093, USA

Curr Topics Behav Neurosci (2018) 37: 13–42
DOI 10.1007/7854_2017_481
Published Online: 05 January 2018

Contents

1 Introduction 14
2 Background 14
2.1 Neurologic Patient H.M 14
2.2 Neurologic Patient E.P 16
2.3 Insights About the Organization of Memory from H.M. and E.P 16
2.4 Animal Model of Medial Temporal Lobe (MTL) Amnesia 19
2.5 Multiple Memory Systems 21
2.6 The Organization of the Medial Temporal Lobe Memory System 23
3 Current Topics 27
3.1 Does Perirhinal Cortex Exclusively Support Memory, or Memory and Higher Order Visual Perception? 27
3.2 Is There an Anatomical Separation Between Recollection and Familiarity? 29
3.3 Is the Organization of Spatial Memory Different Between Humans and Rats, or Is the Difference Between the Working Memory Capacities of the Two Species? 32
References 35

1 Introduction

This first clear insight that the medial temporal lobe (MTL) of the human brain was in fact a system of anatomically connected structures that were organized into a memory system came in 1957 from the observations by Brenda Milner of the noted amnesic patient H.M. Subsequent work in humans, monkeys, and rodents has identified all of the components of the medial temporal lobe that formed the memory system. Currently, work is ongoing to identify the specific contributions each structure in the medial temporal lobe makes towards the formation and storage of long-term declarative memory.

2 Background

2.1 Neurologic Patient H.M

As early as 1899, there was some indication that the hippocampus might be important for memory function. At a medical meeting in St. Petersburg, von Bekhterev (1900) displayed the brain of a patient who had exhibited profound memory problems as the most prominent clinical symptom. The primary brain pathology was described as being a bilateral softening of the hippocampus and medial temporal cortex. During the ensuing decades, a few clinical case studies also suggested a relationship between memory impairment and damage to structures of the medial temporal lobe (Glees and Griffith 1952; Grünthal 1947; Hegglin 1953). However, a clear relationship between memory ability and medial temporal lobe function had to await the clinical descriptions provided by Brenda Milner of the amnesic patient Henry G. Molaison, known now in the scientific literature as patient H.M. (Scoville and Milner 1957).

H.M. had an extensive history of minor and major seizures that were unresponsive to standard antiepileptic medication. He had minor seizures beginning at 10 years of age, and major seizures began to appear when he was 16. The major seizures occurred without warning as generalized convulsions that involved loss of consciousness followed by prolonged periods of sleep. Despite high, near-toxic, doses of medication, the major attacks increased in severity and frequency until eventually, he was unable to work or lead a normal life. A decision was then made with the consent of the family to attempt to relieve the seizures through an experimental surgical intervention. On September 1, 1953 at Hartford Hospital in Hartford Connecticut, William Scoville removed H.M.'s medial temporal lobes bilaterally. The surgery involved tissue removal through a supra-orbital trephine with a fine suction tube and attached cautery while the frontal lobe was carefully retracted. The lesion was designed to extend posteriorly for a distance of 8 cm from the tips of the temporal lobes, with the temporal horns constituting the lateral edges of resection.

The surgery was successful in as much as it reduced the frequency and severity of the seizures. However, H.M. was left with profound amnesia. It is notable that although a number of patients had undergone similar removals prior to H.M., those surgeries were performed in an attempt to relieve severe psychosis rather than to relieve seizures. Because the psychosis remained severe in those patients following surgery, the memory problems that must have surely resulted from the medial temporal lobe resection were not appreciated. For H.M. however, the devastating memory impairment was obvious as soon as he had recovered from surgery (Scoville 1954).

2.1.1 The Anatomy of H.M.'s Lesion

Subsequent standard MRI scans (Corkin et al. 1997), in situ MRI scans (Augustinack et al. 2014) and direct histological analysis (Annese et al. 2014) showed that the lesion was bilaterally symmetrical and included the medial temporal polar cortex, piriform cortex, essentially all of the entorhinal cortex (1–2% sparing) most of the perirhinal cortex and subiculum and amygdaloid complex. The anterior half of the intraventricular aspect of the hippocampal formation was removed. This included the dentate gyrus, hippocampus, and subicular complex. The portion of the hippocampus that was spared was the most posterior aspect where the hippocampus ascends in a dorso-posterior direction. The right hippocampus had slightly more preserved tissue than the left (preserved hippocampus was 45.4 mm in the left hemisphere and 47.2 mm in the right; Annese et al. 2014). It was also notable that while the general size and cortical folds of the cerebral hemispheres were unremarkable for an individual of H.M.'s age, white matter lesions that would be consistent with infarctions were observed. Some cerebellar shrinkage was also apparent and this was probably due to the prolonged exposure to phenytoin sodium, which was administered as part of H.M.'s seizure management. There was also a small focal lesion in the left lateral orbital gyrus and involved both cortex and underlying white matter (Annese et al. 2014).

2.2 *Neurologic Patient E.P*

In November of 1992, when E.P. was 70 years old he became sick with flu-like symptoms that included a fever and lethargy. It was at this point that he also experienced a distinct loss of memory. Over the next several days, the memory problem became worse until he was eventually hospitalized. There he was diagnosed with a viral infection and immediately provided with an antiviral medication for 10 days. Two weeks after admission, a T2-weighted MRI scan revealed that the infection and brain swelling had produced viral encephalitis that was bilateral and symmetrical and encompassed the medial temporal lobe. EP successfully recovered from the infection but the brain damage and associated memory impairment was permanent.

2.2.1 The Anatomy of E.P.'s Lesion

Almost all of the hippocampal formation was removed bilaterally including the cell fields of the dentate gyrus, CA1, CA2, CA3, CA4, and the parasubiculum, presubiculum, and subiculum. Only isolated islands of cells remained in the most causal levels of the hippocampal formation. The amygdala was completely removed bilaterally. Neocortical areas were also bilaterally damaged and this damage was symmetrical. The entorhinal cortex was essentially removed bilaterally with only the layer II cell islands observable at the extreme posterior extent of the structure. The entire perirhinal cortex was ablated bilaterally. Finally, the parahippocampal cortex was mostly present with approximately 25% total damage along the anteroposterior extent. It was notable however that the clear laminar organization of healthy neocortical tissue was ill defined in E.P.'s spared parahippocampal cortical tissue. Damage outside of the medial temporal lobe: The rostral fusiform gyrus was substantially damaged. There was also notable damage to the medial mammillary nucleus, medial septal nucleus, and the claustrum as well as more limited, punctate cell loss in the anterior nucleus of the thalamus and pulvinar. Most of the additional abnormalities involved gliosis in the white matter of the temporal stem (Insausti et al. 2013).

2.3 *Insights About the Organization of Memory from H.M. and E.P*

After his surgery, H.M. essentially became a professional research subject and was studied for a full half-century by Milner, her student Suzanne Corkin, and their various colleagues. He died at the age of 82 on December 2, 2008. It is generally accepted that the evaluations of H.M. initiated the modern era of memory research (Squire 2009). Following his recovery from viral encephalitis, Larry Squire and his colleagues studied E.P. for 14 years beginning in 1994. E.P. died at the age of 84 on March 3, 2008. Although the etiologies of the H.M.'s and E.P.'s brain damage were completely different, the final result was a large bilateral lesion of the medial temporal lobe. H.M.'s lesion more restricted to the MTL that was E.P.'s, but E.P.'s MTL damage more was complete. Nonetheless, the similarity of the damage and the rigor and extent to which these two men were evaluated following their brain damage allowed at least four fundamental principles regarding the organization of memory to be illuminated.

2.3.1 Memory Is a Distinct Cerebral Ability

Before H.M., the idea that memory function was a distinct ability, separable from other functions like intelligence, perception, motivation, or personality was not well accepted. This was due in large part to the systematic and influential work of Karl Lashley. Lashley's work was designed to determine if memory could be localized by making discrete and systematically varied brain lesions in rodents and monkeys and other animals and then testing memory acquisition and retention (Lashley 1929). The conclusion was that memory was not localized but rather was distributed throughout the cortex (termed mass action) and that different parts of the cortex were capable of forming memories when other parts were damaged (termed equipotentiality). Work with H.M. and later E.P. demonstrated that their dramatic memory impairment occurred against a background of preserved intelligence, working memory, personality, perceptual ability, and motivation. The primary impairment was exclusively an impairment of long-term memory. This showed that memory was indeed a discrete brain function and this function was localized to the MTL. The reasons for the discord between Lashley's work and the findings from H. M. and E.P. are discussed below in the section on the development of an animal model of MTL amnesia.

2.3.2 The MTL is Not the Repository of Permanent Long-Term Memory

MTL damage produces an impaired ability to form new memories (anterograde amnesia) and an impaired ability to retrieve some older memories that were acquired before the brain damage (retrograde amnesia). Importantly, the impaired ability to retrieve memories acquired before the brain damage does not extend equally to all memories. Rather, memories that were acquired a short time before the brain damage are disproportionally impaired relative to older memories. That is, while more recent memories appeared to be completely lost, older memories, from say childhood, appeared to be completely normal. This phenomenon is often described as a temporally graded retrograde amnesia (TGRA). These observations had two profound implications to understanding the organization of memory. The first implication is that while the MTL is essential for memory encoding and retrieval, it is not the ultimate repository for permanent long-term memory. The second implication is that when memories are first acquired, they are not stored in a form that will persist as memory indefinitely. Rather these memory representations apparently undergo a process of reorganization that transforms them from MTL-dependent memory into MTL-independent memory. This process is referred to as systems consolidation.

2.3.3 MTL Damage Does Not Impair Working Memory

MTL damage does not prevent the acquisition of limited amounts of information that can be held "in-mind" for brief periods of time. This ability can be described a short-term memory, but more formally as working memory or immediate memory. This is the type of memory that allows us to remember, for example, a phone number long enough to dial it before it is forgotten. This information can be maintained for longer periods of time if it is mentally rehearsed, but when distracted, the information is immediately lost. It is also the type of memory that allows a patient with MTL damage to remember a posed question long enough to comprehend the meaning of the question and to provide a suitable answer. Normally, information contained within working memory can be encoded into long-term memory through focused attention or rehearsal, but with MTL damage this information cannot be encoded into LTM and will be quickly and permanently lost (e.g., Baddeley 2003).

2.3.4 MTL Damage Does Not Impair Many Other Forms of Memory

Despite the pervasive and debilitating memory impairment that follows MTL damage, other forms of memory are spared. The first clear indication of this dissociation came when it was discovered that H.M. was able to learn a mirror drawing skill as well as healthy subjects (Milner 1962). In this task, subjects learn to use a pencil to trace between lines viewed through a mirror. At first, this is difficult because each correct movement must be made in the opposite direction from what is normally required. With practice, subjects steadily improve this skill. H.M. improved at the same rate as healthy subjects despite the fact that after the test he had no memory of even attempting the task. Similarly, E.P. exhibited normal differential delay eyeblink classical conditioning despite not remembering the conditioning session or any of the associated pieces of experimental equipment (Clark and Squire 1998). Because these examples represent an experience-based change in behavior, they could correctly be termed "memory," and thus forms of memory that are independent of the MTL. What gradually became clear was that there are many different forms of memory that depend on different brain systems and operate using different neural computations to accomplish memory (see Multiple Memory Systems below).

2.3.5 MTL Function Supports Declarative Memory

Declarative memory is the type of information that is available to conscious awareness of the memory content. It is what one ordinarily means when talking about memory. In other words, it is memory in the colloquial form. For example, you have a memory for what the word "breakfast" means and you presumably have a memory for what you had for breakfast this morning. This type of memory was termed declarative memory precisely because it could be brought to mind and declared (Squire and Zola 1996). Further, there are two primary forms of declarative memory known as semantic and episodic memory. The example above illustrates both types. Semantic memory describes the general bits of knowledge that have been acquired over a lifetime. This knowledge structure would include things like what different concepts mean. It would also include ideas and facts like what the word "breakfast" means. Episodic memory refers to the memory of autobiographical episodes (i.e., events) and could include all of the rich details that were present at the time of the specific event including what happened, the time and place it happened, and other associated emotions or contextual details including what one might have had for their most recent breakfast. This is the type of memory that allows one to reconstruct their past and by mentally traveling back in time to reexperience the event or episode. Recognition memory is a subcategory of declarative memory and simply refers to the ability to recognize anything that has been previously encountered. Two processes, recollection and familiarity (Atkinson and Juola 1974; Mandler 1980), are thought to support recognition memory. Recollection involves remembering specific details or other contextual information concerning a previously experienced episode. Familiarity simply involves knowing that an object (or anything else) was encountered previously, without having available any additional knowledge concerning the actual event or episode during which the information was acquired. Experimentally, these two processes have been studied using the "remember-know" paradigm where "remember" is a proxy for recollection and "know" is a proxy for familiarity and the possible anatomical distinction between these two processes is a current topic of investigation and is discussed further below.

2.4 Animal Model of Medial Temporal Lobe (MTL) Amnesia

Because early efforts to replicate H.M.'s memory deficit in animals were unsuccessful, there was some skepticism concerning the nature of the actual deficit. Work to establish an animal model began almost immediately when Scoville himself came to Montreal and performed the identical surgery in monkeys that he had done with H.M. (Correll and Scoville 1965). Surprisingly (at the time), these monkeys and others with medial temporal lesions were able to learn tasks that appeared similar to memory tests that H.M. could not perform. For example, H.M. could not succeed on a delayed paired comparison technique that consisted of presenting two visual stimuli in succession, separated by a short time interval (Milner 1972). This observation and others implied that the delay between stimulus presentations was critical for observing memory impairment. However, monkeys with MTL lesions performed normally on visual discrimination problems designed to model the tests presented to H.M. (Orbach et al. 1960). This result persisted even when long delays and distractions were introduced (Orbach et al. 1960). The problem was that at the time, it was not understood that humans and experimental animals often approach seemingly similar tasks using different strategies that involve different memory systems. An important example is that monkeys tend to learn visual discriminations gradually over dozens of trials in a form of habit learning. Habit learning is independent of the MTL and supported by the basal ganglia (Mishkin et al. 1984; Teng et al. 2000). Thus, most of the tasks given to animals with hippocampal lesions were in reality skill-based tasks that patients with MTL damage would have been able to acquire, or they were tasks that animals would learn as a skill whereas humans tended to learn the task by consciously memorizing the material. Accordingly, establishing an animal model would require developing tasks that assess the type of memory impaired in amnesia.

The critical advance in forming a model of human medial temporal lobe amnesia was the establishment of one-trial memory tests for the monkey that assess declarative memory. Importantly, if one wants to relate the animal work to work in humans it is not sufficient to use any convenient test in which the animal must use memory. Instead, one must use "specifically designed animal analogs of those tests that do reveal impairment in human amnesiacs." (Gaffan 1974; p. 1101). In 1978, Mortimer Mishkin trained monkeys on the delayed nonmatching-to-sample (DNMS) task (Mishkin 1978). Here, monkeys were first presented with a sample object and then a choice of that sample object or a new object. The monkey received a reward by selecting the new object. This task exploited the monkey's natural tendency to select the novel object, which meant that animals learned this task quickly (Mishkin and Delacour 1975; Mishkin 1978). After training, the monkeys were prepared with lesions designed to mimic the damage sustained by H. M. Postoperatively the animals reacquired the nonmatching rule, and then the delay between the sample and choice phase was increased progressively from 10 s to 30, 60 and 120 s. In this study, the lesions produced a clear deficit, particularly at the longer delays. The demonstration of delay-dependent impairments was critical for at least two reasons. First, it reproduces the memory impairment phenotype seen in humans (intact working memory and impaired long-term memory). Second, when a brain lesion spares performance at short delays (when the demand on memory is small) and impairs performance selectively at longer delays (when the demand on memory is larger), it rules out a variety of alternative explanations for the impairment (e.g., including the ability to perceptually recognize objects, motivational changes, stress responses, circadian influences, and secondary effects of the lesion including hyperactivity, increased distractibility, motor impairments, and other nonspecific effects).

This study and subsequent studies, which relied especially on the trial unique DNMS task (Zola-Morgan and Squire 1985; Mishkin 1982), document the successful establishment of an animal model of human medial temporal lobe amnesia in the monkey. These findings, and others, led to the conclusion that the hippocampal formation (the CA fields of the hippocampus, the dentate gyrus, the subiculum, and the entorhinal cortex) and the adjacent parahippocampal and perirhinal cortices comprise the major components of the medial temporal lobe memory system (Squire and Zola-Morgan 1991). Large lesions of this system in the monkey produce a pattern of memory impairment that closely resembles the phenotype when similar lesions occur in patients (e.g., patient H.M.; Scoville and Milner 1957; Corkin 1984; Corkin et al. 1997 and patient E.P.; Stefanacci et al. 2000). Subsequent work using the animal model characterized the memory impairment that followed damage to MTL structures. The impairment in monkeys with such lesions exhibits normal skill-based memory and normal habit-like memory (Malamut et al. 1984; Zola-Morgan and Squire 1984) as well as intact short-term memory (Overman et al. 1990). Finally, the impairments in monkeys are long-lasting (Zola-Morgan and Squire 1985) and multimodal (Murray and Mishkin 1984; Suzuki et al. 1993). A detailed description of the contributions of the rodent model is too extensive to be described here, but for review see (Clark and Martin 2005; Martin and Clark 2007).

2.5 Multiple Memory Systems

As early as the 1890s, the idea existed that memory was not a unitary ability. William James, in his classic book Principles of Psychology (1890), wrote separate chapters for describing memory and learned habits. Since then, theories of memory have usually distinguished two forms of memory, one form describing memory in the typical colloquial sense of the word and the other form describing motor memory. For example, theories distinguished between explicit and implicit memory (McDougall 1923), "knowing that" and "knowing how" (Ryle 1949), and declarative and procedural memory (Winograd 1975). The first biological insights into these distinctions came from the study of patient H.M. (Scoville and Milner 1957). H.M. had a profound impairment in declarative memory (memory for facts and events), but nonetheless could learn a motor skill (mirror drawing) as efficiently as controls, while retaining no memory of having practiced the task (Milner 1962). This finding indicated that memory is not a unitary ability. At the time of this finding, the preserved memory ability was thought to be restricted to motor skills, a less cognitive form of memory, while all other memory was still viewed as a single entity. However, we now understand that motor skills were not merely an exception, but rather were the first example of a range of memory abilities that depend on brain systems other than the medial temporal lobe. Subsequent work identified other forms of experience-dependent behaviors that were independent of the medial temporal lobe and conscious awareness. Work in humans identified the phenomenon of priming, which is the improved ability to produce, detect, or classify an item due to a recent encounter with the same or related item (Tulving and Schacter 1990). In addition, the basal ganglia was found to be important for gradual, feedback-guided learning that forms the basis of habit memory (Mishkin et al. 1984; Packard et al. 1989). These tasks must be structured in a way that discourages attempts at memorization (for example, when outcomes of trials are determined probabilistically). In rodents, many discrimination tasks or tasks that require a particular response to be acquired are forms of habit memory and dependent on the basal ganglia (e.g., McDonald and White 1993). Simple forms of classical conditioning (Pavlov 1927), like delay eyeblink classical conditioning, were found to be dependent on the cerebellum and associated brainstem circuitry (Christian and Thompson 2003; Clark et al. 1992; Clark and Lavond 1993, 1994). Classical conditioning of fear responses is critically dependent on the amygdala which is

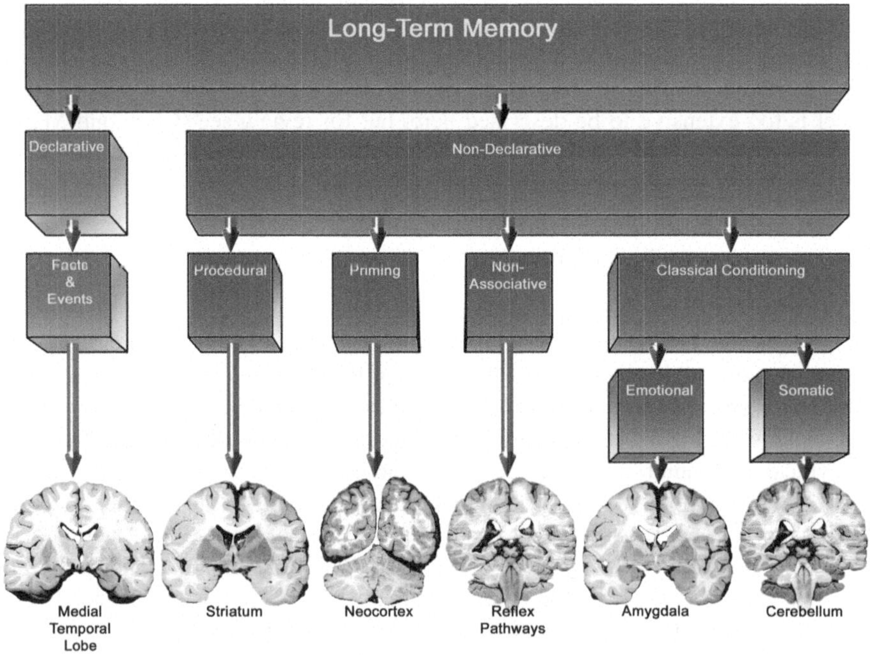

Fig. 1 Memory taxonomy. Long-term memory (purple box) is made up of declarative memory, which includes memory for facts and events (red boxes) and depends on the function of medial temporal lobe (MTL) regions. Nondeclarative memory (blue boxes) is an umbrella term encompassing a collection of learning and memory abilities that are independent the MTL and are principally supported by anatomically distinct and functionally independent brain regions. The critical brain region for each memory system is highlighted in red in coronal sections through the human brain. Adapted from Squire (1992)

thought to be the structure that permanently encodes and stores the hedonic value of the aversive stimulus (Fanselow and Gale 2003). Finally, phylogenetically early forms of behavioral plasticity like habituation and sensitization are also forms of nondeclarative memory. Figure 1 illustrates the categorical taxonomic organization of the mammalian long-term memory systems and the brain structures that support those systems.

2.6 The Organization of the Medial Temporal Lobe Memory System

The system of structures important for declarative memory includes the hippocampus (dentate gyrus, CA fields and subiculum) and the entorhinal, perirhinal, and parahippocampal cortices (Fig. 2; note that in the rat, the region that is synonymous with the parahippocampal cortex is referred to as postrhinal cortex because it is located posterior to the rhinal sulcus). The hippocampus can be conceptualized as residing at the end of a processing hierarchy located in the medial temporal lobe. The hippocampus receives inputs from both the perirhinal and parahippocampal cortices as well as the entorhinal cortex. Guided by the anatomy and physiology, it seems likely that the hippocampus extends and combines functions performed by the structures that project to it (Squire et al. 2007). Additionally, anatomical connections from different regions of neocortex enter the medial temporal lobe at different points. Thus, the higher visual areas TE and TEO project preferentially to the perirhinal cortex. Conversely, spatial information arrives in the medial temporal lobe via the parietal cortex and synapses exclusively in the parahippocampal cortex. Accordingly, it appears that object and spatial information remain segregated in the MTL until combined in the hippocampus (Fig. 3). Consistent with these anatomical facts, damage to parahippocampal cortex was found to impair spatial memory more did than damage to perirhinal cortex (Parkinson et al. 1988; Malkova and Mishkin 1997), and damage to perirhinal cortex impaired performance on the visual object DNMS task more than did damage to parahippocampal cortex (Ramus et al. 1994).

The perirhinal cortex and area TE are immediately adjacent to each other in the temporal lobe and are reciprocally interconnected. These two areas appear to lie at the anatomical border between visual perception and visual memory. Studies of monkeys indicate that perirhinal cortex is important for the memory aspect of recognition memory. Area TE appears to be important for the visual processing that enables the perceptual ability required for successful visual recognition memory (Buffalo et al. 1999, 2000). The functional dissociations that have been reported between the effects of damage to area TE and the effects of damage to perirhinal cortex support these conclusions. For example, monkeys with damage limited to the perirhinal cortex exhibited delay-dependent memory impairment on both visual and tactile versions of the DNMS task (normal performance at short delays when the

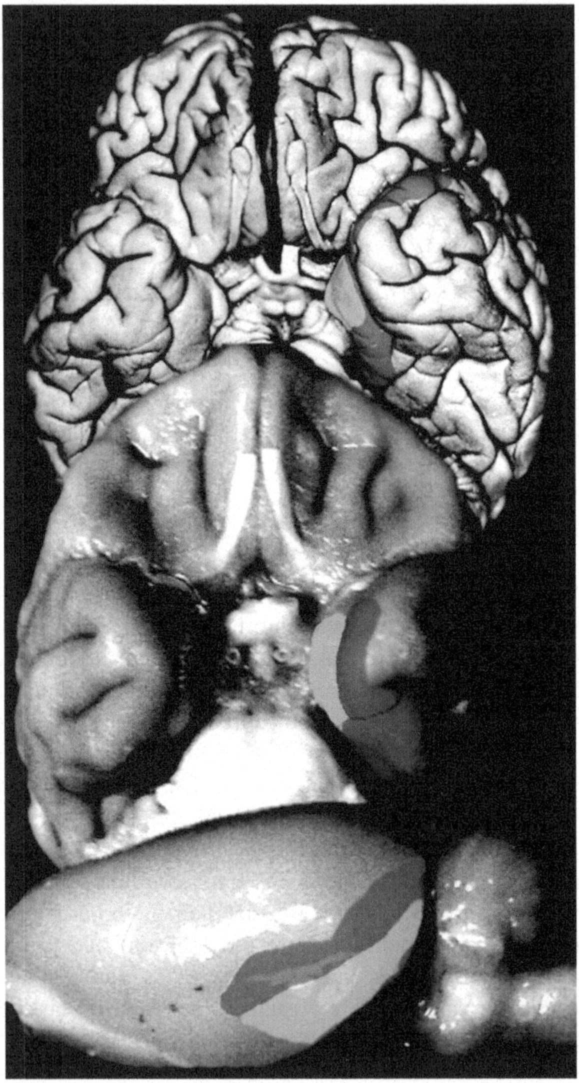

Fig. 2 A ventral view of a human brain (top), a ventral view of a monkey brain (middle), and a lateral view of a rat brain (bottom). The colored structures represent the major cortical components of the medial temporal lobe in all three species: entorhinal cortex (green), perirhinal cortex (red), and parahippocampal cortex (in primates; blue) or postrhinal cortex (in rats; blue). The organization of these structures is highly conserved across these three species

demand on memory is minimal, but impaired performance at longer delays when the demand of memory is greater)—indicating that the primary impairment is mnemonic. By contrast, monkeys with damage restricted to area TE were impaired on visual DNMS but not tactile DNMS (Buffalo et al. 1999). That is, the

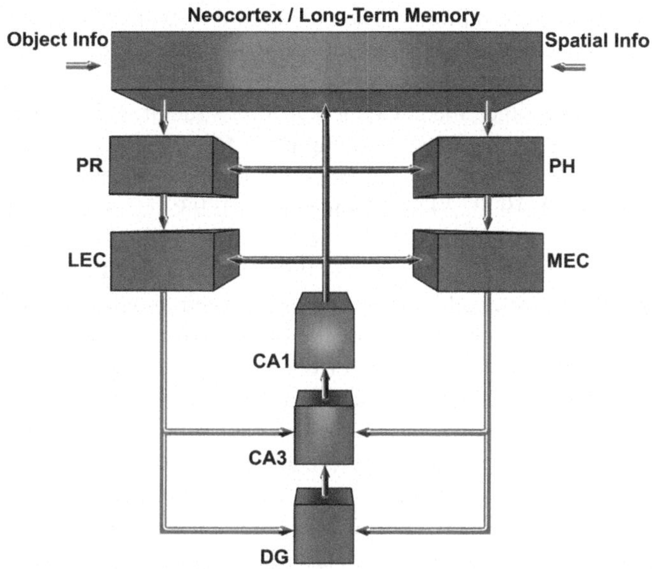

Fig. 3 A schematic view of the medial temporal lobe memory system: The system is hierarchically organized. The hippocampus, defined here as the dentate gyrus (DG), CA3, and CA1, is anatomically situated to receive highly processed information from widespread neocortical regions through four medial temporal cortical areas: the lateral entorhinal cortex (LEC), medial entorhinal cortex (MEC), perirhinal cortex (PR), and parahippocampal cortex (PH; in the rat the postrhinal cortex is used in place of the parahippocampal cortex). Long-term memory (LTM) is thought to be stored in the neocortical areas that were initially involved in processing the various types of sensory information that allowed the encoded experience. Neocortical sensory information is further processed by medial temporal lobe neocortical structures before being relayed to the hippocampus. The DG and CA3 of the hippocampus are anatomically positioned to integrate this information. CA1, primarily through projections to the subiculum (not illustrated), sends this information out of the hippocampus to ultimately contribute the neocortical storage of LTM. In this example, objection information (shown in red) reaches the hippocampus via the PR and LEC. Spatial information (shown in blue) reaches the hippocampus via the PH and MEC. This information is combined in the hippocampus and sent back to neocortex (shown in purple) where it is integrated into LTM. Retrieval of the neocortical LTM is initially dependent on the hippocampus, but through a process of systems consolidation, eventually becomes hippocampus-independent

impairment after TE lesions was unimodal, not multimodal and suggestive of an impairment that is selective to visual perception.

Further, monkeys with perirhinal cortical lesions acquired an automated version of visual DNMS as quickly as normal animals when the delay between sample and choice was only 0.5 s (Buffalo et al. 2000). This finding showed that the ability to perceive the stimuli was unaffected by perirhinal lesions and that monkeys could perform normally when the memory demand was minimal (i.e., the delay was short). In contrast, monkeys with TE lesions were robustly impaired at the 0.5 s delay. Accordingly, monkeys with TE lesions failed even when the memory demands of the task were minimal. The parsimonious explanation of this result is that the monkeys with area TE lesions had difficulty processing the visual stimuli. These findings and others like them (e.g., Buffalo et al. 1999) indicate that perirhinal cortex, like other medial temporal lobe structures, is important for the formation of memory, while area TE is important for visual perceptual processing. Today, it is undisputed that the perirhinal cortex plays a critical role in recognition memory (Buffalo et al. 1999, 2000; Meunier et al. 1993; Eacott et al. 1994; Prusky et al. 2004; Mumby and Pinel 1994; Kornecook et al. 1999; Nemanic et al. 2004; Bussey et al. 1999, 2000; Ennaceur et al. 1996; Winters and Bussey 2005).

Like the perirhinal cortex, the hippocampus has been implicated as important for recognition memory. In fact, the title of the classic paper by Scoville and Milner (1957) was "Loss of recent memory after bilateral hippocampal lesions." The title certainly implies that the memory loss inpatient H.M. was due to the direct damage to the hippocampus. Note however that the last paragraph of the manuscript clearly states correctly, "It is concluded that the anterior hippocampus and hippocampal gyrus, either separately or together, are critically concerned in the retention of current experience." It is interesting that despite the indication that the memory impairment should not be attributed to the hippocampus itself, most subsequent work focused on the hippocampus to the exclusion of the neocortical areas of the medial temporal lobe.

In the 1990s, investigators began using stereotaxic neurosurgical methods to selectively damage the hippocampus in monkeys and to test for recognition memory impairments (Alvarez et al. 1995; Murray and Mishkin 1998; Beason-Held et al. 1999; Zola et al. 2000; Nemanic et al. 2004). The studies concluded that selective hippocampal lesions impair recognition memory (but see Murray and Mishkin 1998). Importantly, Zola et al. (2000) gathered data from 18 monkeys with bilateral lesions of the hippocampus made either by an ischemic procedure, by radio frequency, or by ibotenic acid. Significant recognition memory impairment was observed at all the delays that were tested from 15 s to 40 min. Other work in monkeys using memory tests of spontaneous preference supports the conclusion that selective hippocampal lesions produce robust delay-dependent memory impairment (Zola et al. 2000; Nemanic et al. 2004). These data are congruent with work in humans with damage that included the hippocampus using similar spontaneous recognition memory tests (McKee and Squire 1993; Pascalis et al. 2004). However, work in rodents with selective hippocampal lesions when tested on the DNMS task, has been more mixed. Several studies in the rat have reported that bilateral damage to the hippocampus or the fornix impairs recognition memory (Mumby et al. 1992; Mumby et al. 1995; Wiig and Bilkey 1995; Clark et al. 2001; Prusky et al. 2004). Other studies have failed to find an impairment following bilateral hippocampal or fornix lesions (Aggleton et al. 1986; Rothblat and Kromer 1991; Kesner et al. 1993; Mumby et al. 1996; Duva et al. 1997). A consideration of all the studies suggests that impaired performance on the DNMS task typically occurs following hippocampal damage if the delay is sufficiently long and if the hippocampal lesions are sufficiently large—although these factors alone do not reconcile all the available data (see Clark et al. 2001). I note however that the observed impairment is often relatively mild, although nonetheless significant (Broadbent et al. 2009).

There is a substantial literature reporting that recognition memory impairments following hippocampal damage or disruption in rats and mice when tested using spontaneous preference tasks (Baker and Kim 2002; Broadbent et al. 2004; Clark et al. 2000; de Lima et al. 2006; Gaskin et al. 2003; Hammond et al. 2004; Ainge et al. 2006; Rampon et al. 2000; Rossato et al. 2007). These findings all support the idea that the hippocampus is important for familiarity-based recognition memory. Yet there is also a literature, using spontaneous preference tasks in the rodent suggesting that the hippocampus is not needed for recognition memory (Winters et al. 2004; Forwood et al. 2005; Mumby et al. 2005; O'Brien et al. 2006). It will be important to identify the critical factors that determine when the hippocampus is important for recognition memory and when (or if) normal recognition memory can be accomplished in the absence of the hippocampus (for further discussion of this issue and literature see, Winters et al. 2008). Accordingly, a consensus has not been achieved in the rodent with respect to the role of the hippocampus in recognition memory.

3 Current Topics

3.1 Does Perirhinal Cortex Exclusively Support Memory, or Memory and Higher Order Visual Perception?

As noted above, structures within the medial temporal lobe are critically important for memory. Across several decades, behavioral studies of memory-impaired patients, monkeys, and rodents with bilateral damage to these structures have documented a striking impairment in memory, which occurs against a background of ostensibly preserved perceptual functions (Milner et al. 1968; Squire and Wixted 2011; Mishkin 1982). However, this view has been challenged by a literature suggesting that the perirhinal cortex (a medial temporal lobe neocortical structure) is important not only for memory, but may also have a fundamental role in certain types of high-level visual perception (e.g., Bussey and Saksida 2005; Lee et al. 2005). Specifically, it has been argued that the perirhinal cortex is necessary to resolve visual object discriminations when these discriminations contain a high degree of feature overlap, referred to as feature ambiguity (Bussey et al. 2002; Barense et al. 2005).

This idea emerged from work in the monkey (Eacott et al. 1994). Monkeys with bilateral lesions of the entorhinal and perirhinal cortex were impaired on both a 0 s delay and in a simultaneous matching condition. The authors suggested that these findings might reflect the requirement of the perirhinal cortex to identify stimuli when the stimuli are perceptually similar, because the stimuli used in this study shared many overlapping features. Subsequent work in the monkey was designed specifically to examine the possible contribution of the perirhinal cortex to visual perception. In these studies, various attributes of the stimuli were systematically manipulated during visual discrimination learning tasks to assess the performance of monkeys with lesions of the perirhinal cortex. Impairments were only observed when the visual discriminations involved stimuli with high feature overlap and where good performance appeared to require relatively complex object-level perception (Buckley and Gaffan 1998, 2006; Buckley et al. 2001; Bussey et al. 2002, 2006).

Studies in humans with medial temporal lobe lesions have also addressed this issue, sometimes finding impaired performance and sometimes finding intact performance on these high-level discrimination tasks (Lee et al. 2005; Shrager et al. 2006, respectively). Notably, a comprehensive review (Suzuki 2009) suggested that a reason the issue has been difficult to resolve inpatient studies is that the locus and extent of damage varies among studies, and patients with perceptual impairments might have damage to lateral temporal cortex in addition to medial temporal lobe damage and this extra damage would be expected to impair visual perception. Using the experimental animal, where behavior can be tested after targeted and circumscribed lesions limited to perirhinal cortex, could circumvent these difficulties. However, studies in animals are also problematic. In order to study perceptual ability in animals, the perceptual manipulations must be imbedded within a memory task. In other words, in order to evaluate perceptual function in these studies, animals must typically be trained on a memory task that requires the acquisition of new information in order for the perceptual task to be performed. Accordingly, it is difficult to disambiguate impaired learning and memory from impaired perception using animal studies and difficult to resolve the issue in human studies because of differences in brain lesions and the difficulty in quantifying those lesions even by using modern imaging methods (see Suzuki 2009).

Perhaps the best approach would be to use experimental animals (where the brain lesions can be controlled), but with an experimental design where the influence of any potential mnemonic effect is minimized while still evaluating complex visual perceptual abilities. Accordingly, a novel behavioral paradigm for the rat made it possible to separate the evaluation of memory functions from the evaluation of perceptual functions (Clark et al. 2011). Here, rats were given extensive training on an automated two-choice discrimination task. The extensive training maintained memory performance at a high level during which interpolated probe trials tested visual perceptual ability.

The probe trials were designed to systematically vary the degree of feature ambiguity (i.e., feature overlap) between the two stimuli by morphing the two stimuli into one another across 14 levels of difficulty. As feature ambiguity was increased from very little ambiguity to extreme ambiguity, performance declined in an orderly and monotonic curve that ranged from 87% correct to 50% at the most difficult level. If the perirhinal cortex were critical for feature ambiguous discriminations, then performance should have been intact at the lower morph ambiguity levels and progressively impaired as the stimuli began to share more features and become more difficult/ambiguous. This was not the finding. Bilateral lesions of the perirhinal cortex completely spared the capacity to make these difficult discriminations at every difficulty level (Clark et al. 2011). Control procedures ruled out the possibility the rats were using local cues to solve the discriminations (which would have prevented feature ambiguous, high-level object discriminations from being evaluated). When these same animals were then tested on a recognition memory task, the perirhinal lesions impaired memory—thus confirming the memory impairment caused by these lesions.

Accordingly, the tactic to reduce the possible influence of learning and memory impairment on perceptual performance proved to be successful. Instead of training many discriminations and then presenting a single probe trial for each discrimination, as has been done in the past (Hampton and Murray 2002), animals were trained to learn a single discrimination and then, while maintaining a high level of performance, present 150 probe trials at each of 14 different levels of feature ambiguity. These data suggest that rats with perirhinal cortex lesions exhibited intact performance on every probe trial level because performance did not require any new learning. The basic discrimination was very well learned, and performance remained high throughout testing. These findings provide strong support that the perirhinal cortex is not important for any form of visual perceptual abilities and highlight the value in minimizing the influence of memory impairment when testing perception in the experimental animal. They also explain why prior work in the experimental animal initially reported perceptual impairments. These impairments were likely memory impairments masquerading as perceptual impairments (see Hales et al. 2015 for detailed discussion of these issues). Nonetheless, the topic is still debated and now includes an evaluation of the electrophysiological characteristics of the perirhinal cortex and how both memory and perceptual firing patterns can be inferred (Ahn and Lee 2017).

3.2 Is There an Anatomical Separation Between Recollection and Familiarity?

Recognition memory is commonly viewed as consisting of two components, familiarity and recollection (Mandler 1980). Familiarity involves only knowing that an item was presented without possessing any additional information about the learning episode. Recollection, on the other hand, involves remembering specific contextual details about a prior learning event.

When Brown and Aggleton (2001) proposed a neuroanatomical basis for these two processes, interest in this distinction increased dramatically. Specifically, they proposed that recollection depends on the hippocampus whereas familiarity depends on the adjacent perirhinal cortex. Later work has elaborated on this same proposal (Rugg and Yonelinas 2003; Aggleton and Brown 2006; Eichenbaum et al. 2007), and it has become the basis for the design and analysis of a substantial amount of subsequent experimental work. Importantly, however, alternative formulations have also been suggested regarding the basis of recognition memory and its anatomy (Wixted 2007; Squire et al. 2004).

Much of the discord can be related to the fact that much work taken to support Brown and Aggleton's proposal (2001) have been interpreted in terms of models of recognition memory that are controversial. The implications of these very studies change when the findings are evaluated in terms of an alternative, yet commonly supported model based on signal detection theory.

There have been a large number of studies that have used a variety of neuroscientific methods such as brain lesions, electrophysiological recordings of single neurons, and fMRI, across work with humans, monkeys, and rodents that have evaluated the functional organization of the medial temporal lobe. Much of this work has been originally interpreted in terms of a distinction between recollection and familiarity. However, these same results can also be more simply interpreted with respect to memory strength and indicate that the structures of the medial temporal lobe function in a more integrated and cooperative manner than proposals about the distinction between the hippocampus and perirhinal cortex and recollection and familiarity would suggest.

Receiver Operating Characteristic (ROC) analysis is associated with signal detection theory and was first applied to the analysis of recognition memory more than half century ago (Egan 1958). However, more recently, this analysis has been applied to the neuroanatomical foundations of recollection and familiarity. Briefly, signal detection theory posits that targets (correct items on a recognition memory test) and foils (incorrect or lure items on a recognition memory test) have overlapping distributions of memory strength and the strength and variance of the targets tend to be greater than that of the foils. This model is based on the assumption that familiarity and recollection are both continuous processes that determine the memory strength of a test item (e.g., Mandler 1980; Rotello et al. 2004; Wixted 2007).

A newer two-component theory that also addresses ROC data posits that recollection is a high-threshold component process. In other words, recollection is assumed to support a high-confidence belief that an item has been encountered before (Yonelinas 1994). Additionally, familiarity is thought to be a signal detection process that supports a backup role whenever recollection is absent. Importantly, this high-threshold, signal detection model posits that individual recognition decisions are based on either recollection or on familiarity, but not on a combination of the two.

Using this model (Yonelinas 1994) to evaluate and interpret ROC data, several studies have suggested that hippocampal lesions selectively impair recollection (Aggleton et al. 2005; Daselaar et al. 2006; Fortin et al. 2004; Yonelinas et al. 1998, 2002). The critical point is the asymmetry of the ROC curve is more pronounced in healthy subjects than for amnesic patients with hippocampal damage. The extent to which asymmetry is observed is thought to be the benchmark of recollection in the high-threshold signal detection model (Yonelinas 1994), whereas the more symmetrical ROC produced by amnesic patients is supposed to indicate that their recognition performance is based more on familiarity than is the performance of healthy controls. In fact, familiarity-based performance is often estimated to be normal in the amnesic patients in these studies.

However, in a traditional signal detection theoretical account, a symmetrical ROC only reflects a weak memory, not the absence of recollection. Additionally, an asymmetrical ROC simply implies that the target and foil distribution have an unequal variance, which is tantamount to a strong memory and thus need not suggest that recognition is only supported by recollection (Heathcote 2003; Slotnick and Dodson 2005; Rotello et al. 2005; Smith and Duncan 2004). The analyses of Remember/Know (R/K) judgments have also been used to evaluate the neuroanatomical basis of recollection and familiarity. In the R/K paradigm, subjects must make a judgment as to whether an item is old or new. Then for each old item, they are asked if they "remember" the item (a proxy for recollection) or if they simply "know" that the item had been previously presented (a proxy for familiarity). A number of studies using the R/K method have indicated that hippocampal lesions robustly impair recollection/remember, but impair familiarity/know judgments to a much lesser degree, or in some cases, not at all (Aggleton et al. 2005; Holdstock et al. 2002, 2005; Moscovitch and McAndrews 2002; Yonelinas et al. 2002). However, an obvious concern is that these interpretations would only be valid if the subjective judgments of "Remember" and "Know" are reliable and accurate proxies for recollection and familiarity, respectively.

Importantly, signal detection theory addresses R/K judgments differently (Donaldson 1996; Dunn 2004; Wixted and Stretch 2004). Here, a reduction in "remember" responses combined with little or no reduction in "know" responses is a natural result of strong memories becoming weak memories (see Fig. 2, Squire et al. 2007). In other words, a disproportionate reduction in "remember" judgments need not require a specific loss of recollection. In fact, there is now substantial support for the idea that R/K judgments are an index of memory strength and are not reliable proxies for the qualitatively different processes of recollection and familiarity (Donaldson 1996; Dunn 2004; Rotello et al. 2006; Wixted and Stretch 2004; Wais et al. 2006, 2010).

Accordingly, this collection of findings suggests that the methods that have generally been used to distinguish between recollection and familiarity, might rather distinguish strong memories from weak memories. Further, the functions of the hippocampus and the functions of the perirhinal cortex cannot be crisply dichotomized into the realms of recollection and familiarity, respectively. However, this statement should not be taken to suggest that these two structures function in the same way. For example, neurons in the hippocampus tend to be responsive to more familiar stimuli (Viskontas et al. 2006), whereas neurons in the perirhinal cortex tend to respond to novelty and this response declines as the stimuli become more familiar (Xiang and Brown 1998). At present, it is not clear what data like these indicate about the functional organization of the hippocampus and perirhinal cortex, but they would not appear to relate to anatomical distinctions between recollection and familiarity. A review of the electrophysiological and fMRI literatures instead suggests that both recollection and familiarity signals are apparent in both the hippocampus and perirhinal cortex and that a better approach to understanding these signals would be to evaluate how specific attributes of the stimulus items are encoded into various aspects of the overall memory representation and how these representations may functionally differ between different components of the medial temporal lobe, including the hippocampus and perirhinal cortex (Eichenbaum et al. 1999; Naya et al. 2003; Wood et al. 1999; Lech and Suchan 2013).

3.3 *Is the Organization of Spatial Memory Different Between Humans and Rats, or Is the Difference Between the Working Memory Capacities of the Two Species?*

As described above, declarative memory depends critically on the hippocampus and anatomically associated structures in the MTL. However, these structures, and in particular, the hippocampus, have been strongly associated with the formation and storage of spatial memory (O'Keefe and Nadel 1978; Moser et al. 2008). A current topic of debate relates to the idea that these two perspectives are not fully compatible (Eichenbaum and Cohen 2014; Buffalo 2015). The discord centers on the fundamental distinction between working memory and long-term memory. Working memory is the ability to hold a limited amount of information "in mind" for a limited amount of time after encoding (Baddeley 2003; Cowan 2001; Warrington and Taylor 1973). This information content can be extended in time by active maintenance. The classic example is remembering a new phone number long enough to dial the phone. The average number of digits humans can hold in working memory is 7 (the original length of a phone number), but this number changes depending on the type of information being held. For example, humans can only hold 4 objects, or 1 face in working memory at any given time (Baddeley 2003; Cowan 2001; Warrington and Taylor 1973). Working memory has been thought to be independent of the hippocampus and other MTL structures because it is intact following MTL damage (Milner 1972; Baddeley and Warrington 1970; Clark et al. 2000, 2001; Jeneson and Squire 2012). Thus, tasks that could be maintained by working memory, including spatial tasks, should be intact following MTL damage. However, if MTL structures are necessary to produce the computations required by spatial tasks, then MTL damage should always impair performance on spatial tasks irrespective of the availability of working memory. If true, then for spatial tasks, the distinction between working memory and long-term memory would be irrelevant. Work in humans has been very clear on the issue. For example, work with patients with MTL damage has indicated normal performance on spatial memory tasks under conditions where working memory appears to have supported the performance (Shrager et al. 2008; Jeneson et al. 2010; Kim et al. 2013). Indeed, the noted patient E.P. demonstrated strikingly good performance on spatial navigation when asked about his childhood neighborhoods (Teng and Squire 1999). One type of spatial task is path integration (also known as dead reckoning;

Darwin 1873). In the rodent version of path integration, rats search for food in the dark and then innately attempt to return to the refuge of their start location in order to consume the food. The accuracy with which they return to their start location is the measure of the success of their path integration. Notably, studies of path integration in rats with hippocampal (or entorhinal) damage have reported clear impairments (Maaswinkel et al. 1999; Whishaw et al. 2001; Save et al. 2001; Parron and Save 2004). Importantly, however, all of these studies failed to report the length of time it took for the rats to complete the trials. This means that the rats in these studies may have performed normally whenever the trials were completed quickly because in those instances, performance might have been sustained by their intact working memory.

To address this issue, a new analytical method was developed so that short trials, where performance might be sustained by working, could be analyzed separately from longer trials that would require long-term memory (Kim et al. 2013). Further, the test was administered to both rats with hippocampal damage and humans with hippocampal damage (Kim et al. 2013). The findings from the human section of the study reported that when the trials were simple (that is the target was found quickly), patients with hippocampal damage performed as well as healthy controls. However, rats performed at chance levels even on the simplest trials where the food was found within 3 s or the distance to find the food was 1 m or less, and involved zero turns. These data indicate that rats are unable to use working memory to support performance on this hippocampus-dependent spatial task the way humans do (Kim et al. 2013), and the way rats do on nonspatial, hippocampus-dependent tasks (Clark et al. 2000, 2001).

The implication of this finding and others like them is that spatial memory is organized differently in the rat and human brain. In fact, work with humans clearly suggests that there is nothing unique about spatial memory. Spatial memory is just another form of declarative memory. And like other forms of declarative memory, it is impaired by hippocampal damage, but normal when working memory can be used to support it. Currently, it is not uncommon for scientists who study hippocampal function in the rodent to view the hippocampus exclusively as a spatial processing structure. However, if spatial memory were organized in a fundamentally different way between humans and rats, this would present a very serious challenge for research using rodents as a model system for human hippocampal function (Clark and Squire 2010, 2013).

However, there are at least two other possible explanations for the discrepant finings between humans and rats with respect to spatial memory that do not involve surmising that spatial memory is organized differently between the two species. First, spatial working memory like that required for path integration may be unavailable or perhaps impoverished because the neocortex of the rodent lacks the capacity to construct and maintain a coherent working memory of a spatial environment. Note that in order for subject to use spatial information, a large number of individual components of an environment need to be represented and organized into a meaningful whole. The amount of information necessary for representing a spatial environment may be within the capacity of the grossly expanded neocortical working memory areas of the human brain, but outside of the limits of the rodent neocortical working memory areas. These issues would not apply to nonspatial tasks, because the information that must be maintained in working memory is simpler than in spatial tasks (Clark et al. 2000, 2001).

If working memory capacity was insufficient, then normal rats might accomplish spatial memory tasks, like path integration, by relying exclusively on long-term memory. Rats with hippocampal lesions are impaired at forming long-term memory so they would be impaired on any task, spatial or otherwise, where the information necessary for the task exceeds the capacity of working memory. Similarly, as has been suggested by others, some forms of spatial working memory might depend on the interaction between the medial prefrontal cortex (mPFC) and the hippocampus (Gordon 2011; Hyman et al. 2010; Jones and Wilson 2005; Spellman et al. 2015). Here, hippocampal lesions would disrupt this interaction and impair spatial working memory.

To address these considerations, Sapiurka et al. (2016) tested rats with either mPFC lesions or hippocampal lesions on three tasks of spatial or nonspatial memory: spatial alternation, path integration, and a novel, nonspatial task that required alternation between two different odor-scented cups. The rationale was that if rats do not have sufficient working memory to support spatial tasks, then spatial tasks must always be accomplished by long-term memory. Thus, brain lesions that impair working memory, like lesions of the mPFC, should not affect a spatial task like path integration. Rats were also tested on spatial alternation, a classic working memory task. The results showed that rats with hippocampal lesions were impaired on path integration whereas rats with mPFC were normal (Sapiurka et al. 2016). This finding suggests that path integration in rats is accomplished exclusively by long-term memory and not by working memory. In this study, both groups were impaired on spatial alternation. Here, the interpretation is more complicated. The rats with mPFC lesions were impaired because spatial alternation is a working memory task and mPFC is important for working memory (Horst and Laubach 2009; Hyman et al. 2010; Gordon 2011; Spellman et al. 2015). The rats with hippocampal damage were impaired on spatial alternation because the spatial information exceeded their working memory capacity. Rats with mPFC lesions could not use long-term memory to solve the alternation task because the high interference of repetitive tasks like alternation require working memory (Kane and Engle 2002; Granon et al. 1994). That is, during path integration each trial is unique, whereas each trial of the alternation tasks is dependent on information obtained from the previous trial, and the same response was repeated multiple times in each session (which would cause high interference). Finally, on the nonspatial working memory task (odor alternation), rats with mPFC lesion were again impaired on this working memory task because mPFC supports working memory. However, animals with hippocampal lesions were unimpaired because the odor information was simple enough to be maintained rodent working memory.

In summary, it remains possible that human and rodent spatial memory is fundamentally organized in a biologically different way between the two species. However, the data outlined above suggest that a reasonable, and perhaps parsimonious, alternative perspective is that because of the complex nature of spatial information, limited rodent working memory is unable to support memory tasks that require spatial information and that these tasks necessarily require the use of long-term memory. Rats with hippocampal lesions fail at spatial tasks not because a spatial processing organ has been damaged, but because they are unable to rely on long-term memory. In this regard, the discord between the rodent and human studies is due simply to a difference in the capacity of working memory between the two species (Sapiurka et al. 2016).

References

Aggleton JP, Brown MW (2006) Interleaving brain systems for episodic and recognition memory. Trends Cogn Sci 10:455–463

Aggleton JP, Hunt PR, Rawlins JNP (1986) The effects of hippocampal lesions upon spatial and non-spatial tests of working memory. Behav Brain Res 19:133–146

Aggleton JP, Vann SD, Denby C, Dix S, Mayes AR, Roberts N, Yonelinas AP (2005) Sparing of the familiarity component of recognition memory in a patient with hippocampal pathology. Neuropsychologia 43:1810–1823

Ahn JR, Lee I (2017) Neural correlates of both perception and memory for objects in the rodent perirhinal cortex. Cereb Cortex 24:1–13

Ainge JA, Heron-Maxwell C, Theofilas P, Wright P, de Hoz L, Wood ER (2006) The role of the hippocampus in object recognition in rats: examination of the influence of task parameters and lesion size. Behav Brain Res 167(1):183–195

Alvarez P, Zola-Morgan S, Squire LR (1995) Damage limited to the hippocampal region produces long-lasting memory impairment in monkeys. J Neurosci 15:3796–3807

Annese J, Schenker-Ahmed NM, Bartsch H, Maechler P, Sheh C, Thomas N, Kayano J, Ghatan A, Bresler N, Frosch MP, Klaming R, Corkin S (2014) Postmortem examination of patient H.M.'s brain based on histological sectioning and digital 3D reconstruction. Nat Commun 5:3122

Atkinson RC, Juola JF (1974) Search and decision processes in recognition memory. In: Krantz DH, Atkinson RC, Suppes P (eds) Contemporary developments in mathematical psychology. Freeman, San Francisco, pp 243–290

Augustinack JC, van der Kouwe AJ, Salat DH, Benner T, Stevens AA, Annese J, Fischl B, Frosch MP, Corkin S (2014) H.M.'s contributions to neuroscience: a review and autopsy studies. Hippocampus 24(11):1267–1286

Baddeley A (2003) Working memory: looking back and looking forward. Nat Rev Neurosci 4:829–839

Baddeley AD, Warrington EK (1970) Amnesia and the distinction between long-and short-term memory. J Verbal Learn Verbal Behav 9(2):176–189

Baker KB, Kim JJ (2002) Effects of stress and hippocampal NMDA receptor antagonism on recognition memory in rats. Learn Mem 9:58–65

Barense MD, Bussey TJ, Lee AC, Rogers TT, Davies RR, Saksida LM, Murray EA, Graham KS (2005) Functional specialization in the human medial temporal lobe. J Neurosci 25:10239–10246

Beason-Held LL, Rosene DL, Killiany RJ, Moss MB (1999) Hippocampal formation lesions produce memory impairment in the rhesus monkey. Hippocampus 9:562–574

Broadbent NJ, Squire LR, Clark RE (2004) Spatial memory, recognition memory, and the hippocampus. Proc Natl Acad Sci 101:14515–14520

Broadbent NJ, Gaskin S, Squire LR, Clark RE (2009) Object recognition memory and the rodent hippocampus. Learn Mem 17(1):5–11

Brown MW, Aggleton JP (2001) Recognition memory: what are the roles of the perirhinal cortex and hippocampus? Nat Rev Neurosci 2:51–61

Buckley MJ, Gaffan D (1998) Perirhinal cortex ablation impairs configural learning and paired-associate learning equally. Neuropsychologia 36(6):535–546

Buckley MJ, Gaffan D (2006) Perirhinal cortical contributions to object perception. Trends Cogn Sci 10(3):100–107

Buckley MJ, Booth MC, Rolls ET, Gaffan D (2001) Selective perceptual impairments after perirhinal cortex ablation. J Neurosci 21(24):9824–9836

Buffalo EA (2015) Bridging the gap between spatial and mnemonic views of the hippocampal formation. Hippocampus 25(6):713–718

Buffalo EA, Ramus SJ, Clark RE, Teng E, Squire LR, Zola SM (1999) Dissociation between the effects of damage to perirhinal cortex and area TE. Learn Mem 6:572–599

Buffalo EA, Ramus SJ, Squire LR, Zola SM (2000) Perception and recognition memory in monkeys following lesions of area TE and perirhinal cortex. Learn Mem 7:375–382

Bussey TJ, Saksida LM (2005) Object memory and perception in the medial temporal lobe: an alternative approach. Curr Opin Neurobiol 15(6):730–737

Bussey TJ, Muir JL, Aggleton JP (1999) Functionally dissociating aspects of event memory: the effects of combined perirhinal and postrhinal cortex lesions on object and place memory in the rat. J Neurosci 19:495–502

Bussey TJ, Duck J, Muir JL, Aggleton JP (2000) Distinct patterns of behavioural impairments resulting from fornix transection or neurotoxic lesions of the perirhinal and postrhinal cortices in the rat. Behav Brain Res 111:187–202

Bussey TJ, Saksida LM, Murray EA (2002) Perirhinal cortex resolves feature ambiguity in complex visual discriminations. Eur J Neurosci 15(2):365–374

Bussey TJ, Saksida LM, Murray EA (2006) Perirhinal cortex and feature-ambiguous discriminations. Learn Mem 13(2):103–105

Christian KM, Thompson RF (2003) Neural substrates of eyeblink conditioning: acquisition and retention. Learn Mem 11:427–455

Clark RE, Lavond DG (1993) Reversible lesions of the red nucleus during acquisition and retention of a classically conditioned behavior in rabbits. Behav Neurosci 107:264–270

Clark RE, Lavond DG (1994) Reacquisition of classical conditioning after removal of cerebellar cortex in Dutch Belted rabbits. Behav Brain Res 61:101–106

Clark RE, Martin SJ (2005) Interrogating rodents regarding their object and spatial memory. Curr Opin Neurobiol 15(5):593–598

Clark RE, Squire LR (1998) Classical conditioning and brain systems: a key role for awareness. Science 280:77–81

Clark RE, Squire LR (2010) An animal model of recognition memory and medial temporal lobe amnesia: history and current issues. Neuropsychologia 48(8):2234–2244

Clark RE, Squire LR (2013) Similarity in form and function of the hippocampus in rodents, monkeys, and humans. Proc Natl Acad Sci 110(Suppl 2):10365–10370

Clark RE, Zhang AA, Lavond DG (1992) Reversible lesions of the cerebellar interpositus nucleus during acquisition and retention of a classically conditioned behavior. Behav Neurosci 106:879–888

Clark RE, Zola SM, Squire LR (2000) Impaired recognition memory in rats after damage to the hippocampus. J Neurosci 20:8853–8860

Clark RE, West AN, Zola SM, Squire LR (2001) Rats with lesions of the hippocampus are impaired on the delayed nonmatching-to-sample task. Hippocampus 11(2):176–186

Clark RE, Reinagel P, Broadbent NJ, Flister ED, Squire LR (2011) Intact performance on feature ambiguous discriminations in rats with lesions of the perirhinal cortex. Neuron 70(1):132–140

Corkin S (1984) Lasting consequences of bilateral medial temporal lobectomy: clinical course and experimental findings in H.M. Semin Neurol 4:249–259

Corkin S, Amaral DG, Gonzalez RG, Johnson KA, Hyman BT (1997) H.M.'s medial temporal lobe lesion: findings from magnetic resonance imaging. J Neurosci 17:3964–3980

Correll RE, Scoville WB (1965) Effects of medial temporal lesions on visual discrimination performance. J Comp Physiol Psychol 60:175–181
Cowan N (2001) The magical number 4 in short-term memory: a reconsideration of mental storage capacity. Behav Brain Sci 24:87–185
Darwin C (1873) Origin of certain instincts. Nature 7(179):417–418
Daselaar SM, Fleck MS, Cabeza R (2006) Triple dissociation in the medial temporal lobes: recollection, familiarity, and novelty. J Neurophysiol 96:1902–1911
de Lima MN, Luft T, Roesler R, Schroder N (2006) Temporary inactivation reveals an essential role of the dorsal hippocampus in consolidation of object recognition memory. Neurosci Lett 405(1–2):142–146
Donaldson W (1996) The role of decision processes in remembering and knowing. Mem Cognit 24:523–533
Dunn JC (2004) Remember-know: a matter of confidence. Psychol Rev 111:524–542
Duva CA, Floresco SB, Wunderlich GR, Lao TL, Pinel JPJ, Phillips AG (1997) Disruption of spatial but not object-recognition memory by neurotoxic lesions of the dorsal hippocampus in rats. Behav Neurosci 111:1184–1196
Eacott MJ, Gaffan D, Murray EA (1994) Preserved recognition memory for small sets, and impaired stimulus identification for large sets, following rhinal cortex ablations in monkeys. Eur J Neurosci 6:1466–1478
Egan JP (1958). Recognition memory and the operating characteristic (Tech. Note AFCRC-TN-58-51). Indiana University, Hearing and Communication Laboratory, Bloomington
Eichenbaum H, Cohen NJ (2014) Can we reconcile the declarative memory and spatial navigation views on hippocampal function? Neuron 83(4):764–770
Eichenbaum H, Dudchenko P, Wood E, Shapiro M, Tanila H (1999) The hippocampus, memory, and place cells: is it spatial memory or a memory space? Neuron 23:209–226
Eichenbaum H, Yonelinas AP, Ranganath C (2007) The medial temporal lobe and recognition memory. Annu Rev Neurosci 30:123–152
Ennaceur A, Neave N, Aggleton JP (1996) Neurotoxic lesions of the perirhinal cortex do not mimic the behavioural effects of fornix transection in the rat. Behav Brain Res 80:9–25
Fanselow MS, Gale GD (2003) The amygdala, fear, and memory. Ann N Y Acad Sci 985:125–134
Fortin NJ, Wright SP, Eichenbaum H (2004) Recollection-like memory retrieval in rats is dependent on the hippocampus. Nature 431:188–191
Forwood SE, Winters BD, Bussey TJ (2005) Hippocampal lesions that abolish spatial maze performance spare object recognition memory at delays of up to 48 hours. Hippocampus 15 (3):347–355
Gaffan D (1974) Recognition impaired and association intact in the memory of monkeys after transaction of the fornix. J Comp Physiol Psychol 88(6):1100–1109
Gaskin S, Tremblay A, Mumby DG (2003) Retrograde and anterograde object recognition in rats with hippocampal lesions. Hippocampus 13:962–969
Glees P, Griffith HB (1952) Bilateral destruction of the hippocampus (cornu ammonis) in a case of dementia. Monatsschrift für Psychiatrie und Neurologie 129:193–204
Gordon JA (2011) Oscillations and hippocampal–prefrontal synchrony. Curr Opin Neurobiol 21:486–491
Granon S, Vidal C, Thinus-Blanc C, Changeux JP, Poucet B (1994) Working memory, response selection, and effortful processing in rats with medial prefrontal lesions. Behav Neurosci 108 (5):883
Grünthal E (1947) Über das klinische Bild nach umschriebenem beiderseitigem Ausfall der Ammonshornrinde. Monatsschrift für Psychiatrie und Neurologie 113:1–16
Hales JB, Broadbent NJ, Velu PD, Squire LR, Clark RE (2015) Hippocampus, perirhinal cortex, and complex visual discriminations in rats and humans. Learn Mem 22(2):83–91
Hammond RS, Tull LE, Stackman RW (2004) On the delay-dependent involvement of the hippocampus in object recognition memory. Neurobiol Learn Mem 82:26–34

Hampton RR, Murray EA (2002) Learning of discriminations is impaired, but generalization to altered views is intact, in monkeys (Macaca mulatta) with perirhinal cortex removal. Behav Neurosci 116:363–377
Heathcote A (2003) Item recognition memory and the ROC. J Exp Psychol Learn Mem Cogn 29:1210–1230
Hegglin K (1953) Über einen Fall von isolierter linkseitiger Ammonshornerweichung bei präseniler Dementz Monatsschrift für Psychiatrie und Neurologie 125:170–186
Holdstock JS, Mayes AR, Isaac CL, Cezayirli E, Roberts N, O'Reilly R, Norman K (2002) Under what conditions is recognition spared relative to recall after selective hippocampal damage in humans? Hippocampus 12:341–351
Holdstock JS, Mayes AR, Gong Q, Roberts N, Kapur N (2005) Item recognition is less impaired than recall and associative recognition in a patient with selective hippocampal damage. Hippocampus 15:203–215
Horst NK, Laubach M (2009) The role of rat dorsomedial prefrontal cortex in spatial working memory. Neuroscience 164(2):444–456
Hyman JM, Zilli EA, Paley AM, Hasselmo ME (2010) Working memory performance correlates with prefrontal-hippocampal theta interactions but not with prefrontal neuron firing rates. Front Integr Neurosci 4
Insausti R, Annese J, Amaral DG, Squire LR (2013) Human amnesia and the medial temporal lobe illuminated by neuropsychological and neurohistological findings for patient E.P. Proc Natl Acad Sci 110(21):E1953–E1962
James W (1890) Principles of psychology. Holt, New York
Jeneson A, Squire LR (2012) Working memory, long-term memory, and medial temporal lobe function. Learn Mem 19(1):15–25
Jeneson A, Mauldin KN, Squire LR (2010) Intact working memory for relational information after medial temporal lobe damage. J Neurosci 30(41):13624–13629
Jones M, Wilson M (2005) Theta rhythms coordinate hippocampal-prefrontal interactions in a spatial working memory task. PLoS Biol 2:e402
Kane MJ, Engle RW (2002) The role of prefrontal cortex in working-memory capacity, executive attention, and general fluid intelligence: an individual-differences perspective. Psychon Bull Rev 9(4):637–671
Kesner RP, Bolland BL, Dakis M (1993) Memory for spatial locations, motor responses, and objects: triple dissociation among the hippocampus, caudate nucleus, and extrastriate visual cortex. Exp Brain Res 93:462–470
Kim S, Sapiurka M, Clark RE, Squire LR (2013) Contrasting effects on path integration after hippocampal damage in humans and rats. Proc Natl Acad Sci 110(12):4732–4737
Kornecook TJ, Anzarut A, Pinel JP (1999) Rhinal cortex, but not medial thalamic, lesions cause retrograde amnesia for objects in rats. NeuroReport 10:2853–2858
Lashley KS (1929) Brain mechanisms and intelligence. University of Chicago Press, Chicago
Lech RK, Suchan B (2013) The medial temporal lobe: memory and beyond. Behav Brain Res 254:45–49
Lee AC, Bussey TJ, Murray EA, Saksida LM, Epstein RA, Kapur N, Hodges JR, Graham KS (2005) Perceptual deficits in amnesia: challenging the medial temporal lobe mnemonic view. Neuropsychologia 43:1–11
Maaswinkel H, Jarrard LE, Whishaw IQ (1999) Hippocampectomized rats are impaired in homing by path integration. Hippocampus 9(5):553–561
Malamut BL, Saunders RC, Mishkin M (1984) Monkeys with combined amygdalo-hippocampal lesions succeed in object discrimination learning despite 24-hour intertrial intervals. Behav Neurosci 98:759–769
Malkova L, Mishkin M (1997) Memory for the location of objects after separate lesions of the hippocampus and parahippocampal cortex in Rhesus monkeys. Soc Neurosci Abstr 23:14
Mandler G (1980) Recognizing: the judgment of previous occurrence. Psychol Rev 87:252–271
Martin SJ, Clark RE (2007) The rodent hippocampus and spatial memory: from synapses to systems. Cell Mol Life Sci 64(4):401–431

McDonald RJ, White NM (1993) A triple dissociation of memory systems: hippocampus, amygdala, and dorsal striatum. Behav Neurosci 127(6):835–853
McDougall W (1923) Outline of psychology. Scribners, New York
McKee RD, Squire LR (1993) On the development of declarative memory. J Exp Psychol Learn Mem Cognit 19:397–404
Meunier M, Bachevalier J, Mishkin M, Murray EA (1993) Effects on visual recognition of combined and separate ablations of the entorhinal and perirhinal cortex in rhesus monkeys. J Neurosci 13:5418–5432
Milner B (1962) In: Physiologie de l'hippocampe. Passouant P, editor. Centre National de la Recherche Scientifique, Paris, pp 257–272
Milner B (1972) Disorders of learning and memory after temporal lobe lesions in man. Clin Neurosurg 19:421–446
Milner B, Corkin S, Teuber HL (1968) Further analysis of the hippocampal amnesic syndrome: 14-year followup study of H.M. Neuropsychologia 6:215–234
Mishkin M (1978) Memory in monkeys severely impaired by combined but not by separate removal of amygdala and hippocampus. Nature 273:297–298
Mishkin M (1982) A memory system in the monkey. Philos Trans R Soc Lond B Biol Sci 298:83–95
Mishkin M, Delacour J (1975) An analysis of short-term visual memory in the monkey. J Exp Psychol Anim Behav Process 1:326–334
Mishkin M, Malamut B, Bachevalier J (1984) Memories and habits: two neural systems. In: Lynch G, McGaugh JL, Weinberger NM (eds) Neurobiology of human learning and memory. Guilford, New York, pp 65–77
Moscovitch D, McAndrews MP (2002) Material-specific deficits in remembering in patients with unilateral temporal lobe epilepsy and excisions. Neuropsychologia 40:1335–1342
Moser EI, Kropff E, Moser MB (2008) Place cells, grid cells, and the brain's spatial representation system. Neuroscience 31(1):69
Mumby DG, Pinel JP (1994) Rhinal cortex lesions and object recognition in rats. Behav Neurosci 108:11–18
Mumby DG, Wood ER, Pinel JPJ (1992) Object-recognition memory is only mildly impaired in rats with lesions of the hippocampus and amygdala. Psychobiology 20:18–27
Mumby DG, Pinel JPJ, Kornecook TJ, Shen MJ, Redila VA (1995) Memory deficits following lesions of hippocampus or amygdala in rat: assessment by an object-memory test battery. Psychobiology 23:26–36
Mumby DG, Wood ER, Duva CA, Kornecook TJ, Pinel JPJ, Phillips AG (1996) Ischemia-induced object-recognition deficits in rats are attenuated by hippocampal ablation before or soon after ischemia. Behav Neurosci 110:266–281
Mumby DG, Tremblay A, Lecluse V, Lehmann H (2005) Hippocampal damage and anterograde object-recognition in rats after long retention intervals. Hippocampus 15(8):1050–1056
Murray EA, Mishkin M (1984) Severe tactual as well as visual memory deficits follow combined removal of the amygdala and hippocampus in monkeys. J Neurosci 4:2565–2580
Murray EA, Mishkin M (1998) Object recognition and location memory in monkeys with excitotoxic lesions of the amygdala and hippocampus. J Neurosci 18:6568–6582
Naya Y, Yoshida M, Miyashita Y (2003) Forward processing of long-term associative memory in monkey inferotemporal cortex. J Neurosci 23:2861–2871
Nemanic S, Alvarado MC, Bachevalier J (2004) The hippocampal/parahippocampal regions and recognition memory: insights from visual paired comparison versus object delayed nonmatching in monkeys. J Neurosci 24:2013–2026
O'Keefe J, Nadel L (1978) The hippocampus as a cognitive map. Oxford University Press
O'Brien N, Lehmann H, Lecluse V, Mumby DG (2006) Enhanced context-dependency of object recognition in rats with hippocampal lesions. Behav Brain Res 170(1):156–162
Orbach J, Milner B, Rasmussen T (1960) Learning and retention in monkeys after amygdala-hippocampus resection. Arch Neurol 3:230–251

Overman WH, Ormsby G, Mishkin M (1990) Picture recognition vs. picture discrimination learning in monkeys with medial temporal removals. Exp Brain Res 79:18–24
Packard MG, Hirsh R, White NM (1989) Differential effects of fornix and caudate nucleus lesions on two radial maze tasks: evidence for multiple memory systems. J Neurosci 9:1465–1472
Parkinson JK, Murray EA, Mishkin MA (1988) A selective mnemonic role for the hippocampus in monkeys: memory for the location of objects. J Neurosci 8:4159–4167
Parron C, Save E (2004) Evidence for entorhinal and parietal cortices involvement in path integration in the rat. Exp Brain Res 159(3):349–359
Pascalis O, Hunkin NM, Holdstock JS, Isaac CL, Mayes AR (2004) Visual paired comparison performance is impaired in a patient with selective hippocampal lesions and relatively intact item recognition. Neuropsychologia 42:1293–1300
Pavlov I (1927) Conditioned reflexes: an investigation of the physiological activity of the cerebral cortex. (trans: Anrep GV). Oxford University Press, New York
Prusky GT, Douglas RM, Nelson L, Shabanpoor A, Sutherland RJ (2004) Visual memory task for rats reveals an essential role for hippocampus and perirhinal cortex. PNAS 101(14):5064–5068
Rampon C, Tang YP, Goodhouse J, Shimizu E, Kyin M, Tsien JZ (2000) Enrichment induces structural changes and recovery from nonspatial memory deficits in CA1 NMDAR1-knockout mice. Nat Neurosci 3(3):238–244
Ramus SJ, Zola-Morgan S, Squire LR (1994) Effects of lesions of perirhinal cortex or parahippocampal cortex on memory in monkeys. Soc Neurosci Abs 20:1074
Rossato JI, Bevilaqua LR, Myskiw JC, Medina JH, Izquierdo I, Cammarota M (2007) On the role of hippocampal protein synthesis in the consolidation and reconsolidation of object recognition memory. Learn Mem 14(1):36–46
Rotello CM, Macmillan NA, Reeder JA (2004) Sum-difference theory of remembering and knowing: a two-dimensional signal detection model. Psychol Rev 111:588–616
Rotello CM, Macmillan NA, Reeder JA, Wong M (2005) The remember response: Subject to bias, graded, and not a process-pure indicator of recollection. Psychon Bull Rev 12:865–873
Rotello CM, Macmillan NA., Hicks JL, Hautus M (2006) Interpreting the effects of responsebias on remember-know judgments using signal-detection and threshold models. Mem Cog 34:1598–1614
Rothblat LA, Kromer LR (1991) Object recognition memory in the rat: the role of the hippocampus. Behav Brain Res 42:25–32
Rugg MD, Yonelinas AP (2003) Human recognition memory: a cognitive neuroscience perspective. Trends Cogn Sci 7:313–319
Ryle G (1949) The concept of mind. Hutchinson, San Francisco
Sapiurka M, Squire LR, Clark RE (2016) Distinct roles of hippocampus and medial prefrontal cortex in spatial and nonspatial memory. Hippocampus 26(12):1515–1524
Save E, Guazzelli A, Poucet B (2001) Dissociation of the effects of bilateral lesions of the dorsal hippocampus and parietal cortex on path integration in the rat. Behav Neurosci 115(6):1212–1223
Scoville WB (1954) The limbic lobe in man. J Neurosurg 11:64–66
Scoville WB, Milner B (1957) Loss of recent memory after bilateral hippocampal lesions. J Neurol Neurosurg Psychiat 20:11–21
Shrager Y, Gold JJ, Hopkins RO, Squire LR (2006) Intact visual perception in memory-impaired patients with medial temporal lobe lesions. J Neurosci 26:2235–2240
Shrager Y, Kirwan CB, Squire LR (2008) Neural basis of the cognitive map: path integration does not require hippocampus or entorhinal cortex. Proc Natl Acad Sci 105(33):12034–12038
Slotnick SD, Dodson CS (2005) Support for a continuous (single-process) model of recognition memory and source memory. Mem Cognit 33:151–170
Smith DG, Duncan MJJ (2004) Testing theories of recognition memory by predicting performance across paradigms. J Exp Psychol Learn Mem Cogn 30:615–625
Spellman T, Rigotti M, Ahmari SE, Fusi S, Gogos JA, Gordon JA (2015) Hippocampal-prefrontal input supports spatial encoding in working memory. Nature 522(7556):309–314

Squire LR (1992) Memory and the hippocampus, a synthesis from findings with rats, monkeys, and humans. Psychol Rev 99:195–231
Squire LR (2009) The legacy of patient H.M. for neuroscience. Neuron 15:61(1):6–9
Squire LR, Wixted JT (2011) The cognitive neuroscience of human memory since H.M. Annu Rev Neurosci 34:259–288
Squire LR, Zola SM (1996) Structure and function of declarative and nondeclarative memory systems. Proc Natl Acad Sci 93(24):13515–13522
Squire LR, Zola-Morgan S (1991) The medial temporal lobe memory system. Science 253:1380–1386
Squire LR, Stark CEL, Clark RE (2004) The medial temporal lobe. Annu Rev Neurosci 27:279–306
Squire LR, Wixted JT, Clark RE (2007) Recognition memory and the medial temporal lobe: a new perspective. Nat Rev Neurosci 8(11):872–883
Stefanacci L, Buffalo EA, Schmolck H, Squire LR (2000) Profound amnesia after damage to the medial temporal lobe: a neuroanatomical and neuropsychological profile of patient E. P. J Neurosci 20:7024–7036
Suzuki WA (2009) Perception and the medial temporal lobe: evaluating the current evidence. Neuron 61(5):657–666
Suzuki WA, Zola-Morgan S, Squire LR, Amaral DG (1993) Lesions of the perirhinal and parahippocampal cortices in the monkey produce long-lasting memory impairment in the visual and tactual modalities. J Neurosci 13:2430–2451
Teng E, Squire LR (1999) Memory for places learned long ago is intact after hippocampal damage. Nature 12;400(6745):675–677
Teng E, Stefanacci L, Squire LR, Zola SM (2000) Contrasting effects on discrimination learning after hippocampal lesions and conjoint hippocampal-caudate lesions in monkeys. J Neurosci 20 (10):3853–3863
Tulving E, Schacter DL (1990) Priming and human memory systems. Science 247:301–306
Viskontas IV, Knowlton BJ, Steinmetz PN, Fried I (2006) Differences in mnemonic processing by neurons in the human hippocampus and parahippocampal regions. J Cogn Neurosci 18 (10):1654–1662
von Bekhterev M (1900) Demonstration eines Gehirns mit Zerstörung der vorderen und inneren Theile der Hirnrinde beider Schläfenlappen. Neurologisches Zeitblatt 19:990–991
Wais PE, Wixted JT, Hopkins RO, Squire LR (2006) The hippocampus supports both the recollection and the familiarity components of recognition memory. Neuron 49(3):459–466
Wais PE, Squire LR, Wixted JT (2010) In search of recollection and familiarity signals in the hippocampus. J Cogn Neurosci 22(1):109–123
Warrington EK, Taylor AM (1973) Immediate memory for faces: long- or short-term memory? Q J Exp Psychol 25(3):316–322
Whishaw IQ, Hines DJ, Wallace DG (2001) Dead reckoning (path integration) requires the hippocampal formation: Evidence from spontaneous exploration and spatial learning tasks in light (allothetic) and dark (idiothetic) tests. Behav Brain Res 127(1–2):49–69
Wiig KA, Bilkey DK (1995) Lesions of rat perirhinal cortex exacerbate the memory deficit observed following damage to the fimbria-fornix. Behav Neurosci 109:620–630
Winograd T (1975) Frame representations and the declarative-procedural controversy. In: Bobrow D et al (eds) Representation and understanding: studies in cognitive science. Academic, New York, pp 185–210
Winters BD, Bussey TJ (2005) Transient inactivation of perirhinal cortex disrupts encoding, retrieval, and consolidation of object recognition memory. J Neurosci 25:52–61
Winters BD, Forwood SE, Cowell RA, Saksida LM, Bussey TJ (2004) Double dissociation between the effects of peri-postrhinal cortex and hippocampal lesions on tests of object recognition and spatial memory: heterogeneity of function within the temporal lobe. J Neurosci 24:5901–5908

Winters BD, Saksida LM, Bussey TJ (2008) Object recognition memory: neurobiological mechanisms of encoding, consolidation and retrieval. Neurosci Biobehav Rev 32(5):1055–1070

Wixted JT (2007) Dual-process theory and signal-detection theory of recognition memory. Psychol Rev 114:152–176

Wixted JT, Stretch V (2004) In defense of the signal detection interpretation of remember/know judgments. Psychon Bull Rev 11:616–641

Wood ER, Dudchenko PA, Eichenbaum H (1999) The global record of memory in hippocampal neuronal activity. Nature 397:613–616

Xiang JZ, Brown MW (1998) Differential neuronal encoding of novelty, familiarity and recency in regions of the anterior temporal lobe. Neuropharmacology 37:657–676

Yonelinas AP (1994) Receiver-operating characteristics in recognition memory: evidence for a dual-process model. J Exp Psychol Learn Mem Cogn 20:1341–1354

Yonelinas AP, Kroll NE, Quamme JR, Lazzara MM, Sauve MJ, Widaman KF, Knight RT (2002) Effects of extensive temporal lobe damage or mild hypoxia on recollection and familiarity. Nat Neurosci 5:1236–1241

Zola SM, Squire LR, Teng E, Stefanacci L, Buffalo EA, Clark RE (2000) Impaired recognition memory in monkeys after damage limited to the hippocampal region. J Neurosci 20:451–463

Zola-Morgan S, Squire LR (1984) Preserved learning in monkeys with medial temporal lesions: Sparing of motor and cognitive skills. J Neurosci 4:1072–1085

Zola-Morgan S, Squire LR (1985) Medial temporal lesions in monkeys impair memory on a variety of tasks sensitive to human amnesia. Behav Neurosci 99:22–34

Neural Activity Patterns Underlying Spatial Coding in the Hippocampus

Marielena Sosa, Anna K. Gillespie and Loren M. Frank

Abstract The hippocampus is well known as a central site for memory processing—critical for storing and later retrieving the experiences events of daily life so they can be used to shape future behavior. Much of what we know about the physiology underlying hippocampal function comes from spatial navigation studies in rodents, which have allowed great strides in understanding how the hippocampus represents experience at the cellular level. However, it remains a challenge to reconcile our knowledge of spatial encoding in the hippocampus with its demonstrated role in memory-dependent tasks in both humans and other animals. Moreover, our understanding of how networks of neurons coordinate their activity within and across hippocampal subregions to enable the encoding, consolidation, and retrieval of memories is incomplete. In this chapter, we explore how information may be represented at the cellular level and processed via coordinated patterns of activity throughout the subregions of the hippocampal network.

Keywords hippocampus · learning · memory · oscillations · LFP · network activity · spatial coding · place cells · theta · gamma · sharp-wave ripples

Contents

1 Introduction 44
2 Anatomical Organization of the Hippocampal Network 47
2.1 Hippocampal Inputs 49

Marielena Sosa, Anna K. Gillespie—These authors contributed equally to this chapter.

M. Sosa · A.K. Gillespie · L.M. Frank (✉)
Kavli Institute for Fundamental Neuroscience and Department of Physiology,
University of California, San Francisco, USA
e-mail: loren@phy.ucsf.edu

L.M. Frank
Howard Hughes Medical Institute, Maryland, USA

Curr Topics Behav Neurosci (2018) 37: 43–100
DOI 10.1007/7854_2016_462
Published Online: 25 November 2016

2.2 DG to CA3 ... 50
2.3 CA3 to CA1 ... 51
2.4 CA2 ... 51
2.5 CA1 ... 51
2.6 Subiculum ... 52
3 Electrophysiological Signatures of the Hippocampus ... 52
3.1 Place Cells ... 53
3.2 Theta Oscillations ... 53
3.3 Sharp-Wave Ripples (SWRs) ... 56
3.4 Network Activity Outside of Theta and SWRs ... 56
3.5 Gamma Oscillations ... 57
4 Entorhinal Cortex ... 58
4.1 Cell Characteristics of the EC ... 58
4.2 EC Network Activity ... 59
5 Dentate Gyrus ... 60
5.1 Cell Characteristics of the DG ... 60
5.2 DG Network Activity During the Locomotor State ... 61
5.3 DG Network Activity During SWRs ... 62
6 CA3 and CA1 ... 63
6.1 Functional Roles of CA3 ... 63
6.2 Functional Roles of CA1 ... 64
6.3 Cell Characteristics of CA3 and CA1 ... 65
6.4 CA3 and CA1 Network Activity During the Locomotor State ... 66
6.5 CA3 and CA1 Network Activity During SWRs ... 69
7 CA2 ... 72
7.1 Cell Characteristics of CA2 ... 73
7.2 CA2 Network Activity ... 74
8 Subiculum ... 74
8.1 Cell Characteristics of the Subiculum ... 74
8.2 Subiculum Network Activity ... 75
9 Hippocampal Function Along the Dorsoventral Axis ... 75
9.1 Cell Characteristics Across the Dorsoventral Axis ... 76
9.2 Network Activity Across the Dorsoventral Axis ... 77
10 Hippocampal Output ... 78
11 Conclusion ... 79
References ... 79

1 Introduction

Decades of study have established the hippocampus as a critical center for memory processing in the brain. The hippocampus, along with several associated brain regions, processes the events of daily life and facilitates the long-term storage of these experiences. The link between the medial temporal lobes and memory was indicated first by Scoville and Milner in 1957 as a result of their evaluation of patient now known as H.M. At the age of 27, H.M. underwent bilateral medial temporal lobectomy in a medical effort to alleviate his intractable epilepsy. While the surgical procedure reduced his seizures, H.M. was also rendered unable to form new episodic memories (Scoville and Milner 1957). Subsequent studies of additional patients with more restricted medial temporal lobe lesions, as well as studies in nonhuman primates, identified the hippocampus and parahippocampal gyrus as the most critical regions for memory function (for review, see Squire and Wixted 2011).

While the hippocampus was becoming established as a memory formation center in primates, a seminal series of studies in rodents revealed that hippocampal neurons were remarkably well tuned to spatial location, suggesting a critical role in encoding space. The first evidence for this view of hippocampal function emerged in 1971, when O'Keefe and Dostrovsky reported that a subset of hippocampal neurons fired when rats occupied a specific location in an environment (O'Keefe and Dostrovsky 1971). These neurons became known as place cells, and were shown to be ubiquitous throughout the hippocampus (Muller et al. 1987; Jung and McNaughton 1993; O'Keefe 1976). Moreover, lesion studies of the rat hippocampus revealed a specific deficit in navigation-based memory tasks, further corroborating a role for the hippocampus in spatial processing (Mishkin 1978; Olton and Papas 1979; Morris et al. 1982). Based on these findings, O'Keefe and Nadel proposed that hippocampal neural activity constituted a cognitive map of space, in which individual place cells function to map out the animal's location in reference to its spatial environment (O'Keefe and Nadel 1978).

Since this proposal, many studies have demonstrated that hippocampal neural activity can represent far more than simply spatial location, including aspects of contextual information, object recognition, and time (Eichenbaum et al. 1987; Young et al. 1994; Pastalkova et al. 2008; Hok et al. 2007; Moita et al. 2003; Manns and Eichenbaum 2009). For example, beyond providing a framework for linking locations together to form spatial trajectories, the hippocampus can associate multiple objects with a context (Komorowski et al. 2009), and further link a series of events in a temporally specific order to represent a complex experience (Allen et al. 2016). These observations have led to a proposed expansion of the original spatial cognitive map theory, describing the hippocampal network as a more general relational processing system which enables the rapid association of spatial, temporal, and conceptual aspects of experience (Eichenbaum and Cohen 2001; Eichenbaum et al. 2012). This perspective serves to unify the general memory function of the hippocampus from human and primate studies with the extensive demonstration of a spatial processing function in rodent research.

While these conceptual advances have been important, a complete understanding of the role of the hippocampus will require knowledge of how hippocampal neurons cooperate at a network level to encode, store, and retrieve as memories the complex relationships and experiences that characterize daily life. The original discovery of place cells marked a critical step toward this understanding, as it pointed to a neural mechanism for encoding discrete experiences in the hippocampus. Since then, spatial encoding has been used as a model for the formation of representations that could underlie memory (Eichenbaum and Cohen 2014; Schiller et al. 2015). In this chapter, we will therefore focus on spatial learning and memory as a means to understand mnemonic processing more broadly. In particular, we will explore how coordinated patterns of network activity both within and across the subregions of the hippocampus contribute to spatial memory processing.

BOX: Measuring neural activity in an awake, behaving animal Recording neural activity during behavior allows us to understand how information is processed during an experience and stored as memories. We will discuss two primary types of data collected during this process: single unit activity and local field potential (LFP).

Single unit activity refers to the action potentials, or spikes, fired by individual neurons. Although spikes can now be recorded in vivo using whole cell patch clamp techniques (Tao et al. 2015) or calcium imaging (Ziv et al. 2013), the predominant method for recording unit activity in vivo is extracellular recording. An action potential alters the ionic charge in the extracellular space, as positive sodium ions flow into the cell and away from the recording electrode. The cell's depolarization is therefore reflected as a sharp negative deflection on the extracellular electrode, the inverse of an intracellular recording. The amplitude of this deflection, or waveform, is primarily a function of the electrode's proximity to the cell, as an electrode closer to the cell will measure a larger voltage change. However, because the cell layers of the hippocampus are so densely populated, it can be challenging to distinguish the activity of a single neuron from the surrounding neurons. To address this, hippocampal electrophysiologists typically use recording probes with several closely spaced electrode sites, such as tetrodes, which consist of four insulated electrode wires twisted together (for review, see Buzsaki 2004). Each wire picks up a cell's spike at a slightly different amplitude due to the different proximity of each wire to the cell. This allows the spikes of the cell to be "clustered" by comparing the recorded amplitudes between pairs of electrode wires (Gray et al. 1995; Jog et al. 2002), thus isolating the spike cluster from those of neighboring cells in amplitude space. In contrast, single site electrodes can be sufficient to isolate cells in less densely packed brain regions such as the cortex.

Once spikes have been clustered to link them to a particular neuron, parameters such as the neuron's firing rate, inter-spike interval, and spike waveform can be analyzed to better understand its activity. In the hippocampus, pyramidal cells and fast-spiking interneurons can be putatively identified by their different waveform shapes and firing rates (Fox and Ranck 1981). Not all neuronal populations have clearly differentiable waveforms, however, so it is difficult to definitively identify cell types using extracellular recording alone. Further analysis often describes how the timing of spikes is modulated by behavioral events or by local network activity, as reflected by local field potentials.

Local field potentials (LFP) are defined as the extracellular voltage fluctuations at lower frequencies relative to spiking, which reflect neural network oscillations (hippocampal spiking is typically filtered between ~600–6000 Hz, LFP between ~1–400 Hz) (for review, see Buzsaki et al. 2012). The LFP signal is dominated by synaptic and dendritic activity near the recording electrode for two main reasons. First, high frequency action potentials are largely removed by the low-pass filter. Second, and more importantly, dendritic post-synaptic currents occur at slower timescales than action potentials, increasing the chance of events coinciding in time. The ionic flux of many coincident small synaptic events accumulates, resulting in relatively large fluctuations in the LFP. In laminar structures in which the dendrites and cell bodies of principal neurons are segregated, such as the hippocampus, synaptic input often aligns spatially and temporally, resulting in characteristic layer-specific LFP activity. The amplitude of the LFP signal is influenced by the scale, anatomical organization, and synchrony of inputs to a particular layer (Kajikawa and Schroeder 2011), as well as the proximity of the electrode to the site of maximal current flow, which can be measured using current source density analysis (CSD; Mitzdorf 1985). CSD utilizes the change in LFP signal across closely spaced recording sites to help identify the location of inward or outward current flow. A CSD sink is a negative deflection that represents predominantly positive ions moving into a cell (i.e., an input generating local action potentials), and a source is a positive deflection that is typically interpreted as reflecting the compensatory exit of those positive ions from another part of the cell.

There are many methods for analyzing LFP signals to gain an understanding of how network-level activity is organized within and across brain regions. To isolate particular rhythms, LFPs are often decomposed into their time and frequency components. Measuring the relative intensity of different frequency components can be done using spectral analysis, and the interaction between different frequencies of oscillation can be described by cross-frequency coupling parameters (Tort et al. 2010). LFP can also be compared across multiple brain areas using a measure called coherence, which describes the coordinated modulation of the phase or amplitude of the LFP signals, and may reflect common driving inputs or information flow between the regions (Fries 2005). Finally, as mentioned above, the phase preference of single unit spiking can be determined to understand how LFP signals modulate the firing of local neuronal ensembles. Together, action potentials from individual cells (single unit activity) combined with coordinated network signals (LFP) enable detailed description of neural activity within and across brain regions.

2 Anatomical Organization of the Hippocampal Network

To fully understand how hippocampal network activity contributes to learning and memory, it is important to have a sense for the underlying anatomy that supports this activity. Others have written excellent and detailed reviews (see van Strien et al. 2009; Witter and Amaral 2004), so our goal here is to highlight the fundamental connections in the hippocampal network that facilitate information processing.

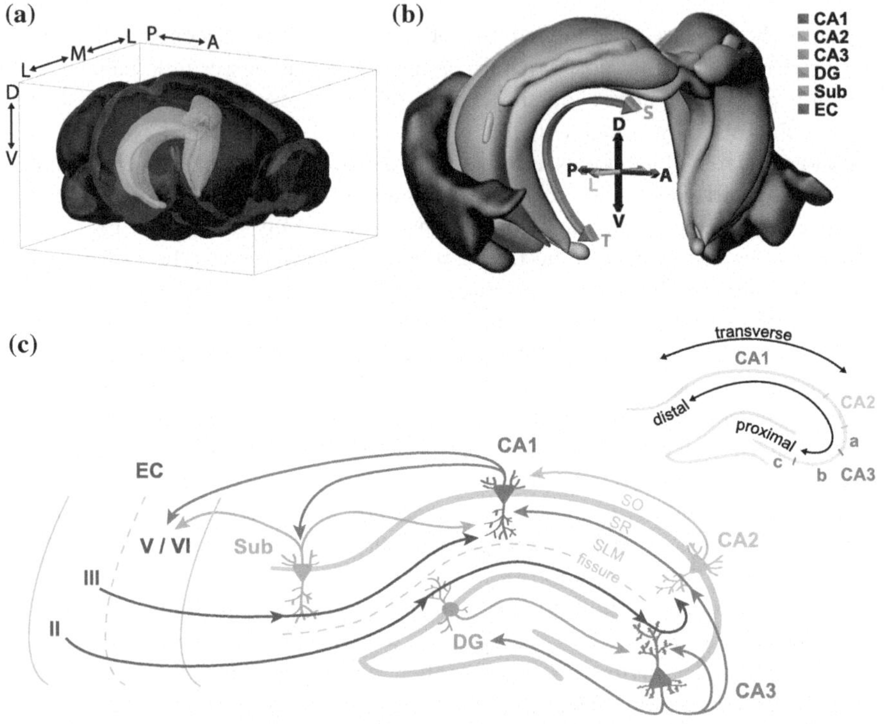

Fig. 1 Anatomical organization of the hippocampal network. **a** Relative location of the hippocampi within the mouse brain. Blue structures highlight the hippocampus proper (CA1, CA2, CA3, and DG) in each hemisphere. The geometry of the hippocampus is very similar in the rat brain. *D* dorsal, *V* ventral, *A* anterior, *P* posterior, *M* medial, *L* lateral. **b** Three-dimensional organization of the hippocampal formation and entorhinal cortex. The hippocampal subregions in each hemisphere are nested such that the DG resides most medially, and the EC wraps around the ventroposterior extent of the hippocampal formation, next to the subiculum (Sub). The curved arrow delineates the septotemporal axis (*S* septal, *T* temporal). Note that in this representation, the transverse axis lies perpendicular to the septotemporal axis, and thus is similar but not exactly analogous to the coronal plane. **a** and **b** are adapted from Brain Explorer 2, © 2015 Allen Institute for Brain Science. Allen Mouse Brain Atlas [Internet]. Available from: http://mouse.brain-map.org . **c** The hippocampal circuit. Major projections into and within the hippocampal circuit are depicted here, following as closely as possibly the true trajectory of axons through the hippocampal layers (*SO* stratum oriens, *SR* stratum radiatum, *SLM* stratum lacunosum moleculare). For example, EC projections target distal apical dendrites of CA1, CA2, and CA3 neurons in SLM, while CA3 targets the proximal apical dendrites of CA2 and CA1 neurons in SR. Minor projections, as well as interneurons and mossy cells, have been omitted for clarity; however, note that these cells are the targets of the depicted CA3 backprojection to the DG hilus. *Arrows* represent synapses, but are not weighted by strength. *Dotted grey lines* represent a subset of layer boundaries, including the hippocampal fissure and the boundary between EC layers II/III and V/VI. The depiction of the EC immediately next to the subiculum is a simplification; note that this exact geometry is only preserved in the horizontal plane of the ventral hippocampus (see panel **b**). Inset: the transverse and proximodistal axes of the hippocampus. Also shown are approximate subdivisions of CA3a, b, and c

The rodent hippocampal formation is a cashew-shaped structure (Fig. 1a) which includes the dentate gyrus (DG), the subiculum, and the hippocampus proper: CA1, CA2 and CA3 (as defined in Witter 1986; Witter et al. 2000). The key axes often used to describe the hippocampus are dorsoventral (often used synonymously with septotemporal, which describes the long axis from the septal, dorsomedial pole of the hippocampus to its temporal pole; Fig. 1b), transverse, and proximodistal (in which proximity is measured relative to DG, Fig. 1c). These axes can delineate anatomical as well as functional gradients, especially along the dorsoventral axis, as we will discuss later. Each hippocampal subregion is organized into layers, formed by the alignment of the principal neurons (Amaral and Witter 1989; Ishizuka et al. 1995). In the hippocampus proper, principal pyramidal neurons are oriented with their basal dendrites in stratum oriens (SO), pyramidal cell bodies in stratum pyramidale (SP), and the apical dendrites in stratum radiatum (SR) and stratum lacunosum moleculare (SLM; Fig. 1c). Various types of interneurons with distinct morphological and functional properties are interspersed throughout each layer (Klausberger and Somogyi 2008). In the DG, the principal granule cell layer is bordered by a molecular layer separating it from the hippocampal fissure. The two "blades" of the granule cell layer surround the hilus, or polymorphic layer, which is composed of interneurons and hilar mossy cells (Freund and Buzsaki 1996; van Strien et al. 2009). The diverse collection of neuronal populations in each hippocampal subregion strongly influences the activity patterns expressed across the hippocampal network.

The hippocampal circuit has canonically been described as a trisynaptic pathway, which involves the perforant path inputs of the entorhinal cortex (EC) to the DG, the mossy fiber projection from the DG to CA3, and the Schaffer collateral projection from CA3 to CA1 (Ramón y Cajal 1893; Lorente de Nó 1934, 1933). However, many local, recurrent, and extrahippocampal connections add complexity to the flow of information through the hippocampus, as we will summarize below.

2.1 *Hippocampal Inputs*

Inputs to the hippocampal formation originate from both cortical and subcortical structures. The hippocampus receives its primary cortical innervation from the entorhinal cortex (Steward and Scoville 1976), via a projection called the perforant pathway (Fig. 1c). EC layer II projects to the apical dendrites of DG granule cells as well as CA3 and CA2 pyramidal cells. While there is some evidence of additional EC input directly onto granule cell bodies (Deller et al. 1996), most EC inputs target the DG molecular layer and SLM of CA2/CA3, with the medial EC (MEC) and the lateral EC (LEC) targeting the more proximal and distal apical dendrites of pyramidal cells, respectively (Witter et al. 1989). The stratification of inputs here may be important for dendritic summation and contribute in specific ways to local LFP (McNaughton and Barnes 1977; Bragin et al. 1995b). In contrast, EC layer III projects to CA1 and the subiculum. In these regions, the subdivision of the MEC and LEC occurs along the proximodistal axis, with the LEC targeting distal CA1 and proximal subiculum, and the MEC targeting proximal CA1 and distal subiculum (van Strien et al. 2009). Direct inputs from sensory and associational cortices primarily target the subiculum, although CA1 has been recently described to receive input directly from the anterior cingulate cortex (Rajasethupathy et al. 2015).

Subcortical inputs to the EC and the hippocampus arise from a variety of structures. The medial septum and diagonal band of Broca (MSDB) send long-range GABAergic and cholinergic afferents to the DG as well as to CA1, CA2 and CA3 (Petsche and Stumpf 1962; Frotscher and Leranth 1985; Freund and Antal 1988; Amaral and Kurz 1985). Additional modulatory inputs come from regions such as the locus coeruleus, the raphe nucleus, and others (Beckstead 1978; Loughlin et al. 1986). CA1 also communicates bidirectionally with the amygdala (Pitkanen et al. 2000; Pikkarainen et al. 1999), which has been long been implicated in emotional forms of learning and conditioned fear memory (Gallagher and Chiba 1996; Paz and Pare 2013; Duvarci and Pare 2014; Janak and Tye 2015). Specifically, the inputs from amygdala to ventral hippocampus have been causally linked to anxiety-like behaviors (Felix-Ortiz and Tye 2014; Felix-Ortiz et al. 2013). CA1 also receives direct input from the nucleus reuniens of the thalamus (Herkenham 1978; Dolleman-Van der Weel and Witter 1996; Vertes et al. 2007) which may influence goal-directed behavior (Ito et al. 2015). For a complete review of hippocampal inputs, see (Witter et al. 1989).

2.2 DG to CA3

Dentate granule cells receive their primary input from the EC and then project to CA3 pyramidal neurons as well as to the other neuronal populations located in the dentate hilus. The DG projection to CA3 is known as the mossy fiber pathway, because of the extensive arborization of granule cell axons and the high density of elaborate postsynaptic spines known as thorny excrescences, which give a "mossy" appearance (Gonzales et al. 2001). In addition to its complex spine structure, CA3 is characterized by heavily recurrent connectivity, meaning that CA3 cells often project onto other CA3 cells (Ishizuka et al. 1990). While recurrence also exists in other hippocampal subregions, it is substantially more prominent in CA3. Specifically, CA3c (Fig. 1c) projects recurrently to the same septotemporal levels of CA3c, while CA3b and CA3a project more extensively within CA3, both across the transverse axis and throughout the septotemporal axis. CA3 is therefore hypothesized to help coordinate activity across the septotemporal extent of the hippocampus (Ishizuka et al. 1990; Li et al. 1994). Furthermore, CA3 projects back to the DG hilus, most strongly from dorsal CA3c and ventral CA3. This backprojection primarily targets excitatory mossy cells and inhibitory interneurons in the hilus, and is therefore hypothesized to indirectly provide both excitation and inhibition of granule cells (Scharfman 2007).

2.3 CA3 to CA1

By far the most intensively studied hippocampal projections are the Schaffer collateral projections from CA3 pyramidal cells to CA1, both ipsilaterally and contralaterally through the hippocampal commissure. The Schaffer collaterals synapse primarily onto the apical dendrites of CA1 pyramidal cells in SR (Fig. 1c), and are stratified by origin: CA3c projects to superficial SR, CA3b to deep SR, and CA3a to SO. Distal CA3 projects to proximal CA1, and proximal CA3 projects to distal CA1 (Laurberg 1979; Ishizuka et al. 1990). Single CA1 pyramidal cells and interneurons receive convergent inputs from both EC layer III and CA3 (Kajiwara et al. 2008; Megias et al. 2001). Although modulated by neural state, the CA3 drive of CA1 is generally thought to be stronger than that of the EC (Spruston 2008).

2.4 CA2

CA2 has received relatively little attention until recently, leaving the functional role of its connections in the hippocampal circuit unclear. CA2 receives input from EC layer II and CA3, as well as strong innervation from the supramammillary nucleus of the hypothalamus (Ishizuka et al. 1990; Chevaleyre and Siegelbaum 2010; Hitti and Siegelbaum 2014; Zhao et al. 2007). Furthermore, neurons in CA2 are extensively recurrently connected and send a strong projection from CA2 to CA1, synapsing primarily in SO and to a lesser degree in SR, and a backprojection from CA2 to CA3 (Hitti and Siegelbaum 2014; Tamamaki et al. 1988; Ishizuka et al. 1990; Cui et al. 2013). Other connections have been more controversial. Notably, individual studies have reported a CA2 to EC layer II projection (Rowland et al. 2013) and a DG to CA2 projection (Kohara et al. 2014), while others do not observe such projections (e.g., Cui et al. 2013).

2.5 CA1

CA1 sends its strongest outputs to the subiculum and to the deep layers of the EC (layers V and VI). The projection to the subiculum is segregated such that proximal CA1 projects most strongly to distal subiculum while distal CA1 projects to proximal subiculum (Amaral et al. 1991). In addition, both CA1 and CA3 project directly to the MSDB (Toth et al. 1993; Toth and Freund 1992; Gulyas et al. 2003; Meibach and Siegel 1977), while other direct projections predominantly from ventral CA1 disperse hippocampal output widely across the brain (Cenquizca and Swanson 2007). Within CA1, local connectivity may have an influence on network patterns and on information processing before signals are sent outward. Specifically, CA1 pyramidal cells synapse laterally onto CA1 interneurons

(Takacs et al. 2012; Amaral et al. 1991), which in turn can even project back to CA3 SR and SO as well as to the DG hilus (Sik et al. 1994, 1995). Furthermore, CA1 axons projecting forward to the subiculum extend collaterals that loop back into CA1 SO (Amaral et al. 1991), providing a small amount of recurrent connectivity within CA1.

2.6 *Subiculum*

The inputs and outputs of the subiculum differ substantially along the dorsal–ventral axis as well as the proximal–distal axis. Dorsal subiculum mostly innervates neocortical regions and receives most inputs from CA1 as well as perirhinal cortex, prefrontal cortex, visual cortex, and MSDB. The ventral subiculum receives the majority of its non-CA1 input from subcortical structures, including hypothalamic nuclei, MSDB, and the amygdala, and returns projections to these regions as well as to the nucleus reuniens of the thalamus and the nucleus accumbens (Witter 2006; Ishizuka 2001; Witter and Amaral 2004). Recently, the early demonstration of a subicular backprojection to CA1 (Kohler 1985; Finch et al. 1983) was confirmed elegantly using Cre-dependent rabies tracing (Sun et al. 2014). Interestingly, both glutamatergic and GABAergic subicular pyramidal neurons innervate all layers of CA1, and the subicular neurons that backproject are the same neurons that receive direct input from CA1. These same cells also receive input from entorhinal cortex, visual cortex, and the MSDB, and both CA1 pyramidal cells and interneurons are targets of this backprojection (Sun et al. 2014). This newly elaborated circuit may provide an important substrate for feedback and fine tuning of hippocampal processing.

3 Electrophysiological Signatures of the Hippocampus

To understand how neural activity in the hippocampal circuit enables the encoding, consolidation, and retrieval of memories, rodent studies of the hippocampus often use place cells as a model for how information can be represented on the single cell and cell ensemble levels. For each subregion of the hippocampal network, we will describe how space is represented at the level of individual neurons, and then how these spatial representations are structured within rhythmic network activity. We will focus primarily on two network patterns that have been implicated in the encoding and retrieval of mnemonic information: theta oscillations and sharp-wave ripples. By understanding how network patterns organize the firing of place cells, we may begin to understand how neural networks may organize information into memories useful for guiding subsequent behavior.

3.1 *Place Cells*

The most striking feature of hippocampal neurons is their spatial specificity. Their stable, location-based receptive fields, called place fields, are now known to be characteristic of the majority of excitatory hippocampal neurons in all subregions (O'Keefe and Dostrovsky 1971; Jung and McNaughton 1993; Muller and Kubie 1987; Thompson and Best 1990). It is important to note that the characterization of place cells is generally done during locomotion, including much of the information we will discuss in the following sections. However, a wealth of evidence suggests that neural activity maintains its place representations outside of locomotion, across behavioral states (e.g., de Lavilleon et al. 2015; Pavlides and Winson 1989; Kay et al. 2016). Place fields develop over the first few minutes of exploration in a new environment and become more refined with experience (Hill 1978; Wilson and McNaughton 1994; Frank et al. 2004). Although exact definitions vary, a cell's place field is generally defined as the region in which its firing exceeds 1 Hz or a certain proportion of the cell's peak firing rate in the environment, such that the place cell fires maximally when the animal is centered in its place field and sparsely or not at all in distant regions of the environment (O'Keefe 1976; Muller et al. 1987). Some place cells have multiple fields, especially in large environments (Fenton et al. 2008; Park et al. 2011). In different environments, different subsets of neurons will become active; this shift in ensemble place activity is called global remapping (Muller et al. 1987; Markus et al. 1995; Lever et al. 2002). A local alteration in an environment (e.g., elimination or addition of a visual cue) might induce rate remapping, in which the active place cell ensemble remains the same, but the firing rates of the ensemble change (Leutgeb et al. 2005b; Anderson and Jeffery 2003; Allen et al. 2012). Rate remapping is thus hypothesized to contribute to the representation of new information within a pre-established spatial framework, while global remapping reflects the creation of an independent spatial representation. Overall, these encoding mechanisms show how neural ensemble activity relates to a representation of the animal's experience, providing a means to investigate how experience is processed within the hippocampal circuit.

3.2 *Theta Oscillations*

During movement, place cells fire at specific times relative to a network rhythm known as theta. Theta is a low frequency oscillation ($\sim$8 Hz, or more broadly 5–12 Hz) which dominates the local field potential (LFP) during locomotion (Fig. 2) and during periods of active engagement in the environment, such as rearing, exploring objects, and preparation for movement (Green and Arduini 1954; Vanderwolf 1969; Grastyan et al. 1959; Foster et al. 1989). An extensive body of literature has thus described theta as the critical marker of an active, location-encoding behavioral state in rodents. Moreover, as theta is known to

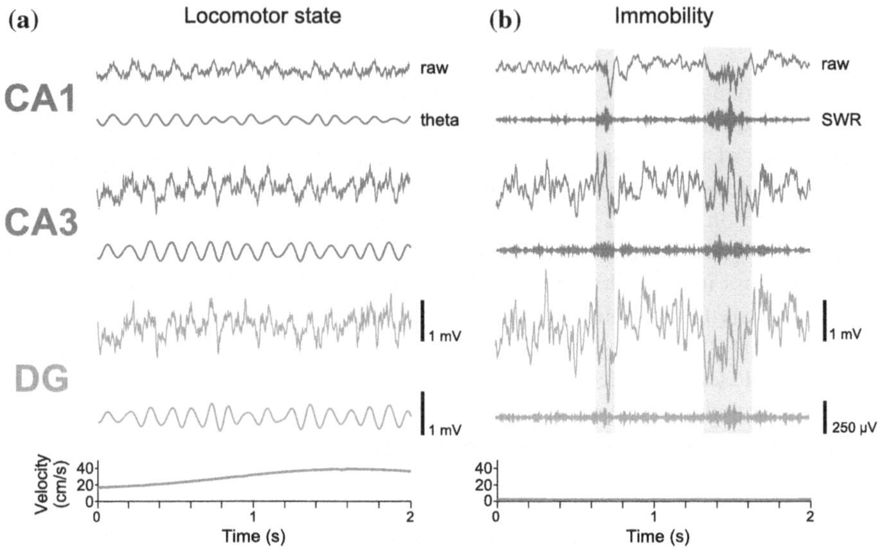

Fig. 2 LFP signatures of the hippocampal network. To illustrate the distinct features of hippocampal LFP, we show raw and filtered LFP detected simultaneously in a single rat from tetrodes located in the principal layer of each major subregion of the hippocampus. During the locomotor state (*left*), persistent theta oscillation dominates the raw LFP signal in all three subregions. As seen in both the raw and theta filtered (5–11 Hz) traces, theta amplitude is smallest in CA1, larger in CA3, and largest in the DG. Graph of the rat's velocity (*bottom*) shows that during this period, the animal is constantly in motion. In contrast, traces on the right show LFP data acquired during awake immobility. Instead of the highly regular rhythmic activity during locomotion, the LFP signal is far more varied, but also increases in amplitude from CA1 to the DG. SWRs detected in CA1 are highlighted in pink, and are easily distinguishable as periods of increased power in the ripple filtered trace (150–250 Hz). The high frequency component of the SWRs are most dominant in CA1, while the sharp wave component is more visible in CA3 and the DG. Note that a substantial amount of time during immobility does not contain SWRs, during which the rat presumably still maintains spatial representations but perhaps through alternate coding mechanisms. During this period, the velocity plot (*bottom*) shows that the rat is motionless

coordinate place cell firing in this state, it has long been thought to be an important contributor to hippocampal processing (for review, see Buzsaki 2002).

It is worth noting that while theta has been observed in mammals other than rodents (Winson 1972; Arnolds et al. 1980), theta is substantially less prominent in bats, cats, monkeys, and humans during analogous periods of locomotion and decision-making (Watrous et al. 2013; Jutras et al. 2013; Ulanovsky and Moss 2007; Kemp and Kaada 1975). For this reason, the specific significance of theta defined as a 5–12 Hz rhythm is not clear. It is possible that an alternative low frequency signal, or irregular but time-locked activity, may perform similar roles in other species (Watrous et al. 2013; Yartsev et al. 2011; Ulanovsky and Moss 2007). To encompass the general behavioral state marked by movement and active sensory engagement across species, we will refer to such periods of activity as the

locomotor state. However, as most studies of hippocampal activity to date were conducted in rodents, we will discuss theta in the locomotor state as one potential mechanism for binding spatial and mnemonic representations.

Hippocampal theta is dependent on activity in the MSDB. Lesions of the MSDB abolish theta activity throughout the entire neocortex (Petsche and Stumpf 1962; Mitchell et al. 1982; Stewart and Fox 1990), and theta suppression from MSDB inactivation has been directly linked to spatial working memory impairment (Mitchell et al. 1982; Givens and Olton 1990; Winson and Abzug 1978; Mizumori et al. 1989). However, other regions such as the supramammillary nuclei may also contribute to the pacing of theta (Kocsis and Vertes 1997; Pan and McNaughton 2002). In addition, lesions of the EC drastically reduce theta power in the hippocampus, suggesting that entorhinal input may be critical for supporting a strong hippocampal theta signal (Buzsaki et al. 1983). Interestingly, while MSDB inactivation abolishes theta entirely, place coding in the hippocampus is somewhat abnormal but not absent (Brandon et al. 2014; Mizumori et al. 1989). The functional role of theta in establishing hippocampal representations thus remains somewhat unclear.

In the rodent hippocampus, theta phase and amplitude are variable across layers (Buzsaki et al. 1986; Bullock et al. 1990) and change along the dorsoventral axis within the same layer (Patel et al. 2012). Theta phase is most consistent just above SP, in SO or corpus callosum, where phase changes within the layer are minimal along the dorsoventral axis (Lubenov and Siapas 2009). The phases of theta represent varying levels of excitation and inhibition. At the trough, for example, inhibition from local inhibitory interneurons is thought to be least, permitting strong firing in local pyramidal cells (Csicsvari et al. 1999b). During the locomotor state, the phase of theta at which a neuron fires is also governed by the animal's proximity to the center of the place field, a phenomenon known as phase precession (O'Keefe and Recce 1993; Skaggs et al. 1996). As an animal enters a place field, that neuron will begin to fire in the later phases of theta, toward the peak of the oscillation. As the animal runs through the neuron's place field, each spike of the neuron tends to align to a progressively earlier phase of the theta cycle, firing near the trough of theta in the place field center and on the descending phases as the rat moves past the place field center (O'Keefe and Recce 1993; Skaggs et al. 1996). This results in each cycle of theta containing a range of place cell activity ordered by location. During most theta cycles, the place cells that fire are those with fields spanning from just ahead to just behind the current position of the animal, such that this sequence of locations is compressed into one theta cycle (Itskov et al. 2008). This has been suggested as a strategy for compressing experience from the timescale of behavior down to the millisecond timescale of spike-timing-dependent plasticity (Skaggs et al. 1996), potentially enabling the rapid encoding of experience in novel environments (Cheng and Frank 2008) and the efficient transmission of spatial information to downstream brain regions (Skaggs et al. 1996; Olypher et al. 2003; Ego-Stengel and Wilson 2007). However, some theta cycles contain a range of place cells representing locations more distant from the animal, and these have been proposed to play a role in planning future trajectories (Gupta et al. 2012;

Wikenheiser and Redish 2015). In general, place cell activity coupled to theta is thought to provide an ongoing representation of location and potentially represent immediate upcoming plans during active behavior.

3.3 Sharp-Wave Ripples (SWRs)

In contrast to the overt rhythmicity of theta during the locomotor state, the hippocampal LFP signal is far more irregular during times of awake immobility (Fig. 2) and slow wave sleep. These periods are punctuated by hippocampal sharp-wave ripples (SWRs) (for review, see Buzsaki 2015). SWRs are perhaps the most synchronous events in the healthy brain, with an estimated 50,000–100,000 neurons discharging over ~50–150 ms in the hippocampus and EC (Chrobak and Buzsaki 1996; Csicsvari et al. 1999a). SWRs have been seen in mammals ranging from mice to humans (Buzsaki et al. 2003; Skaggs et al. 2007; Ulanovsky and Moss 2007; Axmacher et al. 2008), suggesting that their function likely to be conserved. In the rodent, SWRs are characterized by a high frequency ripple oscillation (150–250 Hz) predominantly in the CA1 subregion (Buzsaki et al. 1992) as well as a sharp wave: a simultaneous large negative deflection of the LFP signal detectable throughout most of the hippocampus (Buzsaki 1986). During SWRs, ensembles of place cells become sequentially active in a time-compressed manner, often recapitulating prior experience at high speed (Pavlides and Winson 1989; Lee and Wilson 2002; Skaggs and McNaughton 1996; Buzsaki 1989; Wilson and McNaughton 1994). These sequential reactivations are known as replay events, and are hypothesized to be a key mechanism of hippocampal memory. Complementary to the rapid encoding enabled by compressed place cell sequences within theta cycles, replay during SWRs has been linked to memory consolidation and retrieval. Multiple studies have demonstrated that disrupting SWRs, either during awake immobility or sleep, is sufficient to impair performance on memory-dependent tasks (Girardeau et al. 2009; Ego-Stengel and Wilson 2010; Jadhav et al. 2012; Nokia et al. 2012). Furthermore, SWRs may contribute directly to forming new associations, potentially linking outcomes, such as reward, with the route that leads to them and thus guiding future behavior (Singer and Frank 2009; Foster and Wilson 2006; Lansink et al. 2008, 2009). Distinct patterns of activity concurrent with SWRs have been detected in numerous distant brain areas, suggesting that these events coordinate memory processes across the entire brain (Logothetis et al. 2012).

3.4 Network Activity Outside of Theta and SWRs

While theta and SWRs have been extensively studied as network patterns that facilitate spatial encoding, consolidation and retrieval, relatively little is known about how information in encoded outside of these patterns. During immobility and periods of slow movement, such as when an animal is consuming reward or simply sitting quietly, SWRs only comprise a small fraction (<10 %; Suzuki and Smith 1987; Buzsaki 1989, 2015; Kay et al. 2016) of the ongoing network activity. Recently, a subpopulation of principal neurons in CA2 has been shown to encode location in the absence of locomotion, firing specifically outside of SWRs. Along with neurons in CA1 and CA3, these neurons fire during a transient ~200 ms network pattern with opposite polarity to that of sharp waves. These findings indicate a distinct hippocampal subnetwork dedicated to coding the animal's current location during immobility and even sleep (Kay et al. 2016). In particular, spatial coding during sleep occurred during periods marked by small amplitude LFP activity distinct from the more commonly studied slow wave sleep and REM sleep states (Vanderwolf 1969; Jarosiewicz et al. 2002; Louie and Wilson 2001; Montgomery et al. 2008; Grosmark et al. 2012). Overall, these observations illustrate that that our understanding of the breadth of brain states relevant to hippocampal processing remains incomplete.

3.5 *Gamma Oscillations*

During both the locomotor state and quiescent brain states, an additional rhythm can be seen in the 20–110 Hz range, known as gamma. In the cortex, gamma oscillations have been proposed to play a role in binding neural ensembles, contributing to information transfer and spike-timing dependent plasticity (Gerstner et al. 1996; Markram et al. 1997; Fell and Axmacher 2011), and it is likely that they play similar roles in the hippocampus. However, unlike cortical gamma, hippocampal gamma has been subdivided into several frequency bands, each of which associates with specific states and subregions and is driven by distinct inputs and mechanisms (Buzsaki and Wang 2012; Belluscio et al. 2012). Slow (or low) gamma (20–50 Hz) is thought to be driven predominantly by CA3 (Bragin et al. 1995a; Colgin et al. 2009), while fast gamma (50–90 Hz) is thought to be driven by the MEC (Colgin et al. 2009). An even higher gamma band (90–150 Hz), sometimes called the epsilon band, has also been suggested (Csicsvari et al. 1999a; Sullivan et al. 2011; Freeman 2007). During the locomotor state, all three gamma bands can be observed nested within theta cycles, generally with a single gamma band predominating in each cycle. Thus individual theta cycles tend to exhibit either slow or fast gamma in an interleaved fashion, likely dependent on the cognitive demands experienced by the animal. The frequency of gamma coupled to theta may influence the function and content of theta sequences (Colgin et al. 2009; Zheng et al. 2016). During SWRs, slow gamma in particular is transiently increased in power and coherence throughout the hippocampal circuit, and has been proposed as a clocking mechanism to coordinate accurate replay (Carr et al. 2012; Pfeiffer and Foster 2015).

In the remainder of this chapter, we will focus on the unique contributions of each hippocampal subregion to the patterns of cellular and network activity associated with spatial mnemonic processing in the locomotor state and during SWRs.

4 Entorhinal Cortex

The EC is the main conduit of information to the hippocampus, sending projections to every subregion. The EC receives and integrates sensory information from the primary sensory cortices and head direction information from the thalamic nuclei via the pre- and para-subiculum. In turn, the EC receives direct feedback from the hippocampus. In rodents, the EC is comprised of medial and lateral subregions with distinct functional roles and anatomical connectivity, which are likely preserved in primates although the anatomical delineation between regions is less clear (Witter 1993; Kerr et al. 2007). The MEC conveys mostly spatial information, while encoding in the LEC tends to correspond more to objects, object–context associations, cues, and odors (Deshmukh and Knierim 2013).

4.1 Cell Characteristics of the EC

Within the MEC, distinct subpopulations of cells have been shown to represent major features of any environment. Grid cells are the most common, comprising 30 % of MEC cells, and are found in layers II and III (Zhang et al. 2013). The firing field of a grid cell forms a triangular lattice spanning the entire environment (Hafting et al. 2005). Grid cells are organized into modules that share similar grid scale and orientation; those located in dorsal MEC have smaller grid spacing than those in ventral MEC (Stensola et al. 2012). The other two major MEC populations are border cells, which exhibit firing fields specific to the edge of an environment (Savelli et al. 2008; Solstad et al. 2008; Lever et al. 2009), and head direction cells, which show preference for the animal facing a certain direction independent of location (Sargolini et al. 2006; Taube 2007). These populations can overlap, as some neurons, especially in layers III and V of the MEC, show both grid and head direction tuning (Sargolini et al. 2006). While grid, head direction, and border cells have been the most extensively characterized populations, they comprise only 50 % of the neurons in the MEC, and it is unclear what the function of the remaining neural population may be (Zhang et al. 2013; Sasaki et al. 2015). Recently discovered speed cells may comprise part of this population (Kropff et al. 2015), and may be important for the constant updating of an animal's location on the grid cell map, a process known as path integration (for review, see McNaughton et al. 2006). Together, MEC cells likely provide the animal with a spatial map of the environment and a continuous representation of self-location and transmit that information to the hippocampus.

In contrast to the thoroughly studied MEC, the role of the LEC is much less clear. Current evidence suggests that the LEC may be responsible for encoding the objects, odors, and local cues that occur within an environment rather than mapping space on a global scale (Neunuebel et al. 2013). While objects and local cues can also influence the development of place fields in the MEC, LEC neurons generally lack the spatial tuning frequently seen in MEC neurons (Deshmukh and Knierim 2011; Yoganarasimha et al. 2011). Furthermore, lesions of the LEC have been shown to impair the association of objects with environmental contexts, despite sparing normal object recognition and context recognition (Wilson et al. 2013). This finding, along with others, suggests that the LEC plays a role in linking items and cues with the environment in which they were experienced (Neunuebel et al. 2013). Together, the MEC and LEC are thought to provide the "where" and "what" of an experience to the hippocampus, where it can then be compared to previous experience, integrated with existing frameworks, and stored. In particular, the EC may establish location-based representations of stimuli that could be the basis for the spatial encoding seen in hippocampal cells.

4.2 EC Network Activity

During the locomotor state, theta is prominent in the MEC and strongly entrains neuronal spiking (Deshmukh et al. 2010). Grid cells show theta phase precession (Hafting et al. 2008), and the integrity of grid cell firing is dependent on theta oscillations (Koenig et al. 2011). Surprisingly, although the MEC was initially suspected to be the primary driver of place cell activity, lesions of the MEC do not abolish hippocampal place fields (Van Cauter et al. 2008; Hales et al. 2014). However, MEC lesions do disrupt hippocampal theta phase precession and reduce the spatial specificity and stability of place fields (Van Cauter et al. 2008; Hales et al. 2014; Schlesiger et al. 2015), concomitantly impairing spatial navigation-dependent behavior (Hales et al. 2014). This evidence suggests that the MEC provides critical spatial and temporal cues to refine hippocampal representations of location. Conversely, grid cell integrity is heavily dependent on hippocampal feedback, as inactivation of the hippocampus abolishes grid cell periodicity (Hafting et al. 2008; Bonnevie et al. 2013). In parallel to the reduced spatial encoding observed in the LEC compared to the MEC, the power of the theta rhythm is lower in the LEC than in the MEC or hippocampus, and entrainment of LEC neurons by theta is less prominent. This suggests that theta may be particularly important for coordinating spatial processing across the MEC and hippocampus, and less so for nonspatial object information in the LEC (Deshmukh et al. 2010).

Cells in the layers of the EC that project to the hippocampus are thought to be relatively inactive during SWRs themselves (Chrobak and Buzsaki 1994). However, several studies indicate that the EC, together with the rest of the neocortex, experiences periods of higher firing (up states) and periods of relative inactivity (down states) governed by the neocortical slow oscillation which may influence SWR activity in the hippocampus (Steriade et al. 1993; Sirota et al. 2003; Battaglia et al. 2004a; Isomura et al. 2006). SWRs are more likely to occur during up states (Battaglia et al. 2004a; Sullivan et al. 2011), suggesting that the overall cortical state may influence the ability of the hippocampus to generate SWRs, and this modulation may be conveyed by EC inputs.

5 Dentate Gyrus

Despite its prominent place in the hippocampal circuit, the DG has been one of the least studied of the hippocampal subregions with respect to patterns of network activity. However, it has garnered attention due to several unique characteristics. Most notably, the DG is one of the only regions in the brain which supports persistent neurogenesis throughout life. The regular addition of new neurons to DG circuitry has major implications for network activity in the region (Ge et al. 2008; Schmidt-Hieber et al. 2004) and for behavior dependent on the DG (Dupret et al. 2008; Garthe et al. 2009; Jessberger et al. 2009; Shors et al. 2001; Wojtowicz et al. 2008). A second key feature of the DG is the highly sparse firing of its principal cells, which have very low spontaneous firing rates (Amaral et al. 1990; Jung and McNaughton 1993) and of which only a very small fraction are active in any given environment (Guzowski et al. 1999). It has been proposed that the sparse firing of distributed ensembles and the addition of newborn neurons into those ensembles make the DG uniquely suited to perform pattern separation, a process by which similar experiences are disambiguated and encoded by orthogonal representations (Marr 1971; Clelland et al. 2009).

5.1 Cell Characteristics of the DG

The principal cell type in the DG is the granule cell (GC). These small, tightly packed cells make up the cell layer 'blades' of the DG. Neurogenesis in the subgranular zone lining the border between the GC layer and hilus consistently adds newborn GCs to the GC layer. These newborn cells migrate and integrate into existing GC layer circuitry over a 4–8 week period, during which they show increased excitability compared to mature GCs (Esposito et al. 2005; Ge et al. 2008; Schmidt-Hieber et al. 2004; Gu et al. 2012; Li et al. 2012; Marin-Burgin et al. 2012; Danielson et al. 2016). GCs tend to fire very sparsely (generally less than 0.5 Hz in awake recordings), and have narrow, asymmetric waveforms (Jung and McNaughton 1993). Recordings from putative GCs suggest that they have spatially and directionally specific place fields that are highly stable, although smaller than those found in CA3 and CA1 (Jung and McNaughton 1993; Leutgeb et al. 2007). Moreover, these putative GCs may have more discontiguous sub-place fields than CA3 and CA1 pyramidal cells, although due to dense cellular packing of the DG, this conclusion is confounded by the challenge of identifying the cell type being recorded. It is possible that the cells with multiple sub-place fields may be the hyperexcitable newborn GCs or mossy cells, another excitatory neuronal population in the DG (Danielson et al. 2016; Neunuebel and Knierim 2012).

Small, subtle changes in an environment are sufficient to prompt global remapping of DG ensembles and thus change DG input to CA3, in contrast to CA3 ensembles which adjust slightly but do not remap (Danielson et al. 2016; Leutgeb et al. 2007; Neunuebel and Knierim 2014). This falls in line with the idea of pattern separation, showing that the DG can amplify differences between similar experiences. However, it remains challenging to differentiate the contribution of young and mature GCs to ensemble representations. New GCs seem to be important for pattern separation, as blocking neurogenesis impairs the ability to perform pattern separation, while stimulating neurogenesis enhances it (Nakashiba et al. 2012; Tronel et al. 2012; Creer et al. 2010; Sahay et al. 2011; Clelland et al. 2009). This finding is somewhat contradicted, however, by the hyperactive nature of young GCs, which would seem to undermine the activation of orthogonal ensembles capable of distinguishing similar experiences (Johnston et al. 2016; Danielson et al. 2016). However, young GCs have also been shown to more effectively recruit feedback inhibition than mature GCs (Temprana et al. 2015), which may offset their hyperactivity (McAvoy et al. 2015).

The sparse firing of granule cells is enforced by high levels of GABAergic inhibition from local inhibitory interneurons. Various interneuron subtypes provide both feedback and feedforward inhibition onto GCs by targeting GC bodies or dendrites, respectively (Savanthrapadian et al. 2014). In addition to interneurons, the hilus also contains excitatory mossy cells (Henze and Buzsaki 2007; Scharfman and Myers 2012). As mossy cells are relatively rare (1:100 ratio of mossy cells to granule cells; (Henze and Buzsaki 2007), and because it is unclear how best to distinguish them from GCs, they have not been well characterized electrophysiologically (Neunuebel and Knierim 2012). Mossy cells receive inputs either directly from the EC or indirectly through GCs, and synapse onto hilar interneurons and remote GCs (Buckmaster et al. 1996; Larimer and Strowbridge 2008). This may allow them to integrate activity across the septotemporal axis of the DG by transferring excitation between GCs, or by suppressing the activity of distant GC populations via feedforward inhibition (Larimer and Strowbridge 2008; Henze and Buzsaki 2007). Together, the interactions between sparsely firing mature GCs, excitable newborn GCs, mossy cells, and interneurons may underlie the DG's ability to integrate entorhinal inputs into distinct representations of mnemonic experience.

5.2 DG Network Activity During the Locomotor State

Large, clear theta oscillations can be observed in the DG during the locomotor state (Bragin et al. 1995b) (Fig. 2). Theta entrains the spiking of both GCs and interneurons (Skaggs et al. 1996), although newborn GCs may be more weakly modulated (Rangel et al. 2013). Spatially modulated GCs also exhibit theta phase precession (Skaggs et al. 1996). Theta phase, as well as the coherence of DG theta with the rest of the hippocampus, varies by layer, which may be due to the stratification of inputs from different EC layers (Montgomery et al. 2009). Theta power and coherence measures also fluctuate based on the activity being performed, although the significance of these observations remains unclear.

During the locomotor state, gamma oscillations nested within the theta rhythm are larger in the hilus than anywhere else in the hippocampus (Bragin et al. 1995a; Buzsaki 2002; Montgomery and Buzsaki 2007). Like theta, the power and coherence of DG gamma with the rest of the circuit fluctuates with the cognitive demands of activity performed (Montgomery and Buzsaki 2007), but the exact role of DG gamma is not known. A study using current source density analysis (CSD; see Box) showed the largest gamma current sink in the middle third of the DG molecular layer, where axons from the MEC terminate. This current sink disappeared upon lesion of the EC, further suggesting that DG gamma activity is primarily driven by the EC during locomotor state (Bragin et al. 1995a). This study did not differentiate slow gamma from fast gamma; however, as the EC is thought to promote fast rather than slow gamma in the hippocampus (Colgin et al. 2009), the CSD finding may predominantly reflect fast gamma in the DG during the locomotor state.

5.3 DG Network Activity During SWRs

Since SWRs are thought to originate in CA3 and proceed to CA1, as described later in this chapter, most studies have focused on the SWR-related activity that occurs in those subregions. However, several pieces of evidence suggest that the DG also participates in SWR-associated activity. First, granule cell activity has been observed during SWRs (Penttonen et al. 1997) including reactivation during sleep (Shen et al. 1998), potentially driven by the CA3 backprojection (Scharfman 2007, 1994). Second, state-dependent activity in the DG may affect SWR generation. During slow wave sleep, DG activity can be categorized into "up" and "down" states which correlate with those seen in neocortex (Isomura et al. 2006; Sullivan et al. 2011). As mentioned above, SWRs are more likely to occur during up states than down states (Battaglia et al. 2004a; Isomura et al. 2006; Sullivan et al. 2011). This suggests state-dependent modulation of SWR generation, however, it is unclear whether the DG contributes to this modulation directly, or whether both the DG and CA3 are influenced by EC up/down states in parallel. Finally, a recent study have shown that slow gamma activity in the DG increases during SWRs (Gillespie et al. 2016). This is similar to an SWR-associated transient slow gamma increase observed in CA3 (Carr et al. 2012), which may serve as a critical timing mechanism to organize replay activity during SWRs (Pfeiffer and Foster 2015) as we will further discuss below. In the DG, the power of slow gamma transiently increases during SWRs, as does coherence in this frequency band between

DG-CA3 and DG-CA1. These results suggest that slow gamma activity engages all subregions of the hippocampus proper during SWRs, potentially coordinating information flow through the circuit. Disruption of DG circuitry, caused by the loss of hilar interneuron populations, results in impaired SWR-associated slow gamma activity throughout the hippocampal circuit, further indicating that the DG may be actively engaged during SWRs (Gillespie et al. 2016).

Interestingly, there is another pattern of activity, called dentate spikes (DSs), that also occurs in the DG during awake immobility and slow wave sleep (Bragin et al. 1995b; Penttonen et al. 1997). DSs are brief, large-amplitude LFP deflections seen in the hilus and granule cell layer. Two types have been described, one (DS1) which has a broad waveform, shows a phase reversal in the outer molecular layer, and contains some fast gamma activity, and another (DS2) which shows a single narrow LFP peak with a phase reversal in the inner molecular layer (Bragin et al. 1995b). Lesions of the EC eliminate both DS types, and CSD analysis of these events as well as the location of their phase reversals suggests LEC and MEC drive of DS1 and DS2, respectively (Bragin et al. 1995b). Although DSs and SWRs appear during the same behavioral state, they do not coincide. Instead, DSs seem to have the opposite effect on the hippocampus from SWRs; rather than inducing ensemble activity downstream, DSs seem to transiently suppress CA3 and CA1 activity (Bragin et al. 1995b; Penttonen et al. 1997; Buzsaki et al. 2003). Although behavioral correlates of DSs are not well understood, this observation suggests that they may enable a transient blockade of hippocampal output via CA1.

6 CA3 and CA1

CA3 and CA1 are by far the most well-studied subregions of the hippocampus. As activity patterns are highly coordinated across CA3 and CA1, we will discuss their network activity in parallel, while highlighting the distinctions that embody each region's unique role in hippocampal processing.

6.1 Functional Roles of CA3

CA3 has been functionally implicated in rapid task acquisition (Lee and Kesner 2003; Nakazawa et al. 2003, 2008; Kesner 2007; Lee and Kesner 2004; Cravens et al. 2006), as well as task recall (Nakazawa et al. 2002; Kesner 2007; Lee et al. 2005; Schlesiger et al. 2013), the latter of which is likely facilitated by the retrieval of previously learned patterns. For example, lesions of CA3 impair the ability of animals to use partial cues to trigger memory-based performance of a task acquired in the presence of full cues (Gold and Kesner 2005). This ability to recall a whole memory based on a partial cue is known as pattern completion, and may be a critical neural process for comparing current events to past memories and generalizing across similar experiences (for review, see Leutgeb and Leutgeb 2007; Knierim and Neunuebel 2016; Rolls 2007). Pattern completion may be supported by the trait which most often distinguishes CA3 in the current literature: its relatively high level of recurrent connectivity compared to other hippocampal subregions (Ishizuka et al. 1990; Li et al. 1994; Witter 2007). This recurrence has been presented as anatomical evidence that CA3 acts as an autoassociative network (Marr 1971; Guzowski et al. 2004; Treves and Rolls 1991; Rolls 2007; Papp et al. 2007; Gilbert and Brushfield 2009; McClelland and Goddard 1996; Treves and Rolls 1992, 1994). Autoassociation implies that the activation of a subset of neurons within an ensemble can drive sustained activation of the entire ensemble by propagating excitation through reciprocal connections between cells (Lisman 2003). Such an autoassociative network might also exhibit attractor dynamics (Marr 1971; Mcnaughton and Morris 1987; Kali and Dayan 2000; Leutgeb et al. 2005c; Rolls 2007; Lengyel et al. 2005). In CA3, an ensemble of neurons representing a stored memory could act as the attractor basin; when an input is similar enough to the stored memory, the activity in the network settles on that ensemble. When an external input is sufficiently distinct from the stored pattern, it would outweigh the ongoing activity to transition the pattern of activity to a new group of cells, and thus form a distinct memory (Renno-Costa et al. 2014; Leutgeb and Leutgeb 2007; Colgin et al. 2010). While there is not yet definitive evidence that CA3 functions as a true attractor network, the importance of its recurrent collaterals to pattern completion have been supported by recent findings. Results showing that distal CA3 (where the level of recurrence is highest) shows stronger autoassociation than proximal CA3 (where recurrence is lowest) suggest that the contribution of CA3 subregions to pattern completion depends on the local recurrent connectivity (Lee et al. 2015). The autoassociative nature of CA3 is further supported by the stability of CA3 population representations in response to small changes in environmental cues (Neunuebel and Knierim 2014), indicating that small changes in sensory inputs are insufficient to substantially alter the representation.

6.2 Functional Roles of CA1

The CA1 network represents the final stage of hippocampal processing before information is sent to the subiculum and to the rest of the brain. CA1 continually integrates input received from CA3 and the EC during ongoing experience (Bittner et al. 2015; Spruston 2008; Milstein et al. 2015; Piskorowski and Chevaleyre 2012; Kali and Freund 2005) and permits incremental spatial learning and retrieval even in the absence of CA3 input (Nakashiba et al. 2008). One possible function of CA1 is to compare past experiences stored in and retrieved by CA3 with new information from ongoing experiences transmitted by the EC. In this scenario, CA1 would then create a new representation when there is no past experience that directly corresponds to current input (Lee et al. 2004a; Hasselmo and Schnell 1994; Lisman 1999; Vinogradova 2001; Duncan et al. 2012; Meeter et al. 2005; Meeter et al.

2004). CA1 may therefore compile memories by layering newly learned spatial information onto past and current representations of the global environment.

The possibility that CA1 somehow compares stored and new information is consistent with observations that CA1 responds to novelty (Lisman and Otmakhova 2001; Li et al. 2003; Kumaran and Maguire 2007), specifically by signaling the presence of a novel experience and potentially enhancing the incorporation of novel information into an existing framework (Larkin et al. 2014). In particular, CA1 place cells change their firing rates in response to novel or changing objects (Lenck-Santini et al. 2005; Deshmukh and Knierim 2013; Fyhn et al. 2002; Larkin et al. 2014) and novel spatial environments (Karlsson and Frank 2008; Nitz and McNaughton 2004; VanElzakker et al. 2008). CA1 network patterns, including SWRs and gamma, are also modulated by novelty both during the novel experience and during sleep afterward (Cheng and Frank 2008; Karlsson and Frank 2009; Eschenko et al. 2008; O'Neill et al. 2008; Singer and Frank 2009; Dupret et al. 2010; Kemere et al. 2013; Ramadan et al. 2009). Importantly, novelty-induced increases in firing rate and SWR reactivation appear to be specific to CA1, and not CA3 (Karlsson and Frank 2008), suggesting that the recognition of novelty is a function that emerges uniquely in CA1 or in conjunction with the EC, rather than through input to CA1 from CA3 (Larkin et al. 2014).

6.3 Cell Characteristics of CA3 and CA1

In support of their complementary functional roles, CA3 and CA1 exhibit small but important differences in how their principal cells represent space and other variables. Place cells in the two regions have similar spatial coverage and firing rates (Olton et al. 1978; O'Keefe and Dostrovsky 1971; Best and Ranck 1982) but CA3 is thought to be more strictly responsive to spatial location than CA1 (Barnes et al. 1990; Lee et al. 2004b; Leutgeb et al. 2005a; Knierim et al. 2006; Vazdarjanova and Guzowski 2004; Leutgeb et al. 2004, 2005b). Once CA3 places fields are established in a given environment, firing rates and spatial coverage remain stable over time (Mankin et al. 2012, 2015). CA3 ensembles also show higher sensitivity to absolute location than CA1 ensembles, as distinct populations of CA3 cells can represent distinct spatial locations, even if the local environments in those locations are visually identical (Leutgeb et al. 2004). In contrast, the fields of CA1 cells show prolonged susceptibility to modulation by sensory cues and changes in the environment (e.g., Leutgeb et al. 2004; Vazdarjanova and Guzowski 2004), and thus are more likely to globally remap their firing fields within the same environment than CA3 cells. In addition to encoding spatial information, CA1 place cells can integrate nonspatial information into their firing patterns, usually via rate remapping, to odors, objects, goals, and conditioned stimuli (Eichenbaum et al. 1987; Kobayashi et al. 1997; Kennedy and Shapiro 2009; Hok et al. 2007; Moita et al. 2003; Manns and Eichenbaum 2009; Komorowski et al. 2009; McKenzie et al. 2014; Dupret et al. 2010). The sensitivity of CA1 ensembles is further reflected in novel environments, in which CA1 firing rates start high and then decline along with the proportion of active CA1 cells as an environment becomes familiar. This tunes the population representation to a subset of CA1 neurons (Karlsson and Frank 2008).

During early exposure to an environment, place cells in both CA3 and CA1 may fire in any direction of movement through their place field (Muller et al. 1987). However, over the course of experience on a stereotyped path, such as a linear track, cells tend to develop a directional preference (McNaughton et al. 1983; Frank et al. 2004; Battaglia et al. 2004b). This directional bias is informative, as the ordered firing of unidirectional place cells enables the decoding of not only the spatial trajectory of an animal, but also the animal's direction of movement. In addition, cells in both CA3 and CA1 are capable of developing path equivalence through experience, in which cells fire similarly in geometrically or behaviorally similar areas of a spatial maze. This path equivalence reflects an ability of CA3/CA1 neurons to generalize across related locations and episodes, rather than an inability to distinguish locations (Singer et al. 2010).

Interestingly, despite the flow of information in the canonical trisynaptic loop, the DG is not required for the spatial specificity of place fields in CA3 (McNaughton et al. 1989). Likewise, neither inactivation of CA3 nor the EC is sufficient to abolish place fields in CA1, yet both result in more diffuse, less spatially tuned place fields in CA1 (Mizumori et al. 1989; Brun et al. 2002, 2008a; Van Cauter et al. 2008; Nakashiba et al. 2008). Together, these results suggest that although CA1 place fields can be derived from either EC or CA3 input, both projections are required for robust spatial specificity. These findings also support a role for CA3 in providing a stable spatial framework onto which other types of information can be layered via the more malleable encoding seen in CA1.

6.4 CA3 and CA1 Network Activity During the Locomotor State

During locomotion, CA3 and CA1 cell activity is tightly coupled to the theta rhythm, with both regions exhibiting temporally compressed place cell sequences via theta phase precession that emerge rapidly during novel experience (Feng et al. 2015). As mentioned previously, theta phase precession is thought to be a mechanism by which online spatial encoding is compressed onto a time scale conducive to neural plasticity (Skaggs et al. 1996). Theta phase precession has likewise been proposed to promote synaptic plasticity between CA3 and CA1 cells, such that synapses are strengthened between cells with overlapping place fields via the repeated coincident firing of those cells during experience (Mehta et al. 2000, 2002; O'Neill et al. 2008; Isaac et al. 2009). Moreover, the theta phase at which a place cell fires indicates how far into the firing field the animal is, providing a temporal code for location. This temporal code, consisting of the precise timing of spikes relative to theta, exists independently from the rate code of the local network. The rate code is defined as the collective firing rates of the local ensemble, which represent both location as well as other nonspatial features due to rate remapping (Mehta et al. 2002; Jensen and Lisman 2000; Huxter et al. 2003). The addition of a temporal code to this framework thus could support precise spatial coding despite firing rates that may be highly variable (Mehta et al. 2002; Ahmed and Mehta 2009; Hopfield 1995).

The phases of theta have functional implications in terms of inputs to the circuit, possibly segregating encoding and retrieval (Hasselmo 2005; Hasselmo et al. 2002; Mizuseki et al. 2009). At the trough of theta, Schaffer collateral synapses are highly susceptible to long-term potentiation (Hyman et al. 2003; Kwag and Paulsen 2009) and receive maximal excitation from the EC, potentially facilitating the encoding of new information (Brankack et al. 1993; Kamondi et al. 1998; Mizuseki et al. 2009; Colgin et al. 2009). At the peak of theta, EC input to CA1 decreases and gives way to maximal CA3 input, but CA3 synapses onto CA1 neurons are more likely to be depressed than potentiated (Hyman et al. 2003; Kwag and Paulsen 2009). This may allow for retrieval without corrupting or restoring the retrieved information (Hasselmo 2005; Hasselmo et al. 2002; Mizuseki et al. 2009). Importantly, the theta referred to in these studies was recorded from the hippocampal fissure, which is $\sim 180°$ out of phase with the theta recorded in the CA1 pyramidal cell layer (Hasselmo 2005). Recent behavioral evidence supports this hypothesis of input segregation, as inhibition of CA1 at the peak of theta enhanced spatial working memory performance when delivered during the encoding phase of the task, while inhibition during the trough improved performance during the retrieval phase of the task (Siegle and Wilson 2014). This balance of encoding and retrieval within single theta cycles could be due to the dendritic integration of CA3 and EC inputs on CA1 neurons, regulated by waxing and waning inhibition at theta frequencies (Milstein et al. 2015).

Another critical function of theta in CA3 and CA1 may be the exploration of future trajectories and goals. Early in learning, an animal will often pause at decision points on a maze and visually survey possible routes before choosing a trajectory (termed vicarious trial and error, or VTE) (Muenzinger 1938; Tolman 1938). During periods of VTE behavior, the neural representation of location sweeps ahead of the animal as distant place cells activate (Johnson and Redish 2007). These sweeps of activity within a theta cycle, called theta sequences, are distinct from the majority of theta content because they activate representations outside the animal's current position. Similarly, nonlocal theta sequences can be predictive, reflecting the upcoming behavior of the animal. When deciding between possible reward locations, place cells representing the chosen goal location become active during theta cycles ahead of their place fields, despite often substantial distances from the animal's current location (Wikenheiser and Redish 2015).

During ongoing locomotion, place cells may also fire at different rates in the same location depending on the animal's future destination (prospective coding) or past locations (retrospective coding) (Frank et al. 2000; Wood et al. 2000; Ferbinteanu and Shapiro 2003; Ainge et al. 2007; Ito et al. 2015). These types of coding suggest that place field activity can encode not only absolute location, but also the animal's position relative to an ongoing trajectory (Frank et al. 2000) and thus reflects both upcoming and past experience. Interestingly, the prospective coding phenomenon has also been observed in cells which are modulated by time rather than location (for review see Eichenbaum 2014). These "time cells" were first observed to fire at temporally specific intervals while an animal ran on a treadmill in a singular location, forming sequences which predicted the animal's upcoming trajectory (Pastalkova et al. 2008). Time cells exhibit theta phase precession and can also have place properties, suggesting that time and place encoding coexist within theta sequences (MacDonald et al. 2011; Kraus et al. 2013). Time cell coupling to the theta rhythm has also been described in the context of odor memory in head-fixed animals, indicating that temporal encoding also exists in the absence of movement (MacDonald et al. 2013). Together the current evidence points to a role for theta in exploring future possibilities and temporally organizing spatial experience.

As mentioned earlier, gamma band activity can be found nested within theta oscillations during the locomotor state (Belluscio et al. 2012; Colgin et al. 2009; Csicsvari et al. 2003; Bragin et al. 1995a). The frequency of both slow and fast gamma bands increases with increasing speed of locomotion, as do theta frequencies, indicating locomotor-driven coupling (Ahmed and Mehta 2012). In both CA1 and CA3, slow gamma shows a subtle increase in frequency at higher velocities, while the frequency of fast gamma is strongly modulated by speed (Ahmed and Mehta 2012; Zheng et al. 2015). In addition to modulating oscillation frequency, movement speed differentially alters the power of slow and fast gamma in rats. With increasing speed, slow gamma power decreases while fast gamma power increases, suggesting continuous modulation of the circuit as the behavioral state of the animal changes (Kemere et al. 2013). However, both slow and fast gamma power were positively modulated by speed in mice, suggesting that gamma power modulation varies across species (Chen et al. 2011). The shift to higher frequencies and power of fast gamma is mirrored by an increase in MEC firing rates at faster speeds (Zheng et al. 2015), suggesting predominant engagement of CA1 by the EC at high speeds and by CA3 at low speeds (Kemere et al. 2013; Zheng et al. 2015).

The coupling of theta with slow and fast gamma has been proposed to underlie dynamic switching between difference sources of information in the hippocampus, modulated by behavioral and cognitive demands. Fast gamma, driven by the MEC, is thought to convey information about current location and state, and coincides with spiking activity enriched for place cells with place fields near the animal's current position. In contrast, slow gamma, driven by CA3 (Scheffer-Teixeira et al. 2012; Belluscio et al. 2012; Colgin et al. 2009), may be more likely to coincide with place field spiking that represents trajectories extending beyond the current location (Zheng et al. 2016). The theta cycles containing slow gamma are likely to correspond with the nonlocal representations during VTE or trajectory planning, as mentioned above (Johnson and Redish 2007; Wikenheiser and Redish 2015). As further evidence for a role in active information processing, the phase coupling between theta and both slow and fast gamma is increased during awake locomotion as compared to REM sleep (Montgomery et al. 2008). Together, the evidence suggests that dynamic coupling between network oscillations may reflect changing cognitive demand on the hippocampal circuit during active learning and navigation (Montgomery and Buzsaki 2007; Axmacher et al. 2010; Bott et al. 2015; Igarashi et al. 2014; Colgin 2015; Tort et al. 2009).

6.5 CA3 and CA1 Network Activity During SWRs

CA1 and CA3 are the main contributors to the network activity involved in SWRs and play distinct but highly intertwined roles in supporting these events. SWRs are an intrinsic hippocampal pattern, frequently occurring in hippocampal slice preparations in which major hippocampal afferents, such as those from the EC, are generally disrupted (e.g., Maier et al. 2003). Since the discovery of replay during SWRs, hundreds of studies have investigated their origins and functional contributions to memory processes like consolidation and retrieval.

During SWRs, ensembles of neurons are reactivated in a precise, time-compressed sequence that recapitulates experience. Place cell reactivation was originally demonstrated during sleep after the animal had traversed the reactivated cell's place field (Pavlides and Winson 1989). Ensemble reactivation during sleep was then demonstrated by the finding that pairs of cells with overlapping place fields in a previously explored environment reactivated together more frequently than pairs of cells with distant place fields (Wilson and McNaughton 1994). With improvements in analysis techniques, it became possible to observe the reactivation of more specific neuronal sequences during sleep that recapitulated awake experience (Skaggs and McNaughton 1996; Kudrimoti et al. 1999; Nadasdy et al. 1999; Lee and Wilson 2002). These replay events were shown to occur specifically during SWRs (Kudrimoti et al. 1999; Lee and Wilson 2002). Replay was also suggested to occur during periods of awake immobility (Pavlides and Winson 1989; Kudrimoti et al. 1999) and then confirmed during pauses in awake behavior (Foster and Wilson 2006; Jackson et al. 2006; O'Neill et al. 2006; Diba and Buzsaki 2007). Some studies also reported that SWRs can even occur during movement, particularly in more novel environments (O'Neill et al. 2006; Cheng and Frank 2008). The time-compressed representation of prior experience during SWRs made these network events compelling candidates for neural mechanisms of memory processes.

In support of this theory, several studies have provided causal evidence for the essential role of SWRs in learning and memory. During rest immediately following training sessions on a spatial memory task, disruption of SWRs impaired subsequent task performance and delayed task acquisition (Girardeau et al. 2009; Ego-Stengel and Wilson 2010). These results suggest a critical role for SWRs during sleep in memory consolidation. Disruption of SWRs during awake immobility also had a detrimental effect on spatial task acquisition, impairing the component of the task that most relied on linking experiences across time and making choices based on immediate past experience (Jadhav et al. 2012). This result implicates SWRs in retrieval during ongoing decision-making. Conversely, successful performance of working memory tasks can be correlated with enhanced content or increased incidence of SWRs (Dupret et al. 2010; Eschenko et al. 2006, 2008; Molle et al. 2009; Ramadan et al. 2009) or even predicted by increased place cell reactivation during SWRs (Singer et al. 2013). In addition, experiences such as the exploration of novel environments or receipt of reward increase SWR incidence both in the awake state and during sleep afterwards (Kudrimoti et al. 1999; Karlsson and Frank 2009; Eschenko et al. 2008; Cheng and Frank 2008; Singer and Frank 2009; Wu and Foster 2014), suggesting that particularly salient experiences increase SWR activity. The increase in awake SWRs may facilitate the association of novel spatial trajectories with their outcomes, while the increase in sleep SWRs may support the consolidation of those experiences through persistent reactivation and communication with the neocortex (Carr et al. 2011; O'Neill et al. 2010). Interestingly, SWRs during wakefulness tend to be more accurate in replicating past experience, while SWRs during sleep show lower fidelity (Karlsson and Frank 2009). This may relate to their proposed functional differences: awake SWRs may be critical for the rapid, accurate retrieval of stored experiences to evaluate a current decision, while sleep SWRs may rely on less accurate replay to more flexibly integrate new experiences with existing memory frameworks (Roumis and Frank 2015).

The content of SWR replay is directly dependent on prior experience (Silva et al. 2015), but can also be influenced by many factors with implications for replay function. During sleep SWRs, replay content can be biased by current odor or sound cues, despite the animal not attending to the stimulus (Bendor and Wilson 2012). During awake SWRs, replay often begins at the animal's current location (Csicsvari et al. 2007; Karlsson and Frank 2009; Davidson et al. 2009). The length of a SWR event has been shown to correlate with the length of the trajectory replayed, and multiple SWRs can chain together with replay spanning across the chain (Davidson et al. 2009). Replay events can be either forward, exactly as the event was initially experience, or reverse, "rewinding" through the steps of a trajectory. The distinction between these two options, especially on mazes in which the animal can traverse each section in both directions, relies on the directionality of the cells involved (Diba and Buzsaki 2007; Foster and Wilson 2006). Because place field activity on a linear track becomes more unidirectional with experience (McNaughton et al. 1983; Frank et al. 2004; Battaglia et al. 2004b), unique ensembles distinguish trajectories in the two directions, differentiating forward and reverse replay (Diba and Buzsaki 2007; Foster and Wilson 2006; Gupta et al. 2010; Csicsvari et al. 2007). Forward replay is more common before an animal embarks on a trajectory, suggesting a potential role for planning or evaluating choices. In contrast, reverse replay is observed more after trajectories are completed, which may be important for associating the location of a reward with the steps taken to reach it (Diba and Buzsaki 2007). While functional and correlative studies have pointed to a role for SWRs in planning and reward associations (Pfeiffer and Foster 2013; Singer et al. 2013; Singer and Frank 2009), none have been able to address directional specificity, so the potential implications of forward and reverse replay remain unclear. In addition to the replay of previous and upcoming trajectories, replay of remote environments and distant locations has also been observed during the awake state (Karlsson and Frank 2009; Davidson et al. 2009). The diversity of SWR content likely enables the flexible consolidation and possibly retrieval functions of hippocampal replay.

Several lines of evidence indicate that SWRs are generated in CA3. During SWRs, synchronous neuronal discharge is seen first in the CA3a and b subregions, and later in CA3c and CA1, suggesting a flow of activity (Csicsvari et al. 2000). More than 10 % of the CA3 population must burst synchronously in order to significantly increase firing rates in CA1 (Csicsvari et al. 2000). This burst in CA3 activity results in the sharp wave of the SWR, a large deflection lasting $\sim$200 ms (Buzsaki 1986) in the LFP signal thought to be caused by the massive depolarization of CA1 pyramidal cells from the influx of CA3 excitation. The sharp wave lasts the duration of the SWR ($\sim$100 ms or more; Buzsaki et al. 1992). Once discharged by CA3, the spiking of CA1 pyramidal cells then drives the firing of basket and chandelier interneurons in CA1 (Csicsvari et al. 1999b). The fast feedback between excitation and inhibition in the local network results in the characteristic high frequency ripple oscillation detected in CA1, with principal cells firing at the trough of the oscillation and interneurons firing at the peaks (Buzsaki et al. 1992; Ylinen et al. 1995; Csicsvari et al. 1999b; Cutsuridis and Taxidis 2013). Pharmacological blockade of GABAergic signaling eliminates SWRs, demonstrating a dependence on local inhibition to pace and sustain the high frequency oscillation (Stark et al. 2014). Recently, the sharp wave depolarization of CA1 dendrites was shown to be critical for the long term potentiation of CA3-to-CA1 synapses between cell pairs involved in SWR replay events (Sadowski et al. 2016), although the role of local inhibition in this plasticity remains unknown.

Strikingly, CA1 can generate and maintain brief, high frequency events even when CA3 inputs are removed (Nakashiba et al. 2009). However, these SWR-like events oscillate at a lower frequency than normal SWRs and do not show the ordered reactivation of ensembles indicative of replay (Nakashiba et al. 2009; Stark et al. 2014). This further suggests that CA3 provides important excitatory drive to CA1 that activates the pyramidal cell ensembles underlying meaningful replay events.

Surprisingly, although ensembles active during SWRs often include neurons from the CA3 and CA1 regions of both hemispheres, ripple oscillations themselves are not always coherent across subregions or hemispheres (Csicsvari et al. 1999a; Sullivan et al. 2011; Ylinen et al. 1995). This suggests that a separate network oscillation is necessary for coordinating place cell spiking activity across regions. Recently, a transient increase in slow gamma (20–50 Hz) power was observed in CA1 and CA3 during SWRs which may be an important organizational signal. Specifically, slow gamma coherence between CA3 and CA1 increases during awake SWRs and is positively correlated with higher fidelity of replay, suggesting that slow gamma may coordinate replay activity between the two regions (Carr et al. 2012). As mentioned previously, slow gamma is thought to originate in CA3 (Colgin et al. 2009) and shows highest power in the CA1 SR, where input from

CA3 reaches CA1. Evidence of slow gamma in the DG also coherent with CA3 and CA1 during SWRs (Gillespie et al. 2016), suggests that slow gamma activity might coordinate SWR activity throughout the entire hippocampus. With an established role in binding ensembles at a time scale optimal for plasticity (Kopell et al. 2000; Axmacher et al. 2006; Bibbig et al. 2001; Wespatat et al. 2004; Isaac et al. 2009), slow gamma may provide the temporal organization critical for replay during SWRs (Carr et al. 2012; Colgin 2012). Furthermore, recent evidence shows that spatial trajectories during replay events do not proceed at a constant rate, but instead alternate between virtual movement and stillness in a manner time-locked to the slow gamma rhythm (Pfeiffer and Foster 2015). Together, these findings suggest that slow gamma activity in CA1 and CA3 plays an important organizational role during SWRs.

7 CA2

Relative to CA3 and CA1, CA2 has received relatively little attention in the hippocampal literature. CA2 was first characterized as a distinct hippocampal subregion due to the absence of both thorny excrescences on dendrites and afferent mossy fibers from the DG, thus distinguishing it from CA3, while its enlarged somata distinguished it from CA1 (Lorente de Nó 1934; Ishizuka et al. 1995). More recently, CA2 has been distinguished by a variety of molecular markers (San Antonio et al. 2014; Lee et al. 2010; Vellano et al. 2011; Lein et al. 2004, 2005; Zhao et al. 2001), distinct reciprocal connections with the supramammillary nucleus of the hypothalamus (Cui et al. 2013), and the expression of receptors for the neuromodulators vasopressin and adenosine (Young et al. 2006; Ochiishi et al. 1999). Partially because of these markers, there have been some recent advances in our understanding of the physiology and function of CA2. However, molecularly defined CA2 neurons are intermingled with overlapping cells from proximal CA1 and CA3a (Lein et al. 2005; Hitti and Siegelbaum 2014; Dudek et al. 2016), making it a challenge to study CA2 neurons specifically.

Electrophysiological properties of CA2 remained almost completely unexplored until recently, but it is now clear that CA2 is more than simply a relay center between CA3 and CA1 (Sekino et al. 1997; Bartesaghi and Gessi 2004; Jones and McHugh 2011; Mercer et al. 2007). Plasticity at the CA3-CA2 synapse is surprisingly hard to induce, and CA2 cells do not show changes in synaptic strength with conventional protocols of long term potentiation and depression used in CA3 and CA1 (Simons et al. 2009; Zhao et al. 2007). This is partially because CA2 cells have more negative resting membrane potentials than CA3/CA1 neurons, thus requiring greater input current to fire spikes necessary for activity-dependent synaptic plasticity (Zhao et al. 2007). In addition, CA2 interneurons provide feedback and feedforward inhibition to CA3 and CA1, which may be important for circuit function (Mercer et al. 2007, 2012; Valero et al. 2015). CA2 pyramidal neurons are also unique in their dendritic integration. Unlike CA1 neurons, CA2 neurons are more strongly excited by EC layer II synapses onto their distal dendrites than by CA3 synapses onto their proximal dendrites (Chevaleyre and Siegelbaum 2010). CA2 could thus act as a strong relay from EC to CA1, which may account for the persistence of spatial encoding in CA1 in the absence of CA3 input (Chevaleyre and Siegelbaum 2010; Nakashiba et al. 2008). This idea is supported by slice physiology demonstrating strong excitation of CA1 pyramidal cells at synapses from CA2 (Chevaleyre and Siegelbaum 2010; Kohara et al. 2014).

One of the most intriguing functional roles for CA2 is its contribution to social recognition memory (Hitti and Siegelbaum 2014; Stevenson and Caldwell 2014). This was originally proposed based on strong vasopressinergic afferents to CA2 from the paraventricular nucleus of the hypothalamus (Cui et al. 2013) and the potentiation of CA2 responses to vasopressin (Pagani et al. 2015), given vasopressin's known role in social behavior (Wersinger et al. 2002, 2004; DeVito et al. 2009). This role was specifically tested in mice with inhibited CA2 synaptic output. These mice were significantly impaired in their ability to recognize and distinguish familiar companion mice from novel mice. However, spatial working memory and general sociability was left intact (Hitti and Siegelbaum 2014). The relevance of social experience to CA2 is further supported by recent findings showing that social experience prompts CA2 to globally remap in the absence of any environmental change (Alexander et al. 2016). These studies are the first to link CA2 neural activity with behavioral consequences.

7.1 Cell Characteristics of CA2

CA2 place cells were previously thought to be functionally indistinguishable from CA1 place cells (Martig and Mizumori 2011). Recent studies, however, have suggested unique encoding properties of CA2 cells. Compared to CA1 and CA3 place cells, cells recorded in CA2 tend to have higher mean firing rates, larger spatial coverage of the environment and more spatial firing fields per cell (Lu et al. 2015; Mankin et al. 2015). In addition, they exhibit lower preference for spatial context, such that putative CA2 cells distinguish different environments much more weakly than do CA1 and CA3 cells. Notably, a recent study reports that the CA2 population representation of similar environments decorrelates over extended periods of time, indicating that CA2 cells may facilitate the encoding of similar memory episodes occurring at different time points (Mankin et al. 2015).

Recent evidence suggests that the CA2 neurons described by Mankin and colleagues may comprise a subpopulation of the neurons found at the CA2 anatomical locus. In addition to these neurons, there also exists a subpopulation of cells with spatially specific fields that are predominantly active during periods of low velocity and immobility. These neurons located in CA2 encode the animal's location in the absence of movement, and continue to encode for location during periods of sleep characterized by desynchronized activity (Kay et al. 2016).

7.2 CA2 Network Activity

Very little is known about network activity in CA2. The frequency of theta detected in CA2 is similar to CA1, and CA2 neurons exhibit comparable levels of theta modulation to CA1 neurons (Kay et al. 2016; Mankin et al. 2015), with slightly slower theta phase precession in cells with enlarged place fields (Mankin et al. 2015). During SWRs, recent findings have highlighted a potentially unique activity pattern in CA2. In a study of a small population of cells, Valero et al. (2015) demonstrated that CA2 neurons hyperpolarize during SWRs in vivo while CA1 and CA3 neurons are depolarized and excited. In line with this evidence, it has been hypothesized that CA2 would be largely suppressed during strong activation of CA3 as is thought to occur during SWRs (Jones and McHugh 2011). Notably, the subpopulation of neurons located in CA2 that were uniquely active during immobility were found to be either unmodulated or negatively modulated by SWRs, raising the possibility that lack of participation in SWRs is a defining property of CA2 neurons (Kay et al. 2016).

8 Subiculum

The subiculum is a major site of hippocampal output and is responsible for distributing information received from CA1 pyramidal cells to the neocortex and subcortical structures, as its projections are generally much more numerous than those directly from CA1 (Meibach and Siegel 1977; Rosene and Van Hoesen 1977; Swanson and Cowan 1977; O'Mara 2006). Lesions of the subiculum result in comparable impairment of learning and memory performance to lesions of the hippocampus proper (Morris et al. 1990). Lesions of both hippocampus and subiculum cause a more severe impairment than either alone, suggesting that the subiculum adds a layer of functionality rather than simply relaying hippocampal output (Morris et al. 1990; Potvin et al. 2006, 2007). The dorsal subiculum is thought to be particularly important for spatial memory, and evidence exists to suggest that subicular neurons integrate many layers of information from CA1 and project rich representations to downstream areas. In addition, the ventral subiculum, with substantial inputs from and outputs to subcortical areas, plays a key role in the inhibition of the hypothalamic--pituitary–adrenal axis, thus serving as the interface between hippocampal mnemonic activity and the limbic stress response (O'Mara 2006).

8.1 Cell Characteristics of the Subiculum

Historically, subicular principal cells have often been categorized into two populations: bursting cells and regular firing cells (Witter 2006). It has been suggested that bursting cells are more likely to target more spatial areas, like the MEC, while regular spiking cells may send more projections to the amygdala and LEC (Kim and Spruston 2012). While some evidence suggests that subicular neurons lie on a spectrum of "burstiness" regardless of their location in the subiculum (Kim et al. 2012), other findings indicate that burstiness, as well as firing rates and spatial tuning, depends on the cell's position within the subiculum relative to CA1 (Staff et al. 2000; Sharp and Green 1994; Sharp 2006; Jarsky et al. 2008). Indeed, firing patterns of subicular neurons differ substantially along the proximal-distal axis (Sharp 1996), with pyramidal cells in the proximal subiculum (nearest CA1) exhibiting lower firing rates and relatively smaller firing fields, similar to CA1 place cells, than neurons in the distal subiculum (Kim et al. 2012). The dispersed but rate-modulated firing fields of distal subicular neurons allow them to encode more spatial information than CA1 place cells, and may enable the efficient transfer of this information to the neocortex (Kim et al. 2012). Further in support of this idea, many CA1 neurons are thought to converge onto single subicular neurons, potentially explaining their complex rather than simply location-modulated firing fields. This may allow the subiculum to integrate dispersed CA1 information into a more compressed, rich representation for broad dispersal throughout the brain (O'Mara 2005; Deadwyler and Hampson 2004).

8.2 Subiculum Network Activity

Although generally understudied, several findings inform our understanding of network activity in the subiculum during various behavioral states. During the locomotor state, subicular neurons show robust phase precession similar to CA1 (Kim et al. 2012). During awake immobility and rest, SWRs can be detected in subiculum at the same time as in CA1 (Chrobak and Buzsaki 1994, 1996; Bohm et al. 2015). SWRs differentially modulate populations of subicular neurons; bursting cells tend to be activated during SWRs, while regular-firing cells are suppressed by SWRs (Bohm et al. 2015; Eller et al. 2015). The isolated subiculum has been shown to generate both slow and fast gamma oscillations in vitro (Jackson et al. 2011), suggesting that the subiculum may participate in gamma activity during the locomotor state and SWRs. Beyond this, little is understood of how the subiculum engages in and contributes to network activity in order to facilitate widespread distribution of hippocampal output.

9 Hippocampal Function Along the Dorsoventral Axis

The vast majority of our knowledge about hippocampal activity comes from studies of the dorsal hippocampus (dHPC). This is primarily because it is easily accessible for electrophysiological recording, whereas recording from a deep brain structure such as ventral hippocampus (vHPC) is more technically challenging. Thus far, the hypothesized functional distinction between dHPC and vHPC (which assigns them roles in spatial in spatial and emotional processing, respectively) comes mostly from anatomical connectivity, lesion studies, differences in the spatial specificity of each region's place fields, and differential genetic expression (Moser and Moser 1998; Dong et al. 2009; Fanselow and Dong 2010; Strange et al. 2014). In particular, genetic markers are expressed in a gradient along the dorsoventral axis and highly differentiates ventral pyramidal neurons from dorsal pyramidal neurons (Cembrowski et al. 2016). Dorsal HPC receives more visual cortical inputs via the EC, while vHPC receives more olfactory and gustatory inputs. In terms of outputs, dHPC projects directly to the deep layers of the EC and mostly indirectly, via the subiculum, to other cortical areas such as the retrosplenial cortex and prefrontal cortices. In contrast, vHPC projects not only to the deep layers of the EC, but also heavily and directly to the medial prefrontal cortex, orbitofrontal cortex, olfactory and auditory cortices, amygdala (Cenquizca and Swanson 2007), nucleus accumbens (Groenewegen et al. 1987; Brog et al. 1993), and hypothalamus (Cenquizca and Swanson 2006). These distinct anatomical outputs suggest that vHPC dominates the hippocampal innervation of areas associated with processing emotional information, such as anxiety and reward. Lesion studies further implicate the dHPC in spatial learning to a greater extent than vHPC (Moser et al. 1993, 1995; Pothuizen et al. 2004; Ferbinteanu and McDonald 2001; Bannerman et al. 1999; Richmond et al. 1999), which is implicated in anxiety (Kjelstrup et al. 2002; Bannerman et al. 2004; Henke 1990; Weeden et al. 2015; Wang et al. 2013; Zhang et al. 2001). However, the results of these studies may depend on the animal's training protocol, such that the vHPC may be sufficient for spatial learning over longer time periods while dHPC may be required for rapid acquisition of spatial tasks (de Hoz et al. 2003; Loureiro et al. 2012; Bast 2007; Bast et al. 2009). Ventral HPC may also be important for developing representations of environmental context over time (Komorowski et al. 2013) and transferring spatial knowledge across contexts to facilitate learning in new environments based on prior experience (de Hoz and Martin 2014).

9.1 Cell Characteristics Across the Dorsoventral Axis

The landmark electrophysiological finding that distinguished dHPC and vHPC is that place fields increase gradually in size along the dorsoventral axis, spatially restricted in dHPC and reaching vast spatial coverage at the ventral pole (Poucet et al. 1994; Jung et al. 1994; Kjelstrup et al. 2008; Royer et al. 2010; Komorowski et al. 2013; Ciocchi et al. 2015). This is consistent with an increase in the spacing of grid cell fields along the dorsoventral axis of entorhinal cortex (Stensola et al. 2012; Brun et al. 2008b). From these findings it was hypothesized that vHPC could not accurately encode for spatial information using cells with such little spatial specificity. Recently, however, a study of ventral CA1 in mice suggested that the distributed representations of space in vHPC may actually be ideal for transmitting spatial information to downstream regions (Keinath et al. 2014). While single cell spatial selectivity is lower in ventral CA1 due to large place field sizes, the population of ventral CA1 cells encodes an animal's location just as accurately as the dorsal CA1 population. This is possible via population coding: with some overlap in cells' place fields, each spatial location will be represented by a specific subset of cells firing at specific rates (Keinath et al. 2014; Kim et al. 2012; Osborne et al. 2008; Olypher et al. 2003). In addition to spatial information, non-spatial correlates seem to increase in strength in vHPC, in contrast to dHPC where representation of variables other than place is limited. For example, some ventral CA3 cells are active in motivationally related, but spatially distinct, areas of an environment such as reward sites (Royer et al. 2010; Ciocchi et al. 2015). vHPC cells also tend to represent contextual similarity between locations (Royer et al. 2010) in a manner that develops with experience (Komorowski et al. 2013). In addition, place cells in vHPC have been shown to track changes in olfactory stimuli (Keinath et al. 2014; Petrulis et al. 2005) and may represent locations associated with elevated anxiety (Royer et al. 2010; Ciocchi et al. 2015). While the functional implications of these representations remain largely unexplored, optogenetic manipulation studies point to a role in anxiety behaviors for the ventral HPC (Kheirbek et al. 2013) and its projections to the medial prefrontal cortex (Padilla-Coreano et al. 2016) and a role for vHPC projections to the nucleus accumbens in promoting reward associations (Britt et al. 2012).

9.2 Network Activity Across the Dorsoventral Axis

Although few studies have recorded simultaneous network activity in dHPC and vHPC, several differences have been observed which may have functional relevance. Notably, the theta rhythm is weaker in power in vHPC and less modulated by run speed than in dHPC (Patel et al. 2012; Royer et al. 2010; Schmidt et al. 2013). However, vHPC place cells still exhibit theta phase precession even though they are large, with smaller incremental phase changes such that phase precession persists across the extent of the place field (Kjelstrup et al. 2008). vHPC theta is also shifted 180° relative to dHPC theta, suggesting that theta may be a wave that propagates along the dorsoventral axis (Patel et al. 2012; Lubenov and Siapas 2009). How this "traveling wave" might coordinate activity across the axis in the context of learning and memory remains an open question.

SWRs in vHPC are largely similar in their fundamental characteristics to SWRs in dHPC, with the exception that SWRs in vHPC are typically smaller in amplitude and slightly lower in oscillation frequency (Patel et al. 2013). This is likely due to the more diffuse pyramidal cell layers and lower burstiness of vHPC cells (Royer et al. 2010; Fanselow and Dong 2010). Ventral HPC SWRs tend to occur independently of dHPC SWRs in sleep, although they can also propagate along the entire dorsoventral axis in either direction. The degree of propagation, or synchrony, between dHPC and vHPC may depend on the amplitude of the SWR (Patel et al. 2013), indicating that larger SWRs engage a greater portion of the hippocampus and therefore may reflect increased cognitive demand during behavior. Interestingly, cells of ventral CA1 with tri-directional projections to the medial prefrontal cortex, amygdala, and nucleus accumbens are activated during SWRs in sleep to a greater degree than cells that project to only one or two of those regions (Ciocchi et al. 2015). This suggests that SWRs in vHPC are indeed important for integrating information and synchronizing transfer to distant hippocampal targets. The characteristics and function of vHPC SWRs in the awake state have yet to be explored.

10 Hippocampal Output

Neural activity time-locked to both hippocampal theta and SWRs has been observed in the downstream targets of the hippocampus, suggesting that these network patterns facilitate the integration of spatial information with other modalities. Theta has been posited as a coordinator of brain regions particularly during periods of attention and working memory. Theta power is positively correlated not only with velocity (Montgomery et al. 2009) but also with working memory demands (Belchior et al. 2014; Richard et al. 2013; Schmidt et al. 2013; Tesche and Karhu 2000). Moreover, theta coherence increases between hippocampus and its projection targets (such as mPFC) during coding phases of working memory tasks (Benchenane et al. 2010; Harris and Gordon 2015; Backus et al. 2016). In addition, phase precession relative to hippocampal theta has been observed in the prefrontal cortex (Jones and Wilson 2005; Siapas et al. 2005) as well as in spatially modulated, reward-predictive cells of the nucleus accumbens (van der Meer and Redish 2011; Malhotra et al. 2012). These results point to theta as a mechanism for broadcasting spatial information to downstream regions and coordinating spatial and non-spatial representations across distant brain areas.

There is also substantial evidence that SWRs engage not only the hippocampus but also its projection targets. SWRs can be detected in the deep layers of the EC following the occurrence of CA1 SWRs (Chrobak and Buzsaki 1994, 1996). Strikingly, SWRs also modulate cell ensembles in distant brain regions, including prefrontal cortex (Siapas and Wilson 1998; Wierzynski et al. 2009; Jadhav et al. 2016), visual cortex (Ji and Wilson 2007), nucleus accumbens (Lansink et al. 2008, 2009), and the ventral tegmental area (Gomperts et al. 2015). Activity time-locked to SWRs recorded elsewhere in the brain suggests that the information contained in SWRs is being communicated or integrated with other memory-related modalities (Chrobak and Buzsaki 1996; Siapas and Wilson 1998; Wierzynski et al. 2009; Sirota et al. 2003). SWR-triggered whole brain MRI in monkeys shows widespread activation of cortical areas and coincident suppression of subcortical areas (Logothetis et al. 2012), and the modulation of cohesive functional networks that have been associated with memory processes (Kaplan et al. 2016). These findings point to a pivotal role for SWRs in coordinating memory processes across the entire brain.

11 Conclusion

Since the identification of the hippocampus as a critical neural center for spatial mnemonic processing, our understanding of memory has expanded in stride with developments in neural recording techniques and subsequent discoveries of hippocampal network activity. In particular, by identifying the spatial aspect of memory as a key to decoding hippocampal activity, and beginning to tease apart the unique contributions of each hippocampal subregion to spatial representations, we can begin to understand the patterns of coordinated neural activity that underlie memory function. In this chapter, we have highlighted two key patterns of activity, theta and sharp-wave ripples, which are largely distinct during exploratory activity and quiescence and specialized within each subregion, but that both coordinate the activation and reactivation of neuronal ensembles with high temporal precision. Although our understanding of these patterns focuses largely on spatial memory, we are gradually appreciating the ability of the hippocampus to build relational maps between diverse types of information. Dense, large scale single unit and LFP recordings, optical recording methods, optogenetics, and many other developing techniques will continue to expand our ability to characterize and manipulate hippocampal network activity in order to further our grasp on hippocampal mnemonic processing.

Acknowledgements We wish to thank members of the Frank lab for their careful reviews and constructive comments on this book chapter. In particular, we thank Kenny Kay for the hippocampal recording data displayed in Fig. 2.

References

Ahmed OJ, Mehta MR (2009) The hippocampal rate code: anatomy, physiology and theory. Trends Neurosci 32(6):329–338

Ahmed OJ, Mehta MR (2012) Running speed alters the frequency of hippocampal gamma oscillations. J Neurosci 32(21):7373–7383

Ainge JA, Tamosiunaite M, Woergoetter F, Dudchenko PA (2007) Hippocampal CA1 place cells encode intended destination on a maze with multiple choice points. J Neurosci 27(36):9769–9779

Alexander GM, Farris S, Pirone JR, Zheng C, Colgin LL, Dudek SM (2016) Social and novel contexts modify hippocampal CA2 representations of space. Nat Commun 7:10300

Allen K, Rawlins JN, Bannerman DM, Csicsvari J (2012) Hippocampal place cells can encode multiple trial-dependent features through rate remapping. J Neurosci 32(42):14752–14766

Allen TA, Salz DM, McKenzie S, Fortin NJ (2016) Nonspatial sequence coding in CA1 neurons. J Neurosci 36(5):1547–1563

Amaral DG, Dolorfo C, Alvarez-Royo P (1991) Organization of CA1 projections to the subiculum: a PHA-L analysis in the rat. Hippocampus 1(4):415–435

Amaral DG, Ishizuka N, Claiborne B (1990) Neurons, numbers and the hippocampal network. Prog Brain Res 83:1–11

Amaral DG, Kurz J (1985) An analysis of the origins of the cholinergic and noncholinergic septal projections to the hippocampal formation of the rat. J Comp Neurol 240(1):37–59

Amaral DG, Witter MP (1989) The three-dimensional organization of the hippocampal formation: a review of anatomical data. Neuroscience 31(3):571–591

Anderson MI, Jeffery KJ (2003) Heterogeneous modulation of place cell firing by changes in context. J Neurosci 23(26):8827–8835

Arnolds DE, Lopes da Silva FH, Aitink JW, Kamp A, Boeijinga P (1980) The spectral properties of hippocampal EEG related to behaviour in man. Electroencephalogr Clin Neurophysiol 50(3–4):324–328

Axmacher N, Elger CE, Fell J (2008) Ripples in the medial temporal lobe are relevant for human memory consolidation. Brain 131(Pt 7):1806–1817

Axmacher N, Henseler MM, Jensen O, Weinreich I, Elger CE, Fell J (2010) Cross-frequency coupling supports multi-item working memory in the human hippocampus. Proc Natl Acad Sci USA 107(7):3228–3233

Axmacher N, Mormann F, Fernandez G, Elger CE, Fell J (2006) Memory formation by neuronal synchronization. Brain Res Rev 52(1):170–182

Backus AR, Schoffelen JM, Szebenyi S, Hanslmayr S, Doeller CF (2016) Hippocampal-prefrontal theta oscillations support memory integration. Curr Biol 26(4):450–457

Bannerman DM, Rawlins JN, McHugh SB, Deacon RM, Yee BK, Bast T, Zhang WN, Pothuizen HH, Feldon J (2004) Regional dissociations within the hippocampus–memory and anxiety. Neurosci Biobehav Rev 28(3):273–283

Bannerman DM, Yee BK, Good MA, Heupel MJ, Iversen SD, Rawlins JN (1999) Double dissociation of function within the hippocampus: a comparison of dorsal, ventral, and complete hippocampal cytotoxic lesions. Behav Neurosci 113(6):1170–1188

Barnes CA, McNaughton BL, Mizumori SJ, Leonard BW, Lin LH (1990) Comparison of spatial and temporal characteristics of neuronal activity in sequential stages of hippocampal processing. Prog Brain Res 83:287–300

Bartesaghi R, Gessi T (2004) Parallel activation of field CA2 and dentate gyrus by synaptically elicited perforant path volleys. Hippocampus 14(8):948–963

Bast T (2007) Toward an integrative perspective on hippocampal function: from the rapid encoding of experience to adaptive behavior. Rev Neurosci 18(3–4):253–281

Bast T, Wilson IA, Witter MP, Morris RG (2009) From rapid place learning to behavioral performance: a key role for the intermediate hippocampus. PLoS Biol 7(4):e1000089

Battaglia FP, Sutherland GR, McNaughton BL (2004a) Hippocampal sharp wave bursts coincide with neocortical “up-state” transitions. Learn Mem 11(6):697–704

Battaglia FP, Sutherland GR, McNaughton BL (2004b) Local sensory cues and place cell directionality: additional evidence of prospective coding in the hippocampus. J Neurosci 24 (19):4541–4550

Beckstead RM (1978) Afferent connections of the entorhinal area in the rat as demonstrated by retrograde cell-labeling with horseradish peroxidase. Brain Res 152(2):249–264

Belchior H, Lopes-Dos-Santos V, Tort AB, Ribeiro S (2014) Increase in hippocampal theta oscillations during spatial decision making. Hippocampus 24(6):693–702

Belluscio MA, Mizuseki K, Schmidt R, Kempter R, Buzsaki G (2012) Cross-frequency phase-phase coupling between theta and gamma oscillations in the hippocampus. J Neurosci 32 (2):423–435

Benchenane K, Peyrache A, Khamassi M, Tierney PL, Gioanni Y, Battaglia FP, Wiener SI (2010) Coherent theta oscillations and reorganization of spike timing in the hippocampal- prefrontal network upon learning. Neuron 66(6):921–936

Bendor D, Wilson MA (2012) Biasing the content of hippocampal replay during sleep. Nat Neurosci 15(10):1439–1444

Best PJ, Ranck JB Jr (1982) Reliability of the relationship between hippocampal unit activity and sensory-behavioral events in the rat. Exp Neurol 75(3):652–664

Bibbig A, Faulkner HJ, Whittington MA, Traub RD (2001) Self-organized synaptic plasticity contributes to the shaping of gamma and beta oscillations in vitro. J Neurosci 21(22):9053–9067

Bittner KC, Grienberger C, Vaidya SP, Milstein AD, Macklin JJ, Suh J, Tonegawa S, Magee JC (2015) Conjunctive input processing drives feature selectivity in hippocampal CA1 neurons. Nat Neurosci 18(8):1133–1142

Bohm C, Peng Y, Maier N, Winterer J, Poulet JF, Geiger JR, Schmitz D (2015) Functional diversity of subicular principal cells during hippocampal ripples. J Neurosci 35(40):13608–13618

Bonnevie T, Dunn B, Fyhn M, Hafting T, Derdikman D, Kubie JL, Roudi Y, Moser EI, Moser MB (2013) Grid cells require excitatory drive from the hippocampus. Nat Neurosci 16(3):309–317

Bott JB, Muller MA, Jackson J, Aubert J, Cassel JC, Mathis C, Goutagny R (2015) Spatial reference memory is associated with modulation of theta-gamma coupling in the dentate gyrus. Cereb Cortex

Bragin A, Jando G, Nadasdy Z, Hetke J, Wise K, Buzsaki G (1995a) Gamma (40–100 Hz) oscillation in the hippocampus of the behaving rat. J Neurosci 15(1 Pt 1):47–60

Bragin A, Jando G, Nadasdy Z, van Landeghem M, Buzsaki G (1995b) Dentate EEG spikes and associated interneuronal population bursts in the hippocampal hilar region of the rat. J Neurophysiol 73(4):1691–1705

Brandon MP, Koenig J, Leutgeb JK, Leutgeb S (2014) New and distinct hippocampal place codes are generated in a new environment during septal inactivation. Neuron 82(4):789–796

Brankack J, Stewart M, Fox SE (1993) Current source density analysis of the hippocampal theta rhythm: associated sustained potentials and candidate synaptic generators. Brain Res 615 (2):310–327

Britt JP, Benaliouad F, McDevitt RA, Stuber GD, Wise RA, Bonci A (2012) Synaptic and behavioral profile of multiple glutamatergic inputs to the nucleus accumbens. Neuron 76 (4):790–803

Brog JS, Salyapongse A, Deutch AY, Zahm DS (1993) The patterns of afferent innervation of the core and shell in the "accumbens" part of the rat ventral striatum: immunohistochemical detection of retrogradely transported fluoro-gold. J Comp Neurol 338(2):255–278

Brun VH, Leutgeb S, Wu HQ, Schwarcz R, Witter MP, Moser EI, Moser MB (2008a) Impaired spatial representation in CA1 after lesion of direct input from entorhinal cortex. Neuron 57 (2):290–302

Brun VH, Otnass MK, Molden S, Steffenach HA, Witter MP, Moser MB, Moser EI (2002) Place cells and place recognition maintained by direct entorhinal-hippocampal circuitry. Science 296 (5576):2243–2246

Brun VH, Solstad T, Kjelstrup KB, Fyhn M, Witter MP, Moser EI, Moser MB (2008b) Progressive increase in grid scale from dorsal to ventral medial entorhinal cortex. Hippocampus 18(12):1200–1212

Buckmaster PS, Wenzel HJ, Kunkel DD, Schwartzkroin PA (1996) Axon arbors and synaptic connections of hippocampal mossy cells in the rat in vivo. J Comp Neurol 366(2):271–292

Bullock TH, Buzsaki G, McClune MC (1990) Coherence of compound field potentials reveals discontinuities in the CA1-subiculum of the hippocampus in freely-moving rats. Neuroscience 38(3):609–619

Buzsaki G (1986) Hippocampal sharp waves: their origin and significance. Brain Res 398(2):242–252

Buzsaki G (1989) Two-stage model of memory trace formation: a role for "noisy" brain states. Neuroscience 31(3):551–570

Buzsaki G (2002) Theta oscillations in the hippocampus. Neuron 33(3):325–340

Buzsaki G (2004) Large-scale recording of neuronal ensembles. Nat Neurosci 7(5):446–451

Buzsaki G (2015) Hippocampal sharp wave-ripple: a cognitive biomarker for episodic memory and planning. Hippocampus

Buzsaki G, Anastassiou CA, Koch C (2012) The origin of extracellular fields and currents–EEG, ECoG, LFP and spikes. Nat Rev Neurosci 13(6):407–420

Buzsaki G, Buhl DL, Harris KD, Csicsvari J, Czeh B, Morozov A (2003) Hippocampal network patterns of activity in the mouse. Neuroscience 116(1):201–211

Buzsaki G, Czopf J, Kondakor I, Kellenyi L (1986) Laminar distribution of hippocampal rhythmic slow activity (RSA) in the behaving rat: current-source density analysis, effects of urethane and atropine. Brain Res 365(1):125–137
Buzsaki G, Horvath Z, Urioste R, Hetke J, Wise K (1992) High-frequency network oscillation in the hippocampus. Science 256(5059):1025–1027
Buzsaki G, Leung LW, Vanderwolf CH (1983) Cellular bases of hippocampal EEG in the behaving rat. Brain Res 287(2):139–171
Buzsaki G, Wang XJ (2012) Mechanisms of gamma oscillations. Annu Rev Neurosci 35:203–225
Carr MF, Jadhav SP, Frank LM (2011) Hippocampal replay in the awake state: a potential substrate for memory consolidation and retrieval. Nat Neurosci 14(2):147–153
Carr MF, Karlsson MP, Frank LM (2012) Transient slow gamma synchrony underlies hippocampal memory replay. Neuron 75(4):700–713
Cembrowski MS, Bachman JL, Wang L, Sugino K, Shields BC, Spruston N (2016) Spatial gene-expression gradients underlie prominent heterogeneity of CA1 pyramidal neurons. Neuron
Cenquizca LA, Swanson LW (2006) Analysis of direct hippocampal cortical field CA1 axonal projections to diencephalon in the rat. J Comp Neurol 497(1):101–114
Cenquizca LA, Swanson LW (2007) Spatial organization of direct hippocampal field CA1 axonal projections to the rest of the cerebral cortex. Brain Res Rev 56(1):1–26
Chen Z, Resnik E, McFarland JM, Sakmann B, Mehta MR (2011) Speed controls the amplitude and timing of the hippocampal gamma rhythm. PLoS ONE 6(6):e21408
Cheng S, Frank LM (2008) New experiences enhance coordinated neural activity in the hippocampus. Neuron 57(2):303–313
Chevaleyre V, Siegelbaum SA (2010) Strong CA2 pyramidal neuron synapses define a powerful disynaptic cortico-hippocampal loop. Neuron 66(4):560–572
Chrobak JJ, Buzsaki G (1994) Selective activation of deep layer (V-VI) retrohippocampal cortical neurons during hippocampal sharp waves in the behaving rat. J Neurosci 14(10):6160–6170
Chrobak JJ, Buzsaki G (1996) High-frequency oscillations in the output networks of the hippocampal-entorhinal axis of the freely behaving rat. J Neurosci 16(9):3056–3066
Ciocchi S, Passecker J, Malagon-Vina H, Mikus N, Klausberger T (2015) Brain computation. Selective information routing by ventral hippocampal CA1 projection neurons. Science 348 (6234):560–563
Clelland CD, Choi M, Romberg C, Clemenson GD Jr, Fragniere A, Tyers P, Jessberger S, Saksida LM, Barker RA, Gage FH, Bussey TJ (2009) A functional role for adult hippocampal neurogenesis in spatial pattern separation. Science 325(5937):210–213
Colgin LL (2012) Slow gamma takes the reins in replay. Neuron 75(4):549–550
Colgin LL (2015) Theta-gamma coupling in the entorhinal-hippocampal system. Curr Opin Neurobiol 31:45–50
Colgin LL, Denninger T, Fyhn M, Hafting T, Bonnevie T, Jensen O, Moser MB, Moser EI (2009) Frequency of gamma oscillations routes flow of information in the hippocampus. Nature 19;462(7271):353–357
Colgin LL, Leutgeb S, Jezek K, Leutgeb JK, Moser EI, McNaughton BL, Moser MB (2010) Attractor-map versus autoassociation based attractor dynamics in the hippocampal network. J Neurophysiol 104(1):35–50
Cravens CJ, Vargas-Pinto N, Christian KM, Nakazawa K (2006) CA3 NMDA receptors are crucial for rapid and automatic representation of context memory. Eur J Neurosci 24(6):1771–1780
Creer DJ, Romberg C, Saksida LM, van Praag H, Bussey TJ (2010) Running enhances spatial pattern separation in mice. Proc Natl Acad Sci USA 107(5):2367–2372
Csicsvari J, Hirase H, Czurko A, Mamiya A, Buzsaki G (1999a) Fast network oscillations in the hippocampal CA1 region of the behaving rat. J Neurosci 19(16):RC20
Csicsvari J, Hirase H, Czurko A, Mamiya A, Buzsaki G (1999b) Oscillatory coupling of hippocampal pyramidal cells and interneurons in the behaving rat. J Neurosci 19(1):274–287
Csicsvari J, Hirase H, Mamiya A, Buzsaki G (2000) Ensemble patterns of hippocampal CA3-CA1 neurons during sharp wave-associated population events. Neuron 28(2):585–594

Csicsvari J, Jamieson B, Wise KD, Buzsaki G (2003) Mechanisms of gamma oscillations in the hippocampus of the behaving rat. Neuron 37(2):311–322

Csicsvari J, O'Neill J, Allen K, Senior T (2007) Place-selective firing contributes to the reverse-order reactivation of CA1 pyramidal cells during sharp waves in open-field exploration. Eur J Neurosci 26(3):704–716

Cui Z, Gerfen CR, Young WS 3rd (2013) Hypothalamic and other connections with dorsal CA2 area of the mouse hippocampus. J Comp Neurol 521(8):1844–1866

Cutsuridis V, Taxidis J (2013) Deciphering the role of CA1 inhibitory circuits in sharp wave-ripple complexes. Front Syst Neurosci 7:13

Danielson NB, Kaifosh P, Zaremba JD, Lovett-Barron M, Tsai J, Denny CA, Balough EM, Goldberg AR, Drew LJ, Hen R, Losonczy A, Kheirbek MA (2016) Distinct contribution of adult-born hippocampal granule cells to context encoding. Neuron

Davidson TJ, Kloosterman F, Wilson MA (2009) Hippocampal replay of extended experience. Neuron 63(4):497–507

de Hoz L, Knox J, Morris RG (2003) Longitudinal axis of the hippocampus: both septal and temporal poles of the hippocampus support water maze spatial learning depending on the training protocol. Hippocampus 13(5):587–603

de Hoz L, Martin SJ (2014) Double dissociation between the contributions of the septal and temporal hippocampus to spatial learning: the role of prior experience. Hippocampus 24 (8):990–1005

de Lavilleon G, Lacroix MM, Rondi-Reig L, Benchenane K (2015) Explicit memory creation during sleep demonstrates a causal role of place cells in navigation. Nat Neurosci 18(4):493–495

Deadwyler SA, Hampson RE (2004) Differential but complementary mnemonic functions of the hippocampus and subiculum. Neuron 42(3):465–476

Deller T, Martinez A, Nitsch R, Frotscher M (1996) A novel entorhinal projection to the rat dentate gyrus: direct innervation of proximal dendrites and cell bodies of granule cells and GABAergic neurons. J Neurosci 16(10):3322–3333

Deshmukh SS, Knierim JJ (2011) Representation of non-spatial and spatial information in the lateral entorhinal cortex. Front Behav Neurosci 5:69

Deshmukh SS, Knierim JJ (2013) Influence of local objects on hippocampal representations: landmark vectors and memory. Hippocampus 23(4):253–267

Deshmukh SS, Yoganarasimha D, Voicu H, Knierim JJ (2010) Theta modulation in the medial and the lateral entorhinal cortices. J Neurophysiol 104(2):994–1006.

DeVito LM, Konigsberg R, Lykken C, Sauvage M, Young WS 3rd, Eichenbaum H (2009) Vasopressin 1b receptor knock-out impairs memory for temporal order. J Neurosci 29 (9):2676–2683

Diba K, Buzsaki G (2007) Forward and reverse hippocampal place-cell sequences during ripples. Nat Neurosci 10(10):1241–1242

Dolleman-Van der Weel MJ, Witter MP (1996) Projections from the nucleus reuniens thalami to the entorhinal cortex, hippocampal field CA1, and the subiculum in the rat arise from different populations of neurons. J Comp Neurol 364(4):637–650

Dong HW, Swanson LW, Chen L, Fanselow MS, Toga AW (2009) Genomic-anatomic evidence for distinct functional domains in hippocampal field CA1. Proc Natl Acad Sci USA 106 (28):11794–11799

Dudek SM, Alexander GM, Farris S (2016) Rediscovering area CA2: unique properties and functions. Nat Rev Neurosci 17(2):89–102

Duncan K, Ketz N, Inati SJ, Davachi L (2012) Evidence for area CA1 as a match/mismatch detector: a high-resolution fMRI study of the human hippocampus. Hippocampus 22(3):389–398

Dupret D, O'Neill J, Pleydell-Bouverie B, Csicsvari J (2010) The reorganization and reactivation of hippocampal maps predict spatial memory performance. Nat Neurosci 13(8):995–1002

Dupret D, Revest JM, Koehl M, Ichas F, De Giorgi F, Costet P, Abrous DN, Piazza PV (2008) Spatial relational memory requires hippocampal adult neurogenesis. PLoS ONE 3(4):e1959

Duvarci S, Pare D (2014) Amygdala microcircuits controlling learned fear. Neuron 82(5):966–980
Ego-Stengel V, Wilson MA (2007) Spatial selectivity and theta phase precession in CA1 interneurons. Hippocampus 17(2):161–174
Ego-Stengel V, Wilson MA (2010) Disruption of ripple-associated hippocampal activity during rest impairs spatial learning in the rat. Hippocampus 20(1):1–10
Eichenbaum H (2014) Time cells in the hippocampus: a new dimension for mapping memories. Nat Rev Neurosci 15(11):732–744
Eichenbaum H, Cohen NJ (2001) From conditioning to conscious recollection. Oxford University Press, New York
Eichenbaum H, Cohen NJ (2014) Can we reconcile the declarative memory and spatial navigation views on hippocampal function? Neuron 83(4):764–770
Eichenbaum H, Kuperstein M, Fagan A, Nagode J (1987) Cue-sampling and goal-approach correlates of hippocampal unit activity in rats performing an odor-discrimination task. J Neurosci 7(3):716–732
Eichenbaum H, Sauvage M, Fortin N, Komorowski R, Lipton P (2012) Towards a functional organization of episodic memory in the medial temporal lobe. Neurosci Biobehav Rev 36 (7):1597–1608
Eller J, Zarnadze S, Bauerle P, Dugladze T, Gloveli T (2015) Cell type-specific separation of subicular principal neurons during network activities. PLoS ONE 10(4):e0123636
Eschenko O, Molle M, Born J, Sara SJ (2006) Elevated sleep spindle density after learning or after retrieval in rats. J Neurosci 26(50):12914–12920
Eschenko O, Ramadan W, Molle M, Born J, Sara SJ (2008) Sustained increase in hippocampal sharp-wave ripple activity during slow-wave sleep after learning. Learn Mem 15(4):222–228
Esposito MS, Piatti VC, Laplagne DA, Morgenstern NA, Ferrari CC, Pitossi FJ, Schinder AF (2005) Neuronal differentiation in the adult hippocampus recapitulates embryonic development. J Neurosci 25(44):10074–10086
Fanselow MS, Dong HW (2010) Are the dorsal and ventral hippocampus functionally distinct structures? Neuron 65(1):7–19
Felix-Ortiz AC, Beyeler A, Seo C, Leppla CA, Wildes CP, Tye KM (2013) BLA to vHPC inputs modulate anxiety-related behaviors. Neuron 79(4):658–664
Felix-Ortiz AC, Tye KM (2014) Amygdala inputs to the ventral hippocampus bidirectionally modulate social behavior. J Neurosci 34(2):586–595
Fell J, Axmacher N (2011) The role of phase synchronization in memory processes. Nat Rev Neurosci 12(2):105–118
Feng T, Silva D, Foster DJ (2015) Dissociation between the experience-dependent development of hippocampal theta sequences and single-trial phase precession. J Neurosci 35(12):4890–4902
Fenton AA, Kao HY, Neymotin SA, Olypher A, Vayntrub Y, Lytton WW, Ludvig N (2008) Unmasking the CA1 ensemble place code by exposures to small and large environments: more place cells and multiple, irregularly arranged, and expanded place fields in the larger space. J Neurosci 28(44):11250–11262
Ferbinteanu J, McDonald RJ (2001) Dorsal/ventral hippocampus, fornix, and conditioned place preference. Hippocampus 11(2):187–200
Ferbinteanu J, Shapiro ML (2003) Prospective and retrospective memory coding in the hippocampus. Neuron 40(6):1227–1239
Finch DM, Nowlin NL, Babb TL (1983) Demonstration of axonal projections of neurons in the rat hippocampus and subiculum by intracellular injection of HRP. Brain Res 271(2):201–216
Foster DJ, Wilson MA (2006) Reverse replay of behavioural sequences in hippocampal place cells during the awake state. Nature 440(7084):680–683
Foster TC, Castro CA, McNaughton BL (1989) Spatial selectivity of rat hippocampal neurons: dependence on preparedness for movement. Science 244(4912):1580–1582
Fox SE, Ranck JB Jr (1981) Electrophysiological characteristics of hippocampal complex-spike cells and theta cells. Exp Brain Res 41(3–4):399–410
Frank LM, Brown EN, Wilson M (2000) Trajectory encoding in the hippocampus and entorhinal cortex. Neuron 27(1):169–178

Frank LM, Stanley GB, Brown EN (2004) Hippocampal plasticity across multiple days of exposure to novel environments. J Neurosci 24(35):7681–7689
Freeman WJ (2007) Definitions of state variables and state space for brain-computer interface: Part 1. Multiple hierarchical levels of brain function. Cogn Neurodyn 1(1):3–14
Freund TF, Antal M (1988) GABA-containing neurons in the septum control inhibitory interneurons in the hippocampus. Nature 336(6195):170–173
Freund TF, Buzsaki G (1996) Interneurons of the hippocampus. Hippocampus 6(4):347–470
Fries P (2005) A mechanism for cognitive dynamics: neuronal communication through neuronal coherence. Trends Cogn Sci 9(10):474–480
Frotscher M, Leranth C (1985) Cholinergic innervation of the rat hippocampus as revealed by choline acetyltransferase immunocytochemistry: a combined light and electron microscopic study. J Comp Neurol 239(2):237–246
Fyhn M, Molden S, Hollup S, Moser MB, Moser E (2002) Hippocampal neurons responding to first-time dislocation of a target object. Neuron 35(3):555–566
Gallagher M, Chiba AA (1996) The amygdala and emotion. Curr Opin Neurobiol 6(2):221–227
Garthe A, Behr J, Kempermann G (2009) Adult-generated hippocampal neurons allow the flexible use of spatially precise learning strategies. PLoS ONE 4(5):e5464
Ge S, Sailor KA, Ming GL, Song H (2008) Synaptic integration and plasticity of new neurons in the adult hippocampus. J Physiol 586(16):3759–3765
Gerstner W, Kempter R, van Hemmen JL, Wagner H (1996) A neuronal learning rule for sub-millisecond temporal coding. Nature 383(6595):76–81
Gilbert PE, Brushfield AM (2009) The role of the CA3 hippocampal subregion in spatial memory: a process oriented behavioral assessment. Prog Neuropsychopharmacol Biol Psych 33(5):774–781
Gillespie AK, Jones EA, Lin Y-H, Karlsson MP, Kay K, Yoon SY, Tong LM, Nova P, Carr JS, Frank LM, Huang Y (2016) Apolipoprotein E4 causes age-dependent disruption of slow gamma oscillations during hippocampal sharp-wave ripples. Neuron 90(4):740–51
Girardeau G, Benchenane K, Wiener SI, Buzsaki G, Zugaro MB (2009) Selective suppression of hippocampal ripples impairs spatial memory. Nat Neurosci 12(10):1222–1223
Givens BS, Olton DS (1990) Cholinergic and GABAergic modulation of medial septal area: effect on working memory. Behav Neurosci 104(6):849–855
Gold AE, Kesner RP (2005) The role of the CA3 subregion of the dorsal hippocampus in spatial pattern completion in the rat. Hippocampus 15(6):808–814
Gomperts SN, Kloosterman F, Wilson MA (2015) VTA neurons coordinate with the hippocampal reactivation of spatial experience. eLife 4
Gonzales RB, DeLeon Galvan CJ, Rangel YM, Claiborne BJ (2001) Distribution of thorny excrescences on CA3 pyramidal neurons in the rat hippocampus. J Comp Neurol 430(3):357–368
Grastyan E, Lissak K, Madarasz I, Donhoffer H (1959) Hippocampal electrical activity during the development of conditioned reflexes. Electroencephalogr Clin Neurophysiol 11(3):409–430
Gray CM, Maldonado PE, Wilson M, McNaughton B (1995) Tetrodes markedly improve the reliability and yield of multiple single- unit isolation from multi-unit recordings in cat striate cortex. J Neurosci Methods 63(1–2):43–54
Green JD, Arduini AA (1954) Hippocampal electrical activity in arousal. J Neurophysiol 17 (6):533–557
Groenewegen HJ, Vermeulen-Van der Zee E, te Kortschot A, Witter MP (1987) Organization of the projections from the subiculum to the ventral striatum in the rat. A study using anterograde transport of Phaseolus vulgaris leucoagglutinin. Neuroscience 23(1):103–120
Grosmark AD, Mizuseki K, Pastalkova E, Diba K, Buzsaki G (2012) REM sleep reorganizes hippocampal excitability. Neuron 75(6):1001–1007
Gu Y, Arruda-Carvalho M, Wang J, Janoschka SR, Josselyn SA, Frankland PW, Ge S (2012) Optical controlling reveals time-dependent roles for adult-born dentate granule cells. Nat Neurosci 15(12):1700–1706

Gulyas AI, Hajos N, Katona I, Freund TF (2003) Interneurons are the local targets of hippocampal inhibitory cells which project to the medial septum. Eur J Neurosci 17(9):1861–1872
Gupta AS, van der Meer MA, Touretzky DS, Redish AD (2010) Hippocampal replay is not a simple function of experience. Neuron 65(5):695–705
Gupta AS, van der Meer MA, Touretzky DS, Redish AD (2012) Segmentation of spatial experience by hippocampal theta sequences. Nat Neurosci 15(7):1032–1039
Guzowski JF, Knierim JJ, Moser EI (2004) Ensemble dynamics of hippocampal regions CA3 and CA1. Neuron 44(4):581–584
Guzowski JF, McNaughton BL, Barnes CA, Worley PF (1999) Environment-specific expression of the immediate-early gene Arc in hippocampal neuronal ensembles. Nat Neurosci 2 (12):1120–1124
Hafting T, Fyhn M, Bonnevie T, Moser MB, Moser EI (2008) Hippocampus-independent phase precession in entorhinal grid cells. Nature 453(7199):1248–1252
Hafting T, Fyhn M, Molden S, Moser MB, Moser EI (2005) Microstructure of a spatial map in the entorhinal cortex. Nature 436(7052):801–806
Hales JB, Schlesiger MI, Leutgeb JK, Squire LR, Leutgeb S, Clark RE (2014) Medial entorhinal cortex lesions only partially disrupt hippocampal place cells and hippocampus-dependent place memory. Cell Rep 9(3):893–901
Harris AZ, Gordon JA (2015) Long-range neural synchrony in behavior. Annu Rev Neurosci 38:171–194
Hasselmo ME (2005) What is the function of hippocampal theta rhythm?-Linking behavioral data to phasic properties of field potential and unit recording data. Hippocampus 15(7):936–949
Hasselmo ME, Bodelon C, Wyble BP (2002) A proposed function for hippocampal theta rhythm: separate phases of encoding and retrieval enhance reversal of prior learning. Neural Comput 14 (4):793–817
Hasselmo ME, Schnell E (1994) Laminar selectivity of the cholinergic suppression of synaptic transmission in rat hippocampal region CA1: computational modeling and brain slice physiology. J Neurosci 14(6):3898–3914
Henke PG (1990) Hippocampal pathway to the amygdala and stress ulcer development. Brain Res Bull 25(5):691–695
Henze DA, Buzsaki G (2007) Hilar mossy cells: functional identification and activity in vivo. Prog Brain Res 163:199–810
Herkenham M (1978) The connections of the nucleus reuniens thalami: evidence for a direct thalamo-hippocampal pathway in the rat. J Comp Neurol 177(4):589–610
Hill AJ (1978) First occurrence of hippocampal spatial firing in a new environment. Exp Neurol 62 (2):282–297
Hitti FL, Siegelbaum SA (2014) The hippocampal CA2 region is essential for social memory. Nature 508(7494):88–92
Hok V, Lenck-Santini PP, Roux S, Save E, Muller RU, Poucet B (2007) Goal-related activity in hippocampal place cells. J Neurosci 27(3):472–482
Hopfield JJ (1995) Pattern recognition computation using action potential timing for stimulus representation. Nature 376(6535):33–36
Huxter J, Burgess N, O'Keefe J (2003) Independent rate and temporal coding in hippocampal pyramidal cells. Nature 425(6960):828–832
Hyman JM, Wyble BP, Goyal V, Rossi CA, Hasselmo ME (2003) Stimulation in hippocampal region CA1 in behaving rats yields long-term potentiation when delivered to the peak of theta and long-term depression when delivered to the trough. J Neurosci 23(37):11725–11731
Igarashi KM, Lu L, Colgin LL, Moser MB, Moser EI (2014) Coordination of entorhinal-hippocampal ensemble activity during associative learning. Nature 510 (7503):143–147
Isaac JT, Buchanan KA, Muller RU, Mellor JR (2009) Hippocampal place cell firing patterns can induce long-term synaptic plasticity in vitro. J Neurosci 29(21):6840–6850
Ishizuka N (2001) Laminar organization of the pyramidal cell layer of the subiculum in the rat. J Comp Neurol 435(1):89–110

Ishizuka N, Cowan WM, Amaral DG (1995) A quantitative analysis of the dendritic organization of pyramidal cells in the rat hippocampus. J Comp Neurol 362(1):17–45
Ishizuka N, Weber J, Amaral DG (1990) Organization of intrahippocampal projections originating from CA3 pyramidal cells in the rat. J Comp Neurol 295(4):580–623
Isomura Y, Sirota A, Ozen S, Montgomery S, Mizuseki K, Henze DA, Buzsaki G (2006) Integration and segregation of activity in entorhinal-hippocampal subregions by neocortical slow oscillations. Neuron 52(5):871–882
Ito HT, Zhang SJ, Witter MP, Moser EI, Moser MB (2015) A prefrontal-thalamo-hippocampal circuit for goal-directed spatial navigation. Nature 522(7554):50–55
Itskov V, Pastalkova E, Mizuseki K, Buzsaki G, Harris KD (2008) Theta-mediated dynamics of spatial information in hippocampus. J Neurosci 28(23):5959–5964
Jackson J, Goutagny R, Williams S (2011) Fast and slow gamma rhythms are intrinsically and independently generated in the subiculum. J Neurosci 31(34):12104–12117
Jackson JC, Johnson A, Redish AD (2006) Hippocampal sharp waves and reactivation during awake states depend on repeated sequential experience. J Neurosci 26(48):12415–12426
Jadhav SP, Kemere C, German PW, Frank LM (2012) Awake hippocampal sharp-wave ripples support spatial memory. Science 336(6087):1454–1458
Jadhav SP, Rothschild G, Roumis DK, Frank LM (2016) Coordinated excitation and inhibition of prefrontal ensembles during awake hippocampal sharp-wave ripple events. Neuron 90(1):113–127
Janak PH, Tye KM (2015) From circuits to behaviour in the amygdala. Nature 517(7534):284–292
Jarosiewicz B, McNaughton BL, Skaggs WE (2002) Hippocampal population activity during the small-amplitude irregular activity state in the rat. J Neurosci 22(4):1373–1384
Jarsky T, Mady R, Kennedy B, Spruston N (2008) Distribution of bursting neurons in the CA1 region and the subiculum of the rat hippocampus. J Comp Neurol 506(4):535–547
Jensen O, Lisman JE (2000) Position reconstruction from an ensemble of hippocampal place cells: contribution of theta phase coding. J Neurophysiol 83(5):2602–2609
Jessberger S, Clark RE, Broadbent NJ, Clemenson GD Jr, Consiglio A, Lie DC, Squire LR, Gage FH (2009) Dentate gyrus-specific knockdown of adult neurogenesis impairs spatial and object recognition memory in adult rats. Learn Mem 16(2):147–154
Ji D, Wilson MA (2007) Coordinated memory replay in the visual cortex and hippocampus during sleep. Nat Neurosci 10(1):100–107
Jog MS, Connolly CI, Kubota Y, Iyengar DR, Garrido L, Harlan R, Graybiel AM (2002) Tetrode technology: advances in implantable hardware, neuroimaging, and data analysis techniques. J Neurosci Methods 117(2):141–152
Johnson A, Redish AD (2007) Neural ensembles in CA3 transiently encode paths forward of the animal at a decision point. J Neurosci 27(45):12176–12189
Johnston ST, Shtrahman M, Parylak S, Goncalves JT, Gage FH (2016) Paradox of pattern separation and adult neurogenesis: A dual role for new neurons balancing memory resolution and robustness. Neurobiol Learn Mem 129:60–68
Jones MW, McHugh TJ (2011) Updating hippocampal representations: CA2 joins the circuit. Trends Neurosci 34(10):526–535. (S0166-2236(11)00123-8)
Jones MW, Wilson MA (2005) Theta rhythms coordinate hippocampal-prefrontal interactions in a spatial memory task. PLoS Biol 3(12):e402
Jung MW, McNaughton BL (1993) Spatial selectivity of unit activity in the hippocampal granular layer. Hippocampus 3(2):165–182
Jung MW, Wiener SI, McNaughton BL (1994) Comparison of spatial firing characteristics of units in dorsal and ventral hippocampus of the rat. J Neurosci 14(12):7347–7356
Jutras MJ, Fries P, Buffalo EA (2013) Oscillatory activity in the monkey hippocampus during visual exploration and memory formation. Proc Natl Acad Sci USA 110(32):13144–13149
Kajikawa Y, Schroeder CE (2011) How local is the local field potential? Neuron 72(5):847–858
Kajiwara R, Wouterlood FG, Sah A, Boekel AJ, Baks-te Bulte LT, Witter MP (2008) Convergence of entorhinal and CA3 inputs onto pyramidal neurons and interneurons in hippocampal area CA1–an anatomical study in the rat. Hippocampus 18(3):266–280

Kali S, Dayan P (2000) The involvement of recurrent connections in area CA3 in establishing the properties of place fields: a model. J Neurosci 20(19):7463–7477
Kali S, Freund TF (2005) Distinct properties of two major excitatory inputs to hippocampal pyramidal cells: a computational study. Eur J Neurosci 22(8):2027–2048
Kamondi A, Acsady L, Wang XJ, Buzsaki G (1998) Theta oscillations in somata and dendrites of hippocampal pyramidal cells in vivo: activity-dependent phase-precession of action potentials. Hippocampus 8(3):244–261
Kaplan R, Adhikari MH, Hindriks R, Mantini D, Murayama Y, Logothetis NK, Deco G (2016) Hippocampal sharp-wave ripples influence selective activation of the default mode network. Curr Biol 26(5):686–691
Karlsson MP, Frank LM (2008) Network dynamics underlying the formation of sparse, informative representations in the hippocampus. J Neurosci 28(52):14271–14281
Karlsson MP, Frank LM (2009) Awake replay of remote experiences in the hippocampus. Nat Neurosci 12(7):913–918
Kay K, Sosa M, Chung JE, Karlsson MP, Larkin MC, Frank LM (2016) A hippocampal network for spatial coding during immobility and sleep. Nature 531(7593):185–190
Keinath AT, Wang ME, Wann EG, Yuan RK, Dudman JT, Muzzio IA (2014) Precise spatial coding is preserved along the longitudinal hippocampal axis. Hippocampus 24(12):1533–1548
Kemere C, Carr MF, Karlsson MP, Frank LM (2013) Rapid and continuous modulation of hippocampal network state during exploration of new places. PLoS ONE 8(9):e73114
Kemp IR, Kaada BR (1975) The relation of hippocampal theta activity to arousal, attentive behaviour and somato-motor movements in unrestrained cats. Brain Res 95(2–3):323–342
Kennedy PJ, Shapiro ML (2009) Motivational states activate distinct hippocampal representations to guide goal-directed behaviors. Proc Natl Acad Sci USA 106(26):10805–10810
Kerr KM, Agster KL, Furtak SC, Burwell RD (2007) Functional neuroanatomy of the parahippocampal region: the lateral and medial entorhinal areas. Hippocampus 17(9):697–708
Kesner RP (2007) Behavioral functions of the CA3 subregion of the hippocampus. Learn Mem 14 (11):771–781
Kheirbek MA, Drew LJ, Burghardt NS, Costantini DO, Tannenholz L, Ahmari SE, Zeng H, Fenton AA, Hen R (2013) Differential control of learning and anxiety along the dorsoventral axis of the dentate gyrus. Neuron 77(5):955–968
Kim SM, Ganguli S, Frank LM (2012) Spatial information outflow from the hippocampal circuit: distributed spatial coding and phase precession in the subiculum. J Neurosci 32(34):11539–11558
Kim Y, Spruston N (2012) Target-specific output patterns are predicted by the distribution of regular-spiking and bursting pyramidal neurons in the subiculum. Hippocampus 22(4):693–706
Kjelstrup KB, Solstad T, Brun VH, Hafting T, Leutgeb S, Witter MP, Moser EI, Moser MB (2008) Finite scale of spatial representation in the hippocampus. Science 321(5885):140–143
Kjelstrup KG, Tuvnes FA, Steffenach HA, Murison R, Moser EI, Moser MB (2002) Reduced fear expression after lesions of the ventral hippocampus. Proc Natl Acad Sci USA 99(16):10825–10830
Klausberger T, Somogyi P (2008) Neuronal diversity and temporal dynamics: the unity of hippocampal circuit operations. Science 321(5885):53–57
Knierim JJ, Lee I, Hargreaves EL (2006) Hippocampal place cells: parallel input streams, subregional processing, and implications for episodic memory. Hippocampus 16(9):755–764
Knierim JJ, Neunuebel JP (2016) Tracking the flow of hippocampal computation: Pattern separation, pattern completion, and attractor dynamics. Neurobiol Learn Mem 129:38–49
Kobayashi T, Nishijo H, Fukuda M, Bures J, Ono T (1997) Task-dependent representations in rat hippocampal place neurons. J Neurophysiol 78(2):597–613
Kocsis B, Vertes RP (1997) Phase relations of rhythmic neuronal firing in the supramammillary nucleus and mammillary body to the hippocampal theta activity in urethane anesthetized rats. Hippocampus 7(2):204–214

Koenig J, Linder AN, Leutgeb JK, Leutgeb S (2011) The spatial periodicity of grid cells is not sustained during reduced theta oscillations. Science 332(6029):592–595 (332/6029/592)
Kohara K, Pignatelli M, Rivest AJ, Jung HY, Kitamura T, Suh J, Frank D, Kajikawa K, Mise N, Obata Y, Wickersham IR, Tonegawa S (2014) Cell type-specific genetic and optogenetic tools reveal hippocampal CA2 circuits. Nat Neurosci 17(2):269–279
Kohler C (1985) Intrinsic projections of the retrohippocampal region in the rat brain. I. The subicular complex. J Comp Neurol 236(4):504–522
Komorowski RW, Garcia CG, Wilson A, Hattori S, Howard MW, Eichenbaum H (2013) Ventral hippocampal neurons are shaped by experience to represent behaviorally relevant contexts. J Neurosci 33(18):8079–8087
Komorowski RW, Manns JR, Eichenbaum H (2009) Robust conjunctive item-place coding by hippocampal neurons parallels learning what happens where. J Neurosci 29(31):9918–9929
Kopell N, Ermentrout GB, Whittington MA, Traub RD (2000) Gamma rhythms and beta rhythms have different synchronization properties. Proc Natl Acad Sci USA 97(4):1867–1872
Kraus BJ, Robinson RJ 2nd, White JA, Eichenbaum H, Hasselmo ME (2013) Hippocampal "time cells": time versus path integration. Neuron 78(6):1090–1101
Kropff E, Carmichael JE, Moser MB, Moser EI (2015) Speed cells in the medial entorhinal cortex. Nature 523(7561):419–424
Kudrimoti HS, Barnes CA, McNaughton BL (1999) Reactivation of hippocampal cell assemblies: effects of behavioral state, experience, and EEG dynamics. J Neurosci 19(10):4090–4101
Kumaran D, Maguire EA (2007) Which computational mechanisms operate in the hippocampus during novelty detection? Hippocampus 17(9):735–748
Kwag J, Paulsen O (2009) Bidirectional control of spike timing by GABA(A) receptor-mediated inhibition during theta oscillation in CA1 pyramidal neurons. NeuroReport 20(13):1209–1213
Lansink CS, Goltstein PM, Lankelma JV, Joosten RN, McNaughton BL, Pennartz CM (2008) Preferential reactivation of motivationally relevant information in the ventral striatum. J Neurosci 28(25):6372–6382
Lansink CS, Goltstein PM, Lankelma JV, McNaughton BL, Pennartz CM (2009) Hippocampus leads ventral striatum in replay of place-reward information. PLoS Biol 7(8):e1000173
Larimer P, Strowbridge BW (2008) Nonrandom local circuits in the dentate gyrus. J Neurosci 28 (47):12212–12223
Larkin MC, Lykken C, Tye LD, Wickelgren JG, Frank LM (2014) Hippocampal output area CA1 broadcasts a generalized novelty signal during an object-place recognition task. Hippocampus 24(7):773–783
Laurberg S (1979) Commissural and intrinsic connections of the rat hippocampus. J Comp Neurol 184(4):685–708
Lee AK, Wilson MA (2002) Memory of sequential experience in the hippocampus during slow wave sleep. Neuron 36(6):1183–1194
Lee H, Wang C, Deshmukh SS, Knierim JJ (2015) Neural population evidence of functional heterogeneity along the CA3 transverse axis: pattern completion versus pattern separation. Neuron 87(5):1093–1105
Lee I, Jerman TS, Kesner RP (2005) Disruption of delayed memory for a sequence of spatial locations following CA1- or CA3-lesions of the dorsal hippocampus. Neurobiol Learn Mem 84 (2):138–147
Lee I, Kesner RP (2003) Differential roles of dorsal hippocampal subregions in spatial working memory with short versus intermediate delay. Behav Neurosci 117(5):1044–1053
Lee I, Kesner RP (2004) Differential contributions of dorsal hippocampal subregions to memory acquisition and retrieval in contextual fear-conditioning. Hippocampus 14(3):301–310
Lee I, Rao G, Knierim JJ (2004a) A double dissociation between hippocampal subfields: differential time course of CA3 and CA1 place cells for processing changed environments. Neuron 42(5):803–815
Lee I, Yoganarasimha D, Rao G, Knierim JJ (2004b) Comparison of population coherence of place cells in hippocampal subfields CA1 and CA3. Nature 430(6998):456–459

Lee SE, Simons SB, Heldt SA, Zhao M, Schroeder JP, Vellano CP, Cowan DP, Ramineni S, Yates CK, Feng Y, Smith Y, Sweatt JD, Weinshenker D, Ressler KJ, Dudek SM, Hepler JR (2010) RGS14 is a natural suppressor of both synaptic plasticity in CA2 neurons and hippocampal-based learning and memory. Proc Natl Acad Sci USA 107(39):16994–16998

Lein ES, Callaway EM, Albright TD, Gage FH (2005) Redefining the boundaries of the hippocampal CA2 subfield in the mouse using gene expression and 3-dimensional reconstruction. J Comp Neurol 485(1):1–10

Lein ES, Zhao X, Gage FH (2004) Defining a molecular atlas of the hippocampus using DNA microarrays and high-throughput in situ hybridization. J Neurosci 24(15):3879–3889

Lenck-Santini PP, Rivard B, Muller RU, Poucet B (2005) Study of CA1 place cell activity and exploratory behavior following spatial and nonspatial changes in the environment. Hippocampus 15(3):356–369

Lengyel M, Kwag J, Paulsen O, Dayan P (2005) Matching storage and recall: hippocampal spike timing-dependent plasticity and phase response curves. Nat Neurosci 8(12):1677–1683

Leutgeb JK, Leutgeb S, Moser MB, Moser EI (2007) Pattern separation in the dentate gyrus and CA3 of the hippocampus. Science 315(5814):961–966

Leutgeb JK, Leutgeb S, Treves A, Meyer R, Barnes CA, McNaughton BL, Moser MB, Moser EI (2005a) Progressive transformation of hippocampal neuronal representations in "morphed" environments. Neuron 48(2):345–358

Leutgeb S, Leutgeb JK (2007) Pattern separation, pattern completion, and new neuronal codes within a continuous CA3 map. Learn Mem 14(11):745–757

Leutgeb S, Leutgeb JK, Barnes CA, Moser EI, McNaughton BL, Moser MB (2005b) Independent codes for spatial and episodic memory in hippocampal neuronal ensembles. Science 309 (5734):619–623

Leutgeb S, Leutgeb JK, Moser MB, Moser EI (2005c) Place cells, spatial maps and the population code for memory. Curr Opin Neurobiol 15(6):738–746

Leutgeb S, Leutgeb JK, Treves A, Moser MB, Moser EI (2004) Distinct ensemble codes in hippocampal areas CA3 and CA1. Science 305(5688):1295–1298

Lever C, Burton S, Jeewajee A, O'Keefe J, Burgess N (2009) Boundary vector cells in the subiculum of the hippocampal formation. J Neurosci 29(31):9771–9777

Lever C, Wills T, Cacucci F, Burgess N, O'Keefe J (2002) Long-term plasticity in hippocampal place-cell representation of environmental geometry. Nature 416(6876):90–94

Li S, Cullen WK, Anwyl R, Rowan MJ (2003) Dopamine-dependent facilitation of LTP induction in hippocampal CA1 by exposure to spatial novelty. Nat Neurosci 6(5):526–531

Li XG, Somogyi P, Ylinen A, Buzsaki G (1994) The hippocampal CA3 network: an in vivo intracellular labeling study. J Comp Neurol 339(2):181–208

Li Y, Aimone JB, Xu X, Callaway EM, Gage FH (2012) Development of GABAergic inputs controls the contribution of maturing neurons to the adult hippocampal network. Proc Natl Acad Sci USA 109(11):4290–4295

Lisman J (2003) Long-term potentiation: outstanding questions and attempted synthesis. Philos Trans R Soc Lond B Biol Sci 358(1432):829–842

Lisman JE (1999) Relating hippocampal circuitry to function: recall of memory sequences by reciprocal dentate-CA3 interactions. Neuron 22(2):233–242

Lisman JE, Grace AA (2005) The hippocampal-VTA loop: controlling the entry of information into long-term memory. Neuron 46(5):703–713

Lisman JE, Otmakhova NA (2001) Storage, recall, and novelty detection of sequences by the hippocampus: elaborating on the SOCRATIC model to account for normal and aberrant effects of dopamine. Hippocampus 11(5):551–568

Logothetis NK, Eschenko O, Murayama Y, Augath M, Steudel T, Evrard HC, Besserve M, Oeltermann A (2012) Hippocampal-cortical interaction during periods of subcortical silence. Nature 491(7425):547–553

Lorente de Nó R (1933) Studies on the structure of the cerebral cortex. I. The area entorhinalis. J Psychol Neurol 45:381–438

Lorente de Nó R (1934) Studies on the structure of the cerebral cortex. II. Continuation of the study of the ammonic system. J Psychol Neurol 46:113–177

Loughlin SE, Foote SL, Bloom FE (1986) Efferent projections of nucleus locus coeruleus: topographic organization of cells of origin demonstrated by three-dimensional reconstruction. Neuroscience 18(2):291–306

Louie K, Wilson MA (2001) Temporally structured replay of awake hippocampal ensemble activity during rapid eye movement sleep. Neuron 29(1):145–156

Loureiro M, Lecourtier L, Engeln M, Lopez J, Cosquer B, Geiger K, Kelche C, Cassel JC, Pereira de Vasconcelos A (2012) The ventral hippocampus is necessary for expressing a spatial memory. Brain Struct Funct 217(1):93–106

Lu L, Igarashi KM, Witter MP, Moser EI, Moser MB (2015) Topography of place maps along the CA3-to-CA2 axis of the hippocampus. Neuron 87(5):1078–1092

Lubenov EV, Siapas AG (2009) Hippocampal theta oscillations are travelling waves. Nature 459 (7246):534–539

MacDonald CJ, Carrow S, Place R, Eichenbaum H (2013) Distinct hippocampal time cell sequences represent odor memories in immobilized rats. J Neurosci 33(36):14607–14616

MacDonald CJ, Lepage KQ, Eden UT, Eichenbaum H (2011) Hippocampal "time cells" bridge the gap in memory for discontiguous events. Neuron 71(4):737–749

Maier N, Nimmrich V, Draguhn A (2003) Cellular and network mechanisms underlying spontaneous sharp wave-ripple complexes in mouse hippocampal slices. J Physiol 550(Pt 3):873–887

Malhotra S, Cross RW, van der Meer MA (2012) Theta phase precession beyond the hippocampus. Rev Neurosci 23(1):39–65

Mankin EA, Diehl GW, Sparks FT, Leutgeb S, Leutgeb JK (2015) Hippocampal CA2 activity patterns change over time to a larger extent than between spatial contexts. Neuron 85(1):190–201

Mankin EA, Sparks FT, Slayyeh B, Sutherland RJ, Leutgeb S, Leutgeb JK (2012) Neuronal code for extended time in the hippocampus. Proc Natl Acad Sci USA 109(47):19462–19467

Manns JR, Eichenbaum H (2009) A cognitive map for object memory in the hippocampus. Learn Mem 16(10):616–624

Marin-Burgin A, Mongiat LA, Pardi MB, Schinder AF (2012) Unique processing during a period of high excitation/inhibition balance in adult-born neurons. Science 335(6073):1238–1242

Markram H, Lubke J, Frotscher M, Sakmann B (1997) Regulation of synaptic efficacy by coincidence of postsynaptic APs and EPSPs. Science 275(5297):213–215

Markus EJ, Qin YL, Leonard B, Skaggs WE, Mcnaughton BL, Barnes CA (1995) Interactions between location and task affect the spatial and directional firing of hippocampal-neurons. J Neurosci 15(11):7079–7094

Marr D (1971) Simple memory: a theory for archicortex. Philos Trans R Soc Lond B Biol Sci 262 (841):23–81

Martig AK, Mizumori SJ (2011) Ventral tegmental area disruption selectively affects CA1/CA2 but not CA3 place fields during a differential reward working memory task. Hippocampus 21 (2):172–184

McAvoy K, Besnard A, Sahay A (2015) Adult hippocampal neurogenesis and pattern separation in DG: a role for feedback inhibition in modulating sparseness to govern population-based coding. Front Syst Neurosci 9:120

McClelland JL, Goddard NH (1996) Considerations arising from a complementary learning systems perspective on hippocampus and neocortex. Hippocampus 6(6):654–665

McKenzie S, Frank AJ, Kinsky NR, Porter B, Riviere PD, Eichenbaum H (2014) Hippocampal representation of related and opposing memories develop within distinct, hierarchically organized neural schemas. Neuron 83(1):202–215

McNaughton BL, Barnes CA (1977) Physiological identification and analysis of dentate granule cell responses to stimulation of the medial and lateral perforant pathways in the rat. J Comp Neurol 175(4):439–454

McNaughton BL, Barnes CA, Meltzer J, Sutherland RJ (1989) Hippocampal granule cells are necessary for normal spatial learning but not for spatially-selective pyramidal cell discharge. Exp Brain Res 76(3):485–496

McNaughton BL, Barnes CA, O'Keefe J (1983) The contributions of position, direction, and velocity to single unit activity in the hippocampus of freely-moving rats. Exp Brain Res 52 (1):41–49

McNaughton BL, Battaglia FP, Jensen O, Moser EI, Moser MB (2006) Path integration and the neural basis of the 'cognitive map'. Nat Rev Neurosci 7(8):663–678

Mcnaughton BL, Morris RG (1987) Hippocampal Synaptic Enhancement and Information-Storage within a Distributed Memory System. Trends Neurosci 10(10):408–415

Meeter M, Murre JM, Talamini LM (2004) Mode shifting between storage and recall based on novelty detection in oscillating hippocampal circuits. Hippocampus 14(6):722–741

Megias M, Emri Z, Freund TF, Gulyas AI (2001) Total number and distribution of inhibitory and excitatory synapses on hippocampal CA1 pyramidal cells. Neuroscience 102(3):527–540

Mehta MR, Lee AK, Wilson MA (2002) Role of experience and oscillations in transforming a rate code into a temporal code. Nature 417(6890):741–746

Mehta MR, Quirk MC, Wilson MA (2000) Experience-dependent asymmetric shape of hippocampal receptive fields. Neuron 25(3):707–715

Meibach RC, Siegel A (1977) Efferent connections of the hippocampal formation in the rat. Brain Res 124(2):197–224

Mercer A, Eastlake K, Trigg HL, Thomson AM (2012) Local circuitry involving parvalbumin-positive basket cells in the CA2 region of the hippocampus. Hippocampus 22 (1):43–56

Mercer A, Trigg HL, Thomson AM (2007) Characterization of neurons in the CA2 subfield of the adult rat hippocampus. J Neurosci 27(27):7329–7338

Milstein AD, Bloss EB, Apostolides PF, Vaidya SP, Dilly GA, Zemelman BV, Magee JC (2015) Inhibitory gating of input comparison in the CA1 microcircuit. Neuron 87(6):1274–1289

Mishkin M (1978) Memory in monkeys severely impaired by combined but not by separate removal of amygdala and hippocampus. Nature 273(5660):297–298

Mitchell SJ, Rawlins JN, Steward O, Olton DS (1982) Medial septal area lesions disrupt theta rhythm and cholinergic staining in medial entorhinal cortex and produce impaired radial arm maze behavior in rats. J Neurosci 2(3):292–302

Mitzdorf U (1985) Current source-density method and application in cat cerebral cortex: investigation of evoked potentials and EEG phenomena. Physiol Rev 65(1):37–100

Mizumori SJ, McNaughton BL, Barnes CA, Fox KB (1989) Preserved spatial coding in hippocampal CA1 pyramidal cells during reversible suppression of CA3c output: evidence for pattern completion in hippocampus. J Neurosci 9(11):3915–3928

Mizuseki K, Sirota A, Pastalkova E, Buzsaki G (2009) Theta oscillations provide temporal windows for local circuit computation in the entorhinal-hippocampal loop. Neuron 64(2):267–280

Moita MA, Rosis S, Zhou Y, LeDoux JE, Blair HT (2003) Hippocampal place cells acquire location-specific responses to the conditioned stimulus during auditory fear conditioning. Neuron 37(3):485–497

Molle M, Eschenko O, Gais S, Sara SJ, Born J (2009) The influence of learning on sleep slow oscillations and associated spindles and ripples in humans and rats. Eur J Neurosci 29(5):1071–1081

Montgomery SM, Betancur MI, Buzsaki G (2009) Behavior-dependent coordination of multiple theta dipoles in the hippocampus. J Neurosci 29(5):1381–1394

Montgomery SM, Buzsaki G (2007) Gamma oscillations dynamically couple hippocampal CA3 and CA1 regions during memory task performance. Proc Natl Acad Sci USA 104(36):14495–14500 0701826104 [pii]

/pnas.0701826104

Montgomery SM, Sirota A, Buzsaki G (2008) Theta and gamma coordination of hippocampal networks during waking and rapid eye movement sleep. J Neurosci 28(26):6731–6741

Morris RG, Garrud P, Rawlins JN, O'Keefe J (1982) Place navigation impaired in rats with hippocampal lesions. Nature 297(5868):681–683

Morris RG, Schenk F, Tweedie F, Jarrard LE (1990) Ibotenate lesions of hippocampus and/or subiculum: dissociating components of allocentric spatial learning. Eur J Neurosci 2(12):1016–1028

Moser E, Moser MB, Andersen P (1993) Spatial learning impairment parallels the magnitude of dorsal hippocampal lesions, but is hardly present following ventral lesions. J Neurosci 13 (9):3916–3925

Moser MB, Moser EI (1998) Functional differentiation in the hippocampus. Hippocampus 8 (6):608–619

Moser MB, Moser EI, Forrest E, Andersen P, Morris RG (1995) Spatial learning with a minislab in the dorsal hippocampus. Proc Natl Acad Sci USA 92(21):9697–9701

Muenzinger KF (1938) Vicarious trial and error at a point of choice. I. A general survey of its relation to learning efficiency. J Genet Psychol 12:75–86

Muller RU, Kubie JL (1987) The effects of changes in the environment on the spatial firing of hippocampal complex-spike cells. J Neurosci 7(7):1951–1968

Muller RU, Kubie JL, Ranck JB Jr (1987) Spatial firing patterns of hippocampal complex-spike cells in a fixed environment. J Neurosci 7(7):1935–1950

Nadasdy Z, Hirase H, Czurko A, Csicsvari J, Buzsaki G (1999) Replay and time compression of recurring spike sequences in the hippocampus. J Neurosci 19(21):9497–9507

Nakashiba T, Buhl DL, McHugh TJ, Tonegawa S (2009) Hippocampal CA3 output is crucial for ripple-associated reactivation and consolidation of memory. Neuron 62(6):781–787

Nakashiba T, Cushman JD, Pelkey KA, Renaudineau S, Buhl DL, McHugh TJ, Rodriguez Barrera V, Chittajallu R, Iwamoto KS, McBain CJ, Fanselow MS, Tonegawa S (2012) Young dentate granule cells mediate pattern separation, whereas old granule cells facilitate pattern completion. Cell 149(1):188–201

Nakashiba T, Young JZ, McHugh TJ, Buhl DL, Tonegawa S (2008) Transgenic inhibition of synaptic transmission reveals role of CA3 output in hippocampal learning. Science 319 (5867):1260–1264

Nakazawa K, Quirk MC, Chitwood RA, Watanabe M, Yeckel MF, Sun LD, Kato A, Carr CA, Johnston D, Wilson MA, Tonegawa S (2002) Requirement for hippocampal CA3 NMDA receptors in associative memory recall. Science 297(5579):211–218

Nakazawa K, Sun LD, Quirk MC, Rondi-Reig L, Wilson MA, Tonegawa S (2003) Hippocampal CA3 NMDA receptors are crucial for memory acquisition of one-time experience. Neuron 38 (2):305–315

Neunuebel JP, Knierim JJ (2012) Spatial firing correlates of physiologically distinct cell types of the rat dentate gyrus. J Neurosci 32(11):3848–3858

Neunuebel JP, Knierim JJ (2014) CA3 retrieves coherent representations from degraded input: direct evidence for CA3 pattern completion and dentate gyrus pattern separation. Neuron 81 (2):416–427

Neunuebel JP, Yoganarasimha D, Rao G, Knierim JJ (2013) Conflicts between local and global spatial frameworks dissociate neural representations of the lateral and medial entorhinal cortex. J Neurosci 33(22):9246–9258

Nitz D, McNaughton B (2004) Differential modulation of CA1 and dentate gyrus interneurons during exploration of novel environments. J Neurophysiol 91(2):863–872

Nokia MS, Mikkonen JE, Penttonen M, Wikgren J (2012) Disrupting neural activity related to awake-state sharp wave-ripple complexes prevents hippocampal learning. Front Behav Neurosci 6:84

O'Keefe J (1976) Place units in the hippocampus of the freely moving rat. Exp Neurol 51(1):78–109

O'Keefe J, Dostrovsky J (1971) The hippocampus as a spatial map. Preliminary evidence from unit activity in the freely-moving rat. Brain Res 34(1):171–175

O'Keefe J, Nadel L (1978) The hippocampus as a cognitive map. Oxford University Press, London

O'Keefe J, Recce ML (1993) Phase relationship between hippocampal place units and the EEG theta rhythm. Hippocampus 3(3):317–330
O'Mara S (2005) The subiculum: what it does, what it might do, and what neuroanatomy has yet to tell us. J Anat 207(3):271–282
O'Mara S (2006) Controlling hippocampal output: the central role of subiculum in hippocampal information processing. Behav Brain Res 174(2):304–312
O'Neill J, Pleydell-Bouverie B, Dupret D, Csicsvari J (2010) Play it again: reactivation of waking experience and memory. Trends Neurosci 33(5):220–229
O'Neill J, Senior T, Csicsvari J (2006) Place-selective firing of CA1 pyramidal cells during sharp wave/ripple network patterns in exploratory behavior. Neuron 49(1):143–155
O'Neill J, Senior TJ, Allen K, Huxter JR, Csicsvari J (2008) Reactivation of experience-dependent cell assembly patterns in the hippocampus. Nat Neurosci 11(2):209–215
Ochiishi T, Saitoh Y, Yukawa A, Saji M, Ren Y, Shirao T, Miyamoto H, Nakata H, Sekino Y (1999) High level of adenosine A1 receptor-like immunoreactivity in the CA2/CA3a region of the adult rat hippocampus. Neuroscience 93(3):955–967
Olton DS, Branch M, Best PJ (1978) Spatial correlates of hippocampal unit activity. Exp Neurol 58(3):387–409
Olton DS, Papas BC (1979) Spatial memory and hippocampal function. Neuropsychologia 17 (6):669–682
Olypher AV, Lansky P, Muller RU, Fenton AA (2003) Quantifying location-specific information in the discharge of rat hippocampal place cells. J Neurosci Methods 127(2):123–135
Osborne LC, Palmer SE, Lisberger SG, Bialek W (2008) The neural basis for combinatorial coding in a cortical population response. J Neurosci 28(50):13522–13531
Padilla-Coreano N, Bolkan SS, Pierce GM, Blackman DR, Hardin WD, Garcia-Garcia AL, Spellman TJ, Gordon JA (2016) Direct ventral hippocampal-prefrontal input is required for anxiety-related neural activity and behavior. Neuron 89(4):857–866
Pagani JH, Zhao M, Cui Z, Avram SK, Caruana DA, Dudek SM, Young WS (2015) Role of the vasopressin 1b receptor in rodent aggressive behavior and synaptic plasticity in hippocampal area CA2. Mol Psychiatry 20(4):490–499
Pan WX, McNaughton N (2002) The role of the medial supramammillary nucleus in the control of hippocampal theta activity and behaviour in rats. Eur J Neurosci 16(9):1797–1809
Papp G, Witter MP, Treves A (2007) The CA3 network as a memory store for spatial representations. Learn Mem 14(11):732–744
Park E, Dvorak D, Fenton AA (2011) Ensemble place codes in hippocampus: CA1, CA3, and dentate gyrus place cells have multiple place fields in large environments. PLoS ONE 6(7): e22349
Pastalkova E, Itskov V, Amarasingham A, Buzsaki G (2008) Internally generated cell assembly sequences in the rat hippocampus. Science 321(5894):1322–1327
Patel J, Fujisawa S, Berenyi A, Royer S, Buzsaki G (2012) Traveling theta waves along the entire septotemporal axis of the hippocampus. Neuron 75(3):410–417
Patel J, Schomburg EW, Berenyi A, Fujisawa S, Buzsaki G (2013) Local generation and propagation of ripples along the septotemporal axis of the hippocampus. J Neurosci 33 (43):17029–17041
Pavlides C, Winson J (1989) Influences of hippocampal place cell firing in the awake state on the activity of these cells during subsequent sleep episodes. J Neurosci 9(8):2907–2918
Paz R, Pare D (2013) Physiological basis for emotional modulation of memory circuits by the amygdala. Curr Opin Neurobiol 23(3):381–386
Penttonen M, Kamondi A, Sik A, Acsady L, Buzsaki G (1997) Feed-forward and feed-back activation of the dentate gyrus in vivo during dentate spikes and sharp wave bursts. Hippocampus 7(4):437–450
Petrulis A, Alvarez P, Eichenbaum H (2005) Neural correlates of social odor recognition and the representation of individual distinctive social odors within entorhinal cortex and ventral subiculum. Neuroscience 130(1):259–274

Petsche H, Stumpf C (1962) The origin of theta-rhytm in the rabbit hippocampus. Wien Klin Wochenschr 74:696–700
Pfeiffer BE, Foster DJ (2013) Hippocampal place-cell sequences depict future paths to remembered goals. Nature 497(7447):74–79
Pfeiffer BE, Foster DJ (2015) Place cells. Autoassociative dynamics in the generation of sequences of hippocampal place cells. Science 349(6244):180–183
Pikkarainen M, Ronkko S, Savander V, Insausti R, Pitkanen A (1999) Projections from the lateral, basal, and accessory basal nuclei of the amygdala to the hippocampal formation in rat. J Comp Neurol 403(2):229–260
Piskorowski RA, Chevaleyre V (2012) Synaptic integration by different dendritic compartments of hippocampal CA1 and CA2 pyramidal neurons. Cell Mol Life Sci 69(1):75–88
Pitkanen A, Pikkarainen M, Nurminen N, Ylinen A (2000) Reciprocal connections between the amygdala and the hippocampal formation, perirhinal cortex, and postrhinal cortex in rat—A review. Ann Ny Acad Sci 911:369–391
Pothuizen HH, Zhang WN, Jongen-Relo AL, Feldon J, Yee BK (2004) Dissociation of function between the dorsal and the ventral hippocampus in spatial learning abilities of the rat: a within-subject, within-task comparison of reference and working spatial memory. Eur J Neurosci 19(3):705–712
Potvin O, Allen K, Thibaudeau G, Dore FY, Goulet S (2006) Performance on spatial working memory tasks after dorsal or ventral hippocampal lesions and adjacent damage to the subiculum. Behav Neurosci 120(2):413–422
Potvin O, Dore FY, Goulet S (2007) Contributions of the dorsal hippocampus and the dorsal subiculum to processing of idiothetic information and spatial memory. Neurobiol Learn Mem 87(4):669–678
Poucet B, Thinus-Blanc C, Muller RU (1994) Place cells in the ventral hippocampus of rats. NeuroReport 5(16):2045–2048
Rajasethupathy P, Sankaran S, Marshel JH, Kim CK, Ferenczi E, Lee SY, Berndt A, Ramakrishnan C, Jaffe A, Lo M, Liston C, Deisseroth K (2015) Projections from neocortex mediate top-down control of memory retrieval. Nature 526(7575):653–659
Ramadan W, Eschenko O, Sara SJ (2009) Hippocampal sharp wave/ripples during sleep for consolidation of associative memory. PLoS ONE 4(8):e6697
y Cajal SR (1893) Estructura del asta de Ammon y fascia dentata. Ann Soc Esp Hist Nat 22
Rangel LM, Quinn LK, Chiba AA, Gage FH, Aimone JB (2013) A hypothesis for temporal coding of young and mature granule cells. Front Neurosci 7:75
Renno-Costa C, Lisman JE, Verschure PF (2014) A signature of attractor dynamics in the CA3 region of the hippocampus. PLoS Comput Biol 10(5):e1003641
Richard GR, Titiz A, Tyler A, Holmes GL, Scott RC, Lenck-Santini PP (2013) Speed modulation of hippocampal theta frequency correlates with spatial memory performance. Hippocampus 23 (12):1269–1279
Richmond MA, Yee BK, Pouzet B, Veenman L, Rawlins JN, Feldon J, Bannerman DM (1999) Dissociating context and space within the hippocampus: effects of complete, dorsal, and ventral excitotoxic hippocampal lesions on conditioned freezing and spatial learning. Behav Neurosci 113(6):1189–1203
Rolls ET (2007) An attractor network in the hippocampus: theory and neurophysiology. Learn Mem 14(11):714–731
Rosene DL, Van Hoesen GW (1977) Hippocampal efferents reach widespread areas of cerebral cortex and amygdala in the rhesus monkey. Science 198(4314):315–317
Roumis DK, Frank LM (2015) Hippocampal sharp-wave ripples in waking and sleeping states. Curr Opin Neurobiol 35:6–12
Rowland DC, Weible AP, Wickersham IR, Wu H, Mayford M, Witter MP, Kentros CG (2013) Transgenically targeted rabies virus demonstrates a major monosynaptic projection from hippocampal area CA2 to medial entorhinal layer II neurons. J Neurosci 33(37):14889–14898
Royer S, Sirota A, Patel J, Buzsaki G (2010) Distinct representations and theta dynamics in dorsal and ventral hippocampus. J Neurosci 30(5):1777–1787

Sadowski JH, Jones MW, Mellor JR (2016) Sharp-wave ripples orchestrate the induction of synaptic plasticity during reactivation of place cell firing patterns in the hippocampus. Cell Rep 14(8):1916–1929

Sahay A, Scobie KN, Hill AS, O'Carroll CM, Kheirbek MA, Burghardt NS, Fenton AA, Dranovsky A, Hen R (2011) Increasing adult hippocampal neurogenesis is sufficient to improve pattern separation. Nature 472(7344):466–470

San Antonio A, Liban K, Ikrar T, Tsyganovskiy E, Xu X (2014) Distinct physiological and developmental properties of hippocampal CA2 subfield revealed by using anti-Purkinje cell protein 4 (PCP4) immunostaining. J Comp Neurol 522(6):1333–1354

Sargolini F, Fyhn M, Hafting T, McNaughton BL, Witter MP, Moser MB, Moser EI (2006) Conjunctive representation of position, direction, and velocity in entorhinal cortex. Science 312 (5774):758–762

Sasaki T, Leutgeb S, Leutgeb JK (2015) Spatial and memory circuits in the medial entorhinal cortex. Curr Opin Neurobiol 32:16–23

Savanthrapadian S, Meyer T, Elgueta C, Booker SA, Vida I, Bartos M (2014) Synaptic properties of SOM- and CCK-expressing cells in dentate gyrus interneuron networks. J Neurosci 34 (24):8197–8209

Savelli F, Yoganarasimha D, Knierim JJ (2008) Influence of boundary removal on the spatial representations of the medial entorhinal cortex. Hippocampus 18(12):1270–1282

Scharfman HE (1994) Evidence from simultaneous intracellular recordings in rat hippocampal slices that area CA3 pyramidal cells innervate dentate hilar mossy cells. J Neurophysiol 72 (5):2167–2180

Scharfman HE (2007) The CA3 "backprojection" to the dentate gyrus. Prog Brain Res 163:627–637

Scharfman HE, Myers CE (2012) Hilar mossy cells of the dentate gyrus: a historical perspective. Front Neural Circuits 6:106

Scheffer-Teixeira R, Belchior H, Caixeta FV, Souza BC, Ribeiro S, Tort AB (2012) Theta phase modulates multiple layer-specific oscillations in the CA1 region. Cereb Cortex 22(10):2404–2414

Schiller D, Eichenbaum H, Buffalo EA, Davachi L, Foster DJ, Leutgeb S, Ranganath C (2015) Memory and space: towards an understanding of the cognitive map. J Neurosci 35(41):13904–13911

Schlesiger MI, Cannova CC, Boublil BL, Hales JB, Mankin EA, Brandon MP, Leutgeb JK, Leibold C, Leutgeb S (2015) The medial entorhinal cortex is necessary for temporal organization of hippocampal neuronal activity. Nat Neurosci 18(8):1123–1132

Schlesiger MI, Cressey JC, Boublil B, Koenig J, Melvin NR, Leutgeb JK, Leutgeb S (2013) Hippocampal activation during the recall of remote spatial memories in radial maze tasks. Neurobiol Learn Mem 106:324–333

Schmidt B, Hinman JR, Jacobson TK, Szkudlarek E, Argraves M, Escabi MA, Markus EJ (2013) Dissociation between dorsal and ventral hippocampal theta oscillations during decision-making. J Neurosci 33(14):6212–6224

Schmidt-Hieber C, Jonas P, Bischofberger J (2004) Enhanced synaptic plasticity in newly generated granule cells of the adult hippocampus. Nature 429(6988):184–187

Scoville WB, Milner B (1957) Loss of recent memory after bilateral hippocampal lesions. J Neurol Neurosurg Psychiatry 20(1):11–21

Sekino Y, Obata K, Tanifuji M, Mizuno M, Murayama J (1997) Delayed signal propagation via CA2 in rat hippocampal slices revealed by optical recording. J Neurophysiol 78(3):1662–1668

Sharp PE (1996) Multiple spatial/behavioral correlates for cells in the rat postsubiculum: multiple regression analysis and comparison to other hippocampal areas. Cereb Cortex 6(2):238–259

Sharp PE (2006) Subicular place cells generate the same "map" for different environments: Comparison with hippocampal cells. Behav Brain Res 174(2):206–214

Sharp PE, Green C (1994) Spatial correlates of firing patterns of single cells in the subiculum of the freely moving rat. J Neurosci 14(4):2339–2356

Shen J, Kudrimoti HS, McNaughton BL, Barnes CA (1998) Reactivation of neuronal ensembles in hippocampal dentate gyrus during sleep after spatial experience. J Sleep Res 7(Suppl 1):6–16
Shors TJ, Miesegaes G, Beylin A, Zhao M, Rydel T, Gould E (2001) Neurogenesis in the adult is involved in the formation of trace memories. Nature 410(6826):372–376
Siapas AG, Lubenov EV, Wilson MA (2005) Prefrontal phase locking to hippocampal theta oscillations. Neuron 46(1):141–151
Siapas AG, Wilson MA (1998) Coordinated interactions between hippocampal ripples and cortical spindles during slow-wave sleep. Neuron 21(5):1123–1128
Siegle JH, Wilson MA (2014) Enhancement of encoding and retrieval functions through theta phase-specific manipulation of hippocampus. eLife 3:e03061
Sik A, Penttonen M, Ylinen A, Buzsaki G (1995) Hippocampal CA1 interneurons: an in vivo intracellular labeling study. J Neurosci 15(10):6651–6665
Sik A, Ylinen A, Penttonen M, Buzsaki G (1994) Inhibitory CA1-CA3-hilar region feedback in the hippocampus. Science 265(5179):1722–1724
Silva D, Feng T, Foster DJ (2015) Trajectory events across hippocampal place cells require previous experience. Nat Neurosci 18(12):1772–1779
Simons SB, Escobedo Y, Yasuda R, Dudek SM (2009) Regional differences in hippocampal calcium handling provide a cellular mechanism for limiting plasticity. Proc Natl Acad Sci USA 106(33):14080–14084
Singer AC, Carr MF, Karlsson MP, Frank LM (2013) Hippocampal SWR activity predicts correct decisions during the initial learning of an alternation task. Neuron 77(6):1163–1173
Singer AC, Frank LM (2009) Rewarded outcomes enhance reactivation of experience in the hippocampus. Neuron 64(6):910–921
Singer AC, Karlsson MP, Nathe AR, Carr MF, Frank LM (2010) Experience-dependent development of coordinated hippocampal spatial activity representing the similarity of related locations. J Neurosci 30(35):11586–11604
Sirota A, Csicsvari J, Buhl D, Buzsaki G (2003) Communication between neocortex and hippocampus during sleep in rodents. Proc Natl Acad Sci USA 100(4):2065–2069
Skaggs WE, McNaughton BL (1996) Replay of neuronal firing sequences in rat hippocampus during sleep following spatial experience. Science 271(5257):1870–1873
Skaggs WE, McNaughton BL, Permenter M, Archibeque M, Vogt J, Amaral DG, Barnes CA (2007) EEG sharp waves and sparse ensemble unit activity in the macaque hippocampus. J Neurophysiol 98(2):898–910
Skaggs WE, McNaughton BL, Wilson MA, Barnes CA (1996) Theta phase precession in hippocampal neuronal populations and the compression of temporal sequences. Hippocampus 6(2):149–172
Solstad T, Boccara CN, Kropff E, Moser MB, Moser EI (2008) Representation of geometric borders in the entorhinal cortex. Science 322(5909):1865–1868
Spruston N (2008) Pyramidal neurons: dendritic structure and synaptic integration. Nat Rev Neurosci 9(3):206–221
Squire LR, Wixted JT (2011) The cognitive neuroscience of human memory since H.M. Annu Rev Neurosci 34:259–288
Staff NP, Jung HY, Thiagarajan T, Yao M, Spruston N (2000) Resting and active properties of pyramidal neurons in subiculum and CA1 of rat hippocampus. J Neurophysiol 84(5):2398–2408
Stark E, Roux L, Eichler R, Senzai Y, Royer S, Buzsaki G (2014) Pyramidal cell-interneuron interactions underlie hippocampal ripple oscillations. Neuron 83(2):467–480
Stensola H, Stensola T, Solstad T, Froland K, Moser MB, Moser EI (2012) The entorhinal grid map is discretized. Nature 492(7427):72–78
Steriade M, Nunez A, Amzica F (1993) A novel slow (< 1 Hz) oscillation of neocortical neurons in vivo: depolarizing and hyperpolarizing components. J Neurosci 13(8):3252–3265
Stevenson EL, Caldwell HK (2014) Lesions to the CA2 region of the hippocampus impair social memory in mice. Eur J Neurosci 40(9):3294–3301

Steward O, Scoville SA (1976) Cells of origin of entorhinal cortical afferents to the hippocampus and fascia dentata of the rat. J Comp Neurol 169(3):347–370

Stewart M, Fox SE (1990) Do septal neurons pace the hippocampal theta rhythm? Trends Neurosci 13(5):163–168

Strange BA, Witter MP, Lein ES, Moser EI (2014) Functional organization of the hippocampal longitudinal axis. Nat Rev Neurosci 15(10):655–669

Sullivan D, Csicsvari J, Mizuseki K, Montgomery S, Diba K, Buzsaki G (2011) Relationships between hippocampal sharp waves, ripples, and fast gamma oscillation: influence of dentate and entorhinal cortical activity. J Neurosci 31(23):8605–8616. 31/23/8605

Sun Y, Nguyen AQ, Nguyen JP, Le L, Saur D, Choi J, Callaway EM, Xu X (2014) Cell-type-specific circuit connectivity of hippocampal CA1 revealed through cre-dependent rabies tracing. Cell Rep 7(1):269–280

Suzuki SS, Smith GK (1987) Spontaneous EEG spikes in the normal hippocampus. I. Behavioral correlates, laminar profiles and bilateral synchrony. Electroencephalogr Clin Neurophysiol 67 (4):348–359

Swanson LW, Cowan WM (1977) An autoradiographic study of the organization of the efferent connections of the hippocampal formation in the rat. J Comp Neurol 172(1):49–84

Takacs VT, Klausberger T, Somogyi P, Freund TF, Gulyas AI (2012) Extrinsic and local glutamatergic inputs of the rat hippocampal CA1 area differentially innervate pyramidal cells and interneurons. Hippocampus 22(6):1379–1391

Tamamaki N, Abe K, Nojyo Y (1988) Three-dimensional analysis of the whole axonal arbors originating from single CA2 pyramidal neurons in the rat hippocampus with the aid of a computer graphic technique. Brain Res 452(1–2):255–272

Tao C, Zhang G, Xiong Y, Zhou Y (2015) Functional dissection of synaptic circuits: in vivo patch-clamp recording in neuroscience. Front Neural Circuits 9:23

Taube JS (2007) The head direction signal: origins and sensory-motor integration. Annu Rev Neurosci 30:181–207

Temprana SG, Mongiat LA, Yang SM, Trinchero MF, Alvarez DD, Kropff E, Giacomini D, Beltramone N, Lanuza GM, Schinder AF (2015) Delayed coupling to feedback inhibition during a critical period for the integration of adult-born granule cells. Neuron 85(1):116–130

Tesche CD, Karhu J (2000) Theta oscillations index human hippocampal activation during a working memory task. Proc Natl Acad Sci USA 97(2):919–924

Thompson LT, Best PJ (1990) Long-term stability of the place-field activity of single units recorded from the dorsal hippocampus of freely behaving rats. Brain Res 19;509(2):299–308

Tolman EC (1938) The determiners of behavior at a choice point. Psychol Rev 45:1–41

Tort AB, Komorowski R, Eichenbaum H, Kopell N (2010) Measuring phase-amplitude coupling between neuronal oscillations of different frequencies. J Neurophysiol 104(2):1195–1210

Tort AB, Komorowski RW, Manns JR, Kopell NJ, Eichenbaum H (2009) Theta-gamma coupling increases during the learning of item-context associations. Proc Natl Acad Sci USA 106 (49):20942–20947

Toth K, Borhegyi Z, Freund TF (1993) Postsynaptic targets of GABAergic hippocampal neurons in the medial septum-diagonal band of broca complex. J Neurosci 13(9):3712–3724

Toth K, Freund TF (1992) Calbindin D28 k-containing nonpyramidal cells in the rat hippocampus: their immunoreactivity for GABA and projection to the medial septum. Neuroscience 49 (4):793–805

Treves A, Rolls E (1991) What determines the capacity of autoassociative memories in the brain? Comp Neural Sys 2(4):371–397

Treves A, Rolls ET (1992) Computational constraints suggest the need for two distinct input systems to the hippocampal CA3 network. Hippocampus 2(2):189–199

Treves A, Rolls ET (1994) Computational analysis of the role of the hippocampus in memory. Hippocampus 4(3):374–391

Tronel S, Belnoue L, Grosjean N, Revest JM, Piazza PV, Koehl M, Abrous DN (2012) Adult-born neurons are necessary for extended contextual discrimination. Hippocampus 22(2):292–298

Ulanovsky N, Moss CF (2007) Hippocampal cellular and network activity in freely moving echolocating bats. Nat Neurosci 10(2):224–233
Valero M, Cid E, Averkin RG, Aguilar J, Sanchez-Aguilera A, Viney TJ, Gomez-Dominguez D, Bellistri E, de la Prida LM (2015) Determinants of different deep and superficial CA1 pyramidal cell dynamics during sharp-wave ripples. Nat Neurosci 18(9):1281–1290
Van Cauter T, Poucet B, Save E (2008) Unstable CA1 place cell representation in rats with entorhinal cortex lesions. Eur J Neurosci 27(8):1933–1946
van der Meer MA, Redish AD (2011) Theta phase precession in rat ventral striatum links place and reward information. J Neurosci 31(8):2843–2854
van Strien NM, Cappaert NL, Witter MP (2009) The anatomy of memory: an interactive overview of the parahippocampal-hippocampal network. Nat Rev Neurosci 10(4):272–282
Vanderwolf CH (1969) Hippocampal electrical activity and voluntary movement in the rat. Electroencephalogr Clin Neurophysiol 26(4):407–418
VanElzakker M, Fevurly RD, Breindel T, Spencer RL (2008) Environmental novelty is associated with a selective increase in Fos expression in the output elements of the hippocampal formation and the perirhinal cortex. Learn Mem 15(12):899–908
Vazdarjanova A, Guzowski JF (2004) Differences in hippocampal neuronal population responses to modifications of an environmental context: evidence for distinct, yet complementary, functions of CA3 and CA1 ensembles. J Neurosci 24(29):6489–6496
Vellano CP, Lee SE, Dudek SM, Hepler JR (2011) RGS14 at the interface of hippocampal signaling and synaptic plasticity. Trends Pharmacol Sci 32(11):666–674
Vertes RP, Hoover WB, Szigeti-Buck K, Leranth C (2007) Nucleus reuniens of the midline thalamus: link between the medial prefrontal cortex and the hippocampus. Brain Res Bull 71 (6):601–609
Vinogradova OS (2001) Hippocampus as comparator: role of the two input and two output systems of the hippocampus in selection and registration of information. Hippocampus 11 (5):578–598
Wang ME, Fraize NP, Yin L, Yuan RK, Petsagourakis D, Wann EG, Muzzio IA (2013) Differential roles of the dorsal and ventral hippocampus in predator odor contextual fear conditioning. Hippocampus 23(6):451–466
Watrous AJ, Lee DJ, Izadi A, Gurkoff GG, Shahlaie K, Ekstrom AD (2013) A comparative study of human and rat hippocampal low-frequency oscillations during spatial navigation. Hippocampus 23(8):656–661
Weeden CS, Roberts JM, Kamm AM, Kesner RP (2015) The role of the ventral dentate gyrus in anxiety-based behaviors. Neurobiol Learn Mem 118:143–149
Wersinger SR, Ginns EI, O'Carroll AM, Lolait SJ, Young WS 3rd (2002) Vasopressin V1b receptor knockout reduces aggressive behavior in male mice. Mol Psychiatry 7(9):975–984
Wersinger SR, Kelliher KR, Zufall F, Lolait SJ, O'Carroll AM, Young WS 3rd (2004) Social motivation is reduced in vasopressin 1b receptor null mice despite normal performance in an olfactory discrimination task. Horm Behav 46(5):638–645
Wespatat V, Tennigkeit F, Singer W (2004) Phase sensitivity of synaptic modifications in oscillating cells of rat visual cortex. J Neurosci 24(41):9067–9075
Wierzynski CM, Lubenov EV, Gu M, Siapas AG (2009) State-dependent spike-timing relationships between hippocampal and prefrontal circuits during sleep. Neuron 61(4):587–596
Wikenheiser AM, Redish AD (2015) Hippocampal theta sequences reflect current goals. Nat Neurosci 18(2):289–294
Wilson DI, Langston RF, Schlesiger MI, Wagner M, Watanabe S, Ainge JA (2013) Lateral entorhinal cortex is critical for novel object-context recognition. Hippocampus 23(5):352–366
Wilson MA, McNaughton BL (1994) Reactivation of hippocampal ensemble memories during sleep. Science 265(5172):676–679
Winson J (1972) Interspecies differences in the occurrence of theta. Behav Biol 7(4):479–487
Winson J, Abzug C (1978) Neuronal transmission through hippocampal pathways dependent on behavior. J Neurophys 41(3):716–732

Witter MP (1986) A survey of the anatomy of the hippocampal formation, with emphasis on the septotemporal organization of its intrinsic and extrinsic connections. Adv Exp Med Biol 203:67–82
Witter MP (1993) Organization of the entorhinal-hippocampal system: a review of current anatomical data. Hippocampus 3(Spec no):33–44
Witter MP (2006) Connections of the subiculum of the rat: topography in relation to columnar and laminar organization. Behav Brain Res 174(2):251–264
Witter MP (2007) Intrinsic and extrinsic wiring of CA3: indications for connectional heterogeneity. Learn Mem 14(11):705–713
Witter MP, Amaral DG (2004) Hippocampal formation. In: Paxinos G (ed) The rat nervous system. Elsevier Academic Press, San Diego, pp 635–704
Witter MP, Groenewegen HJ, Lopes da Silva FH, Lohman AH (1989) Functional organization of the extrinsic and intrinsic circuitry of the parahippocampal region. Prog Neurobiol 33(3):161–253
Witter MP, Wouterlood FG, Naber PA, Van Haeften T (2000) Anatomical organization of the parahippocampal-hippocampal network. Ann N Y Acad Sci 911:1–24
Wojtowicz JM, Askew ML, Winocur G (2008) The effects of running and of inhibiting adult neurogenesis on learning and memory in rats. Eur J Neurosci 27(6):1494–1502
Wood ER, Dudchenko PA, Robitsek RJ, Eichenbaum H (2000) Hippocampal neurons encode information about different types of memory episodes occurring in the same location. Neuron 27(3):623–633
Wu X, Foster DJ (2014) Hippocampal replay captures the unique topological structure of a novel environment. J Neurosci 34(19):6459–6469
Yartsev MM, Witter MP, Ulanovsky N (2011) Grid cells without theta oscillations in the entorhinal cortex of bats. Nature 479(7371):103–107
Ylinen A, Bragin A, Nadasdy Z, Jando G, Szabo I, Sik A, Buzsaki G (1995) Sharp wave-associated high-frequency oscillation (200 Hz) in the intact hippocampus: network and intracellular mechanisms. J Neurosci 15(1 Pt 1):30–46
Yoganarasimha D, Rao G, Knierim JJ (2011) Lateral entorhinal neurons are not spatially selective in cue-rich environments. Hippocampus 21(12):1363–1374
Young BJ, Fox GD, Eichenbaum H (1994) Correlates of hippocampal complex-spike cell activity in rats performing a nonspatial radial maze task. J Neurosci 14(11 Pt 1):6553–6563
Young WS, Li J, Wersinger SR, Palkovits M (2006) The vasopressin 1b receptor is prominent in the hippocampal area CA2 where it is unaffected by restraint stress or adrenalectomy. Neuroscience 143(4):1031–1039
Zhang SJ, Ye J, Miao C, Tsao A, Cerniauskas I, Ledergerber D, Moser MB, Moser EI (2013) Optogenetic dissection of entorhinal-hippocampal functional connectivity. Science 340 (6128):1232627
Zhang WN, Bast T, Feldon J (2001) The ventral hippocampus and fear conditioning in rats: different anterograde amnesias of fear after infusion of N-methyl-D-aspartate or its noncompetitive antagonist MK-801 into the ventral hippocampus. Behav Brain Res 126(1–2):159–174
Zhao M, Choi YS, Obrietan K, Dudek SM (2007) Synaptic plasticity (and the lack thereof) in hippocampal CA2 neurons. J Neurosci 27(44):12025–12032
Zhao X, Lein ES, He A, Smith SC, Aston C, Gage FH (2001) Transcriptional profiling reveals strict boundaries between hippocampal subregions. J Comp Neurol 441(3):187–196
Zheng C, Bieri KW, Hsiao YT, Colgin LL (2016) Spatial sequence coding differs during slow and fast gamma rhythms in the hippocampus. Neuron 89(2):398–408
Zheng C, Bieri KW, Trettel SG, Colgin LL (2015) The relationship between gamma frequency and running speed differs for slow and fast gamma rhythms in freely behaving rats. Hippocampus 25(8):924–938
Ziv Y, Burns LD, Cocker ED, Hamel EO, Ghosh KK, Kitch LJ, El Gamal A, Schnitzer MJ (2013) Long-term dynamics of CA1 hippocampal place codes. Nat Neurosci 16(3):264–266

What Versus Where: Non-spatial Aspects of Memory Representation by the Hippocampus

Howard Eichenbaum

Abstract Since the discovery of place cells and other findings indicating strong involvement of the hippocampus in spatial information processing, there has been continued controversy about the extent to which the hippocampus also processes non-spatial aspects of experience. In recent years, many experiments studying the effects of hippocampal damage and characterizing hippocampal neural activity in animals and humans have revealed a clear and specific role of the hippocampus in the processing of non-spatial information. Here this evidence is reviewed in support of the notion that the hippocampus organizes the contents of memory in space, in time, and in networks of related memories.

Keywords Non-spatial memory · Hippocampus · Neural activity · Relational memory · Episodic memory

Contents

1 What Is the Nature of Memory Representation that Supports "Recollection" of Non-spatial Stimuli? 104
1.1 Representations of Episodes as Sequences of Events 107
1.2 Representation of Memories Linked into Relational Networks 109
2 Conclusions—The Role of the Hippocampus in Non-spatial Aspects of Memory 111
References 113

The distinction in hippocampal function between non-spatial and spatial aspects of memory became prominent with the publication of O'Keefe and Nadel's (1978) landmark book in which they argued that the hippocampus is dedicated to cognitive mapping, defined by them as the organization of events in physical space (p. 1). Their principal evidence supporting this hypothesis was a set of tables indicating a

H. Eichenbaum (✉)
Center for Memory and Brain, Boston University, Boston, MA 02215, USA
e-mail: hbe@bu.edu

Curr Topics Behav Neurosci (2018) 37: 101–117
DOI 10.1007/7854_2016_450
Published Online: 28 September 2016

large proportion of then existing publications reporting deficits following hippocampal damage on spatial as compared with a much smaller proportion of studies reporting deficits in non-spatial memory. These findings were complemented by a description of prominent electrophysiological correlates of spatial behavior, including a prominent theta rhythm during movement through space and especially the observation of hippocampal principal neurons that fire when animals occupy a particular location in the environment (place cells), that could reflect a mapping of space.

Since that time, the issue of whether the hippocampus processes non-spatial information has been debated, and has received considerable attention in many experiments. In studies on humans, there is overwhelming evidence that patients with hippocampal damage are impaired in tasks that require memory for specific non-spatial stimuli, such words or pictures, presented visually or orally, and therefore without any demand for remembering where the items were seen or any other aspect of space (Squire et al. 2004). Correspondingly, many functional imaging studies have reported hippocampal activation associated with memory for non-spatial stimuli, most typically when items are associated with memory for the spatial (e.g., Davachi et al. 2003) or non-spatial (Henke et al. 1997; Zeineh et al. 2003; Preston et al. 2004; Qin et al. 2009) context in which they were experienced (for reviews see Davachi 2006; Diana et al. 2007; Eichenbaum et al. 2007).

In animals, the situation is less clear, and here is where the story gets interesting. In O'Keefe and Nadel's (1978) review, few studies reported effects of hippocampal damage on classical or instrumental conditioning or discrimination learning, leading to their conclusion that the hippocampus is not involved in non-spatial memory. This conclusion seemed to be confirmed by later studies using the Morris water maze to show that hippocampal lesions impair learning guided by distal spatial cues but not learning guided by a local visual cue (Morris et al. 1982) and studies using fear conditioning to show that hippocampal lesions impair conditioning to a spatial context but not conditioning to an auditory cue (Phillips and LeDoux 1992). However, there are many exceptions that challenge any simple view about non-spatial conditioning and discrimination learning. For example, a form of classical conditioning where a brief "trace" interval is inserted between the non-spatial conditioning and unconditioned stimuli makes the task hippocampal-dependent (Solomon et al. 1986). In the classic spatial alternation task, animals with hippocampal damage can solve the problem when they can alternate continuously but fail when a brief delay is inserted between alternations (Ainge et al. 2007a, b). In both conditions the spatial cues and any demands for spatial cognition per se are identical—they differ only in the demand to remember across a delay. Showing how complicated the findings can be in any particular formal behavioral paradigm, one study described impairment, no effect, or even facilitation of simple odor discrimination learning in rats with hippocampal system damage (Eichenbaum et al. 1988).

In studies on non-spatial recognition memory in animals, the story is, again, complicated. This story begins with Mishkin's (1978) discovery that large medial temporal lesions impair visual object recognition in the delayed non-matching to sample (DNMS) task, where a single novel object is viewed as the sample, then following a delay, at test monkeys are rewarded for choosing a different novel stimulus over the sample. However, in contrast to the devastating effects of large medial temporal lobe removal, lesions limited to the hippocampus in monkeys had a much less severe effect (Zola et al. 2000) or no effect (Murray and Mishkin 1998). Similarly, in rats, selective hippocampal lesions do not impair object-cued DNMS performance (Mumby 2001; but see Clark et al. 2001), whereas damage to the neighboring perirhinal cortex results in severe DNMS deficits in both monkeys and rats (see Eichenbaum et al. 2007; Squire et al. 2007).

The studies showing remarkably little impairment in non-spatial recognition memory are countered by a different pattern of results revealed using a naturalistic test of recognition memory that measures preferential exploration of objects. In humans and monkeys, subjects are initially allowed to visually explore a novel picture, then following a delay, are shown the same picture and another novel picture. Without any reinforcement, subjects preferentially view the novel picture, and, importantly, this novelty preference depends on the pictures being presented in the same background visual context showing that the picture memory is context dependent (Bachevalier et al. 2015). Hippocampal damage severely impairs this preferential viewing effect (Nemanic et al. 2004; Zola et al. 2000), and the deficit occurs only in context-dependent recognition (Bachevalier et al. 2015). In rodents, subjects initially explore duplicates of a novel three-dimensional object in a familiar environment and then, following a delay, are presented with one of those objects and a new object replacing one of the duplicates. Most studies have reported no effect of hippocampal damage, but hippocampal lesions do impair preferential exploration of a familiar object in a novel place in the familiar environment, or a novel spatial context, and preferential viewing of an earlier explored object over a more recently explored object (Eacott and Norman 2004; Langston and Wood 2010). There are some observations of deficits in novel object preference in mice with knockout of the NMDA receptor (Rampon et al. 2000), and in rats with extensive hippocampal damage (Broadbent et al. 2004), or at very long delays (Clark et al. 2000), and most impressively, during temporary inactivation of the hippocampus (Cohen et al. 2013; Cohen and Stackman 2015). These variable findings on non-spatial recognition memory suggest that there may be more than a single strategy that can support recognition, with one strategy dependent on the hippocampus and another that can be supported by other structures or systems, and whether or not a hippocampal-dependent strategy is critical depends on task demands.

Guided by several findings in humans suggesting a distinction between recognition supported by mere familiarity with a recently experienced stimuli contrasted with recollection of the experience with the stimulus study event (Yonelinas 2001), we addressed the possibility of dual mechanisms supporting recognition by developing a variant of the DNMS task designed to distinguish familiarity-like and recollection-like strategies in rats. Our approach adopted a signal detection analysis in which performance on recognition memory judgments is characterized by the Receiver Operating Characteristic (ROC) function that generates separate indices of the contributions of recollection and familiarity in humans (Yonelinas 2001). To perform this analysis in rats, we modified the standard DNMS protocol to initially present, as a list of sample stimuli, odor cues taken randomly from a pool of familiar odors. Then, following a delay, animals were presented with the "old" stimuli (same odors as the samples) and equal number of "new" stimuli (other odors from the pool) sequentially (Fortin et al. 2004). The resulting ROC curve was very similar to functions observed in humans, indicating the contributions of both familiarity-like and recollection-like processes. Moreover, even when the overall recognition performance (measured by percent correct which combines the contribution of recollection and familiarity) was equivalent in these conditions, normal rats exclusively used recollection whereas and rats with hippocampal damage exclusively relied on familiarity. This double dissociation of strategies unequivocally shows that the hippocampus supports a recollection-like process, while sparing familiarity for the same stimuli. These results provide an explanation for the mixed pattern of findings on hippocampal damage and recognition memory in DNMS and preferential exploration paradigms, suggesting that differences in the memory demands of these tasks drives performance that depends on a hippocampal-dependent recollection-like or can be supported by a hippocampal-independent familiarity-like processes (see Sauvage et al. 2008, 2010). An alternative explanation of the specific role of the hippocampus in recollection is that the hippocampus is necessary only for the development and expression of strong memories (Squire et al. 2007). A potential reconciliation of these perspectives is that recollection involves memories that typically contain vivid associations (e.g., Davachi et al. 2003; Quin et al. 2009).

1 What Is the Nature of Memory Representation that Supports "Recollection" of Non-spatial Stimuli?

Many studies in rodents, monkeys, and humans have described hippocampal neuronal activity associated with a broad range of non-spatial stimuli and behavioral events. In rodents, hippocampal neurons have been reported to fire associated with visual, tactile, olfactory, and auditory cues in a broad range of learning and memory paradigms (Eichenbaum et al. 1999; Eichenbaum 2004, 2010; see below). These findings join with many other reports of hippocampal neural activity associated with combinations of specific stimuli, match/non-match stimulus comparisons, and the locations of these events in animals performing recognition memory tasks (Wood et al. 1999; Weibe and Staubli 1999; Deadwyler et al. 1995; Otto and Eichenbaum 1992). The extent to which non-spatial and spatial cues are represented depends on the context of behavioral demands (Muzzio et al. 2009).

Also, a large fraction of hippocampal neurons in monkeys fire associated with learned associations between specific visual stimuli and eye movement responses (Wirth et al. 2003). Similarly, a large fraction of hippocampal neurons in monkeys respond to visual stimuli modulated by their familiarity in the naturalistic recognition task described above (Jutras and Buffalo 2010). Furthermore, multiple studies have reported that hippocampal neurons in humans also respond to visual stimuli and their responses are modulated by familiarity in recognition tasks (Fried et al. 1997) and distinguish the stimuli that are recalled from those forgotten (Rutishauser et al. 2008). Hippocampal neuronal responses also predict memory for learned verbal paired associates (Cameron et al. 2007). Human hippocampal neurons exhibit sparse and distributed coding of individual remembered stimuli (Wixted et al. 2014) and many hippocampal neurons generalize across closely related stimuli (Quiroga et al. 2005; Krieman et al. 2000a) and fire while the subject is imagining a cued stimulus (Krieman et al. 2000b).

In addition, hippocampal neurons signal learned behavioral actions. Lenck-Santini et al. (2008) described hippocampal neurons that fire during learned "jump" responses, reminiscent of Ranck's (1973) pioneering descriptions of a variety of "behavioral correlates" of hippocampal neurons in rats, later attributed to the location where those behaviors occurred, even though these responses are not explained fully by location alone. These reports of activity associated with specific behavioral actions were preceded by other pioneering observations of hippocampal neuronal activity modeling the conditioned eyeblink response in rabbits (Berger et al. 1983; Hattori et al. 2015), and short-latency auditory-driven responses of hippocampal neurons in rats classically conditioned to seek food upon presentation of a tone stimulus (Olds et al. 1972). In sum, there is considerable variation in the types of non-spatial and spatial information encoded in hippocampal neural activity patterns, leading to the conclusion that hippocampal neuronal activity in rodents and monkeys is "high-dimensional" in the sense that hippocampal neurons exhibit considerable mixed selectivity to multiple relevant non-spatial and spatial dimensions that are salient in a large range of memory tasks (see Mckensie et al. 2014).

The observations discussed above do not directly shed light on how hippocampal neuronal firing patterns specifically support recollection of episodic memories. Next I will further focus on my proposal, made some time ago (Eichenbaum et al. 1999), that the hippocampus contributes to recollective memory by constructing relational representations that bind elements of memories and link memories via their common elements, composing a "memory space" that supports both spatial and non-spatial dimensions of memory organization. According to this view, the representational schemes that underlie relational processing are: (1) the representation of events as objects within the context in which they occur, (2) the representation of episodes as the flow of events across time, and (3) the interleaving of events and episodes into relational networks, supporting the ability to draw novel inferences from memory (Eichenbaum 2004). Here I will review some of the older and more recent findings that support this perspective, focusing on non-spatial aspects of relational memory representations and activity patterns of neurons in the hippocampus that may support these representations.

Representation of events as objects in context. As introduced above, several studies have shown that hippocampal damage causes deficits in the spontaneous preference for exploring familiar objects in new locations or in new environmental contexts, even when preference for novel objects in familiar locations is preserved. Similarly, the hippocampus is essential to recognizing and discriminating objects only when object representations and associations are context dependent (Butterly et al. 2012; Bachevalier et al. 2015). These findings are complemented by several reports that hippocampal neuronal activation that occurs during the exploration of specific objects is embedded within the spatial firing patterns (place fields) of those neurons. For example, following tone-cued fear conditioning, hippocampal neurons come to be driven by the conditioned tone stimulus when the animal is within the place field of that neuron (Moita et al. 2003). Also, in rats performing a variant of the novel object exploration task, hippocampal neurons fired associated with specific objects and their familiarity embedded within the spatial firing patterns (place fields) of these neurons (Manns and Eichenbaum 2009). In rats performing a context-guided object-reward association task, hippocampal neurons fire when animals sample specific objects within particular locations and spatial contexts (Fig. 1; Komorowski et al. 2009). Similarly, after training on somatosensory or auditory discrimination tasks, hippocampal neurons encode tactile and auditory cues along with the locations where they were experienced and rewarded (Itskov et al. 2011, 2012; Vinnik et al. 2012).

Parallel studies using fMRI have shown that specific stimuli are encoded within spatial context in the human hippocampus. For example, when human subjects recall imagined scenes that applied to specific verbal items, the hippocampus was activated only when the subjects recalled the item and scene (Davachi et al. 2003). Also,

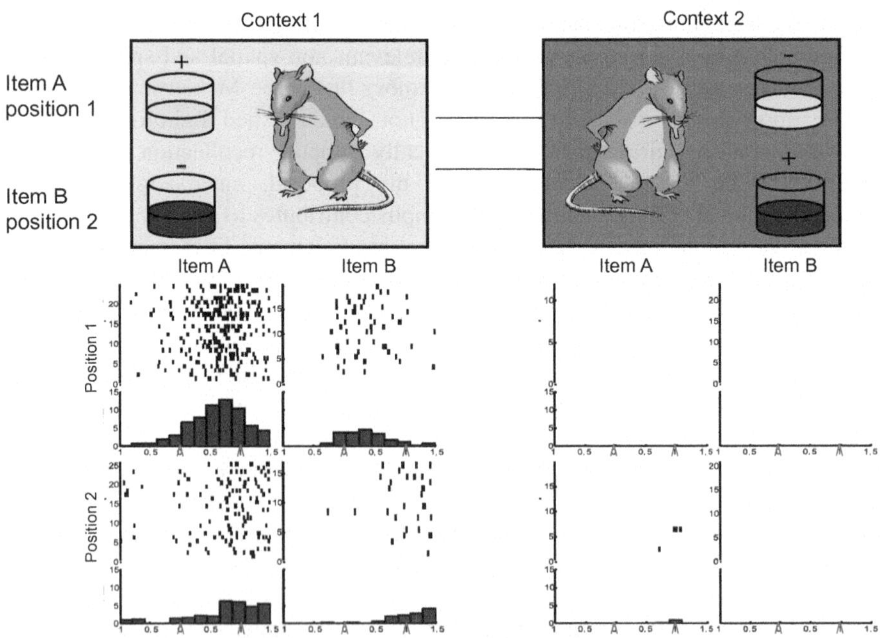

Fig. 1 An example CA1 neuron that fires as a rat samples a specific odor located in a particular position within one spatial context. *Top* The rat performs a context-guided odor association task in which object A, not B, is rewarded in Context 1 whereas object B, not A, is rewarded in Context 2. *Bottom* Rasters and summary histogram of firing rates during object sampling (between *red arrows*) (from Komorowski et al. 2009)

studies on humans indicate that hippocampal damage eliminates the preferential viewing of locations in a scene where objects were previously observed and are now absent (Ryan et al. 2000), and correspondingly, activation of the hippocampus in normal subjects that predicted subsequent successful object-location memory by preferential viewing (Hannula and Ranganath 2009). Also, humans with hippocampal damage are impaired in memory for the locations of multiple objects, and in particular, most errors are due to "swapping" objects between locations (Watson et al. 2013). Together, these findings indicate a specific role for the hippocampus in the organization of non-spatial objects within a spatial context. Notably, even though the representations of non-spatial stimuli described here are embedded within spatial contexts, these paradigms described above do not involve demands for navigation by body movements through space, as is the focus of much current research on the role of the hippocampus in spatial aspects of memory (Hartley et al. 2014). Instead, these tasks simply require identifying objects within their spatial context.

1.1 Representations of Episodes as Sequences of Events

Additional recent evidence indicates that that another non-spatial aspect of memory processing supported by the hippocampus involves the organization of events in time. In humans, there is considerable evidence that hippocampal damage results in deficits in memory for the order of events even when memory for the events is intact, and evidence from imaging studies that the hippocampus is activated during the encoding and retrieval of the order of events in memories, independent of memories for the events themselves (reviewed in Eichenbaum 2013, 2014). Similarly, selective hippocampal lesions result in impairments in memory for the order of studied non-spatial stimuli, even when memory for the stimuli themselves is spared (Fortin et al. 2002; Kesner et al. 2002a, b; Ergorul and Eichenbaum 2004).

There is also growing evidence that memory for the flow of events in experiences is mediated directly by representations of time and order by hippocampal neurons. Thus, recently it has become clear that the same hippocampal principal neurons that are place cells can also represent time independent of place (Eichenbaum 2013). An early study showed that ensemble activity patterns of CA1 neurons gradually change during while rats sampled sequences of odors, and this signal of continuously evolving temporal context predicted success in remembering the odor sequence (Manns et al. 2007). Confirming these findings in a functional imaging study in humans, Ezzyat and Davachi (2014) reported that pattern similarity in hippocampal activation signaled temporal proximity of associated objects and this signal was correlated with memory performance.

In addition, several other studies have identified hippocampal principal neurons that fire at a particular moments in time of a temporally structured event, composing temporal maps of specific experiences. Across these studies, the location of the animal is held constant or firing patterns associated with elapsed time are distinguished from those associated with spatial and behavioral variables, and the firing patterns of these cells are dependent on the critical temporal parameters that characterize the task. Because these properties parallel those of place cells in coding locations in spatially structured experiences, we called these neurons "time cells" (MacDonald et al. 2011), even though these neurons are the same cells that exhibit spatial firing specificity in other circumstances.

Time cells have now been observed in several behavioral paradigms, including during delay periods in maze tasks in which rats alternate goals (Gill et al. 2011; Pastalkova et al. 2008; Kraus et al. 2013; Fig. 2), bridging temporal gaps between associated non-spatial cues (MacDonald et al. 2011), during the delay period in a in non-spatial matching to sample task (MacDonald et al. 2013), and throughout trials in trace eyelid conditioning (Modi et al. 2014). Importantly, in some of these studies, the animal is immobilized and thus space plays no role in ongoing behavior or memory whatsoever (MacDonald et al. 2013; Modi et al. 2014; Naya and Suzuki 2011). The findings of these studies establish a broad scope of temporally structured episodes in which the hippocampus encodes the temporal organization of specific experiences. Confirming these results in a functional imaging study in humans, Hsieh et al. (2014) reported that pattern similarity in hippocampal activation signaled the combination of object and temporal context information in sequence learning. Furthermore, some of the studies in animals have closely linked the emergence of time cells sequences to the encoding of specific memories and to subsequent memory accuracy (Gill et al. 2011; Modi et al. 2014; MacDonald et al. 2013), thus indicating a causal role of time cell firing patterns to memory performance.

The role of the hippocampus in organizing events in time extends even to spatial memories and spatial representations. Thus, for example, while the hippocampus is required for accurate delayed alternation in a T-maze, the hippocampus is not

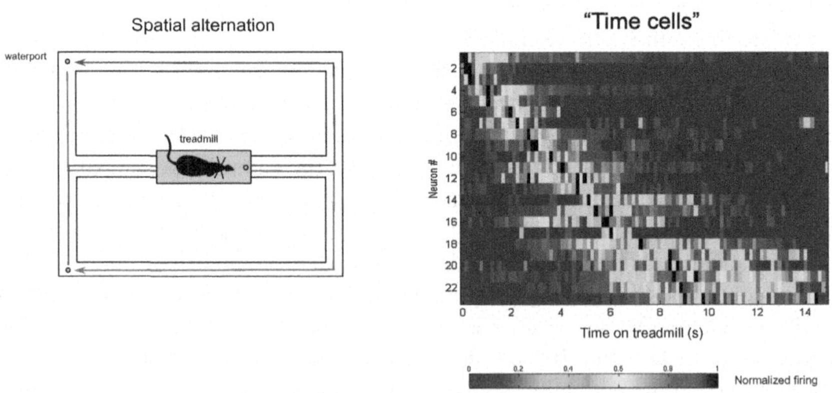

Fig. 2 Hippocampal time cells during the period when a rat runs in place while performing a spatial alternation task. *Left* The spatial alternation task with treadmill in the center of the maze stem. *Red* and *blue lines* indicate alternate right-turn and left-turn paths. *Right* ensemble firing rate mapping where each row represents the normalized average firing rate of a neuron (see cell numbers on *Y*-axis) over the 15 s treadmill run. Note that each cell fires during a specific moment of treadmill running and the entire period of running is filled with time-specific representations (from Kraus et al. 2013)

required for animals to learn to turn in one direction in a T-maze or even to alternate left and right turns if allowed to alternate continuously (O'Keefe and Nadel 1978). Consider that the demands for using spatial representations are equivalent in all these versions of T-maze learning. That is, to distinguish left and right turns at a choice point, and one could perform all these tasks based on the same egocentric or allocentric spatial representations. The difference between these tasks is a demand for memory for time, specifically remembering which turn was last performed, only in delayed alternation and not in learning a consistent turn direction or in continuous alternation. Correspondingly, the spatial firing patterns are different depending on which memory is current—place cells fire differentially depending on whether the animals is in the midst of a left-turn or right-turn trial even in the portion of maze where these routes overlap (Wood et al. 2000). The differentiation of spatial representations that is dependent on the ongoing temporal context occurs both in versions of the task where the hippocampus is required and in those where it is not required, and these context-dependent firing patterns predict accurate memory performance (Robitsek et al. 2013).

Also, the representation of temporally ordered sequences of events by the hippocampus extends to monkeys and humans. In monkeys, hippocampal neuronal activity signals elapsed time in a memory delay between associated objects (Naya and Suzuki 2011). In humans, hippocampal neurons fire in sequence associated with learning (Paz et al. 2010) and memory (Gelbard-Sagiv et al. 2008) of the flow of events experienced in movie clips.

The significance of prominent temporal representation as an aspect of non-spatial coding in the hippocampus is high in two ways. First, as introduced by Tulving (1984) episodic memories are defined by a temporal organization that embodies the temporal organization of events in personal experiences. We know that the hippocampus is critical to episodic memory and to memory for the temporal order of events, even when space is not relevant. Now the existence of time cells provides a mechanism by which the hippocampus organizes memories for events in time. Second, the existence of time cells offers a parallel temporal organizing mechanism to the spatial organizing mechanism offered by place cells. Therefore, the hippocampus could support representations of episodes by mapping objects and events within a framework of space and time, conferring upon those memories connections that reflect the spatial and temporal associations between distinct but related events (Eichenbaum 2013, 2014).

1.2 Representation of Memories Linked into Relational Networks

Since the work of Piaget (1928) we have known that memories are not stored in isolation but rather are integrated into organizations of and personal experiences that I call relational networks. In early work on the role of the hippocampus in building relational networks, Bunsey and Eichenbaum (1996) examined the capacity of rats to link overlapping stimulus associations into relational networks. They trained rats on associations between pairs of odors that shared a common element (A–B and B–C) and then tested for the existence of the relational network that (A–B–C), but assessing knowledge about the indirectly related elements (A–C). Normal rats showed they had developed the relational representation, but animals with hippocampal lesions did not. Subsequently, Dusek and Eichenbaum (1997) showed that normal rats can also learn a series of overlapping stimulus choice problems (choose A over B, B over C, C over D, D over E), and show acquisition of a hierarchical relational representation (A over B over C over D over E) by accurate transitive choices (e.g., B over D). Again the hippocampus was required for the relational judgment. Similarly, rats with hippocampal damage are impaired in learning a circular organization of relations among non-spatial stimuli (A over B over C over A; Dusek and Eichenbaum 1998). Notably, none of these tasks involves spatial organizations.

In more recent work, Tse et al. (2007) showed that when rats learn to find distinct food flavors in specific designated locations within an open field, they develop an organized representation of the spatial relations among the objects in a particular environment and rely on the hippocampus for rapid assimilation of new flavor-place associations within the relational representation (or schema). Mckenzie et al. (2013) explored the neural basis for the development and elaboration of a spatial schema in which rewards could be found at multiple locations. They reported that hippocampal neurons encode the animal's approach to multiple reward locations and rapidly assimilate and reorganize the overall network representation to accommodate the new reward locations. These studies focus on the spatial organization of non-spatial events, and demonstrate a key role for the hippocampus in interleaving non-spatial memories within a spatial relational representation.

In a more ambitious study where rats learned multiple context-dependent object-reward associations, Mckenzie et al. (2014) characterized the neural ensemble representations as a hierarchy of relations among non-spatial and spatial dimensions of events, including the identity of the objects, their reward assignments, the positions within a context in which they were experienced, and the spatial context in which they occurred. These findings revealed the overall structure of the relational representation of all of the events in the task, such that non-spatial features of events (object identifies and reward valences) were embedded within spatial organization (positions) in separate schemas for each context (Fig. 3). Furthermore, after initial learning the initial set of object-reward associations, new object associations were rapidly assimilated into the relational structure that was established by initial learning. In addition, within the overall hierarchical representation, items that had common reward associations in particular positions had strongly similar representations, indicating close associations between objects that were never experienced together. This aspect of the relational representation likely supports the capacity to make novel inferences between those indirectly related objects.

In parallel studies Preston and colleagues (2004) have explored role of the hippocampus in forming relational representations in humans. In their paradigm, subjects learn overlapping pairwise associations between visual objects (e.g., A–B

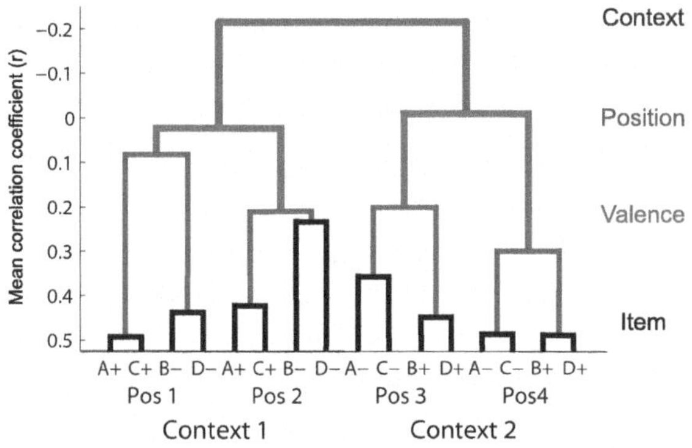

Fig. 3 A dendrogram showing the hierarchy of representational distances measured as correlations between simultaneously recorded neuronal population vectors associated with spatial (context and position) and non-spatial (reward valence and object identity) dimensions in animals performing the same task as show in Fig. 1 where they chose between stimuli A versus B on some trials and between C versus D on other trials. Each type of trial is composed as a specific stimulus (A or B), reward value (+ or −), position (Pos) and context shown on *X*-axis (from Mckenzie et al. 2014)

and B–C) from which they can make inferences between indirectly related elements (A–C), demonstrating the acquisition of a simple relational network that interleaves the overlapping memories and links all three elements (similar to Bunsey and Eichenbaum 1996). Furthermore they showed that, the learning of the second, overlapping pair (B–C) reinstates the hippocampal representation of the earlier learned pair (A–B) and that this content-specific hippocampal activation signaled subsequent success on the inferential judgment (Zeithamova et al. 2012). These findings indicate that the development of relational networks depends upon reinstatement of related networks into which the new information is assimilated, and shows that the subsequent interleaved network supports novel inferences from memory. Recently, Wimmer and Shohamy (2012) have reported parallel findings of hippocampal activation associated with reward values shared between indirectly associated stimuli. In addition, it is notable that the relational network in this paradigm organized neither by space or time, but rather simply by associative links among the elements. These findings extend the scope or relational dimensions supported by the hippocampus to the most fundamental of all dimensions, simple associations between objects.

2 Conclusions—The Role of the Hippocampus in Non-spatial Aspects of Memory

The findings describe above clearly implicate the hippocampus in non-spatial memory processing, including in several studies where space plays no role in memory performance (e.g., Bunsey and Eichenbaum 1996; Alvarez et al. 2002;

Fortin et al. 2004; Naya and Suzuki 2011; Zeithamova et al. 2012). At the same time, the evidence that the hippocampus encodes memories of non-spatial events in isolation from the context and other events is weak. In the view of this investigator, these observations support the idea that the role of the hippocampus is precisely to link event memories with relational representations that are organized by space, time, and associative networks.

Even in seemingly simple memory tasks that require only recognition of individual objects, I believe that relational memory is involved. Consider, for example, the experiment using ROC analysis of odor recognition. It would seem that this task simply requires recognition of individual odors independent of other stimuli, their locations, or their temporal order. But consider further that all of the odors were generally familiar; in each testing session, the sample stimuli were a subset of odors from a pool and odors were presented in different lists many times across testing sessions. Therefore, the memory demand on each test was not, "Have you ever experienced this odor?", but rather, "Was this odor on the list today?". The task thus requires memory for stimuli in the context of the current study list, much as memory for highly familiar words in standard recognition tests in humans demands not a judgment about whether the each word has been seen before, but rather was it on the study list in this experiment (see also Butterly et al. 2012). In my view, the demand for hippocampal function in "recollection" depends on the extent to which performance normally benefits by memory of items in the context of the study experience.

This perspective on the role of the hippocampus in non-spatial memories is supported by a comparison of the kinds of errors made in recognition memory following damage to the hippocampus versus prefrontal cortex in rats. In the ROC paradigm described above, hippocampal and prefrontal lesions both cause selective impairment in the recollection-like component of recognition. However, further inspection of the nature of the impairment indicates a key distinction in these impairments. Hippocampal lesions result in a reduction in the "hit" rate, that is, an increase in errors in which the subject declares stimuli experiences as samples as "new" (Fortin et al. 2004). This kind of error is, of course, what one expects in amnesia caused by hippocampal damage. However, prefrontal lesions do not affect the hit rate, but instead elevate the rate of "false alarms," errors in which items that did not appear in the sample list are declared "old" (Farovik et al. 2008). This kind of error reflects a deficit in distinguishing the source of odor memories as on the current study list as opposed to those on earlier study lists. Thus, prefrontal damage does not cause impairment in odor memory per se, but rather a loss of memory for the study list source or context. These findings indicate that rats do indeed normally benefit from relating individual odor memories to the context of the current study list, and these results suggest that memory for the study items in that context underlies recollection-like performance in rats.

Here I have argued that the role of the hippocampus is to organize memories in context, in order, and in relational networks. The organization can map onto space, but can also map onto time, or onto an associational structure that is neither time nor space. It may be that the hippocampus serves more in the organizational role than the representation of the items themselves, as suggested by some studies that describe hippocampal representation as including the organization alone without item coding (Naya and Suzuki 2011) or as a hierarchical organization with the organization at the top of the hierarchy and object dimensions at the bottom (Mckenzie et al. 2014; Manns and Eichenbaum 2009; Fig. 3). To the extent that organization proves to be the predominant role of the hippocampal representations across paradigms, we might best think of the hippocampus as fundamentally like the orchestra leader whose role it is to organize the performance of musicians who sit in different places and play in a distinct sequence. From this perspective, the division between spatial and non-spatial aspects of hippocampal memory is, in my view, not the most useful compartmentalization of hippocampal function. Rather, I suggest a shift to thinking of a distinction between the contents of memories, the objects and events that occur, and the organization of memories, in dimensions of space, time, associative networks, and perhaps more dimensions by which elements of memories are connected.

Acknowledgments NIMH MH094263, MH51570; MH52090, MH095297; ONR MURI N00014-0-1-0936.

References

Ainge JA, van der Meer MA, Langstron RF, Wood ER (2007a) Exploring the role of context-dependent hippocampal activity in spatial alternation behavior. Hippocampus 17:988–1002

Ainge JA, Tamosiunaite M, Woergoetter F, Dudchencko PA (2007b) Hippocampal CA1 place cells encode intended destination on a maze with multiple choice points. J Neurosci 27:9769–9779

Alvarez P, Wendelken L, Eichenbaum H (2002) Hippocampal formation lesions impair performance in an odor-odor association task independently of spatial context. Neurobiol Learn Mem 8:79–86

Bachevalier J, Nemanic S, Alvarado MC (2015) The influence of context on recognition memory in monkeys: effects of hippocampal, parahippocampal, and perirhinal lesions. Behav Brain Res 285:89–98

Berger TW, Rinaldi PC, Weisz DJ, Thompson RF (1983) Single-unit analysis of different hippocampal cell types during classical conditioning of rabbit nictitating membrane response. J Neurophsiol 50:1197–1219

Broadbent NJ, Squire LR, Clark RE (2004) Spatial memory, recognition memory, and the hippocampus. Proc Natl Acad Sci USA 101:14515–14520

Bunsey M, Eichenbaum H (1996) Conservation of hippocampal memory function in rats and humans. Nature 379:255–257

Butterly DA, Petroccione MA, Smith DM (2012) Hippocampal context processing is critical for interference free recall of odor memories in rats. Hippocampus 22:906–913

Cameron KA, Yashar S, Wilson CL, Fried I (2007) Human hippocampal neurons predict how well word pairs will be remembered. Neuron 30:289–298

Clark RE, Zola SM, Squire LR (2000) Impaired recognition memory in rats after damage to the hippocampus. J Neurosci 20:8853–8860

Clark R, West AN, Zola S, Squire LR (2001) Rats with lesions of the hippocampus are impaired on the delayed nonmatching-to-sample task. Hippocampus 11:176–186
Cohen SJ, Stackman RW Jr (2015) Assessing rodent hippocampal involvement in the novel object recognition task. A review. Behav Brain Res 285:105–117
Cohen SJ, Munchow AH, Rios LM, Zhang G, Asgeirsdottir HN, Stackman RW (2013) The rodent hippocampus is essential for nonspatial object memory. Curr Biol 23:1685–1690
Davachi L (2006) Item, context and relational episodic encoding in humans. Curr Opin Neurobiol 16:693–700
Davachi L, Mitchell J, Wagner AD (2003) Multiple routes to memory: distinct medial temporal lobe processes built item and source memories. Proc Natl Acad Sci 100:2157–2162
Deadwyler SA, Bunn T, Hampson RE (1995) Hippocampal ensemble activity during spatial delayed-nonmatch-to-sample performance in rats. J Neurosci 16:354–372
Diana RA, Yonelinas AP, Ranganath C (2007) Imaging recollection and familiarity in the medial temporal lobe: a three-component model. Trends Cogn Sci 11:379–386
Dusek JA, Eichenbaum H (1997) The hippocampus and memory for orderly stimulus relations. Proc Natl Acad Sci 94:7109–7114
Dusek JA, Eichenbaum H (1998) The hippocampus and transverse patterning guided by olfactory cues. Behav Neurosci 112:762–771
Eacott MJ, Norman G (2004) Integrated memory for object, place, and context in rats, a possible model of episodic-like memory? J Neurosci 24:1948–1953
Eichenbaum H (2004) Hippocampus: cognitive processes and neural representations that underlie declarative memory. Neuron 44:109–120
Eichenbaum H (2010) Spatial and behavioral correlates of hippocampal neuronal activity: a primer for computational analysis. In: Cutsuridis V, Graham B, Cob S, Vida I (eds) Hippocampal microcircuits: a computational modeller's resource book. Springer, NY
Eichenbaum H (2013) Memory on time. Trends Cogn Sci 17:81–88
Eichenbaum H (2014) Time cells in the hippocampus: a new dimension for mapping memories. Nature Rev Neurosci 15:732–744
Eichenbaum H, Fagan A, Mathews P, Cohen N (1988) Hippocampal system dysfunction and odor discrimination learning in rats: impairment or facilitation depending on representational demands. Behav Neurosci 102:331–339
Eichenbaum H, Dudchencko P, Wood E, Shapiro M, Tanila H (1999) The hippocampus, memory, and place cells: is it spatial memory or a memory space? Neuron 23:209–226
Eichenbaum H, Yonelinas AR, Ranganath C (2007) The medial temporal lobe and recognition memory. Ann Rev Neurosci 30:123–152
Ergorul C, Eichenbaum H (2004) The hippocampus and memory for "what," "where," and "when". Learn Mem 11:397–405
Ezzyat Y, Davachi L (2014) Similarity breeds proximity: pattern similarity within and across contexts is related to later mnemonic judgments of temporal proximity. Neuron 81:1179–1189
Farovik A, Dupont LM, Arce M, Eichenbaum H (2008) Medial prefrontal cortex supports recollection, but not familiarity, in the rat. J Neurosci 28:13428–13434
Fortin NJ et al (2002) Critical role of the hippocampus in memory for sequences of events. Nat Neurosci 5:458–462
Fortin NJ, Wright SP, Eichenbaum H (2004) Recollection-like memory retrieval in rats is dependent on the hippocampus. Nature 431:188–191
Fried I, MacDonald KA, Wilson CL (1997) Single neurons activity in human hippocampus and amygdala during recognition of faces and objects. Neuron 18:753–765
Gelbard-Sagiv H, Mukamel R, Harel M, Malach R, Fried I (2008) Internally generated reactivation of single neurons in human hippocampus during free recall. Science 322:96–101
Gill PR et al (2011) Hippocampal episode fields develop with learning. Hippocampus. 21:1240–1249
Hannula DE, Ranganath C (2009) The eyes have it: hippocampal activity predicts expression of relational memory in eye movements. Neuron 63:592–599

Hartley T, Lever C, Burgess N, O'Keefe J (2014) Space in the brain: how the hippocampal formation supports spatial cognition. Phil Trans R Soc B 369:20120510
Hattori S, Chen L, Weiss C, Disterhoft JF (2015) Robust hippocampal responsivity during retrieval of consolidated associative memory. Hippocampus 25:655–669
Henke K, Buck A, Weber B, Wieser HG (1997) Human hippocampus establishes associations in memory. Hippocampus 7:249–256
Hsieh LT, Gruber MJ, Jenkins LJ, Ranganath C (2014) Hippocampal activity patterns carry information about objects in temporal context. Neuron 81:1165–1178
Itskov PM, Vinnik E, Diamond ME (2011) Hippocampal representation of touch-guided behavior in rats: persistent and independent traces of stimulus and reward location. PLoS ONE 6(1): e16462. doi:10.1371/journal.pone.0016462
Itskov PM, Vinnik E, Honey C, Schnupp J, Diamond ME (2012) Sound sensitivity of neurons in rat hippocampus during performance of a sound-guided task. J Neurophysiol 107:1822–1834
Jutras MJ, Buffalo EA (2010) Recognition memory signals in the macaque hippocampus. Proc Natl Acad Sci USA 107:401–406
Kesner RP et al (2002a) The role of the hippocampus in memory for the temporal order of a sequence of odors. Behav Neurosci 116(2):286–290
Kesner RP et al (2002b) The role of the hippocampus in memory for the temporal order of a sequence of odors. Behav Neurosci 116(2):286–290
Komorowski RW, Manns JR, Eichenbaum H (2009) Robust conjunctive item-place coding by hippocampal neurons parallels learning what happens. J Neurosci 29:9918–9929
Kraus BJ, Robinson RJ II, White JA, Eichenbaum H, Hasselmo ME (2013) Hippocampal 'time cells': time versus path integration. Neuron 78:1090–1101
Kreiman G, Koch C, Fried I (2000a) Category specific visual responses of single neurons in the human medial temporal lobe. Nat Neurosci 3:946–953
Kreiman G, Koch C, Fried I (2000b) Imagery neurons in the human brain. Nature 408:357–361
Langston RF, Wood ER (2010) Associative recognition and the hippocampus: differential effects of hippocampal lesions on object-place, object-context and object-place-context memory. Hippocampus 20:1139–1153
Lenck-Santini PP, Fenton AA, Muller RU (2008) Discharge properties of hippocampal neurons during performance of a jump avoidance task. J Neurosci 28:6773–6786
MacDonald CJ, Lepage KQ, Eden UT, Eichenbaum H (2011) Hippocampal "time cells" bridge the gap in memory for discontiguous events. Neuron 71:737–749
MacDonald CJ, Carrow S, Place R, Eichenbaum H (2013) Distinct hippocampal time cell sequences represent odor memories in immobilized rats. J Neurosci 33:14607–14616
Manns J, Eichenbaum H (2009) A cognitive map for object memory in the hippocampus. Learn Mem 16:616–624
Manns JR, Howard M, Eichenbaum H (2007) Gradual changes in hippocampal activity support remembering the order of events. Neuron 56:530–540
McKenzie S, Robinson NTM, Herrera L, Churchill JC, Eichenbaum H (2013) Learning causes reorganization of neuronal firing patterns to represent related experiences within a hippocampal schema. J Neurosci 33:10243–10256
McKenzie S, Frank AJ, Kinsky NR, Porter B, Rivière PD, Eichenbaum H (2014) Hippocampal representation of related and opposing memories develop within distinct, hierarchically-organized neural schemas. Neuron 83:202–215
Mishkin M (1978) Memory in monkeys severely impaired by combined but not by separate removal of amygdala and hippocampus. Nature 273:297–298
Modi MN, Dhawale AK, Bhalla US (2014) CA1 cell activity sequences emerge after reorganization of network correlation structure during associative learning. eLife 3:e01982
Moita MAP, Moisis S, Zhou Y, LeDoux JE, Blair HT (2003) Hippocampal place cells acquire location specific location specific responses to the conditioned stimulus during auditory fear conditioning. Neuron 37:485–497
Morris RGM, Garrud P, Rawlins JP, O'Keefe J (1982) Place navigation impaired in rats with hippocampal lesions. Nature 297:681–683

Mumby DG (2001) Perspectives on object-recognition memory following hippocampal damage, lessons from studies in rats. Behav Brain Res 127:159–181

Murray EA, Mishkin M (1998) Object recognition and location memory in monkeys with excitotoxic lesions of the amygdala and hippocampus. J Neurosci 18:6568–6582

Muzzio IA, Levita L, Kulkarni J, Monaco J, Kentros C, Stead M, Abbott LF, Kandel ER (2009) Attention enhances the retrieval and stability of visuospatial and olfactory representations in the dorsal hippocampus. PLoS Biol 7(6):e1000140. doi:10.1371/journal.pbio.1000140

Naya Y, Suzuki WA (2011) Integrating what and when across the primate medial temporal lobe. Science 333:773–776

Nemanic S, Alvarado MC, Bachevalier J (2004) The hippocampal/parahippocampal regions and recognition memory: insights from visual paired comparison versus object-delayed nonmatching in monkeys. J Neurosci 24:2013–2026

O'Keefe J, Nadel L (1978) The hippocampus as a cognitive map. Oxford University Press, New York

Olds J, Disterhoft JF, Segal M, Kornblith CL, Hirsh R (1972) Learning centers of rat brain mapped by latencies of conditioned unit responses. J Neurophysiol 35:202–219

Otto T, Eichenbaum H (1992) Neuronal activity in the hippocampus during delayed non-match to sample performance in rats: evidence for hippocampal processing in recognition memory. Hippocampus 2:323–334

Pastalkova E. et al (2008) Internally generated cell assembly sequences in the rat hippocampus. Science 321:1322–1327

Paz R, Gelbard-Sagiv H, Mukamel R, Harel M, Malach R, Fried I (2010) A neural substrate in the human hippocampus for linking successive events. Proc Natl Acad Sci USA 107:6046–6051

Phillips RG, LeDoux JE (1992) Differential contribution of amygdala and hippocampus to cued and contextual fear conditioning. Behav Neurosci 106:274–285

Piaget J (1928) Judgment and reasoning in the child. Kegan, Paul, Trench, and Trubner, London

Preston AR, Shrager Y, Dudukovic NM, Gabrieli JD (2004) Hippocampal contribution to the novel use of relational information in declarative memory. Hippocampus 14:148–152

Qin S, Rijpkema M, Tendolkar I, Piekema C, Hermans EJ, Binder M, Petersson KM, Luo J, Fernandez G (2009) Dissecting medial temporal lobe contributions to item and associative memory formation. NeuroImage 46:874–881

Quiroga RQ, Reddy L, Kreiman G, Koch C, Fried I (2005) Invariant visual representation by single neurons in the human brain. Nature 43:1102–1107

Rampon C, Tang Y-P, Goodhouse J, Shimizu E, Kyin M, Tsien J (2000) Enrichment induces structural changes and recovery from non-spatial memory deficits in CA1 NMDAR1-knockout mice. Nat Neurosci 3:238–244

Ranck JB Jr (1973) Studies on single neurons in dorsal hippocampal formation and septum in unrestrained rats. Part I. Behavioral correlates and firing repertoires. Exp Neurol 41:461–531

Robitsek JR, White J, Eichenbaum H (2013) Place cell activation predicts subsequent memory. Behav Brain Res 254:65–72

Rutishauser U, Schuman EM, Mamelak AN (2008) Activity of human hippocampal and amygdala neurons during retrieval of declarative memories. Proc Natl Acad Sci USA. 105:329–334

Ryan JD, Althoff RR, Whitlow S, Cohen NJ (2000) Amnesia is a deficit in relational memory. Psychol Sci 11:6454–6461

Sauvage MM, Fortin NJ, Owens CB, Yonelinas AP, Eichenbaum H (2008) Recognition memory: opposite effects of hippocampal damage on recollection and familiarity. Nat Neurosci 11:16–18

Sauvage MM, Beer Z, Eichenbaum H (2010) Recognition memory: adding a response deadline eliminates recollection but spares familiarity. Learn Mem 17:104–108

Solomon PR, Vander Schaaf ER, Thompson RF, Weisz DJ (1986) Hippocampus and trace conditioning of the rabbit's classically conditioned nictitating membrane response. Behav Neurosci 100:729–744

Squire LR, Stark CEL, Clark RE (2004) The medial temporal lobe. Ann Rev Neurosci 27:279–306

Squire LR, Wixted JT, Clark RE (2007) Recognition memory and the medial temporal lobe, a new perspective. Nat Rev Neurosci 8:872–883

Tse D, Langston RF, Kakeyama M, Bethus I, Spooner PA, Wood ER, Witter MP, Morris RGM (2007) Schemas and memory consolidation. Science 316:76–82

Tulving E (1984) Précis of elements of episodic memory. Behav Brain Sci 7:223–268

Vinnik E, Antopolskiy S, Itskov PM, Diamond ME (2012) Auditory stimuli elicit hippocampal neuronal responses during sleep. Front Syst Neurosci 6:49. doi:10.3389/fnsys.2012.00049

Watson PD, Voss JL, Warren DE, Tranel D, Cohen NJ (2013) Spatial reconstruction by patients with hippocampal damage is dominated by relational memory errors. Hippocampus 23:570–580

Wiebe SP, Stäubli UV (1999) Dynamic filtering f recognition codes in the hippocampus. J Neurosci 19:10562–10574

Wimmer GE, Shohamy D (2012) Preference by association. How memory mechanisms in the hippocampus bias decisions. Science 338:270–273

Wirth S, Yanike M, Frank LM, Smith AC, Brown EN, Suzuki WA (2003) Single neurons in the monkey hippocampus and learning of new associations. Science 300:1578–1581

Wixted JT, Squire LR, Jang Y, Papesh MH, Goldinger SD, Kuhn JR, Smith KA, Treiman DM, Steinmetz PN (2014) Sparse and distributed coding of episodic memory ini neurons of the human hippocampus. Proc Natl Acad Sci USA 111:9621–9626

Wood E, Dudchenko PA, Eichenbaum H (1999) The global record of memory in hippocampal neuronal activity. Nature 397:613–616

Wood E, Dudchenko P, Robitsek JR, Eichenbaum H (2000) Hippocampal neurons encode information about different types of memory episodes occurring in the same location. Neuron 27:623–633

Yonelinas AP (2001) Components of episodic memory, the contribution of recollection and familiarity. Phil Trans Roy Soc Lond Biol Sci 356:1363–1374

Zeineh MM, Engel SA, Thompson PM, Brookheimer SY (2003) Dynamics of the hippocampus during encoding and retrieval of face-name pairs. Science 299:577–580

Zeithamova D, Dominick AL, Preston AR (2012) Hippocampal and ventral medial prefrontal activation during retrieval-mediated learning supports novel inference. Neuron 75:168–179

Zola SM, Squire LR, Teng E, Stefanacci L, Buffalo EA, Clark RE (2000) Impaired recognition memory in monkeys after damage limited to the hippocampal region. J Neurosci 20:451–643

The Functional and Structural Neuroanatomy of Systems Consolidation for Autobiographical and Semantic Memory

Adam J.O. Dede and Christine N. Smith

Abstract It is well established that patients with memory impairment have more difficulty retrieving memories from the recent past relative to the remote past and that damage to the medial temporal lobe (MTL) plays a key role in this pattern of impairment. The precise role of the MTL and how it may interact with other brain regions remains an area of active research. We investigated the role of structures in a memory network that supports remembering. Our chapter focuses on two types of memory: episodic memory and semantic memory. Findings from studies of patients with brain damage and neuroimaging studies in patients and healthy individuals were considered together to identify the functional and structural neuroanatomy of past remembrance.

Keywords Retrograde amnesia · Autobiographical memory · Semantic memory · Connectivity · Neuroimaging · Lesion · Patient

Contents

1 Introduction 120
2 Autobiographical Memory 122
  2.1 Autobiographical Memory in Memory-Impaired Patients 122
  2.2 Possible Semanticization of Episodic Memories 125
  2.3 The Autobiographical Memory Network 126
  2.4 Effects of Lesions in the Autobiographical Memory Network Beyond the MTL 127

A.J.O. Dede
Department of Psychology, University of California San Diego, San Diego
CA 92093, USA

C.N. Smith
Department of Psychiatry, University of California San Diego, San Diego
CA 92093, USA

A.J.O. Dede · C.N. Smith (✉)
Veteran Affairs San Diego Healthcare System, 3350 La Jolla Village Drive (116A), San Diego, CA 92161, USA
e-mail: cnsmith@ucsd.edu

Curr Topics Behav Neurosci (2018) 37: 119–150
DOI 10.1007/7854_2016_452
Published Online: 28 September 2016

2.5 Neuroimaging Studies of Autobiographical Memory 129
2.6 Complicating Factors in Neuroimaging Studies of Autobiographical Memory 131
2.7 Novel Neuroimaging Approaches to the Study of Autobiographical Memory Consolidation 132
2.8 Autobiographical Memory: Conclusions 135
3 Semantic Memory 136
3.1 Semantic Memory in Memory-Impaired Patients 136
3.2 Neuroimaging Studies of Semantic Memory Consolidation 137
3.3 Prospective Studies of Memory Consolidation Across Short Time Scales 139
3.4 Semantic Memory: Conclusions 140
4 Conclusions 141
References 143

1 Introduction

Memory-impaired patients can more easily remember facts and events (declarative memory) formed long before the onset of amnesia compared to those formed more recently (Ribot 1881). This pattern of impaired recent memory and spared remote memory is known as temporally graded retrograde amnesia (TGRA) and this observation has led to the concept of systems consolidation. The idea is that gradual changes occur in the brain systems that support memory retrieval resulting in memories that are more robust to disruption as time passes after learning. In addition, patients exhibiting TGRA often exhibit impairment forming new memories (anterograde amnesia). This pattern of behavior is now thought to occur after damage to structures in the medial temporal lobe (MTL, i.e., the hippocampus and parahippocampal gyrus, which is composed of the perirhinal cortex, entorhinal cortex, and parahippocampal cortex) as well as to the medial diencephalon (McClelland et al. 1995; Kopelman and Bright 2012; Squire 1992). Nevertheless, the precise role of these structures for memory retrieval and how they may interact with other regions of the brain remains a point of controversy.

Although there is agreement that the MTL, and in particular the hippocampus, is always needed for the formation of new declarative memories, it is less clear whether the MTL is always needed to remember information from the past. Declarative memory concerns two types of information: semantic and episodic; and theoretical ideas differ with respect to the function of the MTL for each type of information. Semantic memories are fact-like, divorced from their encoding context, and strongly associated with each other. By contrast, episodic memories are experiential, intrinsically tied to their encoding context, and not strongly associated with each other (Tulving 1983).

Despite these differences, semantic and episodic memory interact substantially to facilitate remembering of autobiographical events. Indeed, the scoring systems used by researchers to evaluate autobiographical memories include separate categories for semantic autobiographical information and episodic autobiographical information. One idea is that semantic information may interact with episodic information at retrieval through an interactive and constructive process. This process results in the retrieved material being consciously registered in the mind of the rememberer (Semon and Simon 1921; Tulving 1983; Bartlett 1932). An implication of this idea is that episodic memories may never be retrieved without semantic information to guide retrieval, and thus, remembered information does not directly represent either semantic or episodic memory but is the result of their interaction at the moment of remembering. Accordingly, to understand fully the nature of systems consolidation for semantic and episodic memory, it is necessary to identify the brain regions important for semantic memory, episodic memory, and the integrative process that links them during remembering. In addition, it will be important to determine how the brain areas involved in these processes and the connections between them change over time.

There are three main theories that have attempted to explain systems consolidation of semantic and episodic information. One theory posits that all kinds of declarative memory (semantic and episodic) become independent of the hippocampus as time passes after learning (standard consolidation theory (SCT); McClelland et al. 1995; Squire and Alvarez 1995). This feat is accomplished by gradual changes in connectivity within the neocortex (Marr 1971). Eventually, connections representing a particular memory in the neocortex are sufficiently well formed that the hippocampus is no longer needed for retrieval.

Another idea is that the episodic component of autobiographical memories is always dependent on the hippocampus (multiple trace theory and transformation hypothesis, Nadel and Moscovitch 1997; Winocur et al. 2010). According to this idea, each retrieval of an episodic memory results in the encoding of a copy of the memory within the hippocampus. With increasing numbers of copies, episodic memories become more robust to disruption over time. In parallel, gradual changes in connectivity within the neocortex affect the integration of new information into semantic memory. Unlike SCT, the transformation hypothesis holds that episodic information is transformed into semantic information as it is incorporated into the neocortex, leaving truly episodic information permanently dependent on the hippocampus. By contrast, semantic memories are thought to eventually become independent of the hippocampus.

Finally, another idea is that the hippocampus is not directly involved in memory encoding and storage at all, but is instead responsible for the construction of atemporal spatial scenes in which the details of memories may be organized (Mullally et al. 2012). According to this view, the organization of information into coherent scenes is critical for the retrieval of episodic information regardless of memory age. The theory is agnostic with respect to semantic information.

Thus, with respect to semantic information, all three theories accommodate the idea that retrieval becomes independent of the hippocampus as time passes after learning. With respect to episodic information, these theories may be summarized as follows: SCT, the hippocampus is important for encoding and time-limited storage; transformation hypothesis, the hippocampus is important for both encoding and storage irrespective of memory age; and spatial view, the hippocampus is important for representing spatial information, a process needed during encoding and retrieval irrespective of memory age. All three theories lack detailed specification of the interactions that may exist between the MTL and other brain areas.

This review primarily considers evidence concerning the process by which autobiographical memories undergo systems consolidation. Beginning with autobiographical memory will facilitate later discussion of semantic memory, episodic memory, and the interaction between them. Our approach is to integrate findings from studies of memory-impaired patients together with findings from neuroimaging studies in order to identify the brain regions relevant for systems consolidation. Accordingly, we begin with studies of autobiographical memory in patients with circumscribed brain damage to several regions, including the MTL, which make up an autobiographical memory network. Next, we turn to neuroimaging studies of autobiographical memory where it is possible to consider how activity and connectivity in these regions changes as memories age in healthy individuals and patients. Finally, we briefly consider evidence from studies of semantic memory and compare it to autobiographical memory. We conclude that the autobiographical memory network may be thought of as two overlapping subnetworks, each responsible for the consolidation of a different type of information (semantic or episodic), and each intrinsically tied to the other by their interaction during retrieval and their mutual dependence on the MTL for the encoding of new information.

2 Autobiographical Memory

2.1 Autobiographical Memory in Memory-Impaired Patients

Studies of memory-impaired patients support the idea that memories become more invulnerable to disruption over time. Although this phenotype had been noted for some time (Ribot 1881), the study of the biological dissociation between recent and remote memory storage came to prominence when Scoville and Milner (1957) reported on the effects of bilateral MTL resection in the famous patient HM. In their initial description, HM's memories for his childhood were described as "vivid and intact" (pg. 14), yet he exhibited a profound deficit in his ability to encode new information. In addition, HM lost the ability to recall information learned within the three years prior to his surgery (e.g., the death of a favorite uncle). Although these reports of HM were suggestive of TGRA, they were based on clinical report rather than quantitative study.

For the early quantitative tests of autobiographical memory, individuals were invited to recall memories from throughout their lives when prompted with concrete nouns (e.g., Tell me a memory from your life that comes to mind when I say, "Tree")(e.g., Tell me a memory from your life that comes to mind when I say

"Tree"; Crovitz and Schiffman 1974). In these studies, the data are the approximate dates of the memories produced by participants. Using this technique, Sagar et al. (1985) found that (unlike controls) HM recalled all of his memories from a time when he was younger than 17 years old (11 years before his surgery). Similarly, this bias toward retrieval of distant memories as opposed to recent memories was also observed for other etiologies that cause memory impairment (i.e., bilateral electroconvulsive therapy, Korsakoff's syndrome, damage to the medial diencephalon, and hypoxia; Zola-Morgan et al. 1983; Beatty et al. 1987). By contrast, this effect was not observed for other etiologies, such as patients with focal brain damage outside the MTL, Parkinson's disease, Alzheimer's disease, and unilateral electroconvulsive therapy (Sagar et al. 1985; Zola-Morgan et al. 1983). These results suggest two ideas. First, they demonstrate that different etiologies may impact retrograde memory in different ways. Second, they provide evidence for systems consolidation in that recent memories were less accessible than remote memories.

Yet, these early studies did not assess the quality of memory. Therefore, these studies could not differentiate between competing views about the fate of truly episodic memories after brain injury or disease. In order to address this issue, researchers measured the quality of memories by counting the number of episodic versus semantic details produced during narrative recollection. In addition, methods were developed to rate the quality of episodic memories holistically through the use of standardized rubrics that seek to evaluate the specificity of memories to time and place, the coherence of narratives, and the richness of imagery.

With increased attention to memory quality came an increased incidence of conflicting reports regarding the status of episodic memory retrieval in patients. Results were largely dichotomous. Patients could either recall richly detailed autobiographical memories from their remote past, or they could recall nothing at all of their remote past. For example, Wilson and Baddeley (1988) reported on a patient (KJ) who suffered from postmeningitic amnesia and was unimpaired at remembering richly detailed episodes from his past. By contrast, Tulving (1985) reported on a patient (KC) who suffered from amnesia pursuant to a closed head motorcycle injury and who exhibited severe impairment remembering any of his past or imagining his future.

Reports of spared or impaired retrieval of remote episodic memories have continued to appear to this day and Lah and Miller (2008) have exhaustively reviewed this literature that encompassed 53 single case studies and 10 group studies. They found that 83 % of studies reported some deficit in autobiographical memory. Of the subset of studies that examined the status of autobiographical memory across different time periods, 46 % found TGRA and 54 % found dense and ungraded retrograde amnesia (RA). These percentages changed very little when we added in the 14 additional studies that were published since Lah and Miller's report (Philippi et al. 2015; Witt et al. 2015; Baird and Samson 2014; St-Laurent et al. 2011; Herfurth et al. 2010; Kurczek et al. 2015; Rosenbaum et al. 2008; Squire et al. 2010; Race et al. 2011; Irish et al. 2011; Thaiss and Petrides 2008; St-Laurent et al. 2014; Kirwan et al. 2008; St-Laurent et al. 2009).

Next, we explore three factors to determine if they can help explain the conflicting results across different patients. First, there may be differences in the procedures used to elicit memories. Second, there may be differences in the way that memories are evaluated. Third, there may be differences in the location and extent of brain damage in different patients.

First, differences in the methods used to elicit memories might explain discordant findings. To elicit memories, participants are typically given a memory prompt. There are many variations on this procedure where some studies provide support by asking specific questions for more detail, whereas others do not. If differences in the methods used to elicit memory were the cause of the discordant findings, then the findings should become concordant if the same method were used. However, this is not the case. For example, the autobiographical memory interview is a standardized tool for assessing episodic memory ability (Kopelman et al. 1989). Using the AMI, patients were reported to have intact autobiographical memory (Bright et al. 2006; Bayley et al. 2003, 2005; Stefanacci et al. 2000) or impaired autobiographical memory (e.g., Philippi et al. 2015; Cipolotti et al. 2001). Thus, one finds considerable variability in the ability to remember the past even when the same method was used.

A second possibility is that individual differences in scoring could have contributed to the different results across studies. For example, if a research group was liberal in their scoring methods then a ceiling effect would occur and both patients and controls would appear to have a similar capacity to retrieve autobiographical memoires. This difficulty would be exacerbated when using measures that are course, when the range of scores is small, and when scoring relies on interpretation of somewhat ambiguous scoring instructions like evaluating whether memories are "specific in time and place" (pg. 732, Kopelman et al. 1989). However, when the same patients were tested using both course and fine-grained rating systems the severity of amnesia was the same (e.g., Hassabis et al. 2007; Bayley et al. 2005; Kirwan et al. 2008). In addition, when the same patients were scored by different laboratories where scorers were blind to group membership the results were the same (Dede et al., under review). Finally, ratings of data sets from different patient groups (e.g., patients with lesions limited to the MTL and patients with lesions to the MTL and lateral temporal cortex, Bright et al. 2006) carried out by a single laboratory would be expected to be similar. This is also not the case.

Third, there is the possibility that anatomical differences between patients may be the cause of the different results across studies. Indeed, it has long been recognized that patients of different etiologies have different RA profiles (e.g., Lidz 1942). Lah and Miller (2008) investigated how the location of anatomical damage affected the probability of observing TGRA by examining the findings from 11 patients or groups of patients with damage limited to the MTL and where memory had been tested for both recent and remote events. Ninety percent of these studies found evidence for TGRA and this percentage increases to 92 % when findings from similar studies published since their report (Squire et al. 2010; Kirwan et al. 2008) are also considered. By contrast, when lesions were not limited to the MTL or when anatomy was not reported, fewer than half of the studies reported TGRA. This result underscores two important points. First, it indicates that obtaining detailed anatomical information from memory-impaired patients is a useful way to clarify the role of damaged areas. Second, the weight of evidence from studies where lesions were restricted to the hippocampus or MTL indicates that these regions are not necessary for the retrieval of truly episodic, remote autobiographical memories.

2.2 *Possible Semanticization of Episodic Memories*

Although the above findings from the study of patients with memory impairment support the idea that the MTL is not necessary for either the storage or retrieval of remote episodic memories, some have questioned whether the episodic content of these memories may ever be independent of the MTL (e.g., Winocur and Moscovitch 2011). According to the transformation hypothesis, semanticization is a necessary consequence of systems consolidation. The idea is that memories transform into semantic scripts as they become independent of the hippocampus. By this view, the experience of mental time travel that accompanies normal autobiographical memory is lost after hippocampal damage. If this is the case, then two predictions can be made: (1) when patients with hippocampal damage are instructed to recollect very detailed memories and fine-grained measures are used, a deficit should be revealed; and (2) when analysis is limited to only those memories that qualify as truly episodic, any deficit observed in patients with MTL damage should be magnified, because they should have no such memories. Neither of these predictions were confirmed.

To test the first prediction, two studies used the same method to assess whether patients could produce very detailed (non-semanticized) memories using scoring methods designed to capture the rich details that accompany autobiographical memories (Kirwan et al. 2008; Rosenbaum et al. 2008). Curiously, these two studies found opposing results: Kirwan et al.: patients were intact relative to controls, Rosenbaum et al.: patients were impaired relative to controls. The key to this puzzle, as reviewed earlier, pertains to whether or not the patients under study had damage limited to the MTL. With the exception of one patient with additional damage to the basal ganglia, the patients examined by Kirwan et al. (2008) had no damage beyond the MTL. By contrast, each patient examined by Rosenbaum et al. (2008) had volume reductions below the 95 % confidence intervals of controls in at least four regions outside of the MTL. In fact, regions outside the MTL best predicted performance for these patients. Specifically, using the data provided in Rosenbaum et al. (2008; Figs. 4 and 5), we used regional volume reductions to predict performance. Performance was best predicted by the volumes of right posterior temporal cortex, left posterior cingulate cortex, and the interaction between the volumes of these two regions ($R^2_{adj} = 0.99$).

To test the second prediction, Kopelman and Bright (2012) reanalyzed the data from their earlier study (Bright et al. 2006) to exclude all but the most episodic and specific memories. Although they had originally reported that MTL patients exhibited weak remote memory, the reanalysis revealed that patients' remote memories were as strong as controls. If the episodic component of memories depended on the hippocampus then the opposite pattern of results would have occurred. That is, the patients' deficit should have increased because none of their memories would have been episodic or specific.

Taken together, these studies suggest that when damage is limited to the MTL, patients exhibit highly detailed memories of the past, their memories are indistinguishable from controls, and their memories are not semanticized (according to the most sensitive tests that exist).

2.3 The Autobiographical Memory Network

The above discussion used evidence from patients with focal brain damage to consider the role of the hippocampus and MTL alone in retrieving autobiographical memories. However, among neuroimaging studies of autobiographical memory, there is now broad agreement that memory retrieval is associated with activity in a network of brain regions (Fig. 1). The network is predominantly left lateralized and is composed of the hippocampus, parahippocampal gyrus, middle temporal gyrus, retrosplenial/posterior cingulate cortex, the temporoparietal junction (angular

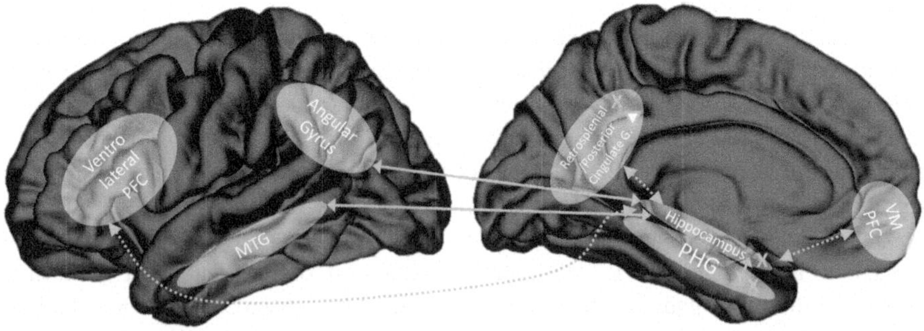

Fig. 1 The autobiographical memory network and memory consolidation. Regions in the autobiographical memory network are shown in *yellow* (cerebellum not shown). *Yellow Xs* indicate regions where damage is associated with temporally-graded retrograde amnesia. *Yellow triangles* indicate regions where brain activity decreases as memories age. *Yellow arrows* indicate that functional connectivity between the regions is associated with autobiographical memory retrieval. *Dashed yellow arrows* indicate functional connectivity that changes as memories age. The *left image* shows lateral surface of the brain and the *right image* shows the medial surface of the brain. The hippocampus is not visible on the surface of the brain, so it is depicted near the parahippocampal gyrus (PHG). *PFC* prefrontal cortex, *MTG* middle temporal gyrus, *VM PFC* ventromedial PFC, *G.* gyrus

gyrus), ventromedial prefrontal cortex (PFC), ventrolateral PFC, and the cerebellum (Svoboda et al. 2006; Cabeza and St Jacques 2007; Maguire 2001).

2.4 *Effects of Lesions in the Autobiographical Memory Network Beyond the MTL*

The patient literature reviewed thus far indicates that the hippocampus is not necessary for the retrieval of remote autobiographical memories. This implies that the information must be stored elsewhere. One prediction is that if the location of autobiographical memory storage outside the hippocampus were to be damaged, then focal RA should result. Focal RA is defined as RA without anterograde amnesia. Reports of focal RA are rare, and it is debated whether or not the phenomenon even exists (for review, see Kopelman 2000). Candidate structures for autobiographical memory storage may include those in the autobiographical memory network. In this section, we will examine the effects of focal damage to three of these regions: lateral temporal cortex, retrosplenial cortex, and medial PFC.

The lateral temporal cortex has long been implicated as a storage site for autobiographical memory. In Penfield's (1938) famous electrical stimulation studies in epilepsy patients, direct stimulation of lateral temporal cortex would sometimes produce vivid recollection-like mental experiences in the patients. Although the terms episodic and semantic were not in use at the time, Penfield's descriptions suggest that these recollections were episodic because the recollections included rich perceptual elements.

Patients with damage to both lateral temporal cortex and MTL exhibit a severe and ungraded RA for both episodic and semantic information (e.g., Bayley et al. 2005; Bright et al. 2006). However, when damage includes lateral temporal cortex alone, episodic memory is relatively spared and semantic memory is impaired (Hodges et al. 1992; Kapur et al. 1994; Irish et al. 2012; Hart and Gordon 1990). Somewhat confusingly, patients with damage thought to be limited to lateral temporal cortex have also been observed with dense RA for autobiographical events (O'Connor et al. 1992; Kapur et al. 1992). One potential difficulty is that damage limited to the lateral temporal cortex may result from several etiologies that are associated with variable patterns of damage. The examples cited here include patients who suffered from trauma (e.g., Kapur et al. 1992), encephalitis (e.g., O'Connor et al. 1992), and semantic dementia (e.g., Irish et al. 2012). These etiologies are all associated with the potential for widespread damage to multiple locations. In addition, the degree and precise location of damage within lateral temporal cortex itself is likely variable between patients. Thus, it may be that patients who exhibit dense RA for autobiographical events have damage that more completely encompasses the critical storage areas in lateral temporal cortex or that extends beyond lateral temporal cortex. To resolve this issue more research will be necessary where the location and extent of damage are measured.

Setting aside specific anatomical considerations, the dynamic interaction between semantic and episodic memory is important when considering the data from patients with damage to lateral temporal cortex. Under Tulving's (1983) conceptualization of semantic and episodic memory, episodic memories cannot be retrieved without a suitable semantic cue, and it has been suggested that semantic information may play a pivotal role in organizing episodic information during retrieval (Irish and Piguet 2013). Thus, the finding that patients with damage to lateral temporal cortex may sometimes exhibit a broad RA for both semantic and episodic information may not represent a true loss of episodic information. Rather, such deficits may be caused by a loss of the semantic information needed to cue episodic information. Indeed, Penfield (1938) found that removal of the very tissue he had stimulated to produce a specific recollection did not destroy the recollection, indicating that the stimulated tissue was not the site of storage. If this is the case, then more consideration of the methods used to cue memories may elucidate further the nature of deficits after lateral temporal lobe damage.

Another region in the autobiographical memory network that has been associated with memory deficits is retrosplenial cortex. Gainotti et al. (1998) found that a patient with a surgical lesion limited to retrosplenial cortex exhibited TGRA that extended approximately 10 years into the past for autobiographical events. Importantly, Gianotti et al. established that the hippocampus (which is adjacent to the retrosplenial cortex) was not damaged. Similar results were reported for a patient with retrosplenial damage following hemorrhage of an arteriovenous malformation (Valenstein et al. 1987). The close proximity of the retrosplenial cortex and the hippocampus makes it difficult to interpret findings from patients whose damage may encompass both regions (Heilman and Sypert 1977; Vann et al. 2009). In addition to TGRA, Gianotti et al. reported that the patient exhibited anterograde amnesia for visual stimuli, but not verbal stimuli. Other reports confirm a specific role for retrosplenial cortex in the episodic aspects of memory (for review see Miller et al. 2014). This modality-specific effect is strikingly different from the global anterograde amnesia exhibited by patients with hippocampal damage. Thus, the retrosplenial cortex appears to play a role in the consolidation of episodic memory and in encoding of new visual memories.

Finally, the medial PFC is part of the autobiographical memory network, but damage to this region does not consistently result in RA (for review, see Nieuwenhuis and Takashima 2011). Instead, medial PFC lesions are associated with confabulation (Schnider 2008), loss of impulse control (e.g., the famous patient Phineas Gage; Harlow 1868), and decreased self-related processing (Philippi et al. 2012). Given these results, it may be that the medial PFC is consistently activated during autobiographical memory retrieval because of the self-referential nature of autobiographical remembering rather than because of any specific role in memory encoding or storage.

Taken together the evidence reviewed here suggests that neither lateral temporal cortex, retrosplenial cortex, nor medial PFC acts as the ultimate site of episodic memory storage. Combined lesions to both MTL and lateral temporal cortex result in dense RA. However, lesions limited to either region alone do not result in dense RA, and it may be that these regions are involved in the dynamic process of remembering rather than in storage. Specifically, semantic information represented in lateral temporal cortex may be necessary for cueing episodic retrieval. Finally, damage to medial PFC results in a wide range of executive impairments, but it does not result in a dense RA.

Although the review provided in this section is not exhaustive, it is important to consider what phenotype would be expected if one of these regions were the ultimate storage site of episodic memory. Such a phenotype would be clear and severe. The idea is not to say that the regions of the autobiographical memory network reviewed here play no role in the long-term storage of episodic information. Rather, our point is that damage to these regions does not result in a clear and severe RA for episodic information, suggesting that the primary site of storage must be elsewhere.

Perhaps one reason there is such difficulty locating the site of autobiographical memory storage is that the contents of memory are distributed throughout the cortex (for an early formulation of this idea, see Lashley 1950). Thus, memories for particular events are represented as connections between the regions that store the elements of content, rather than located in a single storage site. Based on the patient studies reviewed above, hippocampus and retrosplenial cortex likely act as important hubs for these connections during and shortly after encoding. With time, these connections are not needed for retrieval. One possibility is that as time passes after learning, connections between content elements in cortex may be retrieved via cortico-thalamo-cortical loops. Direct cortico-cortical connection is unlikely as it would be difficult to form new long-range connections. If this is true, then damage to this circuit should impair remote memory but not recent memory. Indeed, a patient with damage limited to the medial thalamus (patient JG) exhibited this pattern of RA (Miller et al. 2001, 2003). Moreover, the importance of interactions between MTL and anterior thalamus during memory encoding has been recognized for some time, and damage to anterior thalamus can result in anterograde amnesia (for review see Aggleton and Brown 1999). More investigation into the roles of different nuclei of the thalamus in anterograde and retrograde memory would illuminate this idea.

2.5 Neuroimaging Studies of Autobiographical Memory

Considering the patient literature reviewed above, two predictions can be made in turning to neuroimaging studies of autobiographical memory. First, the involvement of hippocampus and retrosplenial cortex in remembering should decrease with memory age. Second, the role of lateral temporal cortex might increase with memory age, but this finding may reflect this region's role in semantic memory. Although these predictions are relatively straightforward, the findings from neuroimaging studies do not consistently support these predictions, with the exception of retrosplenial cortex.

For neuroimaging studies of autobiographical memory, participants are asked to retrieve autobiographical memories in the scanner. Either before or after scanning, participants are typically asked to provide narrative descriptions of the memories retrieved during scanning. Scoring of these narratives is used to relate neuroimaging data to aspects of memory quality. Even though there is agreement across studies that a specific set of regions is active during memory retrieval (i.e., the autobiographical memory network; Svoboda et al. 2006; Cabeza and St Jacques 2007; Maguire 2001), there is no consensus whether activity in these regions changes as a function of memory age.

Among the 14 studies that have examined activity as a function of memory age, no brain regions exhibited consistent increases or decreases in activity. In fact, the most common finding for these regions was to exhibit no change in activity with memory age, with the exception being posterior cingulate/retrosplenial cortex. Although the hippocampus exhibited changes with memory age in 7 studies, these changes had no consistent direction (Addis et al. 2004; Gilboa et al. 2004; Maguire and Frith 2003; Piefke et al. 2003; Piolino et al. 2004; Rekkas and Constable 2005; Soderlund et al. 2012).

The story is more consistent for the posterior cingulate/retrosplenial cortex. Most studies (7 out of the 12 that examined activity outside of the MTL) reported that activity changed with memory age and all the studies reported higher activity for recent than for remote memory. It is also worth mentioning that, although only 4 studies detected activity in parahippocampal gyrus that changed with memory age (Niki and Luo 2002; Rekkas and Constable 2005; Soderlund et al. 2012; Tsukiura et al. 2002), all but one of these studies found that activity was higher for recent memory than for remote memory.

In summary, the consistent reduction of activity in retrosplenial cortex with memory age is in accord with findings from memory-impaired patients suggesting that this area is more involved in remembering recent memories than remote memories. By contrast, the mixed results for the hippocampus are at odds with findings from patient studies. Further, throughout the rest of the autobiographical network activity does not appear to change with memory age. These results suggest two possibilities. First, although hippocampus and other areas of the autobiographical network remain active for old memories, these regions may not be necessary for remembering. Another possibility is that methodological difficulties may have contributed to the lack of temporal gradients in many areas of the autobiographical network.

Next, we discuss whether three factors could have contributed to the mixed pattern of results across the 14 studies, focusing on the hippocampus. None of the factors appear to explain why these studies fail to detect temporally graded activity. Nevertheless, future studies will benefit from avoiding these difficulties as novel procedures and analysis techniques are brought to bear on this issue. Indeed, carefully designed studies that incorporate new neuroimaging methods are beginning to shed light on the ways in which the functional roles of brain regions may change with memory age, even as overall activity remains stable (see Novel neuroimaging approaches, below).

2.6 Complicating Factors in Neuroimaging Studies of Autobiographical Memory

One methodological concern for autobiographical studies is that remote memories tend to be less vivid, less detailed, and less emotional than recent memories (e.g., Addis et al. 2004; Gilboa et al. 2004; Maguire and Frith 2003). Thus, when one observes differences in brain activity between recent and remote memories, it is difficult to know whether these differences reflect memory age or differences in other qualities that change with memory age. For example, Addis et al. (2004) found that bilateral hippocampal activity was higher for recent memory than for remote memory. However, when the number of details and emotionality were covaried out, only the right hippocampus continued to responds according to memory age.

If differences in memory quality were driving differences in hippocampal activity, one would expect that most of the extant studies should also have identified the hippocampus; this is not the case. Further, one would also expect that the studies that did not detect differential activity would be the same ones that did not report differences in recent and remote memory quality. This is not the case. Taken together, these results suggest that hippocampal activity during episodic memory retrieval does not reliably reflect memory age or memory quality. To clarify the interpretation of memory consolidation studies it will be important for each study to ensure that recent and remote memories are similar in terms of quality before carrying out comparisons for brain activity (e.g., Addis et al. 2004; Gilboa et al. 2004; Bonnici et al. 2012). A good practice is to ensure that memories are similar according to the qualities that are considered central to episodic remembering (i.e., vividness, a sense of mental time travel, and specificity to time and place) and then carry out neuroimaging analyses.

Another methodological concern is how incidental encoding during memory retrieval may affect retrieval-related brain activity. In memory retrieval experiments, the events of the experiment are also encoded into memory. Consequently, subtle differences in activity for recent and remote memory retrieval may be overshadowed by encoding-related activity (Reas and Brewer 2013; Okado and Stark 2003; Buckner et al. 2001; Stark and Squire 2000). Future studies that measure incidental encoding will help disentangle these processes.

A third methodological concern regards the common practice of conducting a pre-scan interview to identify the memories to query during scanning. At first glance, it might appear that this practice would reduce the likelihood of identifying differential activation for recent and remote time periods because memories from both time periods were retrieved recently. However, this is not the case. In fact, the majority of studies that used pre-scan interviews detected differential hippocampal activity for recent and remote memories (even though there was no consistent pattern to the differences). Moreover, hippocampal findings from studies that did not use a pre-scan interview were just as mixed. Thus, even though brain activity can change as a function of the number of recent memory retrievals (Rekkas and Constable 2006; Nadel et al. 2007; Svoboda and Levine 2009), it does not appear that retrieving memories prior to scanning is problematic when detecting activity related to memory age, at least where hippocampus is concerned.

2.7 Novel Neuroimaging Approaches to the Study of Autobiographical Memory Consolidation

The neuroimaging studies discussed above all examined the amplitude of evoked activity for memories as a function of memory age. Alternative ways to investigate how brain signals relate to memory consolidation do not require statistical differences between the evoked neural amplitudes for conditions of interest. For example, some techniques detect correlations between activity in different regions of the brain regardless of the mean amplitude (e.g., connectivity analyses) and other techniques detect patterns of activity across voxels regardless of the mean amplitude of activity (e.g., multi-voxel pattern analysis, MVPA; Norman et al. 2006; Haynes and Rees 2006). Next, we explore findings using these methods as well as studies that have used prospective designs to study autobiographical memory age over shorter timescales (e.g., days to months).

Only one study has examined connectivity as a function of memory age (see below). Instead, most work has examined how connectivity changes according to other factors, such as the stages of retrieval or the quality of memory. For example, within the autobiographical network, St Jacques et al. (2011) found that a subset of regions centered around the MTL (i.e., MTL, posterior cingulate/retrosplenial cortex, tempoparietal junction, and ventromedial PFC) exhibited high interconnectivity during both construction and elaboration stages of autobiographical retrieval. When memories were more easily retrieved, this MTL subnetwork influenced activity in another subnetwork centered around dorsomedial PFC (i.e., dorsomedial and ventrolateral PFC, middle temporal gyrus, tempoparietal junction, and posterior cingulate). Finally, greater connectivity between these two subnetworks during the elaboration stage of retrieval was associated with higher ratings of reliving.

Within its subnetwork, the MTL was more strongly connected with posterior cingulate and ventromedial PFC during autobiographical relative to semantic retrieval (Fuentemilla et al. 2014; Muscatell et al. 2010). Similarly, data collected while participants were at rest revealed that stronger connectivity between MTL and posterior cingulate was associated with more episodic detail production in a subsequent test of autobiographical memory (Sheldon et al. 2016). In addition, findings from complimentary techniques with better temporal resolution, such as magnetoencephalography and intracranial recordings, support the idea that connectivity and coherence between MTL and medial parietal cortex are important for autobiographical retrieval, particularly when memories are associated with visual imagery (Foster et al. 2013; Fuentemilla et al. 2014).

Considering the central role of the MTL in the autobiographical memory network in healthy subjects, it is of interest to examine how damage to the MTL affects connectivity within the network. Maguire et al. (2001) identified similarities and differences in the autobiographical memory network between healthy individuals and an adult patient (Jon) with perinatal injury to the hippocampus. First, Jon had bilateral activations, whereas controls had predominantly left lateralized activations. Second, the most highly connected region of the autobiographical network was the retrosplenial cortex for Jon, but was the hippocampus for controls. A later study examined functional connectivity during rest in a group of three patients of mixed etiology with damage to the MTL (Hayes et al. 2012). They found that retrosplenial connectivity with ventromedial PFC, posterior cingulate, and posterior inferior parietal lobule was higher in the patients relative to controls. By contrast, connectivity between retrosplenial cortex and MTL was lower in patients relative to controls. These findings have been replicated with individuals who have temporal lobe epilepsy (Addis et al. 2007). These results indicate that the brain is capable of compensating for the loss of the hippocampus within the autobiographical memory network and that retrosplenial cortex and medial PFC appear to be important for this compensation. In addition, the fact that activation of the network is more or less intact following hippocampal damage suggests that although the hippocampus may play a role in normal autobiographical remembering, this role is not critical.

One study has examined connectivity for memories that differ according to memory age. Soderlund et al. (2012) used a within-subject test to identify how hippocampal activity and connectivity changed for memories ranging from 1 week to 10 years old. Although they did not detect differences in hippocampal activity as a function of memory age, they did detect differences in hippocampal connectivity. Specifically, hippocampal activity was positively correlated with activity in anterior cingulate, posterior cingulate, anteromedial PFC, and the precuneus for recent memories (1 week, 1 month, and 1 year old), but not for remote memories (10 years old). By contrast, hippocampal activity was negatively correlated with activity in superior temporal gyrus, ventrolateral PFC, and cuneus for remote memories (but not for recent memories). Thus, hippocampal activity was positively associated a set of midline regions for recent memory, but not for remote memory. At the same time, hippocampal activity was negatively associated with set of more lateral regions for remote memory, but not for recent memory. This study illustrates the idea that there can be dissociations between measures of evoked activity and measures of connectivity. Connectivity studies represent fertile ground for detecting how the brain distinguishes between recent and remote memory.

MVPA assesses whether the pattern of activity in a set of voxels carries information capable of differentiating between two or more conditions of interest. Using this technique, Bonnici et al. (2012) demonstrated that regions of the autobiographical memory network could distinguish between 10 individual recent memories (2 weeks old) and it could also distinguish between 10 individual remote memories (10 years old). Specifically, temporal pole and ventromedial PFC contained more information about remote than recent memories. By contrast, MTL (i.e., the hippocampus, entorhinal cortex, and parahippocampal cortex) contained the same amount of information about recent and remote memories. Finally, they observed that voxels in posterior hippocampus (particularly subregions CA3 and dentate gyrus; Bonnici et al. 2013) contained more information about remote than recent memories while voxels in anterior hippocampus contained the same amount of information for recent and remote memories. This study supports the idea that cortical areas become more important for memory storage over time. In addition, it demonstrates the new types of questions that can be posed with MVPA.

In terms of study design, recent studies have employed prospective designs to study autobiographical memory. The benefit of prospective designs is that one has control over the time periods tested and the strength of memory. Although these studies typically measure memory age across short time scales (e.g., days, week, and months) they can detect the autobiographical memory network (Levine et al. 2004; Sheldon and Levine 2013). For example, in a heroic study by Sheldon and Levine (2013), participants recorded narratives soon after significant events in their lives in the 2 years preceding scanning. During scanning, audio clips of these narratives were replayed for recent memories (mean = 31 days old) or remote memories (mean = 565 days old). The authors found that, although evoked activity was no different for recent and remote memories, connectivity of anterior hippocampus differed according to memory age. Hippocampal connectivity with medial parietal regions (posterior cingulate and precuneus) was higher for recent versus remote memories, whereas hippocampal connectivity with PFC and superior temporal gyrus was higher for remote versus recent memories.

Another recent study relied on obtaining photographs of participant's autobiographical events by having them wear a camera (i.e., SenseCam) that took photos about every 30 s. Brain activity was different for the photos when tested 36 h versus 5 months after the photos were obtained. Specifically, activity in hippocampus, parahippocampal gyrus, and ventromedial PFC was higher for recent than for remote memories (Milton et al. 2011). However, according to MVPA analysis of whole-brain activity in response to SenseCam photos that were 1 versus 3 weeks old, there was insufficient information contained in brain activity to distinguish between these conditions (Rissman et al. 2016). Nevertheless, brain activity could distinguish between traditional categories of memory such as hits versus correct rejections or recollection versus familiarity judgments.

2.8 Autobiographical Memory: Conclusions

When memory-impaired patients have damage restricted to the MTL they exhibit TGRA and anterograde amnesia, suggesting that the MTL plays a permanent role in forming new memories and time-limited role in memory retrieval for autobiographical information. For these patients, recent memories are more vulnerable to disruption than remote memories, and remote memories are as detailed as memories from healthy individuals. Thus, the content of autobiographical memory does not reside exclusively in the MTL.

Candidate structures for memory storage might include the large network of regions revealed by neuroimaging studies of autobiographical memory. In patient studies, damage to either the hippocampus or the retrosplenial cortex causes TGRA, but there is no single area where loss results in dense RA for the lifetime. Instead, damage to other regions in the network is associated with impairments in components or features of autobiographical memory. For example, the loss of semantic memory results from lateral temporal cortex damage and the loss of self-referential thinking results from medial PFC damage. Finally, when damage compromises many regions in the network (e.g., patients KC, HC, PH, and GT), ungraded and dense deficits occur. Thus, even if the damaged areas are not involved in memory storage per se, extensive damage to the autobiographical memory network compromises memory retrieval regardless of the age of memory.

The findings from patients suggest that the regions of the autobiographical memory network are primarily associated with retrieval processes rather than memory storage. If this is the case, then so long as the retrieval processes are similar for recent and remote memory, brain activity should be similar for these conditions within the network. Neuroimaging studies of recent and remote memory confirm this idea, as brain activity in the autobiographical memory network does not correspond to the age of memory. An exception to this finding is retrosplenial cortex, which consistently exhibits higher activity for recent memories than remote memories. The hippocampus also exhibits changes in activity with memory age, but there is no consistent direction of change across studies. Activity changes in the hippocampus and other regions may be related to factors that differ across the individual studies such as the quality of memory or the extent of incidental encoding.

Connectivity within the autobiographical memory network changes according to the stage of retrieval (e.g., memory search vs. memory elaboration) and according to features of autobiographical memory (e.g., ease of retrieval, amount of reliving, visual imagery). In particular, connectivity between midline structures (e.g., MTL, retrosplenial and medial parietal cortex, and ventromedial PFC) is associated with features that are characteristic of episodic retrieval. Little is known about how connectivity changes with memory age. The one study that examined brain activity and brain connectivity as memories age indicates that connectivity between the

MTL and these same midline structures changed with memory age. Finally, memory consolidation was more readily detected as changes in connectivity between regions than as changes in activity in individual regions.

3 Semantic Memory

3.1 Semantic Memory in Memory-Impaired Patients

Semantic memory is not tied to any particular encoding context or event, and can be retrieved without interaction with episodic memory. Semantic memory is often assessed by asking participants to demonstrate everyday knowledge (e.g., word meanings, object names, and historical and cultural facts). Such tests draw on retrograde memory, often learned long before testing.

Memory-impaired patients of different etiologies have exhibited both spared and impaired semantic memory. For example, patient RFR exhibited a severe loss of autobiographical memory following encephalitis, but he was relatively unimpaired when asked to provide meanings of words, identify famous names, and rate the familiarity of famous faces (Warrington and McCarthy 1988). By contrast, other patients exhibited pure loss of word meanings without any apparent loss of other functions (Hart and Gordon 1990).

As with autobiographical memory, the locus and extent of brain damage is an important factor in the study of semantic memory impairment. For example, patient RFR had damage that encompassed all of right temporal lobe and left MTL, but left lateral temporal lobe was spared (McCarthy et al. 2005). By contrast, the patients who had impaired knowledge of word meanings had damage limited to left lateral temporal lobe (Hart and Gordon 1990). This pattern extends beyond these examples, and memory-impaired patients whose damage included MTL but spared left lateral temporal cortex consistently exhibit relatively intact retrograde semantic memory (Bayley et al. 2006; Kapur and Brooks 1999; Manns et al. 2003; but see Klooster and Duff 2015). Memory-impaired patients whose damage included left lateral temporal cortex consistently exhibited impaired retrograde semantic memory (Bayley et al. 2005; Gilboa et al. 2005; Bright et al. 2006; Hodges et al. 1992; Barbeau et al. 2012; Gardini et al. 2013; Smith 2014; Levy et al. 2004). These findings indicate that the MTL is not necessary for storage of semantic memory, but that areas within lateral temporal cortex are necessary.

Although the MTL is not necessary for storing semantic memory, it is needed for encoding new semantic memories. For example, patient RFR was retested 14 years after the initial test described above (McCarthy et al. 2005) and he continued to demonstrate intact semantic retrograde memory. By contrast, on a test of words that had entered the lexicon since his injury, RFR was severely impaired. Several other studies have confirmed that the MTL is critical for the encoding of new semantic memories (e.g., Shimamura and Squire 1987), suggesting that MTL damage does not impact maintenance of information encoded before the onset of amnesia.

Thus, as with autobiographical memory, semantic memory relies on the MTL for encoding but retrieval eventually becomes independent of the MTL over time. Unlike autobiographical memory, lateral temporal cortex becomes increasingly important over time as new experiences are transformed into semantic knowledge. Studies examining the extent of RA after damage to the MTL indicate that the hippocampus remains important for the retrieval of semantic information for several years after memory encoding, and that the parahippocampal gyrus remains important for several more years (up to several decades) (Smith et al. 2013; Bayley et al. 2006).

3.2 Neuroimaging Studies of Semantic Memory Consolidation

The functional neuroanatomy of semantic memory has been less well characterized than autobiographical memory. Most studies have investigated brain activity associated with viewing famous faces or names. These studies identified some of the same regions as those involved in the core autobiographical memory network, such as the MTL, lateral temporal cortex, medial and lateral parietal cortex, and PFC (Denkova et al. 2006; Seidenberg et al. 2009; Woodard et al. 2010). In fact, a semantic memory network has been identified by a meta-analysis of 120 neuroimaging studies of semantic memory processing (Binder et al. 2009) and the network overlaps substantially with the autobiographical memory network (Fig. 2).

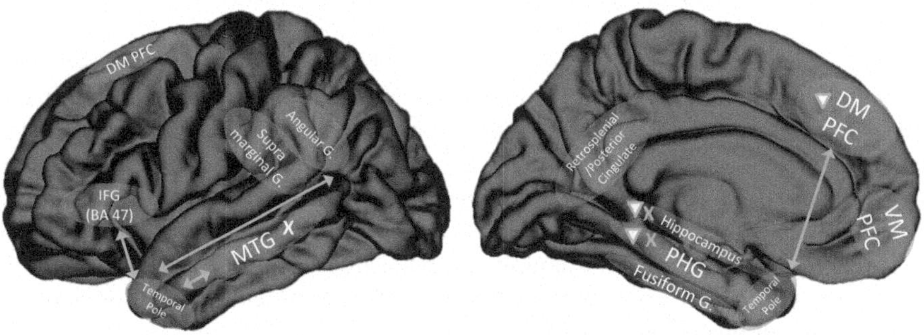

Fig. 2 The semantic memory network and memory consolidation. Regions in the semantic memory network are shown in *blue*. *Blue Xs* indicate regions where damage is associated with temporally graded retrograde amnesia and a *white X* indicates dense, ungraded retrograde amnesia. *Blue triangles* indicate regions where brain activity decreases as memories age. *Blue arrows* indicate that functional connectivity between the regions is associated with semantic memory and processing. The *left image* shows lateral surface of the brain and the *right image* shows the medial surface of the brain. The hippocampus is not visible on the surface of the brain, so it is depicted near the parahippocampal gyrus (PHG). *PFC* prefrontal cortex, *IFG* inferior frontal gyrus, *MTG* middle temporal gyrus, *VM PFC* ventromedial PFC, *DM PFC* dorsomedial PFC, *G.* gyrus; *BA* Brodmann area

In addition to the regions that are shared between the networks [(parahippocampal gyrus, middle temporal gyrus, posterior cingulate/retrosplenial cortex, angular gyrus, inferior frontal gyrus (BA 47), and ventromedial PFC (BA 9,10)], the semantic network also includes dorsomedial PFC (BA 8), supramarginal gyrus (just anterior to the angular gyrus), fusiform gyrus, and temporal pole (Visser et al. 2010).

One idea is that the regions that are common to both networks support the semantic memory component of autobiographical memory. Alternatively, these regions may support cognitive operations that are common for both autobiographical memory and semantic memory such as selecting relevant information from memory stores. Notably absent from this network is the hippocampus, though the studies examined by the meta-analysis examined semantic information learned long ago (e.g., information about word meanings).

Seven studies have examined semantic memory as a function of memory age (two of these studies used the same dataset for MTL analyses, Douville et al. 2005; and non-MTL analysis, Woodard et al. 2007; these two publications are considered one study here). Half of these 6 studies identified activity in the MTL that changed with memory age and all these studies found that activity was higher for recent memories than for remote memories (entorhinal cortex, Haist et al. 2001; hippocampus and parahippocampal gyrus, Douville et al. 2005; hippocampus and entorhinal cortex, Smith and Squire 2009).

Looking outside the MTL for evidence supporting semantic memory consolidation, only two of the six studies identified regions where activity changed with memory age (Smith and Squire 2009; Woodard et al. 2007). For both studies, the activity in dorsomedial PFC (middle frontal gyrus, BA 8) decreased with memory age. In addition, activity in middle temporal gyrus decreased with memory age when participants were asked to recognize famous names (Woodard et al. 2007), but increased with memory age when they identified news events (Smith and Squire 2009).

More work is needed to identify the non-MTL regions that support semantic memory consolidation. Most of the extant studies identified no regions outside the MTL, but the study that identified many regions outside the MTL (Smith and Squire 2009) differed from the other studies. It examined memory age for many items across many time periods instead of for relatively few items across few time periods. Specifically, Smith and Squire (2009) examined 160 news events that occurred 1–30 years prior to scanning separated into 7 time periods. By contrast, the other studies either examined few time periods (one recent time period and one remote time period) or queried few items (as few as 24).

No studies have examined how connectivity changes during semantic memory consolidation. Nevertheless, functional connectivity studies of semantic memory and processing indicate that connectivity associated with the temporal pole is important for semantic retrieval. For example, Maguire et al. (2000) determined that connectivity between temporal pole and middle temporal gyrus was stronger for semantic memory than for autobiographical memory. Using scanning procedures designed to capture signal in the temporal pole, Jackson et al. (2016) found that semantic retrieval was associated with connectivity between temporal pole and middle temporal gyrus. Further, the temporal pole was also connected to dorsomedial PFC, and tempoparietal junction during semantic memory judgments.

3.3 Prospective Studies of Memory Consolidation Across Short Time Scales

The time scales investigated with prospective designs are much shorter than retrospective designs, i.e., ranging from hours to months. The benefit of prospective designs is that one has control over variables such as the time periods tested and the strength of memory. Prospective studies have not examined semantic memory per se, because true semantic memory is thought to form as regularities are extracted across many separate episodic experiences. Instead, prospective studies have examined laboratory-based memory that is typically devoid of the rich, contextual information associated with episodic memory. Thus, we discuss these studies in the semantic memory section because they are distinct from the type of memory normally considered to be episodic memory (i.e., studies of autobiographical memory).

The findings for the hippocampus are mixed across many of the prospective studies, with some studies reporting decreasing hippocampal activity with memory age (Takashima et al. 2006, 2009; Smith et al. 2010; Yamashita et al. 2009), and some studies reporting the opposite pattern (Bosshardt et al. 2005a; Suchan et al. 2008). Other studies found both patterns or no effect memory age (Stark and Squire 2000; Bosshardt et al. 2005b; Janzen et al. 2008; Vilberg and Davachi 2013; Takashima et al. 2007).

A shortcoming of most prospective studies, and a possible explanation for the mixed findings, is that many of these studies have examined only two time periods: recent and remote. In a study that examined 4 time periods across three months (Takashima et al. 2006), there was a steep reduction in hippocampal activity and a parallel increase in ventromedial PFC activity with most of the changes occurring in the first day after learning. Studies that demonstrate systematic differences across multiple time periods should be considered more reliable than studies that find a difference across just two time periods.

Changes in functional connectivity as memories age from hours to weeks have also been observed across many studies and the findings are consistent with the ideas of systems consolidation. Specifically, connectivity between hippocampus and cortex decreased with memory age (Takashima et al. 2006, 2009; Smith et al. 2010), whereas connectivity between cortical regions increased with memory age (Nieuwenhuis et al. 2012; Takashima et al. 2007, 2009; Smith et al. 2010). These findings indicate that changes in connectivity can occur over timeframes as short as days and weeks, but more effort is needed to identify cortical regions that demonstrate a consistent pattern of changes with memory age.

Perhaps the most striking aspect of prospective studies is that the neural reorganization appears to occur over a short time scale (hours to months). By contrast,

TGRA in memory-impaired patients extends over several years. It is possible that the neural changes observed in the days and weeks after learning are the first reflection of a process that takes years to complete. In this case, changes in brain activity and functional connectivity are observed over a short time frame, whereas changes in anatomical connectivity (i.e., the strengthening of connections between the brain regions that contain the contents of memory) take years to form.

Another important observation about the short time frame examined by prospective studies is that this time frame is in good agreement with the findings from the animal literature (Bontempi et al. 1999; Maviel et al. 2004; Frankland et al. 2004). In animal studies of memory consolidation, memory retrieval typically becomes independent of the hippocampus within a few weeks after learning. Thus, the prospective method may provide a way to begin to connect the human and animal literatures.

Prospective studies also allow the assessment of the neural changes associated with factors known to improve memory consolidation (i.e., to attenuate forgetting). For example, study episodes that are spaced over time (spaced learning) reliably create longer lasting memories. By contrast, study episodes that follow each other closely (massed learning) are less effective. Several studies have investigated neural differences between memories studied in a spaced versus massed fashion. Vilberg and Davachi (2013) found that word-picture associations studied in a spaced fashion were more likely to be remembered if activity in the hippocampus and perirhinal cortex was correlated during encoding. By contrast, the correlation between activity in these regions was not important for remembering associations that were learned in a massed fashion, suggesting that connectivity between brain regions supports the improved memory consolidation observed after spaced learning. Takashima et al. (2007) contrasted brain activity during retrieval for face-locations associations learned via massed versus spaced learning. They found that activity in precuneus, temporal pole, temporoparietal junction, middle temporal gyrus, and ventromedial PFC were higher for associations learned in a spaced versus massed fashion, even though memory performance was similar for the two conditions. Furthermore, connectivity between fusiform face area (an area associated with face processing and face memory) and precuneus was higher for associations learned in a spaced versus massed fashion.

3.4 Semantic Memory: Conclusions

Damage limited to the MTL consistently causes anterograde amnesia and TGRA for semantic information. In agreement with these findings, the hippocampus is not a part of the semantic memory network identified by studies of remote semantic memory (e.g., word knowledge). By contrast, the hippocampus is readily identified by neuroimaging studies examining recently learned semantic information (e.g., famous faces, news events), and in these studies MTL activity decreases with memory age. Taken together, these findings are consistent with the idea that the hippocampus plays a permanent role in encoding new semantic memory, but plays a time-limited role in semantic memory retrieval.

The semantic memory network overlaps substantially with the autobiographical memory network including 7 of the same regions (hippocampus, parahippocampal gyrus, middle temporal gyrus, retrosplenial cortex/posterior cingulate cortex, angular gyrus, ventromedial PFC, and ventrolateral PFC). In addition, the semantic network includes four additional regions (dorsomedial PFC, supramarginal gyrus, fusiform gyrus, temporopolar cortex) and the network excludes one region (cerebellum). Although little work has examined how activity in this network changes with the age of memory, it appears that activity in the MTL and dorsomedial PFC consistently decreases with memory age.

Most regions in the network did not exhibit changes in activity with memory age. This finding may indicate that these regions are always involved in semantic memory retrieval. For example, damage to lateral temporal cortex causes severe and ungraded RA, suggesting that this region may store the long-term representation of semantic information. Indeed, semantic memory and semantic processing are both associated with increased connectivity of the temporal pole. It will be important to identify which connections are important for semantic memory consolidation.

There has been an explosion of recent work examining memory consolidation over shorter timescales (days, weeks, and months), when systems consolidation is observed in animals. In humans, the neural changes that emerge within the days and weeks after learning may be the first signs of systems consolidation that takes years to complete.

4 Conclusions

In this section, we compare and contrast memory consolidation for autobiographical memory and semantic memory, identifying three main points that reflect the findings from patient studies and neuroimaging studies.

First, the autobiographical network (Svoboda et al. 2006; Cabeza and St Jacques 2007; Maguire 2001) overlaps substantially with the semantic network (Binder and Desai 2011; Binder et al. 2009). It is perhaps unsurprising that these networks overlap because autobiographical memory requires retrieval of both episodic and semantic components. Indeed, TGRA for both semantic and episodic information is observed after damage limited to MTL. One idea is that there may be a common network for all declarative memory retrieval (Burianova et al. 2010; Burianova and Grady 2007), whereby episodic and semantic retrieval are not distinguished by whether or not regions are active. Instead, episodic and semantic retrieval may be better differentiated by the relative degree of activity and connectivity between regions of the network (Maguire et al. 2000; Heisz et al. 2014).

Although this common memory network was identified by studies of memory retrieval, it is also identified when individuals simply rest quietly (default mode network; Andrews-Hanna et al. 2010, 2014), suggesting that the memory retrieval network may play a role in other cognitive operations. Specifically, the network is active when individuals imagine the future (Addis et al. 2009) or think about events from different perspectives (theory of mind; Spreng and Grady 2010). These findings suggest that the cognitive resources brought to bear during autobiographical and semantic remembering represent a number of subprocesses that interact in different ways depending on the specific task at hand (Addis et al. 2009). Certain regions within the network operate as common connection points across cognitive operations. For example, posterior cingulate and anteromedial PFC may be particularly important because they were the most highly interconnected of all the regions (Andrews-Hanna et al. 2014).

Second, even though semantic and episodic memory are contained within larger one network, focal damage to each of two regions is associated with amnesia for one type of memory but not for the other. Specifically, damage limited to lateral temporal cortex is associated with ungraded RA only for semantic memory, whereas damage limited to retrosplenial cortex is associated with TGRA only for episodic memory. Interestingly, there is no single region within the memory network where damage results in ungraded RA for autobiographical events. By contrast, multiple regions must be damaged for all autobiographical memories to become inaccessible.

Changes in regional brain activity with memory age also distinguish the two types of memory. For autobiographical memory, most studies detected regions outside the MTL where activity changed with memory age, even though the only consistent change was reduction of activity over time in posterior cingulate/retrosplenial cortex. For semantic memory, only a small minority of studies detected regions outside the MTL where activity changed according to memory age. Thus, the cortical regions in the network must be more sensitive to autobiographical memory age than to semantic memory age.

Changes related to memory age within the MTL also differentiate semantic from episodic memory. Specifically, the MTL consistently exhibits less activity for remote semantic memories than for recent ones, but this consistent pattern is not observed for autobiographical memory. On its own, the observation that the MTL is active during remote autobiographical retrieval suggests that the MTL might play a permanent role in retrieving episodic content. However, patients with damage limited to the MTL do not exhibit deficits in remote autobiographical retrieval. Taken together, it is likely the case that although MTL activity sometimes occurs during both recent and remote memory retrieval, it is only necessary for retrieval of recent memories.

One possibility is that, as time passes after learning, episodic memories form multiple, redundant traces supported by different regions of the network making them stronger and more invulnerable to disruption. Analysis of brain connectivity in patients with hippocampal lesions supports the idea that traces stored outside the MTL can support successful autobiographical memory when the MTL is damaged. In accordance with this idea, rodent studies investigating the acute versus chronic effects of hippocampal lesions on remote memory retrieval suggest that the hippocampus may be the default structure that supports memory retrieval, but that other structures can take over this role when the hippocampus is not available (Goshen et al. 2011).

Finally, the memory network is primarily important for supporting memory retrieval rather than for storing the content of memory. Neuroimaging in healthy participants and studies of patients with focal lesions to cortical areas suggest that the content of memory is stored throughout cortex, primarily in secondary sensory cortex and association cortex. For example, Gallant and colleagues (Huth et al. 2016, 2012) have identified that information about semantic meaning is stored throughout cortex. Specific regions of cortex appear to specialize in semantic meaning related to different concepts, such as animate versus inanimate or man-made versus natural. Moreover, patients with focal damage to sensory and association cortex often exhibit RA for specific components of memory, such as amnesia for the colors but not the content of memory or amnesia for animate but not inanimate objects (for review, see Squire and Wixted 2011).

In summary, the results reviewed here tend to agree with SCT (McClelland et al. 1995; Squire and Alvarez 1995). When a memory is created, new connections are formed between cortical regions (e.g., sensory and association cortices), and regions in a memory network initially support these connections. Some regions in the network have a time-limited role in storing the memory trace (e.g., the hippocampus, retrosplenial cortex, and dorsomedial PFC), but others have a permanent role (e.g., lateral temporal cortex for semantic memory). As time passes after learning, connections between cortical storage areas become stronger, making the memory less vulnerable to disruption.

References

Addis DR, Moscovitch M, Crawley AP, McAndrews MP (2004) Recollective qualities modulate hippocampal activation during autobiographical memory retrieval. Hippocampus 14(6):752–762. doi:10.1002/hipo.10215

Addis DR, Moscovitch M, McAndrews MP (2007) Consequences of hippocampal damage across the autobiographical memory network in left temporal lobe epilepsy. Brain 130(Pt 9):2327–2342. Epub 2007 Aug 6

Addis DR, Pan L, Vu MA, Laiser N, Schacter DL (2009) Constructive episodic simulation of the future and the past: distinct subsystems of a core brain network mediate imagining and remembering. Neuropsychologia 47(11):2222–2238. doi:10.1016/j.neuropsychologia.2008.10.02 S0028-3932(08)00422-3 [pii]

Aggleton JP, Brown MW (1999) Episodic memory, amnesia, and the hippocampal-anterior thalamix axis. Behav Brain Sci 22:425–489

Andrews-Hanna JR, Reidler JS, Sepulcre J, Poulin R, Buckner RL (2010) Functional-anatomic fractionation of the brain's default network. Neuron 65(4):550–562. doi:10.1016/j.neuron.2010.02.005

Andrews-Hanna JR, Saxe R, Yarkoni T (2014) Contributions of episodic retrieval and mentalizing to autobiographical thought: evidence from functional neuroimaging, resting-state connectivity, and fMRI meta-analyses. Neuroimage 91:324–335. doi:10.1016/j.neuroimage.2014.01.032

Baird A, Samson S (2014) Music evoked autobiographical memory after severe acquired brain injury: preliminary findings from a case series. Neuropsychol Rehabil 24(1):125–143. doi:10.1080/09602011.2013.858642

Barbeau EJ, Didic M, Joubert S, Guedj E, Koric L, Felician O, Ranjeva JP, Cozzone P, Ceccaldi M (2012) Extent and neural basis of semantic memory impairment in mild cognitive impairment. J Alzheimer's Dis: JAD 28(4):823–837. doi:10.3233/JAD-2011-110989

Bartlett FC (1932) Remembering: a study in experimental and social psychology. The Cambridge psychological library. The University Press, Cambridge

Bayley PJ, Hopkins RO, Squire LR (2003) Successful recollection of remote autobiographical memories by amnesic patients with medial temporal lobe lesions. Neuron 37:135–144

Bayley PJ, Gold JJ, Hopkins RO, Squire LR (2005) The neuroanatomy of remote memory. Neuron 46(5):799–810. doi:10.1016/j.neuron.2005.04.034 S0896-6273(05)00394-6 [pii]

Bayley PJ, Hopkins RO, Squire LR (2006) The fate of old memories after medial temporal lobe damage. J Neurosci 26(51):13311–13317. doi:10.1523/JNEUROSCI.4262-06.2006 26/51/13311 [pii]

Beatty WW, Salmon DP, Bernstein N, Butters N (1987) Remote memories in a patient with amnesia due to hypoxia. Psychol Med 17:657–665

Binder JR, Desai RH (2011) The neurobiology of semantic memory. Trends Cognit Sci 15 (11):527–536

Binder JR, Desai RH, Graves WW, Conant LL (2009) Where is the semantic system? A critical review and meta-analysis of 120 functional neuroimaging studies. Cereb Cortex 19(12):2767–2796

Bonnici HM, Chadwick MJ, Lutti A, Hassabis D, Weiskopf N, Maguire EA (2012) Detecting representations of recent and remote autobiographical memories in vmPFC and hippocampus. J Neurosci 32(47):16982–16991. doi:10.1523/JNEUROSCI.2475-12.2012

Bonnici HM, Chadwick MJ, Maguire EA (2013) Representations of recent and remote autobiographical memories in hippocampal subfields. Hippocampus 23(10):849–854. doi:10.1002/hipo.22155

Bontempi B, Laurent-Demir C, Destrade C, Jaffard R (1999) Time-dependent reorganization of brain circuitry underlying long-term memory storage. Nature 400:671–675

Bosshardt S, Degonda N, Schmidt CF, Boesiger P, Nitsch RM, Hock C, Henke K (2005a) One month of human memory consolidation enhances retrieval-related hippocampal activity. Hippocampus 15(8):1026–1040. doi:10.1002/hipo.20105

Bosshardt S, Schmidt CF, Jaermann T, Degonda N, Boesiger P, Nitsch RM, Hock C, Henke K (2005b) Effects of memory consolidation on human hippocampal activity during retrieval. Cortex 41(4):486–498

Bright P, Buckman JR, Fradera A, Yoshimasu H, Colchester ACF, Kopelman MD (2006) Retrograde amnesia in patients with hippocampal, medial temporal, temporal lobe, or frontal pathology. Learn Mem 13:545–557

Buckner RL, Wheeler ME, Sheridan MA (2001) Encoding processes during retrieval tasks. J Cognit Neurosci 13:406–415

Burianova H, Grady CL (2007) Common and unique neural activations in autobiographical, episodic, and semantic retrieval. J Cognit Neurosci 19(9):1520–1534. doi:10.1162/jocn.2007.19.9.1520

Burianova H, McIntosh AR, Grady CL (2010) A common functional brain network for autobiographical, episodic, and semantic memory retrieval. Neuroimage 49(1):865–874. doi:10.1016/j.neuroimage.2009.08.066

Cabeza R, St Jacques P (2007) Functional neuroimaging of autobiographical memory. TICS 11 (5):219–227. doi:10.1016/j.tics.2007.02.005

Cipolotti L, Shallice T, Chan D, Fox N, Scahill R, Harrison G, Stevens J, Rudge P (2001) Long-term retrograde amnesia…the crucial role of the hippocampus. Neuropsychologia 2 (151):172

Crovitz HF, Schiffman H (1974) Frequency of episodic memories as a function of their age. Bull Psychonomic Soc 4:517–518

Dede AJO, Wixted JT, Hopkins RO, Squire LR (under review) Autobiographical memory, future imagining, and the medial temporal lobe

Denkova E, Botzung A, Manning L (2006) Neural correlates of remembering/knowing famous people: an event-related fMRI study. Neuropsychologia 44(14):2783–2791. doi:10.1016/j.neuropsychologia.2006.05.029

Douville K, Woodard JL, Seidenberg M, Miller SK, Leveroni CL, Nielson KA, Franczak M, Antuono P, Rao SA (2005) Medial temporal lobe activity for recognition of recent and remote famous names: an event-related fMRI study. Neuropsychologia 43(5):693–703. doi:10.1016/j.neuropsychologia.2004.09.005

Foster BL, Kaveh A, Dastjerdi M, Miller KJ, Parvizi J (2013) Human retrosplenial cortex displays transient theta phase locking with medial temporal cortex prior to activation during autobiographical memory retrieval. J Neurosci 33(25):10439–10446. doi:10.1523/JNEUROSCI.0513-13.2013

Frankland PW, Bontempi B, Talton LE, Kaczmarek L, Silva AJ (2004) The involvement of the anterior cingulate cortex in remote contextual fear memory. Science 304(5672):881–883. doi:10.1126/science.1094804 304/5672/881 [pii]

Fuentemilla L, Barnes GR, Duzel E, Levine B (2014) Theta oscillations orchestrate medial temporal lobe and neocortex in remembering autobiographical memories. Neuroimage 85(Pt 2):730–737. doi:10.1016/j.neuroimage.2013.08.029

Gainotti G, Almonti S, Di Betta AM, Silveri MC (1998) Retrograde amnesia in a patient with retrosplenial tumor. Neurocase 4(6):519–526

Gardini S, Cuetos F, Fasano F, Pellegrini FF, Marchi M, Venneri A, Caffarra P (2013) Brain structural substrates of semantic memory decline in mild cognitive impairment. Curr Alzheimer Res 10(4):373–389

Gilboa A, Winocur G, Grady CL, Hevenor SJ, Moscovitch M (2004) Remembering our past: functional neuroanatomy of recollection of recent and very remote personal events. Cereb Cortex 14:1214–1225

Gilboa A, Ramirez J, Kohler S, Westmacott R, Black SE, Moscovitch M (2005) Retrieval of autobiographical memory in Alzheimer's disease: relation to volumes of medial temporal lobe and other structures. Hippocampus 15(4):535–550. doi:10.1002/hipo.20090

Goshen I, Brodsky M, Prakash R, Wallace J, Gradinaru V, Ramakrishnan C, Deisseroth K (2011) Dynamics of retrieval strategies for remote memories. Cell 147(3):678–689. doi:10.1016/j.cell.2011.09.033

Haist F, Bowden Gore J, Mao H (2001) Consolidation of human memory over decades revealed by functional magnetic resonance imaging. Nat Neurosci 4(11):1139–1145. doi:10.1038/nn739 [pii] nn739

Harlow JM (1868) Recovery from the passage of an iron bar through the head. publications of the massachusetts medical society 2 (3)

Hart J Jr, Gordon B (1990) Delineation of single-word semantic comprehension deficits in aphasia, with anatomical correlation. Ann Neurol 27(3):226–231. doi:10.1002/ana.410270303

Hassabis D, Kumaran D, Vann SD, Maguire EA (2007) Patients with hippocampal amnesia cannot imagine new experiences. Proc Natl Acad Sci USA 104(5):1726–1731. doi:10.1073/pnas.0610561104 0610561104 [pii]

Hayes SM, Salat DH, Verfaellie M (2012) Default network connectivity in medial temporal lobe amnesia. J Neurosci 32(42):14622–14629. doi:10.1523/JNEUROSCI.0700-12.2012. 17 Oct 2012

Haynes JD, Rees G (2006) Decoding mental states from brain activity in humans. Nat Rev Neurosci 7(7):523–534. doi:10.1038/nrn1931

Heilman KM, Sypert GW (1977) Korsakoff's syndrome resulting from bilateral fornix lesions. Neurology 27(5):490–493

Heisz JJ, Vakorin V, Ross B, Levine B, McIntosh AR (2014) A trade-off between local and distributed information processing associated with remote episodic versus semantic memory. J Cognit Neurosci 26(1):41–53

Herfurth K, Kasper B, Schwarz M, Stefan H, Pauli E (2010) Autobiographical memory in temporal lobe epilepsy: role of hippocampal and temporal lateral structures. Epilepsy Behav 19 (3):365–371. doi:10.1016/j.yebeh.2010.07.012

Hodges JR, Patterson K, Oxbury S, Funnell E (1992) Semantic dementia: progressive fluent aphasia with temporal lobe atrophy. Brain 115:1783–1806

Huth AG, Nishimoto S, Vu AT, Gallant JL (2012) A continuous semantic space describes the representation of thousands of object and action categories across the human brain. Neuron 76 (6):1210–1224. doi:10.1016/j.neuron.2012.10.014

Huth AG, de Heer WA, Griffiths TL, Theunissen FE, Gallant JL (2016) Natural speech reveals the semantic maps that tile human cerebral cortex. Nature 532(7600):453–458. doi:10.1038/nature17637

Irish M, Piguet O (2013) The pivotal role of semantic memory in remembering the past and imagining the future. Front Behav Neurosci 7:27. doi:10.3389/fnbeh.2013.00027

Irish M, Hornberger M, Lah S, Miller L, Pengas G, Nestor PJ, Hodges JR, Piguet O (2011) Profiles of recent autobiographical memory retrieval in semantic dementia, behavioural-variant frontotemporal dementia, and Alzheimer's disease. Neuropsychologia 49(9):2694–2702. doi:10.1016/j.neuropsychologia.2011.05.017

Irish M, Addis DR, Hodges JR, Piguet O (2012) Exploring the content and quality of episodic future simulations in semantic dementia. Neuropsychologia 50(14):3488–3495. doi:10.1016/j.neuropsychologia.2012.09.012

Jackson RL, Hoffman P, Pobric G, Lambon Ralph MA (2016) The semantic network at work and rest: differential connectivity of anterior temporal lobe subregions. J Neurosci 36(5):1490–1501

Janzen G, Jansen C, van Turennout M (2008) Memory consolidation of landmarks in good navigators. Hippocampus 18(1):40–47. doi:10.1002/hipo.20364

Kapur N, Brooks DJ (1999) Temporally-specific retrograde amnesia in two cases of discrete bilateral hippocampal pathology. Hippocampus 9:247–254

Kapur N, Ellison D, Smith MP, McLellan DL, Burrows EH (1992) Focal retrograde amnesia following bilateral temporal lobe pathology. Brain 115:73–85

Kapur N, Ellison D, Parkin AJ, Hunkin NM, Burrows E, Sampson SA, Morrison EA (1994) Bilateral temporal lobe pathology with sparing of medial temporal lobe structures: lesion profile and pattern of memory disorder. Neuropsychologia 32(1):23–38

Kirwan CB, Bayley PJ, Galvan VV, Squire LR (2008) Detailed recollection of remote autobiographical memory after damage to the medial temporal lobe. Proc Natl Acad Sci USA 105(7):2676–2680. doi:10.1073/pnas.0712155105 0712155105 [pii]

Klooster NB, Duff MC (2015) Remote semantic memory is impoverished in hippocampal amnesia. Neuropsychologia 79 (Pt A):42–52. doi:10.1016/j.neuropsychologia.2015.10.017

Kopelman MD (2000) Focal retrograde amnesia and the attribution of causality: an exceptionally critical review. Cognit Neuropsychol 17(7):585–621. doi:10.1080/026432900750002172

Kopelman MD, Bright P (2012) On remembering and forgetting our autobiographical pasts: retrograde amnesia and Andrew Mayes's contribution to neuropsychological method. Neuropsychologia 50(13):2961–2972. doi:10.1016/j.neuropsychologia.2012.07.028

Kopelman MD, Wilson BA, Baddeley AD (1989) The autobiographical memory interview: a new assessment of autobiographical and personal semantic memory in amnesic patients. J Clin Exp Neuropsy 5:724–744

Kurczek J, Wechsler E, Ahuja S, Jensen U, Cohen NJ, Tranel D, Duff M (2015) Differential contributions of hippocampus and medial prefrontal cortex to self-projection and self-referential processing. Neuropsychologia 73:116–126. doi:10.1016/j.neuropsychologia.2015.05.002

Lah S, Miller L (2008) Effects of temporal lobe lesions on retrograde memory: a critical review. Neuropsychol Rev 18(1):24–52. doi:10.1007/s11065-008-9053-2

Lashley KS (1950) In search of the engram. Sym Soc Exp Biol 4:454–482

Levine B, Turner GR, Tisserand D, Hevenor SJ, Graham SJ, McIntosh AR (2004) The functional neuroanatomy of episodic and semantic autobiographical remembering: a prospective functional MRI study. J Cognit Neurosci 16(9):1633–1646

Levy DA, Bayley PJ, Squire LR (2004) The anatomy of semantic knowledge: Medial vs. lateral temporal lobe. 101:6710–6715

Lidz T (1942) The amnestic syndrome. Arch Neurol Psychiatr 47(4):588–605

Maguire EA (2001) Neuroimaging studies of autobiographical event memory. Philos Trans R Soc Lond B Biol Sci 356(1413):1441–1451. doi:10.1098/rstb.2001.0944

Maguire EA, Frith CD (2003) Lateral asymmetry in the hippocampal response to the remoteness of autobiographical memories. J Neurosci 23:5302–5307

Maguire EA, Mummery CJ, Buchel C (2000) Patterns of hippocampal-cortical interaction dissociate temporal lobe memory subsystems. Hippocampus 10(4):475–482. doi:10.1002/1098-1063(2000)10:4<475:AID-HIPO14>3.0.CO;2-X

Maguire EA, Vargha-Khadem F, Mishkin M (2001) The effects of bilateral hippocampal damage on fMRI regional activations and interactions during memory retrieval. Brain: J Neurol 124(Pt 6):1156–1170

Manns JR, Hopkins RO, Squire LR (2003) Semantic memory and the human hippocampus. Neuron 37:127–133

Marr D (1971) Simple memory: a theory for archicortex. Philos Trans R Soc Lond Ser B 262:23–81

Maviel T, Durkin TP, Menzaghi F, Bontempi B (2004) Sites of neocortical reorganization critical for remote spatial memory. Science 305(5680):96–99. doi:10.1126/science.1098180 305/5680/96 [pii]

McCarthy RA, Kopelman MD, Warrington EK (2005) Remembering and forgetting of semantic knowledge in amnesia: a 16-year follow-up investigation of RFR. Neuropsychologia 43 (3):356–372. doi:10.1016/j.neuropsychologia.2004.06.024

McClelland JL, McNaughton BL, O'Reilly RC (1995) Why there are complementary learning systems in the hippocampus and neocortex: insights from the successes and failures of connectionist models of learning and memory. Psychol Rev 3:419–457

Miller LA, Caine D, Harding A, Thompson EJ, Large M, Watson JD (2001) Right medial thalamic lesion causes isolated retrograde amnesia. Neuropsychologia 39(10):1037–1046

Miller LA, Caine D, Watson JD (2003) A role for the thalamus in memory for unique entities. Neurocase 9(6):504–514. doi:10.1076/neur.9.6.504.29375

Miller AM, Vedder LC, Law LM, Smith DM (2014) Cues, context, and long-term memory: the role of the retrosplenial cortex in spatial cognition. Front Hum Neurosci 8:586. doi:10.3389/fnhum.2014.00586

Milton F, Muhlert N, Butler CR, Smith A, Benattayallah A, Zeman AZ (2011) An fMRI study of long-term everyday memory using SenseCam. Memory 19(7):733–744. doi:10.1080/09658211.2011.552185

Mullally SL, Intraub H, Maguire EA (2012) Attenuated boundary extension produces a paradoxical memory advantage in amnesic patients. Curr Biol 22(4):261–268. doi:10.1016/j.cub.2012.01.001

Muscatell KA, Addis DR, Kensinger EA (2010) Self-involvement modulates the effective connectivity of the autobiographical memory network. Social Cognit Affect Neurosci 5(1):68–76. doi:10.1093/scan/nsp043

Nadel L, Moscovitch M (1997) Memory consolidation, retrograde amnesia and the hippocampal complex. Curr Opin Neurobiol 7:217–227

Nadel L, Campbell J, Ryan L (2007) Autobiographical memory retrieval and hippocampal activation as a function of repetition and the passage of time. Neural Plast 2007:90472

Nieuwenhuis IL, Takashima A (2011) The role of the ventromedial prefrontal cortex in memory consolidation. Behav Brain Res 218(2):325–334. doi:10.1016/j.bbr.2010.12.009

Nieuwenhuis IL, Takashima A, Oostenveld R, McNaughton BL, Fernandez G, Jensen O (2012) The neocortical network representing associative memory reorganizes with time in a process engaging the anterior temporal lobe. Cereb Cortex 22(11):2622–2633. doi:10.1093/cercor/bhr338

Niki K, Luo J (2002) An fMRI study on the time-limited role of the medial temporal lobe in long-term topographical autobiographic memory. J Cognit Neurosci 14:500–507

Norman KA, Polyn SM, Detre GJ, Haxby JV (2006) Beyond mind-reading: multi-voxel pattern analysis of fMRI data. Trends Cognit Sci 10(9):424–430. doi:10.1016/j.tics.2006.07.005

O'Connor M, Butters N, Miliotis P, Eslinger P, Cermak LS (1992) The dissociation of anterograde and retrograde amnesia in a patient with herpes encephalitis. J Clin Exp Neuropsy 14:159–178

Okado Y, Stark CEL (2003) Neural processing associated with true and false memory retrieval. Cogn Affect Behav Neurosci 3(4):323–334

Penfield W (1938) The cerebral cortex in man. Archiv Neurol Psychiatr 40(3):417–442

Philippi CL, Duff MC, Denburg NL, Tranel D, Rudrauf D (2012) Medial PFC damage abolishes the self-reference effect. J Cogn Neurosci 24(2):475–481. doi:10.1162/jocn_a_00138

Philippi N, Botzung A, Noblet V, Rousseau F, Despres O, Cretin B, Kremer S, Blanc F, Manning L (2015) Impaired emotional autobiographical memory associated with right amygdalar-hippocampal atrophy in Alzheimer's disease patients. Front Aging Neurosci 7:21. doi:10.3389/fnagi.2015.00021

Piefke M, Weiss PH, Zilles K, Markowitsch HJ, Fink GR (2003) Differential remoteness and emotional tone modulate the neural correlates of autobiographical memory. Brain: J Neurol 126(Pt 3):650–668

Piolino P, Giffard-Quillon G, Desgranges B, Chetelat G, Baron JC, Eustache F (2004) Re-experiencing old memories via hippocampus: a PET study of autobiographical memory. Neuroimage 22(3):1371–1383. doi:10.1016/j.neuroimage.2004.02.025

Race E, Keane MM, Verfaellie M (2011) Medial temporal lobe damage causes deficits in episodic memory and episodic future thinking not attributable to deficits in narrative construction. J Neurosci 31(28):10262–10269. doi:10.1523/JNEUROSCI.1145-11.2011

Reas ET, Brewer JB (2013) Effortful retrieval reduces hippocampal activity and impairs incidental encoding. Hippocampus 23(5):367–379. doi:10.1002/hipo.22096

Rekkas PV, Constable RT (2005) Evidence that autobiographic memory retrieval does not become independent of the hippocampus: an fMRI study contrasting very recent with remote events. J Cognit Neurosci 17(12):1950–1961

Rekkas PV, Constable RT (2006) Hemodynamic retrieval intensity in hippocampus is decreased by pre-exposure to autobiographic test items. Brain Res Bull 70(4–6):467–473

Ribot T (1881) Les Maladies de la Memoire [English translation: diseases of memory]. Appleton-Century-Crofts, New York

Rissman J, Chow TE, Reggente N, Wagner AD (2016) Decoding fMRI Signatures of Real-world Autobiographical Memory Retrieval. J Cognit Neurosci 28(4):604–620. doi:10.1162/jocn_a_00920

Rosenbaum RS, Moscovitch M, Foster JK, Schnyer DM, Gao F, Kovacevic N, Verfaellie M, Black SE, Levine B (2008) Patterns of autobiographical memory loss in medial-temporal lobe amnesic patients. J Cognit Neurosci 20(8):1490–1506. doi:10.1162/jocn.2008.20105

Sagar HH, Cohen NJ, Corkin S, Growdon JM (1985) Dissociations among processes in remote memory. In: Olton DS, Gamzu E, Corkin S (eds) Memory dysfunctions, 444th edn. Annals of the New York Academy of Sciences, New York, pp 533–535

Schnider A (2008) The confabulating mind. Oxford University Press, USA

Scoville WB, Milner B (1957) Loss of recent memory after bilateral hippocampal lesions. J Neurol Neurosurg Psychiatr 20:11–21

Seidenberg M, Guidotti L, Nielson KA, Woodard JL, Durgerian S, Antuono P, Zhang Q, Rao SM (2009) Semantic memory activation in individuals at risk for developing Alzheimer disease. Neurology 73(8):612–620. doi:10.1212/WNL.0b013e3181b389ad 73/8/612

Semon RW, Simon L (1921) The mneme. G. Allen & Unwin ltd. The Macmillan Company, London, New York

Sheldon S, Levine B (2013) Same as it ever was: vividness modulates the similarities and differences between the neural networks that support retrieving remote and recent autobiographical memories. Neuroimage 83:880–891

Sheldon S, Farb N, Palombo DJ, Levine B (2016) Intrinsic medial temporal lobe connectivity relates to individual differences in episodic autobiographical remembering. Cortex: J devot Stud Nerv Syst Behav 74:206–216. doi:10.1016/j.cortex.2015.11.005

Shimamura AP, Squire LR (1987) A neuropsychological study of fact memory and source amnesia. J Exp Psychol Learn Mem Cognit 13:464–473

Smith CN (2014) Retrograde memory for public events in mild cognitive impairment and its relationship to anterograde memory and neuroanatomy. Neuropsych 28(6):959–972. doi:10.1037/neu0000117

Smith CN, Squire LR (2009) Medial temporal lobe activity during retrieval of semantic memory is related to the age of the memory. J Neurosci 29(4):930–938. doi:10.1523/JNEUROSCI.4545-08.2009 29/4/930 [pii]

Smith JF, Alexander GE, Chen K, Husain FT, Kim J, Pajor N, Horwitz B (2010) Imaging systems level consolidation of novel associate memories: a longitudinal neuroimaging study. Neuroimage 50(2):826–836. doi:10.1016/j.neuroimage.2009.11.053 S1053-8119(09)01238-5 [pii]

Smith CN, Frascino JC, Hopkins RO, Squire LR (2013) The nature of anterograde and retrograde memory impairment after damage to the medial temporal lobe. Neuropsychologia 51 (13):2709–2714. doi:10.1016/j.neuropsychologia.2013.09.015

Soderlund H, Moscovitch M, Kumar N, Mandic M, Levine B (2012) As time goes by: hippocampal connectivity changes with remoteness of autobiographical memory retrieval. Hippocampus 22(4):670–679. doi:10.1002/hipo.20927

Spreng RN, Grady CL (2010) Patterns of brain activity supporting autobiographical memory, prospection, and theory of mind, and their relationship to the default mode network. J Cognit Neurosci 22(6):1112–1123

Squire LR (1992) Memory and the hippocampus: a synthesis from findings with rats, monkeys, and humans. Psychol Rev 99:195–231

Squire LR, Alvarez P (1995) Retrograde amnesia and memory consolidation: a neurobiological perspective. Curr Opin Neurobiol 5:169–177

Squire LR, Wixted JT (2011) The cognitive neuroscience of human memory since H.M. Annu Rev Neurosci 34:259–288. doi:10.1146/annurev-neuro-061010-113720

Squire LR, van der Horst AS, McDuff SGR, Frascino JC, Hopkins RO, Mauldin KN (2010) Role of the hippocampus in remembering the past and imagining the future. Proc Natl Acad Sci USA 107:19044–19048

St Jacques PL, Kragel PA, Rubin DC (2011) Dynamic neural networks supporting memory retrieval. Neuroimage 57(2):608–616

Stark CEL, Squire LR (2000) fMRI activity in the medial temporal lobe during recognition memory as a function of study-test interval. Hippocampus 10:329–337

Stefanacci L, Buffalo EA, Schmolck H, Squire LR (2000) Profound amnesia after damage to the medial temporal lobe: A neuroanatomical and neuropsychological profile of patient E. P. J Neurosci 20(18):7024–7036 20/18/7024 [pii]

St-Laurent M, Moscovitch M, Levine B, McAndrews MP (2009) Determinants of autobiographical memory in patients with unilateral temporal lobe epilepsy or excisions. Neuropsychologia 47 (11):2211–2221. doi:10.1016/j.neuropsychologia.2009.01.032 S0028-3932(09)00045-1 [pii]

St-Laurent M, Moscovitch M, Tau M, McAndrews MP (2011) The temporal unraveling of autobiographical memory narratives in patients with temporal lobe epilepsy or excisions. Hippocampus 21(4):409–421. doi:10.1002/hipo.20757

St-Laurent M, Moscovitch M, Jadd R, McAndrews MP (2014) The perceptual richness of complex memory episodes is compromised by medial temporal lobe damage. Hippocampus 24(5):560–576. doi:10.1002/hipo.22249

Suchan B, Gayk AE, Schmid G, Koster O, Daum I (2008) Hippocampal involvement in recollection but not familiarity across time: a prospective study. Hippocampus 18(1):92–98. doi:10.1002/hipo.20371

Svoboda E, Levine B (2009) The effects of rehearsal on the functional neuroanatomy of episodic autobiographical and semantic remembering: a functional magnetic resonance imaging study. J Neurosci 29(10):3073–3082. doi:10.1523/JNEUROSCI.3452-08.2009 29/10/3073 [pii]

Svoboda E, McKinnon MC, Levine B (2006) The functional neuroanatomy of autobiographical memory: a meta-analysis. Neuropsychologia 44(12):2189–2208. doi:10.1016/j.neuropsychologia.2006.05.023

Takashima A, Petersson KM, Rutters F, Tendolkar I, Jensen O, Zwarts JJ, McNaughton BL, Fernandez G (2006) Declarative memory consolidation in humans: a prospective functional magnetic resonance imaging study. PNAS 103:756–761

Takashima A, Nieuwenhuis IL, Rijpkema M, Petersson KM, Jensen O, Fernandez G (2007) Memory trace stabilization leads to large-scale changes in the retrieval network: a functional MRI study on associative memory. Learn Mem 14(7):472–479. doi:10.1101/lm.605607

Takashima A, Nieuwenhuis IL, Jensen O, Talamini LM, Rijpkema M, Fernandez G (2009) Shift from hippocampal to neocortical centered retrieval network with consolidation. J Neurosci 29:10087–10093

Thaiss L, Petrides M (2008) Autobiographical memory of the recent past following frontal cortex or temporal lobe excisions. Eur J Neurosci 28(4):829–840. doi:10.1111/j.1460-9568.2008.06381.x

Tsukiura T, Fujii T, Okuda J, Ohtake H, Kawashima R, Itoh M, Fukuda H, Yamadori A (2002) Time-dependent contribution of the hippocampal complex when remembering the past: a PET study. NeuroReport 13(17):2319–2323

Tulving E (1983) Elements of episodic memory. Oxford University Press, Cambridge

Tulving E (1985) Memory and consciousness. Can Psychol 26:1–12

Valenstein E, Bowers D, Verfaellie M, Heilman KM, Day A, Watson RT (1987) Retrosplenial amnesia. Brain 110:1631–1646. doi:10.1093/brain/110.6.1631

Vann SD, Aggleton JP, Maguire EA (2009) What does the retrosplenial cortex do? Nat Rev Neurosci 10(11):792–802. doi:10.1038/nrn2733

Vilberg KL, Davachi L (2013) Perirhinal-hippocampal connectivity during reactivation is a marker for object-based memory consolidation. Neuron 79(6):1232–1242. doi:10.1016/j.neuron.2013.07.013

Visser M, Jefferies E, Lambon Ralph MA (2010) Semantic processing in the anterior temporal lobes: a meta-analysis of the functional neuroimaging literature. J Cognit Neurosci 22(6):1083–1094. doi:10.1162/jocn.2009.21309

Warrington EK, McCarthy RA (1988) The fractionation of retrograde amnesia. Brain Cog 7:184–200

Wilson B, Baddeley A (1988) Semantic, episodic, and autobiographical memory in a postmeningitic amnesic patient. Brain Cogn 8(1):31–46

Winocur G, Moscovitch M (2011) Memory transformation and systems consolidation. J Intl Neuropsychol Soc: JINS 17(5):766–780. doi:10.1017/S1355617711000683

Winocur G, Moscovitch M, Bontempi JB (2010) Memory formation and long-term retention in humans and animals: convergence towards a transformation account of hippocampal-neocortical interactions. Neuropsychologia 48:2339–2356

Witt JA, Vogt VL, Widman G, Langen KJ, Elger CE, Helmstaedter C (2015) Loss of autonoetic awareness of recent autobiographical episodes and accelerated long-term forgetting in a patient with previously unrecognized glutamic acid decarboxylase antibody related limbic encephalitis. Front Neurol 6:130. doi:10.3389/fneur.2015.00130

Woodard JL, Seidenberg M, Nielson KA, Miller SK, Franczak M, Antuono P, Douville KL, Rao SM (2007) Temporally graded activation of neocortical regions in response to memories of different ages. J Cognit Neurosci 19(7):1113–1124. doi:10.1162/jocn.2007.19.7.1113

Woodard JL, Seidenberg M, Nielson KA, Smith JC, Antuono P, Durgerian S, Guidotti L, Zhang Q, Butts A, Hantke N, Lancaster M, Rao SM (2010) Prediction of cognitive decline in healthy older adults using fMRI. J Alzheimer's dis 21(3):871–885. doi:10.3233/JAD-2010-091693

Yamashita K, Hirose S, Kunimatsu A, Aoki S, Chikazoe J, Jimura K, Masutani Y, Abe O, Ohtomo K, Miyashita Y, Konishi S (2009) Formation of long-term memory representation in human temporal cortex related to pictorial paired associates. J Neurosci 29(33):10335–10340. doi:10.1523/JNEUROSCI.1328-09.2009

Zola-Morgan S, Cohen NJ, Squire LR (1983) Recall of remote episodic memory in amnesia. Neuropsychologia 21(5):487–500

Memory Reconsolidation

Josue Haubrich and Karim Nader

Abstract Scientific advances in the last decades uncovered that memory is not a stable, fixed entity. Apparently stable memories may become transiently labile and susceptible to modifications when retrieved due to the process of reconsolidation. Here, we review the initial evidence and the logic on which reconsolidation theory is based, the wide range of conditions in which it has been reported and recent findings further revealing the fascinating nature of this process. Special focus is given to conceptual issues of when and why reconsolidation happen and its possible outcomes. Last, we discuss the potential clinical implications of memory modifications by reconsolidation.

Keywords Memory · Consolidation · Reconsolidation · Retrieval · Storage · Forgetting · PTSD · Drug addiction

Contents

1 Reconsolidation and the Dynamic Nature of Memory 152
2 The Discovery of Reconsolidation 152
3 Evidence for Reconsolidation and Alternative Interpretations 155
4 When Does It Occur? Boundary Conditions of Reconsolidation 158
5 Why Does It Occur? Physiological Role of Reconsolidation 161
6 Memory Destabilization 163
7 Common Mechanisms Between Memory Destabilization and Forgetting, Reconsolidation and Maintenance 165
8 Reconsolidation as a Therapeutic Tool 165
9 Conclusion 167
References 167

J. Haubrich · K. Nader (✉)
Department of Psychology, McGill University, Montreal, Canada
e-mail: karim.nader@mcgill.ca

J. Haubrich
Federal University of Rio Grande do Sul, Porto Alegre, Brazil

Curr Topics Behav Neurosci (2018) 37: 151–176
DOI 10.1007/7854_2016_463
Published Online: 25 November 2016

1 Reconsolidation and the Dynamic Nature of Memory

Memory does not consist of a single temporal process. Instead it is a set of distinct phases, each one relying on distinct mechanisms supporting different cognitive demands. The first phase is the encoding (or learning phase) that takes place during a new experience. It allows for the acquisition of information that is being processed by the brain predicting an event. Known since the seminal works of Hebb (1949), "neurons that fire together, wire together." That is, given that, in the brain information flows through specific synaptic connections, in order for it to persist as a long-lasting representation (i.e., a memory trace) the efficiency of such connections must be strengthened and stabilized, otherwise the experience will be forgotten (Ebbinghaus 1885). This happens during the consolidation phase that is triggered after initial acquisition (also referred to as synaptic consolidation) and culminates in memory storage (Glickman 1961; McGaugh 1966). Once stored, to circumvent forgetting it is imperative that the stabilized connections supporting memory are actively sustained by an ongoing process called maintenance (Miller and Springer 1973; Spear 1973). It allows for the retrieval phase that is the recall of the stored information in order to better guide behavior—the main physiological function of memory. Hence, a whole set of different but interconnected processes (acquisition, consolidation, maintenance, and retrieval) play as a well-tuned orchestra that allows animals to adapt their behavior according to their experiences and their environments.

Once it was believed that, following consolidation, memory would be stored in a fixed state for as long as it persisted (McGaugh 1966; Squire et al. 1984). Only the initial consolidation was considered as a neurobiological active phase that required neurons to activate constructive processes such as RNA and protein synthesis, and a number of second messenger pathways involved in AMPA-receptor trafficking (Dudai 2004; Kandel 2001; Martin et al. 2000). Retrieval was viewed by many as simply a passive readout of information encoded in the neuronal circuits supporting an LTM, although there was much data to disagree with it (Lewis 1979; Miller and Marlin 1984; Spear and Mueller 1984). This simplistic view has proven to be wrong and now it is known that retrieval can initiate an active process called reconsolidation. This notion changed the way modern science considers memory—as opposed to a stable, fixed entity, memory is dynamic and susceptible to modifications when recalled. As we will discuss, the discovery of reconsolidation led to major insights regarding the neurobiological nature of memory and has profound clinical implications.

2 The Discovery of Reconsolidation

For more than 100 years, it has been known that new memories go through qualitatively distinct phases over time (Ebbinghaus 1885; Müller and Pilzecker 1900; Ribot 1881). It evolved into the consolidation theory, which in turn became the central tenet around which the memory field evolved. It states that memory is unstable after acquisition and is progressively stabilized over time. This stabilization process is now called synaptic consolidation and encompasses a period of several hours (Dudai and Morris 2000; Kandel 2001; Martin et al. 2000). Once consolidation is completed, memory was thought to be stored in a fixed and stable manner.

The existence of a stabilization period following learning is supported by several lines of evidence. Memory is impaired when amnestic treatments such as protein-synthesis inhibitors (Flexner et al. 1965) and electroconvulsive shock (Duncan 1949) or new competing learning (Gordon and Spear 1973) are presented after learning. In addition, retention can be enhanced by the administration of different compounds after learning (McGaugh and Krivanek 1970). Importantly, all these interventions are only effective if given soon after acquisition but not after a delay, showing that consolidation is limited to the first hours after learning. These findings provided an experimental basis for theories of synaptic consolidation (Glickman 1961; Hebb 1949; McGaugh 1966). Together, these studies led to the interpretation that memory exists in two states—a labile or active state and a consolidated or inactive state. Before the consolidation process is concluded, memory exists in an unstable form called "short-term memory" lasting on the order of hours. During this period, memory is susceptible to interferences that can lead to its impairment or enhancement. Once consolidation is completed and the trace is stabilized as an LTM, such interferences are no longer effective. Thus, if a memory is susceptible to being enhanced or disrupted it is considered in a labile, nonconsolidated state, and if it is insensitive it is by definition consolidated. Several models at different levels of analysis have successfully described changes occurring during the transition of memory from a labile to a fixed trace, such as long-term potentiation (Bliss and Lømo 1973; Martin et al. 2000), and the identification of molecular pathways important for LTM formation (Bourtchuladze et al. 1994; Dash et al. 1990; Kandel 2001; Yin et al. 1994).

The consolidation theory has faced challenges since its inception. One of them arose from initial studies demonstrating that performance can be impaired if amnestic treatments are presented after a reactivation session (Misanin et al. 1968; Schneider and Sherman 1968), what was initially called cue-dependent amnesia. It suggested that following reactivation, memory returned to an unstable state, after which a restabilization process must takes place. This was contrary to the mainstream view that, following consolidation, memory is permanently stored as a fixed, non-labile trace, and all the subsequent processes would be simple passive readouts of the LTM. It led some researchers to disagree with the notion that plasticity was confined only to the birth of a new memory and new theories of memory were proposed (Lewis 1979; Miller and Marlin 1984; Miller and Springer 1973; Spear and Mueller 1984). After the reports of cue-dependent amnesia, it appeared that plasticity was just momentarily on hold when memory was not being used, but could return when it was reactivated.

It is now clear that memory stability is always in check and requires active processes to be sustained. First, we now know that both memory maintenance and forgetting are supported by antagonistic active biological processes that are continuously battling one another (Migues et al. 2010; Pastalkova et al. 2006; Sacktor 2011; Shema et al. 2011). Its mechanisms are starting to be revealed. The constitutively active protein kinase M zeta (PKMζ), an atypical isoform of protein kinase C, is crucial for the maintenance of many types of LTM (Sacktor 2008). It acts by keeping GluA2-containing AMPARs at the postsynaptic density by a continuous regulation of AMPAR trafficking (Migues et al. 2010). This is critical for LTM maintenance and a transient inactivation of PKMζ results in the loss of an established LTM. Forgetting also requires active process to take place (Hardt et al. 2013) such as removal of synaptic GluA2-containing AMPARs (Migues et al. 2016), calcium signaling (Sachser et al. 2016) and hippocampal neurogenesis (Frankland et al. 2013). Second, retrieval is now well known for not being a passive readout of a consolidated LTM, but an active process that can deeply modify memory's stability through the process called reconsolidation. Hence, after consolidation, memory is not stored as if it was written in stone. Instead, it is persistently undergoing active processes that determine its maintenance or forgetting, and each time it is retrieved it can return to a plastic, labile state.

The first insights into the notion that retrieval is not a simple passive readout of an LTM came from a small number of studies showing that an already consolidated memory could be again destabilized and restabilized following a reactivation session. This idea was backed by the same line of evidence that supported the notion of a consolidation process following initial learning. First, performance can be permanently impaired if amnestic treatments such as electroconvulsive shock (Misanin et al. 1968; Schneider and Sherman 1968) or new competing learning (Gordon 1977a) are presented after a reactivation session. Second, performance can be enhanced by some compounds such as strychnine if administered after reactivation (Gordon 1977b). Importantly, all these manipulations are only effective if given soon after reactivation, but not after a delay. These findings implied that after retrieval, memory can return to an unstable, vulnerable state that needs to be restabilized in order to persist. Hence, the consolidation of an LTM is not the end of the road in terms of plasticity. According to the model proposed by Lewis (1979), memory is in essence dynamic and alternates between an active and an inactive state. A memory in the inactive state would be the consolidated LTM that is not being used in the current moment, and by definition it is insensitive to disruption. Recently acquired and reactivated memories, in turn, would be in the active state and thus unstable and sensitive to disruptive agents. This new model endowed memory with a dynamic nature and was able to explain the data supporting consolidation theory as well as the cue-dependent amnesia reports that were inconsistent with the old paradigm.

This dynamic view of memory was seen as somewhat whimsical at a time when the consolidation paradigm prevailed. The subject was, at most, modestly investigated for many years and contemporary textbooks did not even make note of it. The topic reemerged after a seminal work published in 2000. This study (Nader et al. 2000) showed a systematic demonstration of reconsolidation by (i) using a well-defined behavioral protocol (auditory fear conditioning in the rat), (ii) directly targeting a critical brain circuit for fear behavior and its memory consolidation (the basolateral nucleus of amygdala) and (iii) using a drug known to have an amnestic effect upon consolidation through the inhibition of protein synthesis (the antibiotic anisomycin). They found that a reminder is able to bring a consolidated memory back into an unstable state, making it susceptible to disruption by protein-synthesis inhibition in the basolateral amygdala. As in the original articles about cue-dependent amnesia, memory was not impaired if the reactivation session was omitted nor if anisomycin was administered after a delay. Also, post-reactivation short-term memory was intact. Hence, the same lines of evidence that support consolidation as a stabilization process after learning, also support reconsolidation as a necessary restabilization step following a retrieval-induced destabilization.

3 Evidence for Reconsolidation and Alternative Interpretations

Following the publication of Nader et al. (2000), reconsolidation began to be extensively investigated. For instance, reconsolidation has been demonstrated using a wide range of species, memory tasks, and amnestic agents (Table 1). Its occurrence has been consistently demonstrated by the same straightforward lines of evidence that define consolidation. First, performance is reduced if amnestic treatments are given soon after a reactivation (Litvin and Anokhin 2000; Milekic and Alberini 2002; Nader et al. 2000). Second, performance is also impaired if a competing episode of learning takes place shortly after reactivation (Walker et al. 2003). Third, memory can be enhanced by various compounds when they are administered in close temporal proximity to reactivation (Bredy and Barad 2008; Stern and Alberini 2013; Tronson et al. 2006).

Evidence for reconsolidation does not come solely from behavioral studies showing changes in performance. So far, several cellular and molecular correlates of reconsolidation have been reported by studies showing that the molecular correlates of LTM are reversed to the level of untrained animals if reconsolidation is blocked, or are increased after a reactivation (Tronson and Taylor 2007). The effects of manipulations targeting reconsolidation on the brain correlates of LTM are observed at all levels of analysis, including electrophysiological, synaptic, molecular, and systems levels (Table 2). Taking these together, reconsolidation is established as a fundamental memory process. Following the study of Nader et al. (2000) that revitalized this subject, there has been a surge of studies, growing exponentially, and progressively increasing our understanding of this process. It has allowed the development of promising reconsolidation-based treatments for fear and anxiety disorders and addiction. In a testament to the quality of the original articles that started all this progress almost 50 years ago, since the publication of Nader et al. (2000), they have received renewed appreciation and their citation rate is increasing even nowadays.

There are alternative interpretations for the observation of amnesia when a treatment is presented after a reminder—for example, state-dependent learning, new learning, non-specific drug effects, lesions, retrieval impairment, and facilitated

Table 1 Experimental paradigm, amnestic treatment and species

Experimental paradigm	
Auditory fear conditioning	Nader et al. (2000)
Contextual fear conditioning	Debiec et al. (2002)
Episodic memory	Hupbach et al. (2007)
Habituation	Rose and Rankin (2006)
Inhibitory avoidance	Anokhin et al. (2002), Litvin and Anokhin (2000)
Instrumental learning	Sangha et al. (2003); but see Hernandez and Kelley (2004)
Incentive learning	Wang et al. (2005)
Positive reinforcement	Flavell et al. (2011)
Memory for drug reward	Lee et al. (2005), Miller and Marshall (2005), Milton et al. (2008)
Motor sequence learning	Walker et al. (2003)
Object recognition	Kelly et al. (2003)
Spatial memory	Suzuki et al. (2004)
Amnestic treatment	
Antisense	Lee et al. (2004), Taubenfeld et al. (2001)
Inhibition of kinase activity	Duvarci et al. (2005), Kelly et al. (2003)
Inducible knockout mice	Kida et al. (2002)
Interference by new learning	Hupbach et al. (2007), Walker et al. (2003)
Potentiated reconsolidation by increase in kinase activity	Tronson et al. (2006)
Protein-synthesis inhibition	Nader et al. (2000)
Protein knockout mice	Bozon et al. (2003)
Receptor antagonists	Kindt et al. (2009), Suzuki et al. (2004)
RNA synthesis inhibition	Sangha et al. (2003)
Species	
Chick	Anokhin et al. (2002)
Crabs	Pedreira et al. (2002)
Fish	Eisenberg et al. (2003)
Honeybees	Stollhoff et al. (2005)
Humans	Hupbach et al. (2007), Walker et al. (2003)
Mice	Kida et al. (2002)
Nematodes	Rose and Rankin (2006)
Rats	Nader et al. (2000)
Sea slugs	Child et al. (2003), Lee et al. (2012)
Aplysia	Cai et al. (2012)
Sheep	Perrin et al. (2007)

extinction. Given that reconsolidation has been defined by applying the same standards that define consolidation, it is not surprising that most of the alternative interpretations of reconsolidation use the same arguments previously used to

Table 2 Brain correlates of reconsolidation across different levels of analysis

Activation of neurons or brain areas
Arguello, et al. (2013). J Neurosci 33(8), 3646–3658
Barak, S., et al. (2013). Nat Neurosci 16(8), 1111–1117
Bergstrom, et al., (2013). Brain Topogr 26:468–478
Bjorkstrand, J., et al. (2015). PLoS One 10(7), e0129393
Censor, N., et al. (2014). Cortex 58, 281–288
Chia, C., & Otto, T. (2013). Neurobiol Learn Mem 106, 48–55
Diaz-Mataix, L., et al. (2013). Curr Biol 23(6), 467–472
Gräff, J., et al. (2014). Cell 156(1–2), 261–276
Hall, J., et al. (2001b). J Neurosci 21(6), 2186–2193
Lee, J. L., et al. (2004). Science 304(5672), 839–843
Maddox, S. A., et al. (2013b). Learn Mem 20(2), 109–119
Merlo, E., et al. (2005). Learn Mem 12(1), 23–29
Sangha, S., et al. (2003). J Neurosci 23(22), 8034–8040
Schiller, D., et al. (2013). Proc Natl Acad Sci USA 110(50), 20040–20045
Schwabe, L., et al. (2012). Biol Psychiatry 71(4), 380–386
Slaker, M., et al. (2015). J Neurosci 35(10), 4190–4202
Thomas, K. L., et al. (2003). Eur J Neurosci 17(9), 1964–1972
Tronel, S., & Sara, S. J. (2002). Learn Mem 9(3), 105–111
Wirkner, J., et al. (2015). Neurobiol Learn Mem 119, 63–68
Electrophysiological alterations
Clarke, J. R., et al. (2010). Proc Natl Acad Sci USA 107(6), 2652–2657
Diaz-Mataix, L., et al. (2011). J Neurosci 31(26), 9538–9543
Kim, J., et al. (2010). J Neurosci 30(28), 9631–9640
Maddox, S. A., et al. (2013a). PLoS One 8(1), e54463
Maddox, S. A., et al. (2014). Neurobiol Learn Mem 107, 93–100
Thomas, K. L., et al. (2002). Eur J Neurosci 16(9), 1789–1796
Wirkner, J., et al. (2015). Neurobiol Learn Mem 119, 63–68
Lee, S., Kim, J., & Choi, S. (2011). Biochem Biophys Res Co 407(2), 339–342
Changes in molecular pathways
Barnes, P., et al. (2012). Hippocampus 22(2), 149–171
Contreras, M., et al. (2012). Neuropsychopharmacology 37(9), 2101–2108
Hall, J., et al. (2001b). J Neurosci 21(6), 2186–2193
Hall, J., et al. (2001a). Eur J Neurosci 13(7), 1453–1458
Hellemans, K. G., et al. (2006). J Neurosci 26(49), 12694–12699
Jarome, T. J., et al. (2012). Learn Mem 19(7), 300–306
Kelly, A., et al. (2003). J Neurosci 23(12), 5354–5360
Kemenes, G., et al. (2006) J Neurosci 26(23), 6298–6302
Lee, L, (2010). Front Behav Neurosci 4:168
Lubin, F. D., & Sweatt, J. D. (2007). Neuron 55(6), 942–957
Mamiya, N., et al. (2009). J Neurosci 29(2), 402–413
Merlo, E., et al. (2005). Learn Mem 12(1), 23–29
Miller, C. A., & Marshall, J. F. (2005). Neuron 47(6), 873–884
Radiske, A., et al. (2015). J Neurosci 35(16), 6570–6574
Romero-Granados, R., et al. (2010). Hippocampus 20(5), 584–595
von Hertzen, L. S., & Giese, K. P. (2005). J Neurosci 25(8), 1935–1942
Wang, Y., et al. (2012). PLoS One 7(11), e49942
Wells, A. M., et al. (2013). Neuropsychopharmacology 38(5), 753–762
Chen, S., et al. (2014). Elife 3, e03896

(continued)

Table 2 (continued)

Changes in the synapses
Descalzi, G., et al. (2012). Mol Brain 5, ArtID 6, 5
Doyere, V., et al. (2007). Nat Neurosci 10(4), 414–416
Hong, I., et al. (2013). Proc Natl Acad Sci USA 110(20), 8218–8223
Jarome, T. J., et al. (2015). Neuropsychopharmacology 40(13), 3044–3052
Jarome, T. J., et al. (2012). Learn Mem 19(7), 300–306
Li, Y., et al. (2013). Proc Natl Acad Sci USA 110(12), 4798–4803
Lee, S.-H., et al. (2012). Proc Natl Acad Sci USA 109(35), 14200–14205
Monfils, M. H., et al. (2009). Science 324(5929), 951–955
Rehberg, K., et al. (2010). Neurobiol Learn Mem 94(2), 117–126
Ren, Z.-Y., et al. (2013). Neuropsychopharmacology 38(5), 778–790
Rose, J. K., & Rankin, C. H. (2006). J Neurosci 26(45), 11582–11587
Wells, A. M., et al. (2015). Neuropsychopharmacology. doi:10.1038/npp.2015.217
Cai, D., et al. (2012). Curr Biol 22(19), 1783–1788

question the existence of consolidation processes. In common, most of the alternative interpretations are based on negative predictions and have been invalidated for both consolidation (Gold and King 1974) and reconsolidation (Nader and Hardt 2009). Others, such as state-dependent learning, lesions, and non-specific effect of drugs, indeed may occur in specific circumstances but cannot explain the vast literature on reconsolidation. Given the richness of data on reconsolidation (from behavioral to brain correlates across many levels of analysis) alternative interpretations are not capable of explaining this set of data that systematically supports a destabilization–restabilization process triggered by retrieval. For a detailed discussion regarding alternative interpretations, see Nader and Hardt (2009).

The fact that reconsolidation has been demonstrated in a wide range of animal species and memory tasks (Table 1) is compelling evidence that it is a fundamental process. But why would a process that puts memory stability in check be so important? Also, is our brain constantly destabilizing and restabilizing memory or does it happen only in specific circumstances? These are important and complex questions that we are beginning to understand. Interestingly, they appear to be directly related. Within the field there are different perspectives regarding when and why reconsolidation happens. As we will discuss, the findings that reconsolidation are not always triggered by retrieval and may lead to different and interesting outcomes that have given rise to new insights about the neurobiological nature of memory.

4 When Does It Occur? Boundary Conditions of Reconsolidation

Although reconsolidation has been reported across several levels of analysis, like consolidation it is not ubiquitous. Not always will a retrieved memory be reactivated and undergo this process. It happens that there are important parameters that determine whether reconsolidation will take place. These parameters are related to characteristics of the consolidated memory that is being retrieved and the way memory is reactivated. A given factor that constrains reconsolidation occurrence is called a "boundary condition."

Two main properties of a consolidated memory directly determine its probability of being destabilized following retrieval. One is memory strength. It has been reported that very strong training protocols (Eisenberg et al. 2003; Suzuki et al. 2004; Wang et al. 2009; Winters et al. 2009) or protocols that induce asymptotic levels of learning (García-DeLaTorre et al. 2009; Lee 2010; Morris et al. 2006; Rodriguez-Ortiz et al. 2005; Rodriguez-Ortiz et al. 2008) result in a memory that is unlikely to be affected by amnestic agents following a reactivation session. A second property of memory that acts as a boundary condition is its age, given that longer delays between training and reactivation sessions have been shown to create a protective effect against post-reactivation interferences in some studies (Baratti et al. 2008; Eisenberg and Dudai 2004; Frankland et al. 2006; Milekic and Alberini 2002; Suzuki et al. 2004; Haubrich et al. 2015; Bustos et al. 2009).

The characteristics of the retrieval session are also crucial for determining if reconsolidation will occur. Too short reactivation sessions fail to induce reconsolidation (Bustos et al. 2009; Lagasse et al. 2009) whereas long non-reinforced reactivations lead to extinction (Eisenberg et al. 2003; Lee et al. 2006; Mamiya et al. 2009; Pedreira and Maldonado 2003; Perez-Cuesta and Maldonado 2009; Suzuki et al. 2004). Also, it has been reported that reconsolidation is constrained when reactivation occurs in a distinct context (DeVietti and Holiday 1972; Hupbach et al. 2008), when it is predictable (Forcato et al. 2009; Morris et al. 2006; Osan et al. 2011; Pedreira et al. 2004; Robinson et al. 2011), when the stimulus is not directly linked to the memory being assessed (Debiec et al. 2006) or when it occurs during sleep (Diekelmann et al. 2011).

Extinction is sometimes considered a boundary condition per se, given that it is a distinct process that can also be triggered by retrieval (Bouton 2004; Pavlov 1927). It has been suggested that extinction inhibits reconsolidation (Debiec et al. 2002). Supporting this view are reports showing that reconsolidation and extinction are mutually exclusive processes relying on different mechanisms (Eisenberg et al. 2003; Pedreira and Maldonado 2003; Sangha et al. 2003; Suzuki et al. 2004). Hence, as a long non-reinforced retrieval progresses, there appears to be a transition from mere retrieval to reconsolidation and then to extinction. However, it is well established that an extinction procedure results in the consolidation of a new, competing memory trace (Bouton 2004). A more general interpretation of the extinction effect on preventing reconsolidation is that the boundary condition is, in fact, new learning. During retrieval, an animal may be exposed to different amounts of new information. Also, such new information can be directly related to the memory that is being retrieved in varying ways, or be completely unrelated. Hence, there would be a balance determining how this information is stored—by incorporating it into an existing trace through reconsolidation, or through the consolidation of a new memory. Hypothetically, subtle new information, or information directly related to a stored memory, would be best encoded by reconsolidation-mediated updating. On the other hand, large magnitudes of new information, or information that largely opposes that of a stored memory, would be preferentially stored as a new trace (Besnard 2012; Besnard et al. 2012).

The observation of the above-mentioned constraints on reconsolidation led to the idea that old or very strong memories would never be destabilized. However, several reports have shown that this is not always true. The constraints on reconsolidation are not always observed, indicating that there are factors that favor reconsolidation and counteract the boundary conditions. Interestingly, for most of the boundary conditions there have been opposing findings, including the demonstration of reconsolidation of strong (Suzuki et al. 2004; Wang et al. 2009; Robinson and Franklin 2010) and old memories (Brunet et al. 2008; Debiec et al. ; Diergaarde et al. 2006; Lee et al. 2006; Robinson and Franklin 2010; Romero-Granados et al. 2010; Wang et al. 2009). These apparently conflicting results in fact show that the boundary conditions can be circumvented. For example, old or strongly trained memories may be rendered labile by longer reactivation sessions, waiting for 1 month to pass (Frankland et al. 2006; Suzuki et al. 2004), damage to the hippocampus (Wang et al. 2009) or by exposure to novel stimuli during reactivation (Lee 2010; Winters et al. 2009). Also, boosting plasticity with pharmacological agents can make an old memory undergo reconsolidation (Gräff et al. 2014). Hence, there are factors that decrease the probability of memory being destabilized but they are not absolute since they can be opposed by several kinds of interventions including time.

In summary, there are conditions that constrain the probability of reconsolidation taking place, but they can be counteracted by several factors (Table 3). So far, the boundary conditions on reconsolidation that have been most consistently reported are the memory's age, memory strength, weak reactivation sessions, new learning, and the absence of a prediction error. In opposition to the boundary conditions, new information, increased intensity of reactivation session, and pharmacological agents that boost plasticity can promote reconsolidation in conditions where it does not usually happen.

Table 3 Boundary conditions

Characteristics of the memory trace	
Memory age	Baratti et al. (2008), Eisenberg and Dudai (2004), Frankland et al. (2006), Milekic and Alberini (2002), Suzuki et al. (2004), Haubrich et al. (2015), Bustos et al. (2009)
Training strength	Eisenberg et al. (2003), Suzuki et al. (2004), Wang et al. (2009), Winters et al. (2009)
Asymptotic learning	García-DeLaTorre et al. (2009), Lee (2010), Morris et al. (2006), Rodriguez-Ortiz et al. (2005), Rodríguez-Ortiz et al. (2008)

(continued)

Table 3 (continued)

Characteristics of the reactivation session	
Short reactivation	Bustos et al. (2009), Lagasse et al. (2009)
Long reactivation (extinction)	Eisenberg et al. (2003), Lee et al. (2006), Mamiya et al. (2009), Pedreira and Maldonado (2003), Perez-Cuesta and Maldonado (2009), Suzuki et al. (2004)
New environment	DeVietti and Holiday (1972), Hupbach et al. (2008)
Predictable reactivated stimulus	Forcato et al. (2009), Morris et al. (2006), Osan et al. (2011), Pedreira et al. (2004), Robinson et al. (2011)
Stimulus not related to memory being studied	Debiec et al. (2006)
Sleep	Diekelmann et al. (2011)
Parameters that can lift a boundary condition	
Longer reactivation session	Frankland et al. (2006), Suzuki et al. (2004)
Presenting novel information during reactivation	Lee (2010), Winters et al. (2009)
Upregulation of neuroplasticity	Gräff et al. (2014)

5 Why Does It Occur? Physiological Role of Reconsolidation

The demonstration of a plastic period following retrieval led to the important question of its physiological function. Studies showing boundary conditions on reconsolidation provided important insights on this issue, leading to two theories that although distinct, are not mutually exclusive (Alberini 2011; Davis et al. 2010). In common, they suggest that a malleable phase following retrieval allows memory to adapt in order to keep its relevance in an environment that is in constant change.

The memory strength boundary condition (the absence of reconsolidation for strong memories) led to the view that reconsolidation takes place in order to enhance weak memories. Accordingly, each time a weak memory is reactivated and reconsolidated, it is strengthened. If learning is already asymptotical, there is no need for reconsolidation to occur. Given that a memory often recalled is probably well fitted to predict important outcomes, a process that selectively strengthens such a memory is very beneficial. Also, infrequently reactivated weak memories may be lost since they are probably not essential. This idea is backed by reports showing that reconsolidation enhances performance (Frenkel et al. 2005; Lee 2008; De Oliveira Alvares et al. 2013; Forcato et al. 2014; Fukushima et al. 2014) and that memory is not destabilized when learning has reached an asymptote (García-DeLaTorre et al. 2009; Lee 2010; Morris et al. 2006; Rodriguez-Ortiz et al. 2005; Rodriguez-Ortiz et al. 2008).

A second theory of reconsolidation's functional role states that it enables the updating of memory with new information (Abraham and Robins 2005; Dudai 2004; Forcato et al. 2010; Hupbach et al. 2007, 2008; Jones et al. 2012; Lee 2009;

Lukowiak et al. 2007; Morris et al. 2006; Sara 2000; Tronson and Taylor 2007). Accordingly, when memory is retrieved, reconsolidation allows the existing memory trace to incorporate relevant new information (Hupbach et al. 2007). Given that memories are usually retrieved in situations that differ from the event that triggered initial learning, the ability to update memory content with new information seems crucial to keep the memory content relevant (Lee 2008). This idea is supported by the findings that old and strong memories that usually are resistant to destabilization undergo reconsolidation if novelty is presented during retrieval. In addition, it explains the incubation effect, a phenomenon whereby responses learned over time are gradually strengthened (Eysenck 1968; Gabriel 1968; Pickens et al. 2009), a process accelerated by repeated reactivations (Forcato et al. 2011). Reconsolidation-mediated updating was reported in human episodic memory (Forcato et al. 2010; Hupbach et al. 2007) and in animal studies using both aversive (Haubrich et al. 2015; Monfils et al. 2009; De Oliveira Alvares et al. 2013; Rao-Ruiz et al. 2011; Sierra et al. 2013) and appetitive paradigms (Olshavsky et al. 2013).

The two theories of reconsolidation's function, memory strengthening, and updating, are not incompatible. A more general hypothesis for reconsolidation function that reconciles both views is that it occurs when there is a mismatch between what memory predicts based on past experiences and what actually takes place during retrieval (Lee 2009; Wang and Morris 2010). This is the case when a water maze platform is moved to a new location, a well-explored context contains new features, or when an animal continues to learn incrementally over successive exposures with a task. Even repeated trials of the same training procedure could induce reconsolidation as long as there is still something new to learn (Morris et al. 2006; Rodriguez-Ortiz et al 2005; Lee 2008). In all cases, the new information would be added to the memory been recalled, updating it (Wang and Morris 2010). Importantly, "new information" might refer to something that indeed was not present during initial learning, or to something that was not properly encoded, since both are the same from the perspective of the content of a memory trace. This view is in accordance with the work of Rescorla and Wagner (1972) who proposed that learning takes place when surprise occurs. It is logical to assume that an efficient biological system would not expend energy re-encoding information that is already properly stored. Hence, reconsolidation should not occur when additional exposures to a task closely match what is encoded. Importantly, additional exposures to a task may not be necessary at all to bring a weakly learned response to asymptotic levels since in some cases there may be sufficient offline reactivation for the brain to identify and strengthen task-relevant information (Inda et al. 2011).

These predictions are supported by several studies. Brain activity is reduced over multiple exposures to the same event (Grill-Spector et al. 2006). Additional training trials in a water maze task prevent memory from becoming labile following a reactivation session that is able to destabilize a weakly trained memory (Morris et al. 2006; Rodriguez-Ortiz et al. 2008). Changing the texture of flooring during a reactivation session is sufficient to destabilize a well-trained object recognition memory that would not undergo reconsolidation in the unaltered context (Winters et al. 2009).

6 Memory Destabilization

Fundamental to the reconsolidation theory is that after retrieval memory becomes destabilized. Initial evidence for reconsolidation was based mostly on studies looking at the effect of agents with disruptive effects on the restabilization phase. More recently, the mechanisms underlying memory destabilization are been revealed (Table 4). For instance, several compounds can prevent the disruptive effect of amnestic agents on reconsolidation, suggesting that they act by preventing memory being destabilized. This is the case with CB1R antagonists (Suzuki et al. 2004), L-type voltage-gated calcium channel blockers (Suzuki et al. 2004) and NR2B-containing NMDAR antagonists (Mamou et al. 2006; Milton et al. 2013). Importantly, in line with evidence that a reactivated memory requires protein synthesis to be restabilized (Nader et al. 2000), memory becomes unstable after reactivation via ubiquitin/proteasome-dependent protein degradation (Lee et al. 2012). In addition, AMPAR trafficking plays a key role on memory destabilization since its blockade by $GluA2_{3Y}$ prevents a reactivated memory from being disrupted by protein-synthesis inhibition (Hong et al. 2013). Supporting the role of AMPARs, it was shown that soon after retrieval there is an acute increase of calcium-permeable (CP-AMPARs) and decrease of calcium-impermeable AMPA receptors (CI-AMPARs), that reveals a transient labile state of synapses (Clem and Huganir 2010; Rao-Ruiz et al. 2011; Hong et al. 2013). In a clear parallel with synaptic restabilization, there is a gradual increase in CI-AMPARs expression in the hours after retrieval (Clem and Huganir 2010; Rao-Ruiz et al. 2011; Hong et al. 2013). Different NMDAR subunits also play distinct roles during destabilization and restabilization phases. Initial destabilization is mediated by the NR2B subunit whereas restabilization requires NR2A subunits (Milton et al. 2013) which are linked to LTP and increased CREB phosphorylation (Liu et al. 2004), two processes linked to the reconsolidation phase (Hall et al. 2001a; Clarke et al. 2010; Maddox et al. 2014).

Given that reconsolidation occurs in order to restabilize a reactivated memory, any factor that prevents memory becoming unstable will act as a boundary condition. Indeed, it has been shown that the memory strength boundary condition is a consequence of a downregulation of NR2B subunits in the BLA (Wang et al. 2009). Hence, the constraining effects upon reconsolidation by all previously mentioned boundary conditions likely reflect changes at molecular, synaptic, and circuit levels.

A memory trace may comprise many distinct associations encoded by different brain systems. It is not yet clear how different memory systems are affected by a reactivation session. Following retrieval, it is possible that destabilization occurs not in the entire network supporting memory, but in a portion of it. In some conditions such as protein-synthesis inhibition after a fear-memory reactivation, it might be expected that a remaining non-labilized portion of memory could sustain a residual performance later in a test. However, an important point is that no amnestic agent is 100 % effective. With our current tools it is hard to disentangle both effects, but probably both happen to some extent. In an elegant demonstration of partial destabilization, Debiec et al. (2006) found that in a second-order auditory fear-conditioning task [where an auditory tone (CS1) is first associated with shock and then paired with another distinct tone (CS2)], the fear response to CS1 was not impaired when an amnestic agent (anisomycin into the BLA) was presented after a CS2 re-exposure. However, fear expression to both CS1 and CS2 was impaired if anisomycin interference was presented after CS1 reactivation. This suggests that these memory traces might exist within a network that can be either partially or entirely labilized depending on the reactivation cue that is used.

Similarly, Winters et al. (2011) evaluated the effectiveness of anisomycin infused into hippocampus or into perirhinal cortex on disrupting object recognition memory reconsolidation following several reactivation protocols. The reactivation protocols included one novel object, a novel contextual feature, or re-exposure to the original objects in the same context. They found that only the reactivation that included a new contextual feature induced memory destabilization and disruption by anisomycin infused into the hippocampus. The authors suggest that a new contextual feature turns memory hippocampus-dependent and so intra-hippocampal infusions become effective. Another interpretation is that the memory trace stored in the hippocampus, in their protocol, was destabilized only when novelty was presented. As mentioned before, novelty is a factor that can promote reconsolidation in situations in which it does not normally occur. Thus, this result demonstrates that blocking the labilized hippocampus-mediated portion impairs the putatively perirhinal-mediated memory for the objects.

The memory strength boundary condition and the putative physiological role of reconsolidation in strengthening memories also offer insights regarding partial memory destabilization. Rodriguez-Ortiz and Bermudez-Rattoni (2007) proposed that as a learning asymptote is approached, progressively less of a trace would return to a vulnerable state during re-exposure to the task, as evidenced by the decreasing impairment induced by post-reactivation application of amnesic treatments. Thus, the neuronal ensemble mediating the memory may only be partially destabilized to update aspects of an experience that are not already well encoded (or that are new). Theoretically, this view predicts that if a reactivation cue is dissimilar to a memory, a small portion of that memory will be destabilized to encode the new information.

Table 4 Mechanisms of memory destabilization

NR2B-containing NMDAR activation	Mamou et al. (2006), Milton et al. (2013)
CB1R activation	Suzuki et al. (2004)
L-VGCC activation	Suzuki et al. (2004)
Protein degradation	Lee (2008)
AMPAR endocytosis	Clem and Huganir (2010), Rao-Ruiz et al. (2011), Hong et al. (2013)

7 Common Mechanisms Between Memory Destabilization and Forgetting, Reconsolidation and Maintenance

As science moves forward, it becomes more clear how the memory phases that follow consolidation (maintenance or forgetting and reconsolidation) are complex and far from passive processes. Interestingly, there are striking similarities between the mechanisms underlying reconsolidation and those that determine if memory will be forgotten or maintained. For instance, protein kinase M zeta (PKMζ), a constitutively active, atypical isoform of protein kinase C, appears to be necessary for synaptic potentiation and long-term memory maintenance (Sacktor and Fento 2012; Sacktor et al. 1993). PKMζ is constitutively active through a positive feedback loop which induces its own local synthesis (Cai et al. 2011; Ogasawara and Kawato 2010). Inhibiting its activity dramatically impairs memory retention (Pastalkova et al. 2006; Serrano et al. 2008), whereas its genetic overexpression has been observed to enhance long-term memory (Shema et al. 2011).

As discussed, memory destabilization is paralleled by well-orchestrated AMPAR trafficking that involves a rapid endocytosis of CI-AMPAR and the insertion of CP-AMPAR. Interestingly, it is thought that PKMζ preserves the postsynaptic density structure by maintaining GluA2-containing AMPAR insertion by opposing its regulated endocytosis (Migues et al. 2010). The peptide $GluA2_{3Y}$, which acts preventing this endocytic process (Ahmadian et al. 2004), also prevents the memory and LTP-impairing effects of ZIP and prevents the natural forgetting of memory (Migues et al. 2010, 2016; Dong et al. 2015).

Given the central role of GluA2-containing AMPAR endocytosis in memory destabilization, it is thought that PKMζ might not just mediate the maintenance of memory storage but could also selectively control the entry of memory into a plastic state following retrieval (Dudai 2009). Supporting this view, blocking GluA2 endocytosis prevents the effects of disruptive agents upon reconsolidation (Hong et al. 2013) and reconsolidation-mediated updating (Rao-Ruiz et al. 2011). The memory strength boundary condition could also be a consequence of an increased expression of PKMζ, preventing the retrieval-induced AMPAR endocytosis that underlies memory destabilization. However, training strength has not always been observed to limit ZIP effectiveness (Kwapis et al. 2009) and further studies need to be conducted to identify the role of PKMζ during reconsolidation. Hence, memory reconsolidation, maintenance, and forgetting share common features, which reflects their role in memory's dynamic nature.

8 Reconsolidation as a Therapeutic Tool

The notion of memory being dynamic rather than fixed gave rise to reconsolidation-based therapeutic strategies to treat memory-related disorders, such as PTSD and drug addiction. For instance, researchers are searching for effective reactivation protocols and amnestic agents capable of disrupting maladaptive memories in humans. In addition, given that reconsolidation allows memory content to be updated, there have been efforts to develop behavioral procedures to reframe maladaptive memory content to a less detrimental form.

The major challenge for the drug-based strategy is that most of the compounds known to be effective in blocking reconsolidation are toxic or need intracranial infusions, and thus are not applicable in humans. However, there have been great successes in identifying safe and effective drugs. For instance, the beta-blocker propranolol (a drug already on market to treat high blood pressure) was shown to selectively and robustly disrupt the emotional component of spider phobia when administrated after a reactivation (Soeter and Kindt 2015). This effect was long-lasting and did not affect the declarative component of the spider memory, only the anxiety and fear responses elicited by phobia-related cues. Also, as Misanin et al. (1968) first showed in rodents, electroconvulsive shock disrupts human reconsolidation (Gahr et al. 2014; Kroes et al. 2014) and can be a valuable tool when other drugs are not effective.

The updating approach was first described using a so-called reactivation–extinction procedure. Extinction is induced by a long re-exposure to a conditioned stimulus in the absence of the unconditioned stimulus, inducing inhibitory new learning (Bouton et al. 2012). However, given that extinction induces new learning but does not disrupt the original CS+US association, relapses are common (Rescorla and Heth 1975). Monfils (2009) showed that if an extinction session is conducted during the reconsolidation window, a fear response is attenuated permanently. The conclusion is that when preceded by a reactivation, the safe information provided by an extinction session is incorporated into the labile memory, thus changing its content. Subsequent studies found that post-retrieval extinction reduces both fear and reward-related memories in animals and humans (Clem and Huganir 2010; Schiller et al. 2010; Flavell et al. 2011; Rao-Ruiz et al. 2011; Xue et al. 2012; Sartor and Aston-Jones 2014). However, some studies failed to find an updating effect (Auber et al. 2013), possibly due to an unsuccessful reactivation session. Indeed, humans and their memories acquired in real life vary to a great extent in comparison to lab animals and standardized behavioral protocols, thus finding effective reactivation protocols is challenging. Using a distinct updating strategy, Haubrich et al. (2015) found that a fear memory can be reinterpreted as less aversive if reactivated with stimuli of positive valence. Such fear behavior reduction was long-lasting and not amenable to reinstatement and reacquisition, common features of relapse following extinction-based therapies. It shows that memory's emotional valence can be shaped by proper behavioral manipulations, perhaps resulting in effective treatments in the future.

9 Conclusion

The evidence for reconsolidation comes from a wide spectrum of species, amnesic agents and tasks. It encompasses all levels of analysis, from molecular and physiological to behavioral levels, thereby indicating that it is a fundamental property of memory. Hence, opposed to the initial view of memory as a fixed and permanent entity, flexibility is a core characteristic of memory and its maintenance over time is active rather than passive. Reconsolidation remains a topic of intensive research and there is growing interest in applying reconsolidation blockade and its updating properties as a therapeutic tool in several clinical conditions, such as PTSD and drug addiction. Notwithstanding, much about the nature of reconsolidation and memory storage is still unknown. We are just starting to understand how the brain is capable of maintaining memory stability over time and how this stability can be transiently lifted given appropriate learning situations. Apparently conflicting results in the literature show that the mechanisms recruited to destabilize a retrieved memory might vary dramatically depending on training and reactivation characteristics. It is fascinating that the more we solve some details of this "puzzle," the more that new insights and questions emerge. Whatever future research brings to light regarding the nature of memory, the only one thing that seems certain is that memory is dynamic in its essence.

References

Ahmadian G, Ju W, Liu L, Wyszynski M, Lee SH, Dunah AW et al (2004) Tyrosine phosphorylation of GluR2 is required for insulin-stimulated AMPA receptor endocytosis and LTD. EMBO J 23:1040–1050

Alberini CM (2011) The role of reconsolidation and the dynamic process of long-term memory formation and storage. Front Behav Neurosci 5:12

Anokhin KV, Tiunova AA, Rose SP (2002) Reminder effects—reconsolidation or retrieval deficit? Pharmacological dissection with protein synthesis inhibitors following reminder for a passive-avoidance task in young chicks. Eur J Neurosci 15:1759–1765

Arguello AA, Ye X, Bozdagi O, Pollonini G, Tronel S, Bambah-Mukku D, et al (2013) CCAAT enhancer binding protein delta plays an essential role in memory consolidation and reconsolidation. J Neurosci 33:3646–3658

Auber A, Tedesco V, Jones CE, Monfils MH, Chiamulera C (2013) Post-retrieval extinction as reconsolidation interference: methodological issues or boundary conditions? Psychopharmacology 226:631–647

Barak S, Liu F, Hamida S Ben, Yowell Q V, Neasta J, Kharazia V, et al (2013) Disruption of alcohol-related memories by mTORC1 inhibition prevents relapse. Nat Neurosci 16:1111–1117

Baratti CM, Boccia MM, Blake MG, Acosta GB (2008) Reactivated memory of an inhibitory avoidance response in mice is sensitive to a nitric oxide synthase inhibitor. Neurobiol Learn Mem 89:426–440

Barnes P, Kirtley A, Thomas KL (2012) Quantitatively and qualitatively different cellular processes are engaged in CA1 during the consolidation and reconsolidation of contextual fear memory. Hippocampus 22:149–171

Bergstrom HC, McDonald CG, Dey S, Fernandez GM, Johnson LR (2013) Neurons activated during fear memory consolidation and reconsolidation are mapped to a common and new topography in the lateral amygdala. Brain Topogr 26:468–478
Besnard A (2012) A model of hippocampal competition between new learning and memory updating. J Neurosci 32:3281–3283
Besnard A, Caboche J, Laroche S (2012) Reconsolidation of memory: a decade of debate. Prog Neurobiol 99:61–80
Bjorkstrand J, Agren T, Frick A, Engman J, Larsson EM, Furmark T, et al (2015) Disruption of Memory Reconsolidation Erases a Fear Memory Trace in the Human Amygdala: An 18-Month Follow-Up. PLoS One 10:e0129393
Bliss TV, Lømo T (1973) Long-lasting potentiation of synaptic transmission in the dentate area of the anaesthetized rabbit following stimulation of the perforant path. J Physiol 232:331–356
Bourtchuladze R, Frenguelli B, Blendy J, Cioffi D, Schutz G, Silva AJ (1994) Deficient long-term memory in mice with a targeted mutation of the cAMP-responsive element-binding protein. Cell 79:59–68
Bouton ME (2004) Context and behavioral processes in extinction. Learn Mem 11:485–494
Bouton ME, Winterbauer NE, Todd TP (2012) Relapse processes after the extinction of instrumental learning: renewal, resurgence, and reacquisition. Behav Process 90:130–141
Bozon B, Davis S, Laroche S (2003) A requirement for the immediate early gene zif268 in reconsolidation of recognition memory after retrieval. Neuron 40:695–701
Bredy TW, Barad M (2008) The histone deacetylase inhibitor valproic acid enhances acquisition, extinction, and reconsolidation of conditioned fear. Learn Mem 15:39–45
Brunet A, Orr SP, Tremblay J, Robertson K, Nader K, Pitman RK (2008) Effect of post-retrieval propranolol on psychophysiologic responding during subsequent script-driven traumatic imagery in post-traumatic stress disorder. J Psychiatr Res 42:503–506
Bustos SG, Maldonado H, Molina V a (2009) Disruptive effect of midazolam on fear memory reconsolidation: decisive influence of reactivation time span and memory age. Neuropsychopharmacology 34:446–457
Cai D, Pearce K, Chen S, Glanzman DL (2011) Protein kinase M maintains long-term sensitization and long-term facilitation in aplysia. J Neurosci 31:6421–6431
Cai D, Pearce K, Chen S, Glanzman DL (2012) Reconsolidation of long-term memory in Aplysia. Curr Biol 22:1783–1788
Censor N, Dayan E, Cohen LG (2014) Cortico-subcortical neuronal circuitry associated with reconsolidation of human procedural memories. Cortex A J Devoted to Study Nerv Syst Behav 58:281–288
Chia C, Otto T (2013) Hippocampal Arc (Arg3.1) expression is induced by memory recall and required for memory reconsolidation in trace fear conditioning. Neurobiol Learn Mem 106:48–55
Child FM, Epstein HT, Kuzirian AM, Alkon DL (2003) Memory reconsolidation in Hermissenda. Biol Bull 205:218–219
Clem RL, Huganir RL (2010) Calcium-permeable AMPA receptor dynamics mediate fear memory erasure. Science 330(6007):1108–1112
Chen S, Cai D, Pearce K, Sun PY, Roberts AC, Glanzman DL (2014) Reinstatement of long-term memory following erasure of its behavioral and synaptic expression in Aplysia. Elife 3:e03896
Clarke JR, Cammarota M, Gruart A, Izquierdo I, Delgado-Garcia JM (2010) Plastic modifications induced by object recognition memory processing. Proc Natl Acad Sci USA 107:2652–2657
Contreras M, Billeke P, Vicencio S, Madrid C, Perdomo G, Gonzalez M, et al (2012) A role for the insular cortex in long-term memory for context-evoked drug craving in rats. Neuropsychopharmacology 37:2101–2108
Dash PK, Hochner B, Kandel ER (1990) Injection of the cAMP-responsive element into the nucleus of Aplysia sensory neurons blocks long-term facilitation. Nature 345:718–721
Davis S, Renaudineau S, Poirier R, Poucet B, Save E, Laroche S (2010) The formation and stability of recognition memory: What happens upon recall? Front Behav Neurosci 4:177

De Oliveira Alvares L, Crestani AP, Cassini L, Haubrich J, Santana F, Quillfeldt JA (2013) Reactivation enables memory updating, precision-keeping and strengthening: exploring the possible biological roles of reconsolidation. Neuroscience: 1–7. doi:10.1016/j.neuroscience.2013.04.005

Debiec J, LeDoux JE, Nader K (2002) Cellular and systems reconsolidation in the hippocampus. Neuron 36:527–538

Debiec J, Doyere V, Nader K, LeDoux JE, McEwen BS (2006) Directly reactivated, but not indirectly reactivated, memories undergo reconsolidation in the amygdala. PNAS Proc Natl Acad Sci USA 103:3428–3433

Descalzi G, Li X-Y, Chen T, Mercaldo V, Koga K, Zhuo M (2012) Rapid synaptic potentiation within the anterior cingulate cortex mediates trace fear learning. Mol Brain 5:6

Diaz-Mataix L, Debiec J, LeDoux JE, Doyere V (2011) Sensory-specific associations stored in the lateral amygdala allow for selective alteration of fear memories. J Neurosci 31:9538–9543

Diaz-Mataix L, Ruiz Martinez RC, Schafe GE, LeDoux JE, Doyere V (2013) Detection of a temporal error triggers reconsolidation of amygdala-dependent memories. Curr Biol 23:467–472

Diekelmann S, Büchel C, Born J, Rasch B (2011) Labile or stable: opposing consequences for memory when reactivated during waking and sleep. Nat Neurosci 14:381–386

Diergaarde L, Schoffelmeer AN, Vries TJ De (2006) Beta-adrenoceptor mediated inhibition of long-term reward-related memory reconsolidation. Behav Brain Res 170:333–336

Dong Z, Han H, Li H, Bai Y, Wang W, Tu M et al (2015) Long-term potentiation decay and memory loss are mediated by AMPAR endocytosis. J Clin Invest 125:234–247

Doyere V, Debiec J, Monfils MH, Schafe GE, LeDoux JE (2007) Synapse-specific reconsolidation of distinct fear memories in the lateral amygdala. Nat Neurosci 10:414–416

Dudai Y (2004) The neurobiology of consolidations, or, how stable is the engram? Annu Rev Psychol 55:51–86

Dudai Y (2009) Predicting not to predict too much: how the cellular machinery of memory anticipates the uncertain future. Philos Trans R Soc Lond Ser B Biol Sci 364:1255–1262

Dudai Y, Morris R (2000) To consolidate or not to consolidate: what are the questions? In: Bolhius J (ed) Brain, perception, memory: advances in cognitive sciences. Oxford University Press, Oxford, pp 149–162

Duncan CP (1949) The retroactive effect of electroshock on learning. J Comp Physiol Psychol 42:32–44

Duvarci S, Nader K, LeDoux JE (2005) Activation of extracellular signal-regulated kinase-mitogen-activated protein kinase cascade in the amygdala is required for memory reconsolidation of auditory fear conditioning. Eur J Neurosci 21:283–289

Ebbinghaus M (1885) Über das Gedächtnis. Buehler, Leipzig

Eisenberg M, Kobilo T, Berman DE, Dudai Y (2003) Stability of retrieved memory: inverse correlation with trace dominance. Science 301:1102–1104

Eisenberg M, Dudai Y (2004) Reconsolidation of fresh, remote, and extinguished fear memory in Medaka: old fears don't die. Eur J Neurosci 20:3397–3403

Eysenck HJ (1968) A theory of the incubation of anxiety/fear responses. Behav Res Ther 6:309–321

Flavell CR, Barber DJ, Lee JLC (2011) Behavioural memory reconsolidation of food and fear memories. Nat Commun 2:504

Flexner LB, Flexner JB, Stellar E (1965) Memory and cerebral protein synthesis in mice as affected by graded amounts of puromycin. Exp Neurol 13:264–272

Forcato C, Argibay PF, Pedreira ME, Maldonado H (2009) Human reconsolidation does not always occur when a memory is retrieved: the relevance of the reminder structure. Neurobiol Learn Mem 91:50–57

Forcato C, Rodríguez MLC, Pedreira ME, Maldonado H (2010) Reconsolidation in humans opens up declarative memory to the entrance of new information. Neurobiol Learn Mem 93:77–84

Forcato C, Fernandez RS, Pedreira ME (2014) Strengthening a consolidated memory: the key role of the reconsolidation process. J Physiol Paris 108:323–333

Forcato C, Rodríguez MLC, Pedreira ME (2011) Repeated labilization-reconsolidation processesstrengthen declarative memory in humans. PLoS One 6:e23305
Frankland PW, Ding H-KK, Takahashi E, Suzuki A, Kida S, Silva AJ (2006) Stability of recent and remote contextual fear memory. Learn Mem 13:451–457
Frankland PW, Köhler S, Josselyn SA (2013) Hippocampal neurogenesis and forgetting. Trends Neurosci 36:497–503
Frenkel L, Maldonado H, Delorenzi A (2005) Memory strengthening by a real-life episode during reconsolidation: an outcome of water deprivation via brain angiotensin II. Eur J Neurosci 22:1757–1766
Fukushima H, Zhang Y, Archbold G, Ishikawa R, Nader K, Kida S (2014) Enhancement of fear memory by retrieval through reconsolidation. Elife 2014:1–19
Gabriel M (1968) Effects of intersession delay and training level on avoidance extinction and intertrial behavior. J Comp Physiol Psychol 66:412–416
Gahr M, Schonfeldt-Lecuona C, Spitzer M, Graf H (2014) Electroconvulsive therapy and posttraumatic stress disorder: first experience with conversation-based reactivation of traumatic memory contents and subsequent ECT-mediated impairment of reconsolidation. J Neuropsychiatry Clin Neurosci 26:E38–E39
García-DeLaTorre P, Rodriguez-Ortiz CJ, Arreguin-Martinez JL, Cruz-Castañeda P, Bermúdez-Rattoni F (2009) Simultaneous but not independent anisomycin infusions in insular cortex and amygdala hinder stabilization of taste memory when updated. Learn Mem 16:514–519
Glickman S (1961) Perseverative neural processes and consolidation of the memory trace. Psychol Bull 58:218–233
Gold PE, King RA (1974) Retrograde amnesia: storage failure versus retrieval failure. Physiol Rev 81:465–469
Gordon WC (1977a) Similarities of recently acquired and reactivated memories in interference. Am J Psychol 90(2):231–242
Gordon WC (1977b) Susceptibility of a reactivated memory to the effects of strychnine: a time-dependent phenomenon. Physiol Behav 18(1):95–99
Gordon WC, Spear NE (1973) Effect of reactivation of a previously acquired memory on the interaction between memories in the rat. J Exp Psychol 99:349–355
Gräff J, Joseph NF, Horn ME, Samiei A, Meng J, Seo J et al (2014) Epigenetic priming of memory updating during reconsolidation to attenuate remote fear memories. Cell 156:261–276
Grill-Spector K, Henson R, Martin A (2006) Repetition and the brain: Neural models of stimulus-specific effects. Trends Cogn Sci 10:14–23
Hall J, Thomas KL, Everitt BJ (2001a) Fear memory retrieval induces CREB phosphorylation and Fos expression within the amygdala. Eur J Neurosci 13:1453–1458
Hall J, Thomas KL, Everitt BJ (2001b) Cellular imaging of zif268 expression in the hippocampus and amygdala during contextual and cued fear memory retrieval: selective activation of hippocampal CA1 neurons during the recall of contextual memories. J Neurosci 21:2186–2193
Haubrich J, Crestani AP, Cassini LF, Santana F, Sierra RO, Alvares Lde O et al (2015) Reconsolidation allows fear memory to be updated to a less aversive level through the incorporation of appetitive information. Neuropsychopharmacology 40:315–326
Hebb DO (1949) The organization of behavior. Wiley, New York. doi:10.2307/1418888
Hellemans KG, Everitt BJ, Lee JL (2006) Disrupting reconsolidation of conditioned withdrawal memories in the basolateral amygdala reduces suppression of heroin seeking in rats. J Neurosci 26:12694–12699
Hernandez PJ, Kelley AE (2004) Long-term memory for instrumental responses does not undergo protein synthesis-dependent reconsolidation upon retrieval. Learn Mem 11:748–754
Hertzen LS von, Giese KP (2005) Memory reconsolidation engages only a subset of immediate-early genes induced during consolidation. J Neurosci 25:1935–1942
Hong I, Kim J, Kim J, Lee S, Ko H-G, Nader K et al (2013) AMPA receptor exchange underlies transient memory destabilization on retrieval. Proc Natl Acad Sci USA 110:8218–8223

Hupbach A, Gomez R, Hardt O, Nadel L (2007) Reconsolidation of episodic memories: a subtle reminder triggers integration of new information. Learn Mem 14:47–53

Hupbach A, Hardt O, Gomez R, Nadel L (2008) The dynamics of memory: Context-dependent updating. Learn Mem 15:574–579

Inda MC, Muravieva E V, Alberini CM (2011) Memory retrieval and the passage of time: from reconsolidation and strengthening to extinction. J Neurosci 31:1635–1643

Jarome TJ, Kwapis JL, Werner CT, Parsons RG, Gafford GM, Helmstetter FJ (2012) The timing of multiple retrieval events can alter GluR1 phosphorylation and the requirement for protein synthesis in fear memory reconsolidation. Learn Mem 19:300–306

Jarome TJ, Ferrara NC, Kwapis JL, Helmstetter FJ (2015) Contextual Information Drives the Reconsolidation-Dependent Updating of Retrieved Fear Memories. Neuropsychopharmacology 40(13):3044–3052

Jones B, Bukoski E, Nadel L, Fellous J-M (2012) Remaking memories: reconsolidation updates positively motivated spatial memory in rats. Learn Mem 19:91–98

Kandel ER (2001) The molecular biology of memory storage: a dialogue between genes and synapses. Science 294:1030–1038

Kelly A, Laroche S, Davis S (2003) Activation of mitogen-activated protein kinase/extracellular signal-regulated kinase in hippocampal circuitry is required for consolidation and reconsolidation of recognition memory. J Neurosci 23:5354–5360

Kemenes G, Kemenes I, Michel M, Papp A, Muller U (2006) Phase-dependent molecular requirements for memory reconsolidation: differential roles for protein synthesis and protein kinase A activity. J Neurosci 26:6298–6302

Kida S, Josselyn SA, de Ortiz SP, Kogan JH, Chevere I, Masushige S et al (2002) CREB required for the stability of new and reactivated fear memories. Nat Neurosci 5:348–355

Kim J, Song B, Hong I, Kim J, Lee J, Park S, et al (2010) Reactivation of fear memory renders consolidated amygdala synapses labile. J Neurosci 30:9631–9640

Kindt M, Soeter M, Vervliet B (2009) Beyond extinction: erasing human fear responses and preventing the return of fear. Nat Neurosci 12:256–258

Kroes MCW, Tendolkar I, van Wingen GA, van Waarde JA, Strange BA, Fernández G (2014) An electroconvulsive therapy procedure impairs reconsolidation of episodic memories in humans. Nat Neurosci 17:204–206

Kwapis JL, Jarome TJ, Lonergan ME, Helmstetter FJ (2009) Protein kinase Mzeta maintains fear memory in the amygdala but not in the hippocampus. Behav Neurosci 123:844–850

Lagasse F, Devaud JM, Mery F (2009) A switch from cycloheximide-resistant consolidated memory to cycloheximide-sensitive reconsolidation and extinction in Drosophila. J Neurosci 29:2225–2230

Lee JLC (2008) Memory reconsolidation mediates the strengthening of memories by additional learning. Nat Neurosci 11:1264–1266

Lee JLC (2009) Reconsolidation: maintaining memory relevance. Trends Neurosci 32:413–420

Lee JLC (2010) Memory reconsolidation mediates the updating of hippocampal memory content. Front Behav Neurosci 4:168

Lee JLC, Everitt BJ, Thomas KL (2004) Independent cellular processes for hippocampal memory consolidation and reconsolidation. Science 304:839–843

Lee JL, Di Ciano P, Thomas KL, Everitt BJ (2005) Disrupting reconsolidation of drug memories reduces cocaine-seeking behavior. Neuron 47:795–801

Lee JLC, Milton AL, Everitt BJ (2006) Reconsolidation and extinction of conditioned fear: inhibition and potentiation. J Neurosci 26:10051–10056

Lee S, Kim J, Choi S (2011) In vitro synaptic reconsolidation in amygdala slices prepared from rat brains. Biochem Biophys Res Commun 407:339–342

Lee SH, Kwak C, Shim J, Kim JE, Choi SL, Kim HF et al (2012) A cellular model of memory reconsolidation involves reactivation-induced destabilization and restabilization at the sensorimotor synapse in Aplysia. Proc Natl Acad Sci USA 109:14200–14205

Lewis DJ (1979) Psychobiology of active and inactive memory. Psychol Bull 86:1054–1083

Li Y, Meloni EG, Carlezon Jr. WA, Milad MR, Pitman RK, Nader K, et al (2013) Learning and reconsolidation implicate different synaptic mechanisms. Proc Natl Acad Sci USA 110:4798–4803

Litvin OO, Anokhin KV (2000) Mechanisms of memory reorganization during retrieval of acquired behavioral experience in chicks: the effects of protein synthesis inhibition in the brain. Neurosci Behav Physiol 30:671–678

Liu L, Wong TP, Pozza MF, Lingenhoehl K, Wang Y, Sheng M et al (2004) Role of NMDA receptor subtypes in governing the direction of hippocampal synaptic plasticity. Science 304 (5673):1021–1024

Lubin FD, Sweatt JD (2007) The IkappaB kinase regulates chromatin structure during reconsolidation of conditioned fear memories. Neuron 55:942–957

Lukowiak K, Fras M, Smyth K, Wong C, Hittel K (2007) Reconsolidation and memory infidelity in Lymnaea. Neurobiol Learn Mem 87:547–560

Maddox SA, Watts CS, Doyere V, Schafe GE (2013a) A naturally-occurring histone acetyltransferase inhibitor derived from Garcinia indica impairs newly acquired and reactivated fear memories. PLoS One 8:e54463

Maddox SA, Watts CS, Schafe GE (2013b) p300/CBP histone acetyltransferase activity is required for newly acquired and reactivated fear memories in the lateral amygdala. Learn Mem 20:109–119

Maddox SA, Watts CS, Schafe GE (2014) DNA methyltransferase activity is required for memory-related neural plasticity in the lateral amygdala. Neurobiol Learn Mem 107:93–100

Mamiya N, Fukushima H, Suzuki A, Matsuyama Z, Homma S, Frankland PW, et al (2009) Brain region-specific gene expression activation required for reconsolidation and extinction of contextual fear memory. J Neurosci 29:402–413

Mamou CB, Gamache K, Nader K (2006) NMDA receptors are critical for unleashing consolidated auditory fear memories. Nat Neurosci 9:1237–1239

Martin SJ, Grimwood PD, Morris RG (2000) Synaptic plasticity and memory: an evaluation of the hypothesis. Annu Rev Neurosci 23:649–711

McGaugh JL (1966) Time-dependent processes in memory storage. Science 153:1351–1358

McGaugh JL, Krivanek JA (1970) Strychnine effects on discrimination learning in mice: effects of dose and time of administration. Physiol Behav 5:1437–1442

Merlo E, Freudenthal R, Maldonado H, Romano A (2005) Activation of the transcription factor NF-kappaB by retrieval is required for long-term memory reconsolidation. Learn Mem 12:23–29

Migues PV, Hardt O, Wu DC, Gamache K, Sacktor TC, Wang YT et al (2010) PKMzeta maintains memories by regulating GluR2-dependent AMPA receptor trafficking. Nat Neurosci 13:630–634

Migues PV, Liu L, Archbold GEB, Einarsson EO, Wong J, Bonasia K et al (2016) Blocking synaptic removal of GluA2-containing AMPA receptors prevents the natural forgetting of long-term memories. J Neurosci 36:3481–3494

Milekic MH, Alberini CM (2002) Temporally graded requirement for protein synthesis following memory reactivation. Neuron 36:521–525

Miller RR, Marlin NA (1984) The physiology and semantics of consolidation: of mice and men. In: Weingartner H, Parker ES (eds) Memory consolidation: psychobiology of cognition. Erlbaum, Hillsdale, NJ, pp 85–109

Miller CA, Marshall JF (2005) Molecular substrates for retrieval and reconsolidation of cocaine-associated contextual memory. Neuron 47:873–884

Miller RR, Springer AD (1973) Amnesia, consolidation, and retrieval. Psychol Rev 80:69–79

Milton AL, Lee JLC, Butler VJ, Gardner R, Everitt BJ (2008) Intra-amygdala and systemic antagonism of NMDA receptors prevents the reconsolidation of drug-associated memory and impairs subsequently both novel and previously acquired drug-seeking behaviors. J Neurosci 28:8230–8237

Milton AL, Merlo E, Ratano P, Gregory BL, Dumbreck JK, Everitt BJ (2013) Double dissociation of the requirement for GluN2B- and GluN2A-containing NMDA receptors in the destabilization and restabilization of a reconsolidating memory. J Neurosci 33:1109–1115
Misanin JR, Miller RR, Lewis DJ (1968) Retrograde amnesia produced by electroconvulsive shock after reactivation of a consolidated memory trace. Science 160(3827):554–555
Monfils MH, Cowansage KK, Klann E, LeDoux JE (2009) Extinction-reconsolidation boundaries: key to persistent attenuation of fear memories. Science 324(5929):951–955
Morris RGM, Inglis J, Ainge J a, Olverman HJ, Tulloch J, Dudai Y, et al (2006) Memory reconsolidation: sensitivity of spatial memory to inhibition of protein synthesis in dorsal hippocampus during encoding and retrieval. Neuron 50:479–489
Müller GE, Pilzecker A (1900) Experimentelle Beitraege zur Lehre vom Gedaechtnis. Zeitschrift für Psychologie 1:1
Nader K, Hardt O (2009) A single standard for memory: the case for reconsolidation. Nat Rev Neurosci 10:224–234
Nader K, Schafe GE, Le Doux JE (2000) Fear memories require protein synthesis in the amygdala for reconsolidation after retrieval. Nature 406:722–726
Ogasawara H, Kawato M (2010) The protein kinase Mζ network as a bistable switch to store neuronal memory. BMC Syst Biol 4:181
Olshavsky ME, Song BJ, Powell DJ, Jones CE, Monfils M-H, Lee HJ (2013) Updating appetitive memory during reconsolidation window: critical role of cue-directed behavior and amygdala central nucleus. Front Behav Neurosci 7, ArtID 186, 7
Osan R, Tort AB, Amaral OB (2011) A mismatch-based model for memory reconsolidation and extinction in attractor networks. PLoS One 6:e23113
Pastalkova E, Serrano P, Pinkhasova D, Wallace E, Fenton AA, Sacktor TC (2006) Storage of spatial information by the maintenance mechanism of LTP. Science 313:1141–1144
Pavlov IP (1927) Conditioned reflexes: An investigation of the physiological activity of the cerebral cortex. Book 17
Pedreira ME, Maldonado H (2003) Protein synthesis subserves reconsolidation or extinction depending on reminder duration. Neuron 38:863–869
Pedreira ME, Perez-Cuesta LM, Maldonado H (2002) Reactivation and reconsolidation of long-term memory in the crab Chasmagnathus: protein synthesis requirement and mediation by NMDA-type glutamatergic receptors. J Neurosci 22:8305–8311
Pedreira ME, Pérez-Cuesta LM, Maldonado H (2004) Mismatch between what is expected and what actually occurs triggers memory reconsolidation or extinction. Learn Mem 11:579–585
Perez-Cuesta LM, Maldonado H (2009) Memory reconsolidation and extinction in the crab: Mutual exclusion or coexistence? Learn Mem 16:714–721
Perrin G, Ferreira G, Meurisse M, Verdin S, Mouly A-M, Lévy F (2007) Social recognition memory requires protein synthesis after reactivation. Behav Neurosci 121:148–155
Pickens CL, Golden SA, Adams-Deutsch T, Nair SG, Shaham Y (2009) Long-lasting incubation of conditioned fear in rats. Biol Psychiatry 65:881–886
Radiske A, Rossato JI, Kohler CA, Gonzalez MC, Medina JH, Cammarota M (2015) Requirement for BDNF in the reconsolidation of fear extinction. J Neurosci 35:6570–6574
Rao-Ruiz P, Rotaru DC, van der Loo RJ, Mansvelder HD, Stiedl O, Smit AB et al (2011) Retrieval-specific endocytosis of GluA2-AMPARs underlies adaptive reconsolidation of contextual fear. Nat Neurosci 14:1302–1308
Rehberg K, Bergado-Acosta JR, Koch JC, Stork O (2010) Disruption of fear memory consolidation and reconsolidation by actin filament arrest in the basolateral amygdala. Neurobiol Learn Mem 94:117–126
Ren ZY, Liu MM, Xue YX, Ding ZB, Xue LF, Zhai SD, et al (2013) A critical role for protein degradation in the nucleus accumbens core in cocaine reward memory. Neuropsychopharmacology 38:778–790
Rescorla RA,Wagner AR (1972) A theory of Pavlovian conditioning: variations in the effectiveness of reinforcement and nonreinforcement. In Classical Conditioning II. Appleton-Century-Crofts, New York, pp 64–99

Rescorla RA, Heth CD (1975) Reinstatement of fear to an extinguished conditioned stimulus. J Exp Psychol Anim Behav Process 1:88–96

Ribot T (1881) Les Maladies de la Memoire. Appleton-Century-Crofts, New York

Robinson MJ, Franklin KB (2010) Reconsolidation of a morphine place preference: impact of the strength and age of memory on disruption by propranolol and midazolam. Behav Brain Res 213:201–207

Robinson MJF, Ross EC, Franklin KBJ (2011) The effect of propranolol dose and novelty of the reactivation procedure on the reconsolidation of a morphine place preference. Behav Brain Res 216:281–284

Rodriguez-Ortiz CJ, Bermudez-Rattoni F (2007) Memory Reconsolidation or Updating Consolidation? Neural Plast Mem From Genes to Brain Imaging at http://www.ncbi.nlm.nih.gov/pubmed/21204424

Rodriguez-Ortiz CJ, la Cruz V De, Gutiérrez R, Bermudez-Rattoni F (2005) Protein synthesis underlies post-retrieval memory consolidation to a restricted degree only when updated information is obtained. Learn Mem 12:533–537

Rodriguez-Ortiz CJ, Garcia-DeLaTorre P, Benavidez E, Ballesteros MA, Bermudez-Rattoni F (2008) Intrahippocampal anisomycin infusions disrupt previously consolidated spatial memory only when memory is updated. Neurobiol Learn Mem 89:352–359

Romero-Granados R, Fontan-Lozano A, Delgado-Garcia JM, Carrion AM (2010) From learning to forgetting: behavioral, circuitry, and molecular properties define the different functional states of the recognition memory trace. Hippocampus 20:584–595

Rose JK, Rankin CH (2006) Blocking memory reconsolidation reverses memory-associated changes in glutamate receptor expression. J Neurosci 26:11582–11587

Sachser RM, Santana F, Crestani AP, Lunardi P, Pedraza LK, Quillfeldt JA et al (2016) Forgetting of long-term memory requires activation of NMDA receptors, L-type voltage-dependent Ca(2 +) channels, and calcineurin. Sci Rep 6:22771

Sacktor TC (2008) PKMzeta, LTP maintenance, and the dynamic molecular biology of memory storage. Prog Brain Res 169:27–40

Sacktor TC (2011) How does PKMζ maintain long-term memory? Nat Rev Neurosci 12:9–15

Sacktor TC, Fenton AA (2012) Appropriate application of ZIP for PKMζ inhibition, LTP reversal, and memory erasure. Hippocampus 22:645–647

Sacktor TC, Osten P, Valsamis H, Jiang X, Naik MU, Sublette E (1993) Persistent activation of the zeta isoform of protein kinase C in the maintenance of long-term potentiation. Proc. Natl. Acad. Sci. USA 90:8342–8346

Sangha S, Scheibenstock A, Lukowiak K (2003) Reconsolidation of a long-term memory in Lymnaea requires new protein and RNA synthesis and the soma of right pedal dorsal 1. J Neurosci 23:8034–8040

Sara SJ (2000) Retrieval and Reconsolidation: Toward a Neurobiology of Remembering. Learn Mem 7:73–84

Sartor GC, Aston-Jones G (2014) Post-retrieval extinction attenuates cocaine memories. Neuropsychopharmacology 39:1059–1065

Schiller D, Monfils M-H, Raio CM, Johnson DC, Ledoux JE, Phelps EA (2010) Preventing the return of fear in humans using reconsolidation update mechanisms. Nature 463:49–53

Schiller D, Kanen JW, LeDoux JE, Monfils M-H, Phelps EA (2013) Extinction during reconsolidation of threat memory diminishes prefrontal cortex involvement. PNAS Proc Natl Acad Sci United States Am 110:20040–20045

Schneider AM, Sherman W (1968) Amnesia: a function of the temporal relation of footshock to electroconvulsive shock. Science 159(3811):219–21

Schwabe L, Nader K, Wolf OT, Beaudry T, Pruessner JC (2012) Neural signature of reconsolidation impairments by propranolol in humans. Biol Psychiatry 71:380–386

Serrano P, Friedman EL, Kenney J, Taubenfeld SM, Zimmerman JM, Hanna J, et al (2008) PKMζ maintains spatial, instrumental, and classically conditioned long-term memories. PLoS Biol 6:2698–2706

Shema R, Haramati S, Ron S, Hazvi S, Chen A, Sacktor TC et al (2011) Enhancement of consolidated long-term memory by overexpression of protein kinase Mzeta in the neocortex. Science 331:1207–1210

Sierra RO, Cassini LF, Santana F, Crestani AP, Duran JM, Haubrich J, et al (2013) Reconsolidationmay incorporate state-dependency into previously consolidated memories. Learn Mem 20:379–87

Slaker M, Churchill L, Todd RP, Blacktop JM, Zuloaga DG, Raber J, et al (2015) Removal of perineuronal nets in the medial prefrontal cortex impairs the acquisition and reconsolidation of a cocaine-induced conditioned place preference memory. J Neurosci 35:4190–4202

Soeter M, Kindt M (2015) An abrupt transformation of phobic behavior after a post-retrieval amnesic agent. Biol Psychiatry. doi:10.1016/j.biopsych.2015.04.006

Spear NE (1973) Retrieval of memory in animals. Psychol Rev 80:163–194

Spear N, Mueller C (1984) Consolidation as a function of retrieval. In: Weingarten H, Parker E (eds) Memory consolidation: psychobiology of cognition. Erlbaum, London, pp 111–147

Squire LR, Cohen NH, Nadel L (1984) The medial temporal region and memory consolidation: a new hypothesis. In: Weingartner H, Parker E (eds) Memory consolidation. Erlbaum, Hillsdale, NJ, pp 185–210

Stern SA, Alberini CM (2013) Mechanisms of memory enhancement. Wiley Interdiscip Rev Syst Biol Med 5:37–53

Stollhoff N, Menzel R, Eisenhardt D (2005) Spontaneous recovery from extinction depends on the reconsolidation of the acquisition memory in an appetitive learning paradigm in the honeybee (Apis mellifera). J Neurosci 25:4485–4492

Suzuki A, Josselyn SA, Frankland PW, Masushige S, Silva AJ, Kida S (2004) Memory reconsolidation and extinction have distinct temporal and biochemical signatures. J Neurosci 24:4787–4795

Taubenfeld SM, Milekic MH, Monti B, Alberini CM (2001) The consolidation of new but not reactivated memory requires hippocampal C/EBPbeta. Nat Neurosci 4:813–818

Thomas KL, Arroyo M, Everitt BJ (2003) Induction of the learning and plasticity-associated gene Zif268 following exposure to a discrete cocaine-associated stimulus. Eur J Neurosci 17:1964–1972

Thomas KL, Hall J, Everitt BJ (2002) Cellular imaging with zif268 expression in the rat nucleus accumbens and frontal cortex further dissociates the neural pathways activated following the retrieval of contextual and cued fear memory. Eur J Neurosci 16:1789–1796

Tronel S, Sara SJ (2002) Mapping of olfactory memory circuits: Region-specific c-fos activation after odor-reward associative learning or after its retrieval. Learn Mem 9:105–111

Tronson NC, Taylor JR (2007) Molecular mechanisms of memory reconsolidation. Nat Rev Neurosci 8:262–275

Tronson NC, Wiseman SL, Olausson P, Taylor JR (2006) Bidirectional behavioral plasticity of memory reconsolidation depends on amygdalar protein kinase A. Nat Neurosci 9:167–169

Walker MP, Brakefield T, Hobson JA (2003) Dissociable stages of human memory consolidation and reconsolidation. Nature 425:8–12

Wang S-H, Morris RG (2010) Hippocampal-neocortical interactions in memory formation, consolidation, and reconsolidation. Annu Rev Psychol 61:49–79

Wang SH, Ostlund SB, Nader K, Balleine BW (2005) Consolidation and reconsolidation of incentive learning in the amygdala. J Neurosci 25:830–835

Wang S-H, Oliveira Alvares L de, Nader K (2009) Cellular and systems mechanisms of memory strength as a constraint on auditory fear reconsolidation. Nat Neurosci 12: 905–912

Wang Y, Zhang TY, Xin J, Li T, Yu H, Li N, et al (2012) Differential involvement of brain-derived neurotrophic factor in reconsolidation and consolidation of conditioned taste aversion memory. PLoS One 7:e49942

Wells AM, Arguello AA, Xie X, Blanton MA, Lasseter HC, Reittinger AM, et al (2013) Extracellular signal-regulated kinase in the basolateral amygdala, but not the nucleus accumbens core, is critical for context-response-cocaine memory reconsolidation in rats. Neuropsychopharmacology 38:753–762

Wells AM, Xie X, Higginbotham JA, Arguello AA, Healey KL, Blanton M, Fuchs RA, et al (2015) Contribution of an SFK-Mediated Signaling Pathway in the Dorsal Hippocampus to Cocaine-Memory Reconsolidation in Rats. Neuropsychopharmacology 41(3):675–685

Winters BD, Tucci MC, DaCosta-Furtado M (2009) Older and stronger object memories are selectively destabilized by reactivation in the presence of new information. Learn Mem 16: 545–553

Winters BD, Tucci MC, Jacklin DL, Reid JM, Newsome J, et al (2011) On the dynamic nature of the engram: evidence for circuit-level reorganization of object memory traces following reactivation. Neurosci 31:17719–28

Wirkner J, Low A, Hamm AO, Weymar M (2015) New learning following reactivation in the human brain: targeting emotional memories through rapid serial visual presentation. Neurobiol Learn Mem 119:63–68

Xue Y-X, Luo Y-X, Wu P, Shi H-S, Xue L-F, Chen C et al (2012) A memory retrieval-extinction procedure to prevent drug craving and relapse. Science 336:241–245

Yin J C, Wallach J S, Del Vecchio M, Wilder E L, Zhou H, Quinn W G, Tully T (1994) Induction of a dominant negative CREB transgene specifically blocks long-term memory in Drosophila. Cell 79(1):49-58

The Sensory Neocortex and Associative Memory

Dominik Aschauer and Simon Rumpel

Abstract Most behaviors in mammals are directly or indirectly guided by prior experience and therefore depend on the ability of our brains to form memories. The ability to form an association between an initially possibly neutral sensory stimulus and its behavioral relevance is essential for our ability to navigate in a changing environment. The formation of a memory is a complex process involving many areas of the brain. In this chapter we review classic and recent work that has shed light on the specific contribution of sensory cortical areas to the formation of associative memories. We discuss synaptic and circuit mechanisms that mediate plastic adaptations of functional properties in individual neurons as well as larger neuronal populations forming topographically organized representations. Furthermore, we describe commonly used behavioral paradigms that are used to study the mechanisms of memory formation. We focus on the auditory modality that is receiving increasing attention for the study of associative memory in rodent model systems. We argue that sensory cortical areas may play an important role for the memory-dependent categorical recognition of previously encountered sensory stimuli.

Keyword Synaptic plasticity · Structural plasticity · Microcircuit · Neuronal assembly · Tonotopic map · Auditory cortex · Categorical perception · Rodent

Contents

1 Memory 178
2 Sensory Circuits in the Brain 179
3 Plasticity of the Sensory Cortex 180
4 Behavioral Paradigms to Study Associative Memory in the Auditory Modality 182
5 Functional Organization of Auditory Cortex 186

D. Aschauer · S. Rumpel (✉)
Institute of Physiology, Focus Program Translational Neurosciences (FTN), University Medical Center, Johannes Gutenberg University, Mainz, Germany
e-mail: Sirumpel@uni-mainz.de

Curr Topics Behav Neurosci (2018) 37: 177–211
DOI 10.1007/7854_2016_453
Published Online: 27 December 2016

6 Tonotopy in Auditory Cortex 189
7 Auditory Map Plasticity 190
8 Plasticity in Functional Properties of Individual Neurons 191
9 Microcircuit Mechanisms Gating Plasticity 194
10 Synaptic Plasticity During Learning 195
11 Category Learning 196
12 Future Challenges 198
References 200

1 Memory

Imagine you are in a foreign country, scanning for a radio station in your rental car. You do not understand much of what is being said, but suddenly a song you recognize comes on. You have no explicit memory of when and where you heard it for the first time, or whether you have heard it since, but you know you have heard it before.

The ability to memorize past experiences is probably the most central function of our brains. Memories shape our personalities, influence our current perception, and determine the decisions we make. Memory is the process by which acquired knowledge is encoded, stored, and later retrieved. How and where memories are stored in the brain is one of the fundamental questions of neuroscience (Kandel 1991; Hubener and Bonhoeffer 2010; Josselyn 2010). Current models put forward the idea that memory is stored as a change in the structure and strength of neural connections. Pioneering theoretical work from Donald Hebb provided a model of how synaptic plasticity in a neural circuit could promote learning and memory via the formation of functionally linked groups of neurons, so-called 'cell assemblies.' This functional linkage depends on the temporal coherence of pre- and postsynaptic activity. Hebb's theory provides a model for neural circuit formation and the integration of newly acquired information through functional changes (Hebb 1949; Buzsaki 2010; Harris and Mrsic-Flogel 2013). Following the first experimental description of long-lasting changes in synaptic transmission induced by specific patterns of neuronal activity (Bliss and Lomo 1973), a substantial body of evidence has been obtained over the last several decades supporting the idea that synaptic plasticity is correlated with and necessary for the formation of a memory (Morris et al. 1986; McKernan and Shinnick-Gallagher 1997; Rogan et al. 1997; Rumpel et al. 2005; Whitlock et al. 2006; Nabavi et al. 2014) and sufficient for the expression of a memory (Nabavi et al. 2014). However, understanding how synaptic plasticity specifically impacts on the dynamics of neuronal networks and thereby mediates the storage of a memory is still a challenge to experimenters due to the vast scale of the problem and the limitations of the methods available (Singer 1999; Lansner 2009; Wallace and Kerr 2010).

Interestingly, Hebb also put forth the idea that cell assemblies are not restricted to local circuits, but may be distributed across multiple brain (Polk and Farah 1995). The distributed nature of a memory trace (or 'engram') naturally raises the question of what aspect of a complex memory is associated to a particular brain structure. It is well established that different classes of memories are organized across different brain systems (Ashby and O'Brien 2005; Poldrack and Foerde 2008). For example, declarative memories are believed to be mediated by different brain structures than procedural memories. The neocortex has also been proposed as a site of memory storage, with modality-specific memories being distributed across their corresponding sensory areas (Penfield 1959; Marr 1970).

Prior experience shapes perception and is essential for the recognition of a sensory stimulus. Sensory processing involves both (1) the detection of an environmental stimulus through sensory receptors in the periphery, represented by the bottom-up flow of information, as well as (2) the concurrent retrieval of internal representations of the environment, represented by the top-down feedback. Hence, the concept of associative memory in the context of sensory processing refers to the long-term storage of sensory information and the ability to interpret incoming stimuli according to previous experience, a process exemplified by the act of categorization (Miller et al. 2003; Russ et al. 2007; Freedman and Miller 2008). Other forms of short-term perceptual memories are beyond the scope of this chapter, but have been discussed in detail elsewhere (Dick 1974; Desimone 1996; Courtney et al. 1997; Magnussen 2000, Pasternak and Greenlee 2005).

In this chapter, we will discuss the current status of research on primary sensory areas of the neocortex and connected brain regions, focusing on the role these circuits play in the formation of associative long-term memories. Specifically, we will describe the general architecture of sensory cortical circuits, their input and output connections, and how experience is integrated through plasticity mechanisms. We will focus on the auditory modality as a representative and versatile model to study perception, learning, and memory. Particularly in the rodent model, several behavioral paradigms have been developed for the purpose of studying learning-related plasticity at the level of neuronal circuits and behavior. There is a vast body of evidence in the auditory modality describing plasticity of single neuron activity, as well as large-scale cortical maps. Finally, we will describe mechanisms at the circuit and synaptic level that mediate experience-dependent functional adaptations.

2 Sensory Circuits in the Brain

Each major sensory modality has a corresponding primary area in the neocortex. Studies showing specific perceptual impairments following insults to these areas in human patients provide much of the evidence highlighting their importance for conscious perception (Mendez and Geehan 1988; Kentridge et al. 1999). How does sensory information reach the neocortex? The majority of peripheral input to these areas (except for olfaction) is carried along sensory tracts through subcortical nuclei before converging in the thalamus. The thalamus is connected to several brain regions, including the hippocampus and other structures of the limbic system. In addition, dedicated thalamic subnuclei convey information to the corresponding primary sensory region of the cortex (Sherman 2007; Guillery and Sherman 2011; Sherman 2012; Metzger et al. 2013).

Subcortical and subthalamic nuclei of the sensory pathway contribute to the processing of particular characteristics of temporal or spatial structure of a stimulus. Furthermore, anatomical pathways linking sensory input to motor output are also located in subcortical areas, thus decoupling some sensorimotor transformations from cortical control (Kobayashi and Isa 2002; Redgrave et al. 2010). Top-down modulation by brain state, experience, or multimodal context typically grows when ascending the subcortical hierarchy. Top-down feedback not only affects subcortical and subthalamic nuclei, but can even include primary sensory receptors of the periphery (Gilbert and Li 2013; Markovitz et al. 2013). Thus, the streams of sensory information are not merely feedforward pathways, but each area is able to modify activity in areas up- or downstream by integrating immediate sensory and behavioral context. Therefore, each percept can be seen as a result of the bottom-up and top-down flow of information (Gilbert and Sigman 2007; Lee et al. 2014).

Along these lines, neuronal activity in primary sensory cortices can be prominently modulated by other cortical areas. For example, recent findings have highlighted the functional relevance of connections between motor cortex and primary sensory cortical areas in mediating movement-related changes in sensory processing (Zagha et al. 2013; Fu et al. 2014; Manita et al. 2015). Associative areas in the cortex are distinguished from primary sensory areas and motor areas by their primary input, which originates from other cortical areas rather than modality-specific thalamic nuclei, typically conveying information from multiple sensory modalities. Associative areas can also influence processing in primary areas. For example, using optogenetic manipulations in the mouse, it has been demonstrated that cingulate cortex, a region of the frontal cortex, can modulate processing in primary visual cortex, leading to an improvement of visual discrimination (Zhang et al. 2014).

In summary, sensory information from a specific modality is processed in a specialized region of the neocortex, where it is also integrated with inputs representing other modalities, internal attentive states, and prior experiences.

3 Plasticity of the Sensory Cortex

Neuronal circuit plasticity allows us to adapt in an ever-changing environment. The induction of plastic rearrangements affecting the processing of sensory stimuli can be interpreted as evidence for a particular brain region contributing to the formation of a memory. Indeed, there is strong evidence from different model organisms that primary sensory areas of the neocortex exhibit experience-dependent plasticity (Katz and Shatz 1996; Weinberger 2004; Hensch 2005a, b). Hence, plastic adaptations are not restricted to emotional centers, like the limbic system, or to other structures with well-established roles in memory, like the hippocampus.

A classical approach for the study of plasticity is sensory deprivation: Closure of the contralateral eyelid induces ocular dominance plasticity in the visual cortex of adult mice (Sawtell et al. 2003). Studies in somatosensory cortex have utilized similar deprivation paradigms, such as whisker trimming in adult rodents. Here, a high degree of plasticity has also been reported, whereby stimulus-evoked responses to the spared whisker become overrepresented in comparison to undeprived animals (Diamond et al. 1993; Sellien and Ebner 2007).

Further classical examples of cortical plasticity induced by sensory deprivation involve work on the somatosensory system in humans and other primates. In non-human primates, substantial reorganization of somatosensory representations in the neocortex occurs following peripheral denervation by transection of the median nerve (Merzenich et al. 1983). In humans, such cortical plasticity has been observed in amputees, where receptive fields of neighboring cortical areas expand. It is believed that this reorganization underlies the sensation of a phantom limb elicited by the stimulation of other parts of the skin (Ramachandran 1993; Flor et al. 1995; Borsook et al. 1998). In patients suffering from syndactyly, surgical separation of webbed fingers leads to cortical reorganization correlating to the new functional status of the separated digits (Mogilner et al. 1993). Complementary to sensory deprivation, also environmental enrichment leading to increased sensory stimulation was shown to induce plastic rearrangements in the sensory cortex (Polley et al. 2004b).

In addition to gross manipulations of sensory inputs, plasticity in sensory cortex has also been investigated in a more physiological context: since learning is a prerequisite to form a novel memory, experimenters have devised several specific behavioral paradigms focusing on different aspects of learning and memory. In primary somatosensory cortex, repeated associative stimulation of two whiskers leads to the appearance of responses to a previously ineffective stimulus and to modifications of pre-existing stimulus responses (Delacour et al. 1987). Functional representations in barrel cortex increase in spatial extent after pairing whisker stimulation to an electric shock applied to the tail (Siucinska and Kossut 1996). Importantly, the enlargement can be potently reversed by extinction of the association memory through repeated presentations of the whisker stimulation without the foot shock (Siucinska and Kossut 1996). This plasticity mechanism is not restricted to aversive learning—reward learning can also trigger increases of cortical representations (Siucinska and Kossut 2004). Large-scale population recordings demonstrated that increases in response strength after associative fear learning are accompanied by a decrease in numbers of responding neurons in barrel cortex (Gdalyahu et al. 2012). For detailed reviews on the whisker system and barrel cortex, please see (Feldman and Brecht 2005, Margolis et al. 2014). In humans, cortical correlates of associative learning were observed after tactile co-activation on fingertips. Combining psychophysical assessment of discrimination ability with functional magnetic resonance imaging, it was shown that the cortical reorganization leading to an increased spatial separation and enlargement of the areas representing the finger tips is correlated with a lower spatial discrimination threshold (Dinse et al. 2003; Pleger et al. 2003).

Furthermore, plastic changes of neuronal responses related to task-relevant stimuli have been also demonstrated in associative areas of the neocortex. For example, neurons in visual association cortex of the cat modify their responses after associative learning (Morrell et al. 1983). Likewise, neural correlates of associative long-term memory have been described in primate temporal cortex (Miyashita 1988), and orbitofrontal cortex neurons in the rodent fire selectively to a rewarded stimulus after but not before learning (Schoenbaum et al. 1998; Lipton et al. 1999; Schoenbaum et al. 1999).

In operant conditioning, even neurons of premotor (Mitz et al. 1991) and motor cortex (Zach et al. 2008) develop learning-related plasticity. Interestingly, neurons in primary motor cortex can sensitize to stimuli of different modalities, such as color, if they are behaviorally important (Zach et al. 2008). Furthermore, activity of primary motor cortex neurons typically represents distinct movements, but the particular movement they represent can change during learning (Rokni et al. 2007; Huber et al. 2012).

Taken together, there is ample experimental evidence for plasticity in many cortical areas, including primary sensory cortex, that is induced by perturbations of sensory inputs as well as learning.

4 Behavioral Paradigms to Study Associative Memory in the Auditory Modality

For the rest of this chapter, we will focus on experimental paradigms and observations that have been made in small animal models, particularly rodents. These model organisms have become increasingly valuable due to the rapidly developing genetic, behavioral, and neurophysiological toolboxes available for the study of perception and memory. Moreover, we will discuss associative learning and memory in the context of the auditory modality, which allows exceptional control of the sensory stimuli.

A common approach in the investigation of plasticity mechanisms involves associative learning paradigms. Here, a previously behaviorally neutral stimulus (conditional stimulus, CS) is presented in a predictive manner to signal an innately behavioral important stimulus (unconditional stimulus, US) and experimental subjects form and memorize an association between these two initially unrelated stimuli through a process of reinforcement. In a simplistic view, associative learning can be divided into two basic processes. The first process involves the formation of a memory about the properties of the CS, which will be necessary for the recognition of this specific stimulus during a memory test. Eventually, repeated presentations of a CS can trigger processes that lead to improved perceptual ability, such as a lower detection threshold. The second process involves the linkage of the CS with a behavioral response, the so-called conditional response (CR), which is often used as a behavioral readout of learning. It is important to differentiate between these two processes, since both require the involvement of different, though overlapping brain areas. Two major forms of associative learning paradigms are usually distinguished: in classical conditioning, the sequential presentation of a CS preceding a US leads to associative learning (Pavlov and Anrep 1927), while in operant conditioning, subjects learn to emit a response which produces a certain outcome (such as the presentation of a food reward) (Skinner 1938). Forms of associative learning can be observed throughout the animal kingdom, including many invertebrates as well as all classes of vertebrates (Macphail 1982).

A commonly used paradigm is auditory cued fear conditioning (ACFC), where an animal learns to associate a specific sound (CS) with a US, e.g., an electric shock. Relatively few tone-shock pairings are needed in order to form an association, which can be measured by a defensive response to the CS—rodents typically exhibit freezing behavior. Albeit rapidly acquired, this form of memory can last several months to years (LeDoux 2000; Johansen et al. 2011; LeDoux 2014).

Classical ACFC offers the experimenter many degrees of freedom in the choice of stimulus (CS or US) and behavioral (e.g., freezing) and physiological readouts (e.g., heartbeat, blood pressure, release of stress hormones). The paradigm has been exploited extensively to identify and study the involved brain regions and to gain insight into the neural mechanisms underlying learning. It should be noted that the association between the CS and the US is not the only type of learning which is triggered by ACFC. An animal will also express fear when returned to the spatial context where the pairing between the auditory CS and US occurred. This phenomenon, called contextual fear conditioning, (Anagnostaras et al. 2001; Maren et al. 2013; Orsini et al. 2013) requires the hippocampus, which mediates the processing of spatial cues (O'Keefe 1979; Hartley et al. 2014; Geva-Sagiv et al. 2015). Presenting the auditory CS and monitoring a CR in a neutral context can provide a behavioral readout of the association made to the auditory CS.

Numerous studies in recent decades have implicated the amygdala, an almond-shaped small nucleus of the limbic system, in the formation and storage of a fear memory (LeDoux 2000; Johansen et al. 2011). The amygdala is subdivided into several subnuclei, and has been identified as a point of convergence of sensory pathways conveying information about the CS and US. Several behavioral and physiological output responses can be triggered by the amygdala. Previous research has demonstrated that plastic changes in synapses within the amygdala induced by ACFC mediate the formation of a fear memory. This includes LTP-like associative mechanisms enhancing neural responses to the auditory CS (Rogan et al. 1997; Repa et al. 2001), facilitation of synaptic transmission from the medial geniculate body (MGB) to the amygdala (McKernan and Shinnick-Gallagher 1997), increase in postsynaptic receptor trafficking following ACFC (Rumpel et al. 2005), and modulation of a fear memory by optogenetic manipulation of synaptic strength (Nabavi et al. 2014). Importantly, when the lateral amygdala and central nucleus of the amygdala are lesioned or inactivated by application of the GABA agonist muscimol, acquisition of a fear memory is impaired (Maren et al. 2001; Nader et al. 2001; Ciocchi et al. 2010).

The auditory cortex is also involved in ACFC, since auditory information reaches the amygdala via the auditory thalamus (specifically the medial division of the medial geniculate body, MGBm) and cortex. Findings from pre-training lesion studies using pure tones as the CS indicate that projections from the auditory cortex are not necessary for the acquisition of conditioned fear responses (LeDoux et al. 1984). However, similar results were also obtained when direct thalamo-amygdaloid projections were selectively lesioned, leaving thalamo-cortical projections and cortico-amygdaloid projections intact (Romanski and LeDoux 1992). Hence, either pathway is sufficient, but not necessary to support ACFC to pure tones. Whereas pre-training lesions are typically interpreted to assess the necessity of a brain structure for fear memory acquisition, post-training lesions assess the structure's necessity for successful memory recall and expression of fear-related behavioral responses. Post-training lesions showed a stronger effect on behavior when targeting cortical inputs to the amygdala (Campeau and Davis 1995; Boatman and Kim 2006). Interestingly, the ability to form a new association, as tested by a repeated CS-US pairing, was not affected (Boatman and Kim 2006; Peter et al. 2012; Moczulska et al. 2013). Furthermore, lesion studies were designed to not only investigate the acquisition and recall of a fear memory, but also the specificity of the memory. Here, fear-related responses to the conditioned sound (CS^+) were compared to responses to sounds that were not paired with the US to assess generalization and discrimination (Jarrell et al. 1987; Pearce 1994; Armony et al. 1997; Antunes and Moita 2010). It should be noted that the interpretation of results from lesion studies can be difficult, as the finding that a lesion or inactivation of a particular region can be functionally compensated for does not necessarily mean that it is not normally involved in a process.

Interestingly, if sounds with more complex spectrotemporal properties than pure tones of a single frequency are being used as CS, cortical pathways appear to be necessary (Ohl et al. 1999; Kholodar-Smith et al. 2008; Letzkus et al. 2011; Peter et al. 2012), leading to the hypothesis that auditory cortex crucially contributes to the perceptual analysis of complex sound stimuli, for which MGB function alone is not sufficient. Furthermore, auditory cortex is not only necessary for the acquisition of a fear memory of complex sounds, as shown in studies where the lesion was performed prior to the ACFC paradigm, but also for memory recall, as demonstrated by studies where lesions were performed after the ACFC paradigm (Moczulska et al. 2013).

ACFC has a major advantage for the analysis of memory-related processes as it is typically acquired with a single or few tone-shock pairings and thus allows a fast comparison of the involved circuits before and after learning. Carefully measuring changes in perception induced by a learning paradigm often requires the presentation of multiple stimuli over multiple repetitions. However, repeated presentations of a CS without further pairing with the US will lead to a new form of learning—extinction learning—which will lead to altered behavioral responses. By contrast, discrimination learning paradigms, which involve associative memory, typically allow the assessment of behavioral responses over hundreds of stimulus presentations. In a discrimination learning paradigm, neutral stimuli initially serve as a cue and are differentially reinforced in an operant manner. Discrimination learning involves different phases and qualities of learning, including perceptual learning and category learning (De Baene et al. 2008). A hallmark of the ability to form categories is the transfer of the learned features to novel stimuli (Lamberts 2002; Keri 2003). Hence, discrimination tasks offer the flexibility to not only study the subjectively perceived similarity or dissimilarity of stimuli, but also, after learning has occurred, to study the underlying mechanisms of spontaneous categorization of novel stimuli (Scheich et al. 2007).

A number of discrimination tasks, each designed to study specific neurobiological mechanisms, have been successfully implemented across several species, including humans (Keri 2003), non-human primates (Shadlen and Newsome 2001, Romo and Salinas 2003, Insabato et al. 2014), and rodents (Uchida et al. 2006; Stuttgen et al. 2011).

Typically in operant conditioning paradigms, two different sensory stimuli are presented in random sequence. A specific behavioral response is reinforced for each of the two stimuli. Since control of motivation is crucial in animal models, often a small food pellet or drop of water is used in deprived subjects as reward. Punishment is often achieved by a time-out or mild aversive stimulus like an air puff. For example, in a 'go/no-go' paradigm the 'go' cue signals the availability of a reward that can be obtained by approach to a port, whereas the same behavior in a 'no-go' trial can be punished. Thus, the optimal behavioral response in a 'no-go' trial is the suppression of a 'go' response. In a more symmetrical task design, such as the 'two alternative forced choice' task, the appropriate behavioral response to both kinds of trials is similar, and would typically involve the differential usage of two reward ports. Note that these paradigms are also called 'yes/no' tasks in the psychological literature (Stuttgen et al. 2011).

Subjects trained in discrimination paradigms have to successfully pass different phases of training (Irvine et al. 2000; Seger and Miller 2010). During initial learning, the animal adapts its behavior in a task-related manner, learning the association of the sensory cue and the particular type of trial, thereby increasing the amount of rewards obtained (Freedman and Miller 2008). In this phase, the animal has to map the sensory inputs to the appropriate motor behavior. In a final phase, when interested in categorization learning, so-called catch trials can be introduced (Salzman et al. 1990), to probe for spontaneous categorizations of novel stimuli.

When studying decision making and categorization, classically implemented by a 'two alternative forced choice' task, it is essential to study the link between sensory processing and motor representations involved in the initiation of an action, where several possible and overlapping routes connecting involved brain areas exist. In the auditory modality, several inputs to motor areas, either motor cortex or subcortical areas like basal ganglia or superior colliculus, have been anatomically described. In rodents there exists a collicular route, a thalamo-striatal route, and a cortico-cortical route (Pai et al. 2011). The collicular route involves the inferior colliculus (IC) in the mesencephalon, which receives auditory input from the superior olivary nucleus in the brain stem via the lateral lemniscus and can directly drive the superior colliculus, which controls unconscious motor reflexes (e.g., eye movements). The thalamo-striatal route involves direct projections from the MGB in the thalamus to the striatum. The striatum has been shown to be activated by auditory input and is implicated in action selection (Kimchi and Laubach 2009, Cui et al. 2013), reinforcement learning (Reynolds et al. 2001), and decision making (Ding and Gold 2010). The cortico-cortical route involves classical ascending pathways to the auditory cortex, which connects to prefrontal cortex, a non-sensory cortex involved in short-term memory (Yoon et al. 2008) and decision making (Kesner and Churchwell 2011), which in turn sends connections to motor cortex. Additionally, neurons in the primary auditory cortex projecting to the striatum are able to directly influence decisions in an auditory discrimination task where animals have to categorize sounds based on their frequency content (Znamenskiy and Zador 2013). Moreover, synaptic connections between cortico-striatal neurons and their striatal targets conveying task-relevant frequency information are selectively strengthened during learning (Xiong et al. 2015).

In summary, there are a number of well-defined behavioral paradigms that can be cued by auditory stimuli in rodents and allow the study of associative memory. These tasks recruit also other memory-related processes in addition to associative memory. At the circuit level, many brain structures have been identified to mediate these tasks, including circuits that do not directly involve the auditory cortex.

5 Functional Organization of Auditory Cortex

Considering a simplified view of unidirectional flow of sensory information (i.e., by ignoring feedback projections), the auditory cortex is the final target for afferents in the auditory modality. Similar to other sensory cortices, it can be subdivided in rodents into different areas, including primary auditory cortex (AI), secondary auditory cortex (AII), and the anterior auditory field (AAF) (Harel et al. 2000). In comparison with other sensory cortices, the auditory pathway is complex, bearing a high number of subcortical processing stations. Several aspects of a sound, like the spatial origin (Middlebrooks and Knudsen 1984; Brainard and Knudsen 1993; Palmer 2005; Singheiser et al. 2012) or temporal features (Frisina 2001), are analyzed subcortically. Sensory information reaches the cortex via the auditory thalamus, which itself receives inputs from two distinct pathways. The ventral division of the medial geniculate body (MGBv) belongs to the lemniscal pathway and is tonotopically organized. Neurons show narrow frequency tuning, reliably respond to repetitive stimuli, and project mainly to layer 4 of the auditory cortex (Morel and Kaas 1992; Guillery and Sherman 2011; Metzger et al. 2013). By contrast, MGBm belongs to the nonlemniscal pathway, is not tonotopically organized, and is multisensory. Neurons show broad frequency tuning, adapt to repetitive stimuli, and project diffusely, primarily to layer 2/3 of the primary auditory cortex, as well to structures of the limbic system (amygdala, insular cortex, striatum) (Ma and Suga 2009).

Single neurons in the auditory cortex are often characterized by their tuning in response to specific pure tone frequencies, originating from the functional architecture of the cochlea (Narayan et al. 1998). Commonly analyzed features are the characteristic frequency (CF), to which firing can be still observed at the lowest sound pressure levels, and a best frequency (BF) which is the particular pure tone frequency to which a neuron responds with the highest firing rate for a given sound intensity (Read et al. 2002). Prototypically, neurons of the primary auditory cortex have a V-shaped tuning curve, indicating a specific response to a given frequency at low sound pressure levels and an increasingly unselective response at high sound pressure levels which can span over five octaves (Recanzone et al. 2000). Particularly in unanaesthetized animals also other forms of tuning curves displaying for example level tolerance or O-shaped tuning have been described (Sutter 2000; Bizley et al. 2005; Hromadka et al. 2008; Sadagopan and Wang 2008, 2010; Parto Dezfouli and Daliri 2015).

Furthermore, auditory cortex shows a columnar organization with individual neurons sharing similar BFs and aural dominance along the vertical cortical axis (Imig and Adrian 1977; Guo et al. 2012a). Mechanistically, excitatory and inhibitory synaptic currents are co-tuned to sound frequency and intensity, whereby inhibitory currents arrive delayed by a few milliseconds, thereby balancing excitation and sharpening neural responses in time (Wehr and Zador 2003). During development, inhibitory and excitatory frequency tuning profiles are mismatched due to broader inhibitory tuning. Exposure to environmental sound stimuli leads to an increase in synaptic strength and enhanced coupling of excitatory and inhibitory inputs in young animals, and hence, reorganization of synaptic receptive fields of primary auditory cortex neurons (Dorrn et al. 2010).

It has been estimated that a single cortical neuron receives inputs from tens of thousands of synapses (DeFelipe and Farinas 1992), which mainly originate from nearby excitatory neurons, leading to the establishment of local microcircuits. Despite the fact that an individual neuron receives many inputs, the connectivity in local circuits has been shown to be rather sparse: the likelihood of detecting a direct connection between two neighboring neurons is typically less than 10–20 % (White 2007; Oswald and Reyes 2008; Ko et al. 2011; Levy and Reyes 2012; Cossell et al. 2015).

The strength of synapses between pairs of connected neurons varies substantially. The distribution of connection strengths in the cortex is highly skewed with many weak and only few strong synapses and can be approximated well with a log-normal function (Song et al. 2005). A similar observation has been made using chronic in vivo two-photon imaging of dendritic spines, the morphological correlates of the postsynaptic half of excitatoriy synapses impinging on pyramidal cells, in the auditory cortex of transgenic mice expressing a fluorescent protein in a subset of cortical neurons (Loewenstein et al. 2011). Synaptic connections in the living brain display substantial fluctuations in their strengths over the time course of a few days, as monitored in changes in the sizes of dendritic spines. Interestingly, the stationary log-normal distribution of spine sizes can be explained as an outcome of their ongoing multiplicative dynamics, i.e., that the amount of change depends on their initial size (Loewenstein et al. 2011). Interestingly, this form of plasticity can be observed even in the absence of explicit learning experiences. Furthermore, synaptic connections in the auditory cortex undergo turnover (Moczulska et al. 2013). Thus the number of synaptic connections that is observed at a given time point is determined by the rate of synapses being formed and eliminated, which is in balance in adult mice under basal conditions. These observations highlight that local cortical circuits are dynamic structures. However, it should be noted that despite substantial volatility, a fraction of connections will be maintained for years (Yang et al. 2009). Interestingly, turnover of synaptic connections is not entirely stochastic, and features that predict the longevity of synapses can be extracted (Loewenstein et al. 2015).

Excitatory synaptic inputs onto pyramidal neurons of layer 2/3 of the auditory cortex are functionally heterogeneous, as measured by calcium imaging in dendritic spines. Slightly less than half of spines show sound-evoked calcium responses and marked tuning to specific frequencies, as expected due to sparse representations and the high recurrent connectivity in cortical principal neurons (Chen et al. 2011). Neighboring spines on the same dendritic shaft show strong diversity in tuning. Most neighboring spines are tuned to different frequencies, consistent with a global computation strategy in auditory cortex (Chen et al. 2011). This is in contrast to findings from somatosensory cortex, which show marked functional clustering of spine activity (Takahashi et al. 2012). However, the pooled BFs of single spines are biased toward the best frequency of the pyramidal neuron, an important link for the understanding of how information from multiple diverse synaptic inputs is integrated and leads to the transformation of input information to create an output (Branco and Hausser 2010; Cossell et al. 2015).

In contrast to other primary sensory cortices like visual cortex, where many neurons can be driven to fire at high rates after presentation of optimal stimuli, e.g., moving gratings, representations in the auditory cortex have been considered sparse, perhaps as it has been difficult to identify such optimal stimuli (Hromadka and Zador 2009). In awake marmoset monkeys, a highly vocal species, most responsive neurons were strongly driven by amplitude- or frequency-modulated sounds rather than by pure tones (Wang et al. 2005). This finding highlights the different functional roles of auditory cortex for the perception of acoustic stimuli compared to subcortical areas, where the majority of neurons show strong frequency tuning (Ehret and Schreiner 2005; Lumani and Zhang 2010). In primary auditory cortex of the rat, it was shown that strong and persistent firing in response to sounds are present in less than 5 % of neurons, and half of the neurons tested showed no change in firing rate for any stimulus presentation (Hromadka et al. 2008). Moreover, the majority of responses of single neurons were extremely sparse, consisting often of only a single action potential, at least under anesthesia (DeWeese et al. 2003). It has been difficult to predict neuronal responses to a frequency-modulated complex sound based on a neuron's response properties assessed with pure tones (Machens et al. 2004; Bar-Yosef and Nelken 2007; Laudanski et al. 2012; Mizrahi et al. 2014). Much of this complexity in tuning likely arises by nonlinear interactions with other neurons within a local circuit. Considering the vast synaptic connections of single neurons within and outside of cortex (Douglas and Martin 2007), it will be crucial for the understanding of sensory coding to examine how response features of individual neurons are integrated on a population level.

6 Tonotopy in Auditory Cortex

A typical feature of primary sensory cortices is the spatial organization into topographically structured maps in discrete areas of the neocortex which reflects the local relationships of sensory receptors at the peripheral epithelium. These maps are characterized by nearby neurons sharing similar response properties. The auditory cortex shows tonotopic organization, in which neurons are organized along a gradient reflecting the frequency of the pure tone stimulus driving a response most effectively. Tonotopic organization originates in the cochlea where sound pressure waves are decomposed based on their frequency content (Narayan et al. 1998). Tonotopy is mostly conserved throughout the ascending central auditory pathways (Aitkin and Schuck 1985; Ross et al. 1988). Inputs from the MGBv reach primary auditory cortex in a tonotopically organized manner, leading to a fine frequency gradient along the cortical surface (Merzenich et al. 1973; Hackett et al. 2011; Guo et al. 2012b; Issa et al. 2014). The first demonstration of tonotopic organization of auditory cortex was provided in studies on cats from Woolsey and Walzl (1942) and has later been observed in many other species, including humans (Merzenich and Brugge 1973; Merzenich et al. 1973; Merzenich et al. 1976; Oliver et al. 1976; Morel and Kaas 1992; Binder et al. 1994; Pantev et al. 1995; Stiebler et al. 1997; Esser and Eiermann 1999; Harel et al. 2000; Talavage et al. 2000; Wallace et al. 2000; Morosan et al. 2001; Doron et al. 2002; Talavage et al. 2004; Bendor and Wang 2005; Langers and van Dijk 2012; Profant et al. 2013).

In mouse auditory cortex, four low-to-high frequency gradients can be identified (Issa et al. 2014). The individual gradients are likely corresponding to different subfields of the auditory cortex, namely AI, AAF, AII, and in the ultrasonic field (UF). Neurons in AI, AAF, and UF exhibit sensitive firing to low sound pressure levels, narrow tuning curves, and clearly detectable BF (Lee et al. 2004; Guo et al. 2012b). On the other hand, individual neurons of AII, and a separate associative area, the dorsoposterior field (DP), show broad frequency tuning and more complex response characteristics; hence, BFs are difficult to determine (Stiebler et al. 1997). Also, neurons in the DP have longer tone-evoked latencies compared with primary areas (Guo et al. 2012a, b; Joachimsthaler et al. 2014). The representation of stimulus features other than pure-tone tuning in a topographic manner appears to be largely absent or much less pronounced in the rodent auditory cortex (Kilgard and Merzenich 1999; Guo et al. 2012b).

Despite decades of research providing evidence of tonotopic organization of auditory cortex neurons, the presence of a tonotopic gradient on a population level of single cells and its functional relevance for sensory coding has been debated. Opinions are diverse and range from it being an organizational principle essential for sensory processing and perception to being an epiphenomenon only reflective of the structural architecture of the cochlea specifically the apex-to-base selectivity for increasing pure tone frequencies (Kaas 1997; Weinberg 1997; Bandyopadhyay et al. 2010; Rothschild et al. 2010; Pienkowski and Eggermont 2011; Aschauer and Rumpel 2014; Kanold et al. 2014).

7 Auditory Map Plasticity

Since cortical topographic maps have long been observed and well described across different sensory modalities, and tightly conserved in mammalian species, they have served as an attractive read out of cortical functional organization. In the following two sections, we will first discuss evidence for plastic functional reorganization of the auditory cortex at the level of cortical maps, and later at the level of individual neurons.

The first evidence for auditory map plasticity came from a study measuring uptake of radioactively labeled 2-deoxyglucose (2-DG) as proxy for neuronal activity. Pairing pure tone presentation with stimulation of the midbrain reticular formation—which leads to responses of the animal commonly seen with a mild aversive stimulus (bradycardia, rise in blood pressure, freezing)—2-DG uptake was especially high in areas corresponding to the topographic location of neurons with BFs near the CS frequency (Gonzalez-Lima and Scheich 1986). Recanzone and colleagues later demonstrated this phenomenon by training owl monkeys to discriminate small differences in the frequency of sequentially presented pure tones (Recanzone et al. 1993). Measurements of multiunit activity in auditory cortex revealed that cortical areas representing target tone frequencies were expanded in trained animals. Importantly, the amount of expansion was correlated with improved behavioral discrimination.

Following these investigations of tonotopic map plasticity in the context of discrimination learning, significant effort was put into the identification of mechanisms mediating this plasticity. Specifically, the action of the neuromodulator acetylcholine was the focus of numerous studies, first due to the strong projection of cholinergic neurons in the Nucleus Basalis (NB) of Meynert to the neocortex (Mesulam et al. 1983), and second, because of its long known critical role in cognition, learning and memory (Deutsch 1971; Dekker et al. 1991; Hasselmo 1995). In rodents, prolonged pairing of NB stimulation with presentation of a pure tone CS leads to large-scale tonotopic map plasticity (Kilgard and Merzenich 1998), thereby increasing representations of behaviorally relevant stimuli. This expansion leads to a compensatory decrease of spatial representations of other pure tone frequencies (Kilgard and Merzenich 1998). Pairing the stimulation of dopaminergic VTA neurons with pure tone presentations leads to an increase of the area representing the CS frequency, and representations of nearby frequencies are reduced in a compensatory manner—all of which is thwarted by the systemic application of dopamine antagonists (Bao et al. 2001).

In addition to primary neocortical areas, higher order associative areas also show topographic organization. Interestingly, and in contrast to NB pairing, Bao and colleagues found that VTA pairing leads to an apparent overall enlargement of primary auditory cortex, as defined by physiological parameters like sharpness of pure tone tuning. The reason for this is that neurons in the multimodal association area, called ventroposterior field (VPF), which normally do not show distinct frequency tuning, become sharply and almost exclusively tuned to the frequency of the paired CS. Map plasticity has been also monitored in the suprarhinal auditory field (Polley et al. 2006), and the posterior auditory field (PAF) (Puckett et al. 2007). Surprisingly, in contrast to primary auditory cortex, PAF shows a decrease in the number of neurons responding to the CS frequency (Puckett et al. 2007).

Since the first observation that the degree of map plasticity is correlated with improved behavioral discrimination (Recanzone et al. 1993; Rutkowski and Weinberger 2005), it was critically debated whether auditory map plasticity directly serves a specific learning-related function (Ohl and Scheich 2005; Weinberger 2015). Interestingly, only reinforcement, rather than the presentation of a stimulus alone, is sufficient to induce map plasticity (Polley et al. 2006). Furthermore, a large increase in representational area after associative conditioning coincides with resistance to extinction (Bieszczad and Weinberger 2010). Such evidence suggests a relationship between the increase in representational area and the strength of memory, but these findings were largely correlational.

Later studies utilized a more direct approach by analyzing the effect of NB pairing outside of a behavioral context and probing for learning thereafter (Reed et al. 2011). NB pairing before training leads to an accelerated acquisition in the task, which points toward a functional significance of auditory map plasticity. Intriguingly, behavioral performance remains stable, even though representational cortical areas revert to their original sizes over time (Reed et al. 2011). This evidence indicates that map expansion of behaviorally relevant stimuli is not the way by which improved perceptual skills are stored in the neocortex, but is rather functionally significant for learning.

8 Plasticity in Functional Properties of Individual Neurons

The impact of learning paradigms on the functional properties of individual neurons or small local groups of neurons, rather than large-scale tonotopic maps, has also been investigated extensively. One of the first studies providing evidence for learning-induced plasticity in the auditory cortex came from Robert Galambos and colleagues. In this study, recordings of local field potentials, reflecting the summed synaptic currents from multiple nearby neurons, showed that sound-evoked responses in the auditory cortex are larger and more reliable in cats who were subjected to ACFC paradigms, compared with untrained animals (Galambos et al.

1956). This finding was corroborated by multiunit recordings (Buchwald et al. 1966), and also by studies of single units (Weinberger et al. 1984). In contrast to the global increase in the LFP signal, changes in sound-evoked activity of single cells were quite heterogeneous. Many cells increased their stimulus-evoked firing, while others showed a marked decrease or no change at all (Weinberger et al. 1984).

Neurons in the auditory cortex can exhibit plasticity not only in responsiveness, but also in frequency tuning. Interestingly, tuning curves of individual neurons can shift according to the frequency of a pure tone CS. Moreover, neurons can increase their firing to the CS and decrease in response to unpaired pure tone stimuli (Bakin and Weinberger 1990). Relatively few CS-US pairings are sufficient to induce plasticity, which consolidates over time, is temporally remarkably stable, and is retained for at least 10 days (Edeline et al. 1993; Galvan and Weinberger 2002). However, the mentioned plasticity features are only present in a subset of neurons (Bakin and Weinberger 1990). Remarkably, ACFC-induced plasticity in auditory cortex even persists after extensive extinction training, suggesting a role for long-term storage of memories (Quirk et al. 1997). This is in contrast to neurons in the lateral amygdala, which revert their learning-induced increase in firing to the CS to baseline levels after extinction (Quirk et al. 1997; LaBar et al. 1998; Herry et al. 2008). It should be noted that neurons in other regions of the auditory pathway, like MGB and IC, show plasticity in response to ACFC as well (Edeline 1990; Yan and Suga 1998; Maren et al. 2001). Changes in the tuning properties of auditory cortex neurons, in principle, could be fully accounted for by plastic changes in brain areas projecting to the auditory cortex. However, as we will discuss in the following sections, there is direct evidence for structural plastic changes in the circuits of the auditory cortex itself.

Plasticity in auditory tuning is not only present after ACFC, but with other associative learning tasks, like instrumental learning, where the CS does not elicit a fear response (Bakin et al. 1996). Response plasticity is particularly strong if the stimuli are paired with a reinforcer, be it aversive (e.g., foot shock) or rewarding (e.g., food pellet), and mere presentation of the stimulus is not sufficient to induce these changes (Blake et al. 2002). Seeing similar forms of plasticity independent of how they are reinforced is consistent with a hypothesis that changes in tuning curves reflect learning of specific stimulus properties rather than the associative linking of the stimulus to a particular behavior or emotional state. Subjects' performance in a pure tone detection task correlates with changes in frequency tuning of single neurons (Blake et al. 2002), and the strength of plasticity effects is also well correlated with inter-individual differences in learning (Fritz et al. 2003). More insights into the temporal dynamics of tuning shifts were gained in a study where ferrets were trained in an aversive conditioning task with a flexible CS, forcing animals to quickly adapt to the new task requirements (Fritz et al. 2003).

In addition to behavioral relevance, internal states like attentiveness and expectation can significantly alter perception (Petkov et al. 2004; Fritz et al. 2007). If task-relevant stimuli are presented in a similar manner as during training, task performance is significantly better than in the context of flexible task designs. It has been demonstrated that this phenomenon, which reflects a learned anticipation, is reflected in neuronal activity. Namely, even the presentation of behaviorally irrelevant sounds can lead to increased evoked responses if they are presented at a time point where normally the reinforced stimuli occur (Jaramillo and Zador 2011).

The impact the cholinergic modulatory system on the functional properties of single cells or small groups of cells has also been investigated. Sound-evoked firing rates are facilitated following co-activation of NB in auditory cortex (Hars et al. 1993). Repeated presentations of a pure tone stimulus, which normally does not lead to long-lasting changes in tuning properties, induces BF shifts when presented with local application of agonists of muscarinic acetylcholine receptors (McKenna et al. 1989). Furthermore, acetylcholine application can specifically enhance changes in tuning that are observed during ACFC (Ji et al. 2001). This enhancement of plasticity is not seen when acetylcholine is applied in conjunction with atropine, an antagonist of muscarinic acetylcholine receptors (Ji et al. 2001). At the behavioral level, associative learning to sounds is impaired after NB lesions or blockade of acetylcholine signaling in the cortex (Butt and Hodge 1997; Butt et al. 2009).

Additional evidence for the cholinergic system's role in controlling plasticity comes from experiments in echolocating bats, where plastic changes in tuning of auditory cortex neurons during conditioning have been observed (Gao and Suga 2000). While focal electrical stimulations of the auditory cortex alone lead to transient BF shifts (Chowdhury and Suga 2000), co-application of acetylcholine induces more long-lasting BF shifts (Ma and Suga 2005). Interestingly, focal electrical stimulation of the auditory cortex can also change BFs of neurons in the inferior colliculus (IC), indicative of substantial top-down control in the auditory pathway (Gao and Suga 1998). However, in contrast to neurons of auditory cortex, the shifts in the BF of IC neurons typically recover over time, whether or not acetylcholine is exogenously administered (Ma and Suga 2005).

Other neuromodulatory systems have also been implied in the control of plasticity in auditory cortex. The locus coeruleus (LC) is the major source of noradrenergic projections to the cortex. Repeated LC stimulation with simultaneous presentation of a pure tone induced BF shifts in about 30 % of auditory cortex neurons, showing either selective increases or decreases in response to the paired pure tone (Edeline et al. 2011). Similarly, pairing stimulation of the peripheral vagus nerve with pure tone trains of a specific repetition rate increases neuronal spiking to the paired rate of pure tone trains (Engineer et al. 2011; Shetake et al. 2012).

Frequency tuning is not the only functional feature showing plastic changes induced by the learning of a behaviorally relevant association to a stimulus. Encoding of sound intensity by primary auditory cortex neurons was changed when rats were trained to associate specific sound intensities with a food reward in an instrumental or classical conditioning paradigm (Polley et al. 2004a). Additionally, training in a temporal rate discrimination task, where animals successfully learn to find the spatial location of sounds presented with high repetition rates in a maze, leads to enhanced cortical responses to high-rate noise pulses, irrelevant of the sound frequency. Also, these changes take place in neurons irrelevant of their CF (Bao et al. 2004).

In the previous two sections, we summarized evidence for functional plasticity in the auditory cortex, at the level of tonotopic maps as well as individual neuron that is induced by learning. This corroborates the idea that primary sensory cortices play a role in the formation of long-term associative memories and implicates neuromodulatory systems in regulating cortical plasticity.

9 Microcircuit Mechanisms Gating Plasticity

Many recent studies have considered the local circuit mechanisms mediating and controlling learning-induced plasticity in auditory cortex. It has long been known that cortical interneurons play a critical role in regulating the activity of excitatory principal cells. Under normal conditions, excitatory and inhibitory inputs to auditory cortex neurons are largely co-tuned (Wehr and Zador 2003, Tan et al. 2004). After pharmacological blockage of inhibitory connections, sensory tuning functions broaden substantially (Sillito 1975; Chen and Jen 2000; Kurt et al. 2006). A differential regulation of inhibition and excitation has been reported following the pairing of sound CS with NB stimulation, mediating a shift in the neuron's BF (Froemke et al. 2007). Using pharmacologic and optogenetic manipulations, it was demonstrated that interneurons in layer I of primary auditory cortex play a crucial role in the processing of aversive stimuli (Letzkus et al. 2011). The activity of layer 1 interneurons is tightly controlled by cholinergic fibers from NB. After an electric foot shock, the majority of layer I interneurons show an increase in firing, presumably through a direct synaptic connection with fibers releasing acetylcholine. In turn, a different type of interneuron in layer 2/3, tonically active Parvalbumin positive (PV^+) interneurons (Markram et al. 2004), decreases their firing in response to a foot shock, suggesting a synaptic connection from layer 1 interneurons which inhibit PV^+ interneurons in layer 2/3. This leads to a disinhibition of pyramidal cells in layer 2/3, which normally receive perisomatic inhibition from PV^+ cells. Thus, layer 1 interneurons can act as a 'gatekeeper' for the induction of activity-dependent plasticity in the auditory cortex.

It was later demonstrated that a similar disinhibitory mechanism as put forward by Letzkus and colleagues (Letzkus et al. 2011) exists also in visual cortex and medial prefrontal cortex, suggesting this is a general cortical circuit motif (Pi et al. 2013). Precisely, a subgroup of inhibitory neurons, genetically identified by their strong expression of vasoactive intestinal polypeptide (VIP) (Markram et al. 2004), functions in a related manner in response to a reinforcement signal (Pi et al. 2013). VIP-positive interneurons are activated by long-range cholinergic fibers originating from NB in the basal forebrain, which themselves respond vigorously to an aversive stimulus. The VIP-positive interneurons themselves are connected to two other types of interneurons, the PV+, and the somatostatin positive, thereby leading to a disinhibition of excitatory pyramidal neurons. All in all, VIP-positive interneurons perform gain control in local cortical processing, thus enabling plasticity mechanisms during learning (Pi et al. 2013).

Behaviorally relevant stimuli from multiple modalities can impact auditory cortical activity through these non-auditory, reinforcement-related circuits. For example, expression of immediate early genes is often used as a signature of increased neuronal activity. mRNA levels of immediate early genes show low basal expression levels, and are heavily transcribed after a stimulus drives neuronal firing. Many immediate early genes code for transcription factors, like c-fos or zif-268, which in turn control the expression of effector genes and are therefore crucial for network plasticity (Donato et al. 2013). Expression of immediate early genes is strongly increased in auditory cortex after ACFC, but also after the application of a foot shock alone—a stimulus which is purely somatosensory by nature and commonly used as a control in an ACFC experiment (Peter et al. 2012).

10 Synaptic Plasticity During Learning

There is considerable evidence that modulatory inputs play an important role in controlling the extent to which auditory stimuli can induce plastic changes in cortical circuits. Such plastic changes could be a neural correlate of associative memory. Froemke and colleagues demonstrated that receptive field plasticity resulting from pairing a sound with NB stimulation involved the long-lasting modification of cortical inhibitory and excitatory synaptic currents, while inputs from thalamus remained unchanged (Froemke et al. 2007). Evidence from current source density analyses allowing measurements of synaptic activity with laminar resolution suggests a higher degree of synaptic plasticity in layers 2/3 than in the primary input layer 4 (Guo et al. 2013). Pairing NB stimulation with pure tone presentation induces an increase in excitatory currents at the paired input frequency and intensity, and decreases at the original BF and best intensity, whereby the overall amount of excitation remains conserved. In contrast to excitation, NB pairing leads to an initial decrease in inhibition at all frequencies, and selectively recovers at the paired sound frequencies, leading to a rebalancing of excitatory and inhibitory currents. Thus, NB pairing is able to transiently destabilize cortical receptive fields by enhancing responses to behaviorally relevant stimuli (Froemke et al. 2013). Spike-timing-dependent plasticity mechanisms were shown to contribute to the normalization of excitatory–inhibitory balance, since they lead to the modification of inhibitory and excitatory synapses. Thus, regardless of the initial relative tuning of excitation and inhibition, repetitive spike pairing can effectively enhance co-tuning with that of the postsynaptic neuron (D'Amour and Froemke 2015).

As mentioned above, the advent of chronic imaging techniques has allowed unprecedented views on the dynamics of synaptic connections in the living brain (Bhatt et al. 2009; Holtmaat and Svoboda 2009). These techniques have been also applied to investigate the impact of learning experience on synapses in vivo (Xu et al. 2009; Yang et al. 2009; Caroni et al. 2012). When combining imaging of dendritic spines with auditory fear conditioning, a transient shift in the balance of spine formation and elimination was observed (Moczulska et al. 2013). Paired conditioning led to the increased formation of spines in the auditory cortex, which left a trace in the network that persisted for more than a week. Interestingly, this observation resembled previous reports on the effects of motor skill learning on dendritic spines in the primary motor cortex of mice (Xu et al. 2009; Yang et al. 2009), but is in contrast to findings in the frontal association cortex during ACFC. In these studies, an increase in spine elimination was observed during fear memory acquisition, but similar increases in spine formation were only observed during extinction learning (Lai et al. 2012). These results corroborate the notion that even a relatively simple learning paradigm like ACFC affects multiple areas in the brain differentially. However, although the observation of plastic rearrangements of sensory cortical circuits during learning experiences could be interpreted as a signature of the formation of an associative memory, the functional significance is not yet clear.

The study of synaptic dynamics in the brain may nevertheless provide some insight on the processes mediating learning and memory. It is commonly observed in both humans and laboratory models that individuals vary substantially in the amount of time required to become skilled at a task. In a combination of theoretical modeling and psychophysics using a 'go/no-go' discrimination task in mice, it was shown that a reinforcement learning model endowed with a multiplicative learning rule captures the variability in individual learning curves (Bathellier et al. 2013). Considering that synapse turnover also displays multiplicative dynamics (Loewenstein et al. 2011), it is believed that the two processes could be causally related.

11 Category Learning

In previous sections we discussed current evidence for synaptic and functional plasticity in sensory cortices in the context of behavioral paradigms that involve associative memory. Stimulus categorization is a cognitive task requiring integration of information about multiple sensory and mnemonic aspects. (Liberman et al. 1967). Categorical perception is a fundamental aspect of how the world is perceived and it occurs so effortlessly that one does not tend to consider this as a difficult task unless trying to teach a computer to do so. When subjects are asked what sound was first heard on a given day, typical answers are 'my alarm clock' or 'birds' or 'my snoring husband'. The answers are not 'a sound with rhythmic structure in the low frequency bands'. The naming of the sound source, rather than a description of the sound itself, reflects categorical perception. In order to understand processes of perception, it is fundamental to reveal underlying neural mechanisms of how such categories are formed. Despite our knowledge on how the brain detects basic stimulus features, like motion direction in vision, or frequency in audition, much less is known about how the brain learns, stores, and recalls categories (Nelken et al. 2003; Russ et al. 2007; Scheich et al. 2007; Freedman and Miller 2008; Brosch et al. 2015).

In psychological theories that describe underlying mechanisms during a categorization task, it is assumed that an animal has stored the representation of many different categories through previous encounters with members of such categories. If an unfamiliar stimulus is encountered, at least three different tasks have to be carried out. The first task is the perception of the stimulus, where features relevant for the categorization have to be analyzed. Next, the stimulus has to be compared to previously stored representations of all relevant categories, before finally, a decision has to be made as to which category the unfamiliar stimulus is assigned (Lamberts 2002).

Early evidence for the involvement of auditory cortex during categorization comes from studies in Mongolian Gerbils (Ohl et al. 2001, 2003; Scheich et al. 2007; Deliano et al. 2009)—a commonly used model organism for auditory research due their high sensitivity to perceive sounds at low frequencies that are important for human speech, and their behavioral readiness in auditory tasks (Stuermer et al. 2003; Gleich and Strutz 2012). In a psychophysical experiment in which rising or falling frequency-modulated tones had to be categorized, individual animals abruptly developed the ability to spontaneously categorize novel stimuli at some variable time point during the training period. Strikingly, category-specific spatiotemporal activity patterns of cortical neurons emerged at a similar timing as the behavioral switch. Moreover, activity patterns evoked by sounds belonging to the same category shared a high degree of similarity (Ohl et al. 2001). This suggests the formation of an internal representation in primary auditory cortex reflecting the structure of the category, rather than specific features of the individual stimulus (Selezneva et al. 2006). This finding demonstrates that primary auditory cortex does not only play a role in the first phase of categorization—perception—but also in the second phase, where sensory input is integrated into the memory of already stored categories.

A necessary component of categorization is the assignment of many varying inputs to the same category. Interestingly, activity of local neuronal ensembles in the superficial layers of the primary auditory cortex of mice in response to a large set of sound stimuli (pure tones, as well as frequency-modulated complex sounds) reveals only a limited repertoire of sound-evoked population activity patterns (Bathellier et al. 2012). These so-called response modes are superposed on high trial-to-trial variability (Hromadka et al. 2008). When presenting gradually changing mixtures of two sounds that alone would drive either the one or the other response mode, the population response changed abruptly from one mode to the other (Bathellier et al. 2012). These nonlinear 'winner-takes-all' dynamics share properties with artificial neuronal network models of cortical circuits (Hopfield 1982; Amit and Brunel 1997; Maass et al. 2007; Mongillo et al. 2008; Wang 2008). Despite having a restricted repertoire at the local scale, activity patterns recorded from multiple ensembles are sufficiently different to form a global representation that allows prediction of spontaneous categorization behavior in mice. An interesting aspect of these observations may be that the essential units for sensory coding are not single neurons, but neuronal ensembles with nonlinear response properties (Maass and Zador 1999).

Another model used to study categorization in the auditory modality is the processing of conspecific animal vocalizations. Much research has been aimed to identify mechanisms in the auditory pathway that underlie selective processing of vocalization calls. Primary auditory cortical neurons of the common marmoset respond much more vigorously to natural vocalizations than to artificially modified stimuli (Wang et al. 1995). If marmoset vocalizations were presented to a different species, there was no such response selectivity present (Wang and Kadia 2001). This points toward a categorization mechanism, which might be innate or acquired through experience (Cheung et al. 2005). When such vocalizations are presented to naïve individuals who have never been exposed to these vocalizations before, preferences for natural over artificial vocalizations are not present, which again illustrates the importance of experience for this categorization behavior (Schnupp et al. 2006). Thus, it appears that primary auditory cortex is able to represent behaviorally relevant acoustic features through discrete patterns of neuronal activity.

One hallmark of cortical population activity in response to a given stimulus is its high trial-to-trial variability (Hromadka et al. 2008). Interestingly, response variability to a fixed stimulus is often correlated across populations of neurons. These ‘noise correlations’ can be measured via paired recordings, alongside ‘signal correlations,’ which describe the similarity in the average spike response to a stimulus. Noise correlations can be found ubiquitously in the nervous system (deCharms and Merzenich 1996; Romo et al. 2003; Romo and Salinas 2003; Bruno and Sakmann 2006; Ecker et al. 2010). Noise correlations may play a role in sensory coding, since the concomitant responses of multiple sensory neurons must be integrated to generate a percept. The analysis of synaptic currents during NB pairing with a pure tone frequency in rats revealed that both signal and noise correlations in EPSCs can be plastically modified, by increasing in mutual information while trial-to-trial variability is decreased (Froemke et al. 2013). Interestingly, similar observations have been made in the auditory system of starlings during associative learning (Jeanne et al. 2013).

The activity of local populations of neurons, or cell assemblies (Hebb 1949), provides an attractive strategy for sensory coding (Harris et al. 2003; Buzsaki 2010; Harris and Mrsic-Flogel 2013). The activity of single neurons in sensory cortices reflects the detection of specific features of an object, but in order to perceive the whole object, the cooperative interaction of larger cell assemblies can provide the relevant information. Here, single neurons are able to participate in different combinations to form a plastic assembly, incorporating their individual and diverse response characteristics (Freiwald et al. 2001).

12 Future Challenges

In this chapter we reviewed current evidence for the role of sensory cortex in mediating associative memory. In contrast to the traditional view that primary sensory cortices are merely involved in the extraction of specific features of the environment, modern research has provided strong evidence for the importance of its function in higher cognitive tasks as well. Hence, it will be crucial to identify a link between the well-described local circuit phenomena taking place during learning and the remarkable behavioral flexibility of mammals and other animals to adapt to an ever-changing environment. The activity of single cortical neurons has long been shown to code for specific aspects of stimuli, such as frequency in the auditory modality or orientation in the visual modality. However, for the achievement of complex perceptual tasks, like discrimination and categorization, it seems essential to understand the combined activity of larger populations of neurons functioning in cell assemblies (Harris and Mrsic-Flogel 2013; Ohl 2015), since tuning curves of single neurons alone or large-scale topographic maps are not able to account for these processes.

Hebbian theory provides a model of how the combined activity of an interconnected group of neurons, the cell assembly, can form a basis for cognitive processes, like memory or decision making. Through the development of modern techniques, like multielectrode arrays or in vivo imaging, it has become possible to simultaneously record the activity of large populations of neurons with single-cell resolution, and these methods will generate new insights into the mechanisms of coordinated activity in ensembles. Recent findings highlighting cortical circuits as dynamic structures, an emerging question in research of sensory coding, concern the durability of a cell assembly that may convey functional stability (Buzsaki 2010; Lutcke et al. 2013). Most current data on the activity of large populations of neurons is limited to a snap shot in the lifetime of an animal. In the future, it will be necessary to study the stability of representations on a population level, focusing on responses to environmentally relevant stimuli.

However, before even asking how functional ensembles are maintained in a dynamic structure, one may wonder what are the processes that lead to their formation in the first place? Two sources of plasticity could give rise to the formation of new assemblies underlying a new memory or category. Plasticity directly correlated to behaviorally relevant learning experiences could lead to the formation of a new assembly, similar to the Platonic concept of a memory as the stamping of an image into a wax tablet (Plato and Campbell 1883). However, cortical circuit dynamics under basal conditions could lead to the formation of novel neuronal assemblies even before the actual learning experience. Therefore, the additional wave of synaptic rearrangements observed during learning may specifically reflect the selection and stabilization of a group of neurons useful for the adaptation of sensory processing. Future studies utilizing longitudinal functional imaging with cellular resolution in learning animals may allow the disambiguation of these scenarios.

The biggest future challenge, perhaps, is related to the question how memorized or also novel stimuli lead to an actual percept. Most of what we have learned about the processes in sensory cortex during perception stems from studies that have monitored the activity of single or many neurons under different conditions of sensory stimulation or behavioral contexts. As such, they are correlative in nature and allow only limited conclusions on the neuronal mechanisms that underlie a sensory percept. For example, it is unclear if the activity of all responsive neurons in the sensory cortex, or only a subset, is important to evoke a percept. In the future it will be necessary to complement these studies with others of a new kind in which the subjective perception of artificially generated activity patterns is measured, in order to reproduce and thus understand the mechanisms underlying the sensation of a scent, an image, or a song.

Acknowledgments The authors would like to thank Dr. A. Chambers, Dr. M. Stüttgen and Dr. M. Kaschube for comments on the manuscript.

References

Aitkin L, Schuck D (1985) Low frequency neurons in the lateral central nucleus of the cat inferior colliculus receive their input predominantly from the medial superior olive. Hear Res 17(1):87–93

Amit DJ, Brunel N (1997) Model of global spontaneous activity and local structured activity during delay periods in the cerebral cortex. Cereb Cortex 7(3):237–252

Anagnostaras SG et al (2001) Hippocampus and contextual fear conditioning: recent controversies and advances. Hippocampus 11(1):8–17

Antunes R, Moita MA (2010) Discriminative auditory fear learning requires both tuned and nontuned auditory pathways to the amygdala. J Neurosci 30(29):9782–9787

Armony JL et al (1997) Stimulus generalization of fear responses: effects of auditory cortex lesions in a computational model and in rats. Cereb Cortex 7(2):157–165

Aschauer DF, Rumpel S (2014) Measuring the functional organization of the neocortex at large and small scales. Neuron 83(4):756–758

Ashby FG, O'Brien JB (2005) Category learning and multiple memory systems. Trends Cogn Sci 9(2):83–89

Bakin JS et al (1996) Induction of receptive field plasticity in the auditory cortex of the guinea pig during instrumental avoidance conditioning. Behav Neurosci 110(5):905–913

Bakin JS, Weinberger NM (1990) Classical conditioning induces CS-specific receptive field plasticity in the auditory cortex of the guinea pig. Brain Res 536(1–2):271–286

Bandyopadhyay S et al (2010) Dichotomy of functional organization in the mouse auditory cortex. Nat Neurosci 13(3):361–368

Bao S et al (2001) Cortical remodelling induced by activity of ventral tegmental dopamine neurons. Nature 412(6842):79–83

Bao S et al (2004) Temporal plasticity in the primary auditory cortex induced by operant perceptual learning. Nat Neurosci 7(9):974–981

Bar-Yosef O, Nelken I (2007) The effects of background noise on the neural responses to natural sounds in cat primary auditory cortex. Front Comput Neurosci 1:3

Bathellier B et al (2013) A multiplicative reinforcement learning model capturing learning dynamics and interindividual variability in mice. Proc Natl Acad Sci U S A 110(49):19950–19955

Bathellier B et al (2012) Discrete neocortical dynamics predict behavioral categorization of sounds. Neuron 76(2):435–449

Bendor D, Wang X (2005) The neuronal representation of pitch in primate auditory cortex. Nature 436(7054):1161–1165

Bhatt DH et al (2009) Dendritic spine dynamics. Annu Rev Physiol 71:261–282

Bieszczad KM, Weinberger NM (2010) Representational gain in cortical area underlies increase of memory strength. Proc Natl Acad Sci U S A 107(8):3793–3798

Binder JR et al (1994) Functional magnetic resonance imaging of human auditory cortex. Ann Neurol 35(6):662–672
Bizley JK et al (2005) Functional organization of ferret auditory cortex. Cereb Cortex 15 (10):1637–1653
Blake DT et al (2002) Neural correlates of instrumental learning in primary auditory cortex. Proc Natl Acad Sci U S A 99(15):10114–10119
Bliss TV, Lomo T (1973) Long-lasting potentiation of synaptic transmission in the dentate area of the anaesthetized rabbit following stimulation of the perforant path. J Physiol 232(2):331–356
Boatman JA, Kim JJ (2006) A thalamo-cortico-amygdala pathway mediates auditory fear conditioning in the intact brain. Eur J Neurosci 24(3):894–900
Borsook D et al (1998) Acute plasticity in the human somatosensory cortex following amputation. NeuroReport 9(6):1013–1017
Brainard MS, Knudsen EI (1993) Experience-dependent plasticity in the inferior colliculus: a site for visual calibration of the neural representation of auditory space in the barn owl. J Neurosci 13(11):4589–4608
Branco T, Hausser M (2010) The single dendritic branch as a fundamental functional unit in the nervous system. Curr Opin Neurobiol 20(4):494–502
Brosch M et al (2015) Neuronal activity in primate auditory cortex during the performance of audiovisual tasks. Eur J Neurosci 41(5):603–614
Bruno RM, Sakmann B (2006) Cortex is driven by weak but synchronously active thalamocortical synapses. Science 312(5780):1622–1627
Buchwald JS et al (1966) Changes in Cortical and Subcortical Unit Activity during Behavioral Conditioning. Physiol Behav 1(1):11
Butt AE et al (2009) Association learning-dependent increases in acetylcholine release in the rat auditory cortex during auditory classical conditioning. Neurobiol Learn Mem 92(3):400–409
Butt AE, Hodge GK (1997) Simple and configural association learning in rats with bilateral quisqualic acid lesions of the nucleus basalis magnocellularis. Behav Brain Res 89(1–2):71–85
Buzsaki G (2010) Neural syntax: cell assemblies, synapsembles, and readers. Neuron 68(3):362–385
Campeau S, Davis M (1995) Involvement of subcortical and cortical afferents to the lateral nucleus of the amygdala in fear conditioning measured with fear-potentiated startle in rats trained concurrently with auditory and visual conditioned stimuli. J Neurosci 15(3 Pt 2):2312–2327
Caroni P et al (2012) Structural plasticity upon learning: regulation and functions. Nat Rev Neurosci 13(7):478–490
Chen QC, Jen PH (2000) Bicuculline application affects discharge patterns, rate-intensity functions, and frequency tuning characteristics of bat auditory cortical neurons. Hear Res 150 (1–2):161–174
Chen X et al (2011) Functional mapping of single spines in cortical neurons in vivo. Nature 475 (7357):501–505
Cheung SW et al (2005) Plasticity in primary auditory cortex of monkeys with altered vocal production. J Neurosci 25(10):2490–2503
Chowdhury SA, Suga N (2000) Reorganization of the frequency map of the auditory cortex evoked by cortical electrical stimulation in the big brown bat. J Neurophysiol 83(4):1856–1863
Ciocchi S et al (2010) Encoding of conditioned fear in central amygdala inhibitory circuits. Nature 468(7321):277–282
Cossell L et al (2015) Functional organization of excitatory synaptic strength in primary visual cortex. Nature 518(7539):399–403
Courtney SM et al (1997) Transient and sustained activity in a distributed neural system for human working memory. Nature 386(6625):608–611
Cui G et al (2013) Concurrent activation of striatal direct and indirect pathways during action initiation. Nature 494(7436):238–242
D'Amour JA, Froemke RC (2015) Inhibitory and excitatory spike-timing-dependent plasticity in the auditory cortex. Neuron 86(2):514–528

De Baene W et al (2008) Effects of category learning on the stimulus selectivity of macaque inferior temporal neurons. Learn Mem 15(9):717–727
deCharms RC, Merzenich MM (1996) Primary cortical representation of sounds by the coordination of action-potential timing. Nature 381(6583):610–613
DeFelipe J, Farinas I (1992) The pyramidal neuron of the cerebral cortex: morphological and chemical characteristics of the synaptic inputs. Prog Neurobiol 39(6):563–607
Dekker AJ et al (1991) The role of cholinergic projections from the nucleus basalis in memory. Neurosci Biobehav Rev 15(2):299–317
Delacour J et al (1987) "Learned" changes in the responses of the rat barrel field neurons. Neuroscience 23(1):63–71
Deliano M et al (2009) Auditory cortical activity after intracortical microstimulation and its role for sensory processing and learning. J Neurosci 29(50):15898–15909
Desimone R (1996) Neural mechanisms for visual memory and their role in attention. Proc Natl Acad Sci U S A 93(24):13494–13499
Deutsch JA (1971) The cholinergic synapse and the site of memory. Science 174(4011):788–794
DeWeese MR et al (2003) Binary spiking in auditory cortex. J Neurosci 23(21):7940–7949
Diamond ME et al (1993) Experience-dependent plasticity in adult rat barrel cortex. Proc Natl Acad Sci U S A 90(5):2082–2086
Dick AO (1974) Iconic Memory and Its Relation to Perceptual Processing and Other Memory Mechanisms. Percept Psychophys 16(3):575–596
Ding L, Gold JI (2010) Caudate encodes multiple computations for perceptual decisions. J Neurosci 30(47):15747–15759
Dinse HR et al (2003) Pharmacological modulation of perceptual learning and associated cortical reorganization. Science 301(5629):91–94
Donato F et al (2013) Parvalbumin-expressing basket-cell network plasticity induced by experience regulates adult learning. Nature 504(7479):272–276
Doron NN et al (2002) Redefining the tonotopic core of rat auditory cortex: physiological evidence for a posterior field. J Comp Neurol 453(4):345–360
Dorrn AL et al (2010) Developmental sensory experience balances cortical excitation and inhibition. Nature 465(7300):932–936
Douglas RJ, Martin KA (2007) Recurrent neuronal circuits in the neocortex. Curr Biol 17(13): R496–500
Ecker AS et al (2010) Decorrelated neuronal firing in cortical microcircuits. Science 327 (5965):584–587
Edeline JM (1990) Frequency-specific plasticity of single unit discharges in the rat medial geniculate body. Brain Res 529(1–2):109–119
Edeline JM et al (2011) Induction of selective plasticity in the frequency tuning of auditory cortex and auditory thalamus neurons by locus coeruleus stimulation. Hear Res 274(1–2):75–84
Edeline JM et al (1993) Rapid development of learning-induced receptive field plasticity in the auditory cortex. Behav Neurosci 107(4):539–551
Ehret G, Schreiner CE (2005) Spectral and intensity coding in the auditory midbrain. In: Winer JA, Schreiner CE (eds)The inferior colliculus. Springer New York, New York, pp 312–345
Engineer ND et al (2011) Reversing pathological neural activity using targeted plasticity. Nature 470(7332):101–104
Esser KH, Eiermann A (1999) Tonotopic organization and parcellation of auditory cortex in the FM-bat Carollia perspicillata. Eur J Neurosci 11(10):3669–3682
Feldman DE, Brecht M (2005) Map plasticity in somatosensory cortex. Science 310(5749):810–815
Flor H et al (1995) Phantom-limb pain as a perceptual correlate of cortical reorganization following arm amputation. Nature 375(6531):482–484
Freedman DJ, Miller EK (2008) Neural mechanisms of visual categorization: insights from neurophysiology. Neurosci Biobehav Rev 32(2):311–329
Freiwald WA et al (2001) Synchronization and assembly formation in the visual cortex. Prog Brain Res 130:111–140

Frisina RD (2001) Subcortical neural coding mechanisms for auditory temporal processing. Hear Res 158(1–2):1–27
Fritz J et al (2003) Rapid task-related plasticity of spectrotemporal receptive fields in primary auditory cortex. Nat Neurosci 6(11):1216–1223
Fritz JB et al (2007) Auditory attention–focusing the searchlight on sound. Curr Opin Neurobiol 17(4):437–455
Froemke RC et al (2013) Long-term modification of cortical synapses improves sensory perception. Nat Neurosci 16(1):79–88
Froemke RC et al (2007) A synaptic memory trace for cortical receptive field plasticity. Nature 450(7168):425–429
Fu Y et al (2014) A cortical circuit for gain control by behavioral state. Cell 156(6):1139–1152
Galambos R et al (1956) Electrophysiological correlates of a conditioned response in cats. Science 123(3192):376–377
Galvan VV, Weinberger NM (2002) Long-term consolidation and retention of learning-induced tuning plasticity in the auditory cortex of the guinea pig. Neurobiol Learn Mem 77(1):78–108
Gao E, Suga N (1998) Experience-dependent corticofugal adjustment of midbrain frequency map in bat auditory system. Proc Natl Acad Sci U S A 95(21):12663–12670
Gao E, Suga N (2000) Experience-dependent plasticity in the auditory cortex and the inferior colliculus of bats: role of the corticofugal system. Proc Natl Acad Sci U S A 97(14):8081–8086
Gdalyahu A et al (2012) Associative fear learning enhances sparse network coding in primary sensory cortex. Neuron 75(1):121–132
Geva-Sagiv M et al (2015) Spatial cognition in bats and rats: from sensory acquisition to multiscale maps and navigation. Nat Rev Neurosci 16(2):94–108
Gilbert CD, Li W (2013) Top-down influences on visual processing. Nat Rev Neurosci 14(5):350–363
Gilbert CD, Sigman M (2007) Brain states: top-down influences in sensory processing. Neuron 54 (5):677–696
Gleich O, Strutz J (2012) The Mongolian Gerbil as a Model for the Analysis of Peripheral and Central Age-Dependent Hearing Loss. Hearing Loss. S, Naz, InTech
Gonzalez-Lima F, Scheich H (1986) Neural substrates for tone-conditioned bradycardia demonstrated with 2-deoxyglucose. II. Auditory cortex plasticity. Behav Brain Res 20 (3):281–293
Guillery RW, Sherman SM (2011) Branched thalamic afferents: what are the messages that they relay to the cortex? Brain Res Rev 66(1–2):205–219
Guo F et al (2012a) Auditory discrimination training rescues developmentally degraded directional selectivity and restores mature expression of GABA(A) and AMPA receptor subunits in rat auditory cortex. Behav Brain Res 229(2):301–307
Guo W et al (2012b) Robustness of cortical topography across fields, laminae, anesthetic states, and neurophysiological signal types. J Neurosci 32(27):9159–9172
Guo F et al (2013) Tone-detection training enhances spectral integration mediated by intracortical pathways in primary auditory cortex. Neurobiol Learn Mem 101:75–84
Hackett TA et al (2011) Linking topography to tonotopy in the mouse auditory thalamocortical circuit. J Neurosci 31(8):2983–2995
Harel N et al (2000) Three distinct auditory areas of cortex (AI, AII, and AAF) defined by optical imaging of intrinsic signals. Neuroimage 11(4):302–312
Harris KD et al (2003) Organization of cell assemblies in the hippocampus. Nature 424 (6948):552–556
Harris KD, Mrsic-Flogel TD (2013) Cortical connectivity and sensory coding. Nature 503 (7474):51–58
Hars B et al (1993) Basal forebrain stimulation facilitates tone-evoked responses in the auditory cortex of awake rat. Neuroscience 56(1):61–74
Hartley T et al (2014) Space in the brain: how the hippocampal formation supports spatial cognition. Philos Trans R Soc Lond B Biol Sci 369(1635):20120510

Hasselmo ME (1995) Neuromodulation and cortical function: modeling the physiological basis of behavior. Behav Brain Res 67(1):1–27
Hebb DO (1949) The organization of behavior. Wiley & Sons, New York
Hensch TK (2005a) Critical period mechanisms in developing visual cortex. Curr Top Dev Biol 69:215–237
Hensch TK (2005b) Critical period plasticity in local cortical circuits. Nat Rev Neurosci 6 (11):877–888
Herry C et al (2008) Switching on and off fear by distinct neuronal circuits. Nature 454(7204):600–606
Holtmaat A, Svoboda K (2009) Experience-dependent structural synaptic plasticity in the mammalian brain. Nat Rev Neurosci 10(9):647–658
Hopfield JJ (1982) Neural networks and physical systems with emergent collective computational abilities. Proc Natl Acad Sci U S A 79(8):2554–2558
Hromadka T et al (2008) Sparse representation of sounds in the unanesthetized auditory cortex. PLoS Biol 6(1):e16
Hromadka T, Zador AM (2009) Representations in auditory cortex. Curr Opin Neurobiol 19 (4):430–433
Hubener M, Bonhoeffer T (2010) Searching for engrams. Neuron 67(3):363–371
Huber D et al (2012) Multiple dynamic representations in the motor cortex during sensorimotor learning. Nature 484(7395):473–478
Imig TJ, Adrian HO (1977) Binaural columns in the primary field (A1) of cat auditory cortex. Brain Res 138(2):241–257
Insabato A et al (2014) The influence of spatiotemporal structure of noisy stimuli in decision making. PLoS Comput Biol 10(4):e1003492
Irvine DR et al (2000) Specificity of perceptual learning in a frequency discrimination task. J Acoust Soc Am 108(6):2964–2968
Issa JB et al (2014) Multiscale optical Ca2 + imaging of tonal organization in mouse auditory cortex. Neuron 83(4):944–959
Jaramillo S, Zador AM (2011) The auditory cortex mediates the perceptual effects of acoustic temporal expectation. Nat Neurosci 14(2):246–251
Jarrell TW et al (1987) Involvement of cortical and thalamic auditory regions in retention of differential bradycardiac conditioning to acoustic conditioned stimuli in rabbits. Brain Res 412 (2):285–294
Jeanne JM et al (2013) Associative learning enhances population coding by inverting interneuronal correlation patterns. Neuron 78(2):352–363
Ji W et al (2001) Effects of acetylcholine and atropine on plasticity of central auditory neurons caused by conditioning in bats. J Neurophysiol 86(1):211–225
Joachimsthaler B et al (2014) Quantitative analysis of neuronal response properties in primary and higher-order auditory cortical fields of awake house mice (Mus musculus). Eur J Neurosci 39 (6):904–918
Johansen JP et al (2011) Molecular mechanisms of fear learning and memory. Cell 147(3):509–524
Josselyn SA (2010) Continuing the search for the engram: examining the mechanism of fear memories. J Psychiatry Neurosci 35(4):221–228
Kaas JH (1997) Topographic maps are fundamental to sensory processing. Brain Res Bull 44 (2):107–112
Kandel ER (1991) Learning and memory. In: Kandel ER, Schwartz JH, Jessel TM (eds) Principles of neural science. Elsevier, New York, pp 1009–1031
Kanold PO et al (2014) Local versus global scales of organization in auditory cortex. Trends Neurosci 37(9):502–510
Katz LC, Shatz CJ (1996) Synaptic activity and the construction of cortical circuits. Science 274 (5290):1133–1138
Kentridge RW et al (1999) Attention without awareness in blindsight. Proc Biol Sci 266 (1430):1805–1811

Keri S (2003) The cognitive neuroscience of category learning. Brain Res Brain Res Rev 43 (1):85–109
Kesner RP, Churchwell JC (2011) An analysis of rat prefrontal cortex in mediating executive function. Neurobiol Learn Mem 96(3):417–431
Kholodar-Smith DB et al (2008) Fear conditioning to discontinuous auditory cues requires perirhinal cortical function. Behav Neurosci 122(5):1178–1185
Kilgard MP, Merzenich MM (1998) Cortical map reorganization enabled by nucleus basalis activity. Science 279(5357):1714–1718
Kilgard MP, Merzenich MM (1999) Distributed representation of spectral and temporal information in rat primary auditory cortex. Hear Res 134(1–2):16–28
Kimchi EY, Laubach M (2009) Dynamic encoding of action selection by the medial striatum. J Neurosci 29(10):3148–3159
Ko H et al (2011) Functional specificity of local synaptic connections in neocortical networks. Nature 473(7345):87–91
Kobayashi Y, Isa T (2002) Sensory-motor gating and cognitive control by the brainstem cholinergic system. Neural Netw 15(4–6):731–741
Kurt S et al (2006) Differential effects of iontophoretic in vivo application of the GABA(A)-antagonists bicuculline and gabazine in sensory cortex. Hear Res 212(1–2):224–235
LaBar KS et al (1998) Human amygdala activation during conditioned fear acquisition and extinction: a mixed-trial fMRI study. Neuron 20(5):937–945
Lai CS et al (2012) Opposite effects of fear conditioning and extinction on dendritic spine remodelling. Nature 483(7387):87–91
Lamberts K (2002) Feature sampling in categorization and recognition of objects. Q J Exp Psychol A 55(1):141–154
Langers DR, van Dijk P (2012) Mapping the tonotopic organization in human auditory cortex with minimally salient acoustic stimulation. Cereb Cortex 22(9):2024–2038
Lansner A (2009) Associative memory models: from the cell-assembly theory to biophysically detailed cortex simulations. Trends Neurosci 32(3):178–186
Laudanski J et al (2012) Differences between spectro-temporal receptive fields derived from artificial and natural stimuli in the auditory cortex. PLoS ONE 7(11):e50539
LeDoux JE (2000) Emotion circuits in the brain. Annu Rev Neurosci 23:155–184
LeDoux JE (2014) Coming to terms with fear. Proc Natl Acad Sci U S A 111(8):2871–2878
LeDoux JE et al (1984) Subcortical efferent projections of the medial geniculate nucleus mediate emotional responses conditioned to acoustic stimuli. J Neurosci 4(3):683–698
Lee AM et al (2014) Identification of a brainstem circuit regulating visual cortical state in parallel with locomotion. Neuron 83(2):455–466
Lee CC et al (2004) Concurrent tonotopic processing streams in auditory cortex. Cereb Cortex 14 (4):441–451
Letzkus JJ et al (2011) A disinhibitory microcircuit for associative fear learning in the auditory cortex. Nature 480(7377):331–335
Levy RB, Reyes AD (2012) Spatial profile of excitatory and inhibitory synaptic connectivity in mouse primary auditory cortex. J Neurosci 32(16):5609–5619
Liberman AM et al (1967) Perception of the speech code. Psychol Rev 74(6):431–461
Lipton PA et al (1999) Crossmodal associative memory representations in rodent orbitofrontal cortex. Neuron 22(2):349–359
Loewenstein Y et al (2011) Multiplicative dynamics underlie the emergence of the log-normal distribution of spine sizes in the neocortex in vivo. J Neurosci 31(26):9481–9488
Loewenstein Y et al (2015) Predicting the Dynamics of Network Connectivity in the Neocortex. J Neurosci 35(36):12535–12544
Lumani A, Zhang H (2010) Responses of neurons in the rat's dorsal cortex of the inferior colliculus to monaural tone bursts. Brain Res 1351:115–129
Lutcke H et al (2013) Steady or changing? Long-term monitoring of neuronal population activity. Trends Neurosci 36(7):375–384

Ma X, Suga N (2005) Long-term cortical plasticity evoked by electric stimulation and acetylcholine applied to the auditory cortex. Proc Natl Acad Sci U S A 102(26):9335–9340
Ma X, Suga N (2009) Specific and nonspecific plasticity of the primary auditory cortex elicited by thalamic auditory neurons. J Neurosci 29(15):4888–4896
Maass W et al (2007) Computational aspects of feedback in neural circuits. PLoS Comput Biol 3 (1):e165
Maass W, Zador AM (1999) Dynamic stochastic synapses as computational units. Neural Comput 11(4):903–917
Machens CK et al (2004) Linearity of cortical receptive fields measured with natural sounds. J Neurosci 24(5):1089–1100
Macphail EM (1982) Brain and intelligence in vertebrates. Clarendon Press, Oxford
Magnussen S (2000) Low-level memory processes in vision. Trends Neurosci 23(6):247–251
Manita S et al (2015) A Top-Down Cortical Circuit for Accurate Sensory Perception. Neuron 86 (5):1304–1316
Maren S et al (2013) The contextual brain: implications for fear conditioning, extinction and psychopathology. Nat Rev Neurosci 14(6):417–428
Maren S et al (2001) The amygdala is essential for the development of neuronal plasticity in the medial geniculate nucleus during auditory fear conditioning in rats. J Neurosci 21(6):RC135
Margolis DJ et al (2014) Microcircuit dynamics of map plasticity in barrel cortex. Curr Opin Neurobiol 24(1):76–81
Markovitz CD et al (2013) Tonotopic and localized pathways from primary auditory cortex to the central nucleus of the inferior colliculus. Front Neural Circuits 7:77
Markram H et al (2004) Interneurons of the neocortical inhibitory system. Nat Rev Neurosci 5 (10):793–807
Marr D (1970) A theory for cerebral neocortex. Proc R Soc Ser B-Biol Sci 176(1043):161– +
McKenna TM et al (1989) Cholinergic modulation of frequency receptive fields in auditory cortex: I. Frequency-specific effects of muscarinic agonists. Synapse 4(1):30–43
McKernan MG, Shinnick-Gallagher P (1997) Fear conditioning induces a lasting potentiation of synaptic currents in vitro. Nature 390(6660):607–611
Mendez MF, Geehan GR Jr (1988) Cortical auditory disorders: clinical and psychoacoustic features. J Neurol Neurosurg Psychiatry 51(1):1–9
Merzenich MM, Brugge JF (1973) Representation of the cochlear partition of the superior temporal plane of the macaque monkey. Brain Res 50(2):275–296
Merzenich MM et al (1976) Auditory cortex in the grey squirrel: tonotopic organization and architectonic fields. J Comp Neurol 166(4):387–401
Merzenich MM et al (1983) Topographic reorganization of somatosensory cortical areas 3b and 1 in adult monkeys following restricted deafferentation. Neuroscience 8(1):33–55
Merzenich MM et al (1973) Cochleotopic organization of primary auditory cortex in the cat. Brain Res 63:343–346
Mesulam MM et al (1983) Cholinergic innervation of cortex by the basal forebrain: cytochemistry and cortical connections of the septal area, diagonal band nuclei, nucleus basalis (substantia innominata), and hypothalamus in the rhesus monkey. J Comp Neurol 214(2):170–197
Metzger CD et al (2013) Functional mapping of thalamic nuclei and their integration into cortico-striatal-thalamo-cortical loops via ultra-high resolution imaging-from animal anatomy to in vivo imaging in humans. Front Neurosci 7:24
Middlebrooks JC, Knudsen EI (1984) A neural code for auditory space in the cat's superior colliculus. J Neurosci 4(10):2621–2634
Miller EK et al (2003) Neural correlates of categories and concepts. Curr Opin Neurobiol 13 (2):198–203
Mitz AR et al (1991) Learning-dependent neuronal activity in the premotor cortex: activity during the acquisition of conditional motor associations. J Neurosci 11(6):1855–1872
Miyashita Y (1988) Neuronal correlate of visual associative long-term memory in the primate temporal cortex. Nature 335(6193):817–820

Mizrahi A et al (2014) Single neuron and population coding of natural sounds in auditory cortex. Curr Opin Neurobiol 24(1):103–110
Moczulska KE et al (2013) Dynamics of dendritic spines in the mouse auditory cortex during memory formation and memory recall. Proc Natl Acad Sci U S A 110(45):18315–18320
Mogilner A et al (1993) Somatosensory Cortical Plasticity in Adult Humans Revealed by Magnetoencephalography. Proc Natl Acad Sci U S A 90(8):3593–3597
Mongillo G et al (2008) Synaptic theory of working memory. Science 319(5869):1543–1546
Morel A, Kaas JH (1992) Subdivisions and connections of auditory cortex in owl monkeys. J Comp Neurol 318(1):27–63
Morosan P et al (2001) Human primary auditory cortex: cytoarchitectonic subdivisions and mapping into a spatial reference system. Neuroimage 13(4):684–701
Morrell F et al (1983) Conditioning of single units in visual association cortex: cell-specific behavior within a small population. Exp Neurol 80(1):111–146
Morris RG et al (1986) Selective impairment of learning and blockade of long-term potentiation by an N-methyl-D-aspartate receptor antagonist, AP5. Nature 319(6056):774–776
Nabavi S et al (2014) Engineering a memory with LTD and LTP. Nature 511(7509):348–352
Nader K et al (2001) Damage to the lateral and central, but not other, amygdaloid nuclei prevents the acquisition of auditory fear conditioning. Learn Mem 8(3):156–163
Narayan SS et al (1998) Frequency tuning of basilar membrane and auditory nerve fibers in the same cochleae. Science 282(5395):1882–1884
Nelken I et al (2003) Primary auditory cortex of cats: feature detection or something else? Biol Cybern 89(5):397–406
O'Keefe J (1979) A review of the hippocampal place cells. Prog Neurobiol 13(4):419–439
Ohl FW (2015) Role of cortical neurodynamics for understanding the neural basis of motivated behavior—lessons from auditory category learning. Curr Opin Neurobiol 31:88–94
Ohl FW et al (2003) Early and late patterns of stimulus-related activity in auditory cortex of trained animals. Biol Cybern 88(5):374–379
Ohl FW, Scheich H (2005) Learning-induced plasticity in animal and human auditory cortex. Curr Opin Neurobiol 15(4):470–477
Ohl FW et al (2001) Change in pattern of ongoing cortical activity with auditory category learning. Nature 412(6848):733–736
Ohl FW et al (1999) Bilateral ablation of auditory cortex in Mongolian gerbil affects discrimination of frequency modulated tones but not of pure tones. Learn Mem 6(4):347–362
Oliver DL et al (1976) Tonotopic Organization and Connections of Primary Auditory-Cortex in Tree Shrew, Tupaia Glis. Anat Rec 184(3):491
Orsini CA et al (2013) Ensemble coding of context-dependent fear memory in the amygdala. Front Behav Neurosci 7:199
Oswald AM, Reyes AD (2008) Maturation of intrinsic and synaptic properties of layer 2/3 pyramidal neurons in mouse auditory cortex. J Neurophysiol 99(6):2998–3008
Pai S et al (2011) Minimal impairment in a rat model of duration discrimination following excitotoxic lesions of primary auditory and prefrontal cortices. Front Syst Neurosci 5:74
Palmer AR, Kuwada S (2005) Binaural and spatial coding in the inferior colliculus. In: Winer JA, Schreiner CE (eds) The inferior colliculus. Springer New York, New York, pp 377–410
Pantev C et al (1995) Specific tonotopic organizations of different areas of the human auditory cortex revealed by simultaneous magnetic and electric recordings. Electroencephalogr Clin Neurophysiol 94(1):26–40
Parto Dezfouli M, Daliri MR (2015) The effect of adaptation on the tuning curves of rat auditory cortex. PLoS ONE 10(2):e0115621
Pasternak T, Greenlee MW (2005) Working memory in primate sensory systems. Nat Rev Neurosci 6(2):97–107
Pavlov IP, Anrep GV (1927) Conditioned reflexes; an investigation of the physiological activity of the cerebral cortex. London, Oxford University Press, Humphrey Milford
Pearce JM (1994) Similarity and discrimination: a selective review and a connectionist model. Psychol Rev 101(4):587–607

Penfield W (1959) The interpretive cortex; the stream of consciousness in the human brain can be electrically reactivated. Science 129(3365):1719–1725
Peter M et al (2012) Induction of immediate early genes in the mouse auditory cortex after auditory cued fear conditioning to complex sounds. Genes Brain Behav 11(3):314–324
Petkov CI et al (2004) Attentional modulation of human auditory cortex. Nat Neurosci 7(6):658–663
Pi HJ et al (2013) Cortical interneurons that specialize in disinhibitory control. Nature 503 (7477):521–524
Pienkowski M, Eggermont JJ (2011) Cortical tonotopic map plasticity and behavior. Neurosci Biobehav Rev 35(10):2117–2128
Plato, Campbell L (1883) The Theaetetus of Plato, with a revised text and English notes/by Lewis Campbell. The Clarendon press, Oxford
Pleger B et al (2003) Functional imaging of perceptual learning in human primary and secondary somatosensory cortex. Neuron 40(3):643–653
Poldrack RA, Foerde K (2008) Category learning and the memory systems debate. Neurosci Biobehav Rev 32(2):197–205
Polk TA, Farah MJ (1995) Brain localization for arbitrary stimulus categories: a simple account based on Hebbian learning. Proc Natl Acad Sci U S A 92(26):12370–12373
Polley DB et al (2004a) Associative learning shapes the neural code for stimulus magnitude in primary auditory cortex. Proc Natl Acad Sci USA 101(46):16351–16356
Polley DB et al (2004b) Naturalistic experience transforms sensory maps in the adult cortex of caged animals. Nature 429(6987):67–71
Polley DB et al (2006) Perceptual learning directs auditory cortical map reorganization through top-down influences. J Neurosci 26(18):4970–4982
Profant O et al (2013) The response properties of neurons in different fields of the auditory cortex in the rat. Hear Res 296:51–59
Puckett AC et al (2007) Plasticity in the rat posterior auditory field following nucleus basalis stimulation. J Neurophysiol 98(1):253–265
Quirk GJ et al (1997) Fear conditioning enhances different temporal components of tone-evoked spike trains in auditory cortex and lateral amygdala. Neuron 19(3):613–624
Ramachandran VS (1993) Behavioral and magnetoencephalographic correlates of plasticity in the adult human brain. Proc Natl Acad Sci U S A 90(22):10413–10420
Read HL et al (2002) Functional architecture of auditory cortex. Curr Opin Neurobiol 12(4):433–440
Recanzone GH et al (2000) Frequency and intensity response properties of single neurons in the auditory cortex of the behaving macaque monkey. J Neurophysiol 83(4):2315–2331
Recanzone GH et al (1993) Plasticity in the frequency representation of primary auditory cortex following discrimination training in adult owl monkeys. J Neurosci 13(1):87–103
Redgrave P et al (2010) Interactions between the midbrain superior colliculus and the basal ganglia. Front Neuroanat 4:132
Reed A et al (2011) Cortical map plasticity improves learning but is not necessary for improved performance. Neuron 70(1):121–131
Repa JC et al (2001) Two different lateral amygdala cell populations contribute to the initiation and storage of memory. Nat Neurosci 4(7):724–731
Reynolds JN et al (2001) A cellular mechanism of reward-related learning. Nature 413(6851):67–70
Rogan MT et al (1997) Fear conditioning induces associative long-term potentiation in the amygdala. Nature 390(6660):604–607
Rokni U et al (2007) Motor learning with unstable neural representations. Neuron 54(4):653–666
Romanski LM, LeDoux JE (1992) Equipotentiality of thalamo-amygdala and thalamo-cortico-amygdala circuits in auditory fear conditioning. J Neurosci 12(11):4501–4509
Romo R et al (2003) Correlated neuronal discharges that increase coding efficiency during perceptual discrimination. Neuron 38(4):649–657

Romo R, Salinas E (2003) Flutter discrimination: neural codes, perception, memory and decision making. Nat Rev Neurosci 4(3):203–218
Ross LS et al (1988) Origin of ascending projections to an isofrequency region of the mustache bat's inferior colliculus. J Comp Neurol 270(4):488–505
Rothschild G et al (2010) Functional organization and population dynamics in the mouse primary auditory cortex. Nat Neurosci 13(3):353–360
Rumpel S et al (2005) Postsynaptic receptor trafficking underlying a form of associative learning. Science 308(5718):83–88
Russ BE et al (2007) Neural and behavioral correlates of auditory categorization. Hear Res 229(1–2):204–212
Rutkowski RG, Weinberger NM (2005) Encoding of learned importance of sound by magnitude of representational area in primary auditory cortex. Proc Natl Acad Sci U S A 102(38):13664–13669
Sadagopan S, Wang X (2008) Level invariant representation of sounds by populations of neurons in primary auditory cortex. J Neurosci 28(13):3415–3426
Sadagopan S, Wang X (2010) Contribution of inhibition to stimulus selectivity in primary auditory cortex of awake primates. J Neurosci 30(21):7314–7325
Salzman CD et al (1990) Cortical microstimulation influences perceptual judgements of motion direction. Nature 346(6280):174–177
Sawtell NB et al (2003) NMDA receptor-dependent ocular dominance plasticity in adult visual cortex. Neuron 38(6):977–985
Scheich H et al (2007) The cognitive auditory cortex: task-specificity of stimulus representations. Hear Res 229(1–2):213–224
Schnupp JW et al (2006) Plasticity of temporal pattern codes for vocalization stimuli in primary auditory cortex. J Neurosci 26(18):4785–4795
Schoenbaum G et al (1998) Orbitofrontal cortex and basolateral amygdala encode expected outcomes during learning. Nat Neurosci 1(2):155–159
Schoenbaum G et al (1999) Neural encoding in orbitofrontal cortex and basolateral amygdala during olfactory discrimination learning. J Neurosci 19(5):1876–1884
Seger CA, Miller EK (2010) Category learning in the brain. Annu Rev Neurosci 33:203–219
Selezneva E et al (2006) Dual time scales for categorical decision making in auditory cortex. Curr Biol 16(24):2428–2433
Sellien H, Ebner FF (2007) Rapid plasticity follows whisker pairing in barrel cortex of the awake rat. Exp Brain Res 177(1):1–14
Shadlen MN, Newsome WT (2001) Neural basis of a perceptual decision in the parietal cortex (area LIP) of the rhesus monkey. J Neurophysiol 86(4):1916–1936
Sherman SM (2007) The thalamus is more than just a relay. Curr Opin Neurobiol 17(4):417–422
Sherman SM (2012) Thalamocortical interactions. Curr Opin Neurobiol 22(4):575–579
Shetake JA et al (2012) Pairing tone trains with vagus nerve stimulation induces temporal plasticity in auditory cortex. Exp Neurol 233(1):342–349
Sillito AM (1975) The contribution of inhibitory mechanisms to the receptive field properties of neurones in the striate cortex of the cat. J Physiol 250(2):305–329
Singer W (1999) Neuronal synchrony: a versatile code for the definition of relations? Neuron 24 (1):49–65, 111–125
Singheiser M et al (2012) The representation of sound localization cues in the barn owl's inferior colliculus. Front Neural Circuits 6:45
Siucinska E, Kossut M (1996) Short-lasting classical conditioning induces reversible changes of representational maps of vibrissae in mouse SI cortex–a 2DG study. Cereb Cortex 6(3):506–513
Siucinska E, Kossut M (2004) Experience-dependent changes in cortical whisker representation in the adult mouse: a 2-deoxyglucose study. Neuroscience 127(4):961–971
Skinner BF (1938) The behavior of organisms: an experimental analysis. B.F. Skinner Foundation, Cambridge, Massachusetts

Song S et al (2005) Highly nonrandom features of synaptic connectivity in local cortical circuits. PLoS Biol 3(3):e68
Stiebler I et al (1997) The auditory cortex of the house mouse: left-right differences, tonotopic organization and quantitative analysis of frequency representation. J Comp Physiol A 181 (6):559–571
Stuermer IW et al (2003) Intraspecific allometric comparison of laboratory gerbils with Mongolian gerbils trapped in the wild indicates domestication in Meriones unguiculatus (Milne-Edwards, 1867) (Rodentia: Gerbillinae). Zoologischer Anzeiger 242(3):249–266
Stuttgen MC et al (2011) Mapping spikes to sensations. Front Neurosci 5:125
Sutter ML (2000) Shapes and level tolerances of frequency tuning curves in primary auditory cortex: quantitative measures and population codes. J Neurophysiol 84(2):1012–1025
Takahashi N et al (2012) Locally synchronized synaptic inputs. Science 335(6066):353–356
Talavage TM et al (2000) Frequency-dependent responses exhibited by multiple regions in human auditory cortex. Hear Res 150(1–2):225–244
Talavage TM et al (2004) Tonotopic organization in human auditory cortex revealed by progressions of frequency sensitivity. J Neurophysiol 91(3):1282–1296
Tan AY et al (2004) Tone-evoked excitatory and inhibitory synaptic conductances of primary auditory cortex neurons. J Neurophysiol 92(1):630–643
Uchida N et al (2006) Seeing at a glance, smelling in a whiff: rapid forms of perceptual decision making. Nat Rev Neurosci 7(6):485–491
Wallace DJ, Kerr JN (2010) Chasing the cell assembly. Curr Opin Neurobiol 20(3):296–305
Wallace MN et al (2000) Identification and localisation of auditory areas in guinea pig cortex. Exp Brain Res 132(4):445–456
Wang X, Kadia SC (2001) Differential representation of species-specific primate vocalizations in the auditory cortices of marmoset and cat. J Neurophysiol 86(5):2616–2620
Wang X et al (2005) Sustained firing in auditory cortex evoked by preferred stimuli. Nature 435 (7040):341–346
Wang X et al (1995) Representation of a species-specific vocalization in the primary auditory cortex of the common marmoset: temporal and spectral characteristics. J Neurophysiol 74 (6):2685–2706
Wang XJ (2008) Decision making in recurrent neuronal circuits. Neuron 60(2):215–234
Wehr M, Zador AM (2003) Balanced inhibition underlies tuning and sharpens spike timing in auditory cortex. Nature 426(6965):442–446
Weinberg RJ (1997) Are topographic maps fundamental to sensory processing? Brain Res Bull 44 (2):113–116
Weinberger NM (2004) Specific long-term memory traces in primary auditory cortex. Nat Rev Neurosci 5(4):279–290
Weinberger NM (2015) New perspectives on the auditory cortex: learning and memory. Handb Clin Neurol 129:117–147
Weinberger NM et al (1984) Physiological Plasticity of Single Neurons in Auditory-Cortex of the Cat during Acquisition of the Pupillary Conditioned-Response. 1. Primary Field (Ai). Behav Neurosci 98(2):171–188
White EL (2007) Reflections on the specificity of synaptic connections. Brain Res Rev 55(2):422–429
Whitlock JR et al (2006) Learning induces long-term potentiation in the hippocampus. Science 313(5790):1093–1097
Woolsey CN, Walzl EM (1942) Topical projection of nerve fibers from local regions of the cochlea to the cerebral cortex of the cat. Bull Johns Hopkins Hosp 71:315–344
Xiong Q et al (2015) Selective corticostriatal plasticity during acquisition of an auditory discrimination task. Nature 521(7552):348–351
Xu T et al (2009) Rapid formation and selective stabilization of synapses for enduring motor memories. Nature 462(7275):915–919
Yan W, Suga N (1998) Corticofugal modulation of the midbrain frequency map in the bat auditory system. Nat Neurosci 1(1):54–58

Yang G et al (2009) Stably maintained dendritic spines are associated with lifelong memories. Nature 462(7275):920–924
Yoon T et al (2008) Prefrontal cortex and hippocampus subserve different components of working memory in rats. Learn Mem 15(3):97–105
Zach N et al (2008) Emergence of novel representations in primary motor cortex and premotor neurons during associative learning. J Neurosci 28(38):9545–9556
Zagha E et al (2013) Motor cortex feedback influences sensory processing by modulating network state. Neuron 79(3):567–578
Zhang S et al (2014) Selective attention. Long-range and local circuits for top-down modulation of visual cortex processing. Science 345(6197):660–665
Znamenskiy P, Zador AM (2013) Corticostriatal neurons in auditory cortex drive decisions during auditory discrimination. Nature 497(7450):482–485

The Representational Basis of Working Memory

Derek Evan Nee and Mark D'Esposito

Abstract Working memory refers to a system involved in the online maintenance and manipulation of information in the absence of external input. Due to the importance of working memory in higher-level cognition, a wealth of neuroscience studies has investigated its neural basis. These studies have often led to conflicting viewpoints regarding the importance of the prefrontal cortex (PFC) and posterior sensory cortices. Here, we review evidence for each position. We suggest that the relative contributions of the PFC and sensory cortices to working memory can be understood with respect to processing demands. We argue that procedures that minimize processing demands lead to increased importance of sensory representations, while procedures that permit transformational processing lead to representational abstraction that relies on the PFC. We suggest that abstract PFC representations support top-down control over posterior representations while also providing bottom-up inputs into higher-level cognitive processing. Although a number of contemporary studies have studied working memory while using procedures that minimize the role of the PFC, we argue that consideration of the PFC is critical for our understanding of working memory and higher-level cognition more generally.

Keywords Short-term memory · Prefrontal cortex · Abstraction · Multivariate pattern analysis

Contents

1 Introduction 214
2 Working Memory and the Prefrontal Cortex 215

D.E. Nee (✉)
Department of Psychology, Florida State University, 1107 W Call St, Tallahassee, FL 32306-4301, USA
e-mail: nee@psy.fsu.edu

M. D'Esposito
Helen Wills Neuroscience Institute and Department of Psychology, University of California, 132 Barker Hall, Berkeley, CA 94720-3190, USA

Curr Topics Behav Neurosci (2018) 37: 213–230
DOI 10.1007/7854_2016_456
Published Online: 28 September 2016

3 Beyond the Prefrontal Cortex ... 216
4 Multivariate Contributions to Working Memory ... 217
5 Frontal Versus Posterior Contributions to Working Memory ... 220
6 The Role of Abstraction in Working Memory, Control, and Higher-Level Cognition ... 222
7 Conclusion ... 226
References ... 226

1 Introduction

Atkinson and Shiffrin (1968) used the term "working memory" to describe a short-term store—a system involved in the brief maintenance of information that could be distinguished from both a sensory store (i.e., perception) and a long-term store (i.e., long-term memory). However, Baddeley is typically given credit for popularizing the term "working memory" and galvanizing an enormous amount of research into the topic. When Baddeley and Hitch (1974) described working memory, storage was merely one part of the system. A critical aspect—one that Atkinson and Shiffrin (1968) neglected as part of their description—was the processing performed on the storage system. This processing included directing what should and should not be stored, as well as manipulating the contents of the stored information in order to reason and problem solve. One could argue that it is the processing that makes working memory work.

Despite Baddeley's emphasis on the importance of the processing aspects of working memory, the bulk of research in working memory has focused on storage with a minimization of processing. For example, in many neuroscientific investigations, a canonical task is to present participants with a sample stimulus and require a match/non-match decision after a period of several seconds. In such tasks, neural activity during the retention interval is typically isolated (e.g., from encoding and retrieval) and taken as a measure of working memory (Curtis and D'Esposito 2003). While such tasks are useful for examining the retention of information in working memory, minimal processing must be done on an item to make a recognition decision. This track perhaps follows from Baddeley's own focus on storage, which he considered to be the more tractable part of working memory to study (Baddeley 2012). In turn, investigation into the processing aspects of working memory has often proceeded with a minimization of storage demands under the heading of "cognitive control." Hence, it has become sufficient, and perhaps even desirable, to study working memory in the absence of processing.

From the neuroscience perspective, how one defines working memory delineates the scope of the investigation. The narrowest, and perhaps most popular, approach is to investigate the mechanisms involved in actively maintaining information in working memory. Here, "active" is an operational term, denoting that active neural firing is the physical embodiment of working memory (Luck and Vogel 2013). Increasingly, the possibility that synaptic modification may also play a role in maintaining information in mind is gaining interest as a potential alternative, or in the very least, additional mechanism (Jonides et al. 2008; Mongillo et al. 2008; Sugase-Miyamoto et al. 2008; Nee and Jonides 2013a). However, given the difficulty in studying synaptic plasticity, synaptic contributions to working memory remain largely theoretical. As a result, we will focus primarily on active neural firing and its hypothesized correlates (e.g., BOLD signal), resorting to synaptic mechanisms only when theoretically necessary.

A more expansive definition of working memory that includes processing as originally formulated by Baddeley broadens the scope of the investigation considerably. In this case, of interest is not only how information is maintained, but for what it is being maintained. Critically, the former may depend on the latter. As we will review, the nature of how information will be used dictates the form in which it is stored in working memory. It is commonly argued that the importance of working memory can be appreciated by its relationship to higher-order cognitive skills such as reasoning, problem solving, and intelligence (Daneman and Carpenter 1980; Carpenter et al. 1990; Just and Carpenter 1992; Daneman and Merikle 1996; Fukuda et al. 2010). To accomplish higher-order cognitive tasks, it is likely that some transformation of sensory inputs takes place in order to form symbolic codes that can be the basis of complex cognition. However, many contemporary experimental designs minimize the potential for such transformations in an effort to examine pure measures of working memory retention. Whether limiting transformations captures the essential nature of working memory bears consideration.

What follows is a review of neuroscientific contributions to the study of the representational basis of working memory. By "representation" we mean the manner in which information is codified while sustained until its eventual use. We begin with a brief recount of early interest in the role of the prefrontal cortex (PFC) in working memory. We follow with more contemporary investigations that demonstrate the importance of posterior cortical areas in the maintenance of working memory, especially when representations are visual in nature. We continue by detailing efforts to describe the interactions between frontal and posterior areas that may explain discrepancies between early and more contemporary research. We close by suggesting that future investigation into the dynamic nature of working memory may yield additional insights by capturing the way working memory is utilized in higher-level cognition.

2 Working Memory and the Prefrontal Cortex

A commonly held view in the early 1990s was that working memory is dependent upon the PFC in much the same way as long-term memory is dependent upon the hippocampus (Goldman-Rakic 1995). In a landmark review, Goldman-Rakic (1987) marshaled lesion, neuroanatomical, and neurophysiological evidence to advocate the central role of the PFC in working memory. Among the most salient pieces of evidence is the demonstration of so-called delay cells in the monkey PFC whose activity persists during periods when a stimulus is no longer present in the environment until a response contingent upon that stimulus is made (Fuster and Alexander 1971; Kubota and Niki 1971; Funahashi et al. 1989). Many such cells are stimulus-specific, firing in response to certain stimuli, but not others. These observations led to the conclusion that such delay cells are the physiological basis of working memory. This idea was bolstered by demonstrations that "mnemonic scotomas" could be induced by lesions to selective aspects of the PFC (Funahashi et al. 1993), suggesting that specific information maintained in working memory could be localized to distinct areas of the PFC.

A challenge to a PFC-centric view of working memory was the demonstration that delay cells are not restricted to the PFC. For example, similar activity is observed in temporal (Fuster and Jervey 1982; Miyashita and Chang 1988), parietal (Chafee and Goldman-Rakic 1998), and even early visual cortices (Super et al. 2001). However, subsequent research demonstrated that delay-period activity in such posterior regions can be abolished by the insertion of distraction, while activity in the PFC persists (Constantinidis and Steinmetz 1996; Miller et al. 1996). The robustness of PFC activity to disruption led to the conclusion that only the PFC maintains an enduring trace for the purpose of working memory.

3 Beyond the Prefrontal Cortex

Neurophysiological studies in monkeys have been invaluable to our understanding of working memory. However, a drawback of this approach is the ability to examine only circumscribed areas of the brain at a time. This leads to the possibility that critical areas involved in a behavior may be overlooked. Furthermore, training monkeys to perform a cognitive task requires a great deal of effort. This limits the kinds of designs one can examine and makes it difficult to contrast a wide array of tasks. The advent of human neuroimaging, and fMRI in particular, brought the ability to examine the entire brain during a wide variety of tasks. These advantages have led to a great deal more insight into the neural underpinnings of working memory.

Two important conclusions can be drawn from the study of working memory with fMRI. First, working memory does not selectively recruit the PFC. Meta-analyses of neuroimaging activations reveal that working memory recruits a broad constellation of frontal, posterior, and subcortical areas (Wager and Smith 2003; Owen et al. 2005; Rottschy et al. 2012; Nee et al. 2013). The precise localization tends to differ depending on the type of material maintained and processed in working memory. Broadly speaking, spatial working memory recruits dorsal frontal and parietal areas, while identity-based working memory recruits ventral frontal and temporal areas, a segregation that had been predicted by research in monkeys (Goldman-Rakic 1987; Levy and Goldman-Rakic 2000). The second principal finding is that the PFC is not selectively recruited by working memory. Indeed, meta-analyses of a broad array of tasks have often found very little to distinguish working memory from other functions in terms of spatial localization in the PFC and elsewhere (Duncan and Owen 2000; Niendam et al. 2012). As a result of these data, a PFC-centric view of working memory has largely fallen out of favor (Jonides et al. 2005; Postle 2006; D'Esposito 2007; Jonides et al. 2008).

The broadly overlapping activations observed in neuroimaging across tasks led to a revised view of working memory. While the psychological tradition favored "walling off" working memory into boxes distinct from perception and long-term memory (Baddeley 1986, 2012), neuroimaging studies have made it clear that there are no such walls in the brain—at least not at an appreciable spatial scale. Instead, there appears to be considerable economy in the brain. For example, temporal areas involved in the perception of faces (Puce et al. 1995; Kanwisher et al. 1997) also demonstrate sustained activation when faces are held in working memory (Druzgal and D'Esposito 2001, 2003; Postle et al. 2003; Ranganath et al. 2004a; Lepsien and Nobre 2007). Similarly, temporal areas involved in scene perception show sustained activation when scenes are held in working memory (Ranganath et al. 2004a; Lepsien and Nobre 2007). Furthermore, it seems to matter little whether the information maintained in working memory comes from perception or from long-term memory—the same areas are engaged regardless (Ranganath et al. 2004b). Dorsal frontal and parietal areas involved in spatial working memory are virtually indistinguishable for those involved in spatial attention (Tamber-Rosenau et al. 2011; Jerde et al. 2012) and eye movements (Postle et al. 2000) (but see also Courtney et al. 1998). The view that emerges from these data is that working memory is simply the sustained activity in areas involved in information representation for other cognitive activities such as perception, action, and long-term memory (Jonides et al. 2005; Postle 2006; D'Esposito 2007; Jonides et al. 2008; Nee and Jonides 2013a). Hence, the principal aspect that seems to distinguish working memory from other cognitive functions is its persistence in the absence of external input, rather than large-scale spatial localization.

4 Multivariate Contributions to Working Memory

To better understand the representations that underlie working memory, recent neuroimaging studies have turned to multivariate methods. These methods differ from classic neuroimaging analyses that use a single measure to describe the response of a region (e.g., average signal within a region-of-interest—univariate approach). Instead, patterns of activation within and between regions are considered distributed codes that underlie information representation (Norman et al. 2006). Two kinds of inferences can be drawn from these data: (1) what areas of the brain contribute to the representations underlying working memory and (2) how are these codes modulated as a function of cognitive demands. Note that the first kind of inference is similar in principle to univariate approach reviewed above in that it attempts to localize areas of the brain responsible for working memory. However, the univariate approach is somewhat limited in that it can only indicate what areas of the brain are more/less active as a function of task demands. Without careful control conditions, it is difficult to know what leads to activation differences. For example, a prototypical means to examine areas related to working memory is to parametrically vary the number of items to be held in working memory and explore which areas of the brain track these demands (e.g., Cohen et al. 1997). While areas of the brain responsible for storage are hypothesized to increase their activation as they store more and more items, so too will areas not involved in storage such as areas responsible for monitoring the success/failure of retention and areas involved in general arousal. By contrast, the multivariate approach allows the researcher to associate particular patterns of activation to particular stimuli/stimulus classes in order to determine which areas contribute to signal that discriminates different representations. In this way, the activations are directly tied to the representations themselves and are therefore less likely to correspond to ancillary demands (but see Todd et al. 2013 for important considerations).

A number of insights have been gained from the multivariate neuroimaging approach. First, the approach has revealed information representation underlying working memory in areas such as early visual cortex that were previously unappreciated by fMRI studies (Harrison and Tong 2009; Serences et al. 2009). Because early visual areas do not show global increases in activation during working memory retention intervals, these areas were overlooked by univariate methods. However, not only do early visual areas discriminate individual stimuli in working memory, the fidelity with which they do so correlates with the precision with which a visual working memory is maintained (Emrich et al. 2013; Ester et al. 2013). Such data indicate that the presence of working memory representations in early visual areas is not an epiphenomenon, but instead that early visual representation is a means to accurately preserve sensory input over delays.

Some studies indicate that it is sensory cortices, and not the PFC, that maintain information in working memory. As mentioned previously, sustained activity in the PFC during working memory maintenance was taken as a measure of storage of working memory representations. However, an fMRI study examined whether PFC areas that show elevated working memory-related activation could distinguish individual representations in working memory (Riggall and Postle 2012). Participants were presented with a moving dot pattern and were cued to remember either the direction or speed of motion. The authors found no evidence that the PFC could distinguish individual directions or velocities held in working memory, while strong evidence was found in early visual areas. By contrast, the PFC could distinguish the relevant feature held in working memory (i.e., motion or speed). The authors took this as evidence that the PFC represents the task (i.e., attend to motion vs. attend to speed), while visual areas represent the specific memoranda (see also Chen et al. 2012; Sreenivasan et al. 2014). Other studies have similarly demonstrated visual working memory representation in posterior cortices, but not the PFC (Christophel et al. 2012; Linden et al. 2012; Emrich et al. 2013). These findings are a substantial departure from the PFC-centric view of working memory suggested by earlier monkey studies mentioned above, a point that we revisit below.

In addition to regional inference, multivariate methods permit inferences on the nature of representations themselves. In one study (Lewis-Peacock and Postle 2008) patterns of fMRI activation were examined while subjects made perceptual judgments to visually present stimuli. Thereafter, participants learned to relate pairs of these stimuli, forming long-term associations between them. In a final test, participants were given a sample stimulus and were asked to judge whether a probe appearing after a several second delay interval was the learned associate. The researchers then looked during the delay interval for the patterns of activation that had been observed during the perceptual judgments. During this time, patterns of activation underlying working memory shifted from representing the sample stimulus to the associate. These data emphasize two important points: (1) That information in working memory can be identified using the same representational codes observed during perception and (2) that such codes are re-activated during retrieval from long-term memory. These data demonstrate the continuous nature of perception, working memory, and long-term memory in that each uses the same distributed representation (see also Ranganath et al. 2004b for similar conclusions using a univariate approach).

Observing shifting representations in working memory enables the ability to better understand the dynamic nature of processing in working memory. In complex cognitive activities, we often shift attention among multiple representations held in working memory. For example, during mental arithmetic (e.g., 321 × 13), we may hold multiple numbers in mind, but focus on a subset of those numbers at a given time (e.g., 1 × 3). What are the consequences of attending to specific information in working memory? In an fMRI study, Lewis-Peacock et al. (2012) presented participants with two stimuli to maintain in working memory. Thereafter, a cue indicated that subjects would make a judgment based on one of the items. After the judgment, a second cue signaled another judgment to be made on the same item or the other item. Hence, the cue directed attention to particular items in working memory. The authors found that only the attended item could be detected in distributed activation signals. Although subjects could successfully switch between items, demonstrating that information about the unattended item was still present in some manner, the unattended item could not be identified in distributed activation signals (see also Lepsien and Nobre 2007 for similar data using a univariate approach). A follow-up study using EEG reached the same conclusion (LaRocque et al. 2013). These data suggest that active neural firing is only associated with attended information in working memory.

What is the neural status of unattended information? This is currently an active area of research. Univariate analyses indicate that there are at least two distinct states of unattended information each distinguished by their accessibility measured behaviorally and neurally (Nee and Jonides 2011, 2013a, b). However, multivariate analyses have thus far been unable to find evidence of active representation of unattended information of any sort in BOLD and EEG data (but see Lewis-Peacock et al. 2014 for evidence that unattended features of an attended object can persist). Such results have led to speculation that unattended information may maintain some level of accessibility due to rapid synaptic weight changes (LaRocque et al. 2013; Nee and Jonides 2013a; Larocque et al. 2014). By this account, encoding and maintenance of information in working memory leads to synaptic modification that facilitates reactivation of the same circuits involved during encoding. Recently discovered short-term synaptic plasticity may be a candidate neurophysiological mechanism (Erickson et al. 2010). Notably, this mechanism is distinct from long-term plasticity—the mechanism thought to underlie the formation of long-term memories. Furthermore, computational modeling has demonstrated the mechanistic feasibility of synaptic modification as means for working memory storage, although the precise mechanisms differ across models (Mongillo et al. 2008; Sugase-Miyamoto et al. 2008). While intriguing, the contribution of synaptic modification to working memory remains theoretical and requires additional investigation.

5 Frontal Versus Posterior Contributions to Working Memory

As reviewed above, neurophysiological data recorded from monkeys demonstrated that stimulus-selective neural activity related to working memory could be observed in multiple areas of the brain, but that such activity in the PFC seemed to be uniquely resistant to distraction (Miller et al. 1996). By contrast, recent fMRI studies have suggested that visual working memories can be decoded from the activation in posterior visual cortices, but not the PFC (Christophel et al. 2012; Linden et al. 2012; Riggall and Postle 2012; Emrich et al. 2013). At first blush, these data appear to be contradictory. If the PFC is the site of working memory storage, there should not be instances wherein information in working memory is actively maintained elsewhere in the brain, but not the PFC. Conversely, if posterior cortices are the site of storage, how can information in working memory be maintained in these regions when their activity is disrupted?

It is important to note the types of stimuli used to examine working memory in most experimental studies. Human fMRI studies examining visual working memory tend to use stimuli that cannot be easily verbalized in order to minimize potential verbal recoding of stimuli. The resultant stimuli are often drawn from a continuous space (e.g., 360 degrees of motion angles), which is difficult to discretize into categories. Hence, the only means to successfully perform such tasks is to sustain the original percept. By contrast, the stimuli used by Miller et al. (1996) consisted of real-world objects that were distinctive from one another. Since monkeys were the participants, verbal recoding was not a concern. Nevertheless, even with monkeys, the original percept of real-world objects may be recoded into a more meaningful representation. It is possible that the PFC represents such abstracted, meaningful representations, but not continuous representations that are difficult to categorize.

A recent fMRI study examined whether the categorical nature of stimuli could differentiate PFC versus posterior representational bases (Lee et al. 2013). Participants held a sample object in visual working memory in anticipation of one of two probe tasks. In the visual task, the probe consisted of an object fragment and participants decided whether the fragment was part of the sample object. In the non-visual task, the probe consisted of a new object and participants decided whether the probe object came from the same object category as the sample. In both tasks, the same exact stimuli were used. However, in the visual task, the category/meaning of the object was of little consequence, whereas in the non-visual task, the converse was true. The authors found that inferior temporal areas showed representational selectivity in the visual, but not non-visual task, while the PFC showed the opposite effect. These data indicate that the nature of the task and corollary processing has important implications regarding how information is represented (see also Curtis et al. 2004). The same exact stimulus can be represented in working memory in an entirely different manner depending on future processing demands. Moreover, these data indicate that PFC representations are categorical in nature, while temporal representations are visual in nature.

Further support for categorical coding in the PFC comes from a recent study that examined visual working memory for color in monkeys (Lara and Wallis 2014). Twenty color stimuli were drawn from a continuous spectrum that rendered categorical coding of color difficult. In this task, monkey PFC neurons did not show evidence of representing color. Instead, PFC neurons coded for which of the four locations the stimuli had been presented in. Hence, the PFC maintained a representation of the stimuli, but only for the feature that was categorical in nature (i.e., spatial quadrant) and not for the feature that was continuous (i.e., color). Another recent study found evidence for sustained monkey PFC signals reflecting motion direction (Mendoza-Halliday et al. 2014), but only four directions were included in the stimulus set, again affording categorical coding. These data indicate that discrepancies between the earlier monkey data that showed PFC representations maintained in working memory and the more recent human fMRI data that did not are not due to species differences, but instead due to stimuli and the type of representational code to which they are amenable.

Differences between the representational coding of the PFC and temporal cortex are further highlighted in neurophysiological data from monkeys. Freedman et al. (2003) presented monkeys with stimuli generated by morphing six exemplar stimuli drawn from two distinct object classes: dogs and cats. The morphs resulted in some pure stimuli (e.g., 100 % dog) and other stimuli with various levels of mixing (e.g., 60 % dog). Monkeys were trained to classify the stimuli into the two distinct categories. Thereafter, activity in the PFC and inferior temporal cortex was examined during a category-matching working memory task. Activity in inferior temporal cortex showed a continuous pattern: Cells preferring one category showed parametric firing rates as a function of purity of the sample. By contrast, activity in the PFC was categorical in nature with little regard to the purity of the sample.

Hence, the inferior temporal neurons reflected the visual stimulus properties while the PFC reflected the category boundaries (see also Meyers et al. 2008). These data further suggest that the PFC forms abstractions from continuous posterior inputs.

6 The Role of Abstraction in Working Memory, Control, and Higher-Level Cognition

Abstraction has recently become an important theoretical lens through which to understand the representations of the PFC. We described above how the PFC is proposed to form categorical abstractions from continuous posterior representations. Several theories suggest that the PFC itself is organized such that more anterior areas of the PFC form abstracted representations from more posterior areas of the PFC (Koechlin et al. 2003; Badre 2008; Badre and D'Esposito 2009; Nee et al. 2014). Abstractions from categories indicate the contexts under which a particular categorical representation is relevant. Such abstractions are often thought of as rules that guide behavior. For example, a concrete rule might specify an action to make for a particular stimulus (e.g., if phone rings, answer it). A more abstract rule might specify that the action for a particular stimulus depends on the context in which it is encountered (e.g., my office: if phone rings, answer it; colleague's office: if phone rings, don't answer). Additional layers of context may also be added (e.g., if a colleague tells you he is expecting an important call, you might answer the phone for him if he is not around). Data demonstrate that as rules become increasingly abstract, increasingly anterior areas of the PFC are engaged (Koechlin et al. 2003; Badre and D'Esposito 2007; Badre et al. 2010; Nee and Brown 2012, 2013). Furthermore, lesions to more anterior areas of the PFC impair actions based on more abstract rules (Badre et al. 2009; Azuar et al. 2014). These data suggest that abstraction is a fundamental function of the PFC and a way in which it exerts control over behavior.

What are the implications for abstract PFC representation in working memory? One implication is that abstract PFC representations can provide a means to bias posterior representations (Miller and Cohen 2001). In a working memory setting, the storage of an abstract category or set can specify what types of information should be encoded, maintained, or discarded from working memory, thereby providing a means to select relevant posterior representations (Nee et al. 2013). Consistent with this proposal, we noted earlier that activations in the PFC distinguish between broad classes of relevant sensory inputs to be encoded into working memory such as remembering motion versus speed (Riggall and Postle 2012) or faces versus scenes (Chen et al. 2012; Sreenivasan et al. 2014). Further evidence comes from recent data that demonstrate how the PFC influences posterior processing in order to provide the selectivity necessary to choose relevant from irrelevant material. A pair of studies examined the effect of PFC disruption to categorical processing in temporal cortex (Miller et al. 2011; Lee and D'Esposito 2012). Both studies observed that PFC disruption reduced the selectivity with which temporal cortices responded to stimuli drawn from different categories. These results suggest that top-down input from the PFC tunes posterior neurons to respond with greater category selectivity, providing a basis for top-down selection.

A categorical signal arising from the PFC may also provide a useful cue to reactivate dormant representations. Consider again the data from Miller et al. (1996). The requirement to process new inputs disrupted sustained firing in the temporal cortex, but not the PFC. If the PFC maintains a categorical representation abstracted from the original percept, then this information may be useful as a cue to reactivate stimulus-specific posterior representations following disruption. This could be accomplished if some residual trace of the original stimulus is maintained via synaptic modification (Mongillo et al. 2008; Sugase-Miyamoto et al. 2008; Erickson et al. 2010). Assuming that PFC-posterior connections are structured so that the PFC can selectively influence distinct sets of neurons (Goldman and Nauta 1977), a category cue from the PFC could limit the space of potential reactivated memories to the appropriate category. This would provide a useful means to maintain information in working memory in the midst of distraction.

Some evidence for this idea comes from an fMRI study that examined PFC-temporal interactions during a working memory task that included a distractor (Clapp et al. 2010). Participants maintained a scene in working memory. During the delay interval, an intervening stimulus appeared that was either to be processed (interrupting stimulus), ignored (distracting stimulus), or no stimulus appeared (no interference). Connectivity between the PFC and a scene-selective temporal area was disrupted by the interrupting stimulus, compared to the distracting stimulus and no distraction. Critically, the degree to which the PFC and temporal cortex re-coupled following interruption predicted working memory performance. These data suggest that the PFC provided a top-down signal that reactivated working memory-related activity in temporal cortex that aided in working memory retention.

We have thus far argued that abstract, categorical representations maintained by the PFC provide a means to exert top-down control over continuous posterior representations. This categorical account of PFC function makes two salient predictions. First, distracting stimuli drawn from the same category should interfere with working memory more so than distracting stimuli drawn from a different category. This is because targets and distractors drawn from the same category would compete for a common PFC categorical signal. Such effects have been observed both during retention (Jha et al. 2004; Yoon et al. 2006) and retrieval (Atkins et al. 2011). Second, if all stimuli are drawn from the same stimulus class and no categorical boundaries can be drawn, then the PFC will not be able to abstract categorical information. In these cases, we would not expect the PFC to represent stimulus-related information in working memory. As reviewed above, such results have been demonstrated under these conditions (Christophel et al. 2012; Linden et al. 2012; Riggall and Postle 2012; Emrich et al. 2013). A further corollary is that representations solely dependent upon posterior representations should be particularly vulnerable to disruption since the PFC cannot intervene to reactivate disrupted memories. While we do not know of data that directly speak to this issue, working memory for non-categorical visual stimuli has been demonstrated to be incredibly fragile, at times failing to survive even a single intervening stimulus (Phillips and Christie 1977). We would conjecture that such stimuli rely solely on posterior representations, although empirical verification is necessary.

Thus far, we have described evidence that abstracted representations in the PFC can be useful for providing top-down control over posterior cortices. A second use of abstracted representations is bottom-up input into higher-level cognitive processes. As we mentioned earlier, the value of working memory is evident in the relationship between working memory capacity and higher-level cognitive processing such as reasoning, problem solving, and language comprehension (Daneman and Carpenter 1980; Carpenter et al. 1990; Just and Carpenter 1992; Daneman and Merikle 1996; Fukuda et al. 2010). Common to many higher-level cognitive processes is the use of abstract, symbolic codes. For example, logic statements underlying reasoning use symbolic codes to form inferences (e.g., if P then Q; $\sim$Q; therefore, $\sim$P). If we take working memory capacity as the capacity to maintain several symbolic codes, it is clear why working memory is so important: The more abstract, symbolic codes that can be maintained, the more workspace one has for higher-level cognition.

If abstract representations of the PFC form the basis of higher-level cognitive processes, then disrupting these representations should impair the higher-level cognitive functions that are dependent upon them. Badre et al. (2009) examined patients with damage to different areas of the PFC while they performed a series of tasks that required varying levels of rule abstraction. Previous fMRI data demonstrated that lower level rules activated posterior areas of the PFC, while higher-level rules activated anterior areas of the PFC (Badre and D'Esposito 2007). Such data suggest that posterior areas of the PFC represent lower level rules, while anterior areas of the PFC represent higher-level rules. If the lower level rules formed inputs into the higher-level rules, then damage to posterior areas of the PFC should asymmetrically impair processing based on the higher-level rules. This is because higher-level rule processing depends on the lower level inputs. Conversely, damage to anterior areas of the PFC that process higher-level rules should not impair processing based on lower level rules. This is because processing based solely on lower level rules does not need higher-level input. This asymmetric pattern of deficit was precisely what was observed. This pattern has since been replicated using a different design and a similar logic (Azuar et al. 2014). These data indicate that posterior areas of the PFC, that represent abstractions of posterior cortical inputs, are themselves the inputs to higher-level cognitive processes instantiated by anterior areas of the PFC.

The data reviewed above suggest that abstraction is a central function of the PFC that has critical importance for working memory and its use in higher-level cognition. Despite the apparent importance of abstract, categorical representations, many contemporary studies, particularly in the visual domain, utilize continuous stimuli that are not amenable to categorization. One reason for the rise in the use of continuous stimuli is that such stimuli lend themselves to interesting quantitative investigation. For instance, various conclusions have been drawn on the nature of working memory by investigating the precision of working memory representations (Wilken and Ma 2004; Ma et al. 2014). Precision refers to how closely a participant can reproduce a continuous sample after a retention interval (e.g., reproduce the angle with which an orientation was presented). Modeling the mechanisms that give rise to variability in mnemonic precision is therefore thought to be a window into understanding how working memory is information limited (Bays 2014; van den Berg et al. 2014). As mentioned earlier, early visual activity during working memory retention intervals directly relates to the precision of the memoranda (Emrich et al. 2013; Ester et al. 2013), suggesting that mechanisms that limit precision reflect processing in sensory cortices. However, as we reviewed above, it is unclear that the precise sensory representations that these studies rely upon are abstracted into a categorical form. If such representations cannot be abstracted into a categorical form, our account would suggest that they would not form inputs into higher-level cognition.

Evidence for this claim comes from a recent study that examined the relationship between working memory capacity, precision, and intelligence (Fukuda et al. 2010). Here, intelligence is a proxy for higher-level cognition, and capacity and precision are two modeled aspects of working memory (Zhang and Luck 2008). Capacity reflects the number of items that can be maintained in working memory, while precision reflects the fidelity with which representations can be maintained. Consistent with previous data, capacity predicted intelligence ($r = 0.66$). However, precision did not ($r = -0.05$) suggesting that precision is unrelated to higher-level cognitive processing. Moreover, there is even some evidence that precision is negatively related to top-down attentional control (Machizawa and Driver 2011). Hence, while examining the precision of sensory representations may lead to interesting quantitative accounts of sensory cortices, the importance of such representations in the broader scope of working memory remains to be determined.

The review above suggests that abstract PFC representations influence working memory in two directions: (1) top-down as a means to influence posterior representations and (2) bottom-up as input into higher-level cognition. Further progress in understanding these processes will likely rely on both multivariate analyses to identify representations in working memory and connectivity analyses to investigate PFC-posterior and PFC-PFC interactions. Critical to these endeavors will be consideration of the types of processing performed on information in working memory and how this processing transforms the underlying neural code (Stokes et al. 2013). Since such transformations likely occur rapidly, the use of time-resolved techniques such as EEG will be particularly important (Garcia et al. 2013). Hence, a multimodal approach using the time resolution of EEG to track representations and the spatial localization of fMRI to describe network interactions would be particularly informative.

7 Conclusion

We have reviewed evidence for the representational basis of working memory. We have argued that the neural basis for representation depends on the nature of underlying code. While early research argued that the PFC forms the representational basis that underlies working memory, more contemporary research has argued strongly for the contribution of sensory cortices. We have hypothesized that these discrepancies reflect differences in nature of the processing performed on information maintained in working memory resulting in different representational codes. Limiting the kinds of transformations that can be performed on a stimulus and requiring faithful reproduction of concrete properties of a stimulus leads to a sensory representation dependent upon sensory cortices rather than the PFC. However, when sensory information can be categorized into an abstract code, this abstracted code is maintained by the PFC. We have argued that abstract PFC codes are not only a means to store information, but also provide a mechanism for top-down control over posterior cortices. Since these codes are more robust to distraction than sensory codes, they can also provide a means with which to reactivate disrupted sensory traces. Furthermore, we have suggested that abstract codes form the input to higher-level cognition, which underscores the importance of working memory in cognition. While interesting quantitative data can be derived from using tasks that force a continuous, concrete representation, how such representations are utilized beyond literal sensory reproduction is unclear. By contrast, abstract PFC representations provide a basis for storage, and critically, for processing in working memory. Hence, future research into how abstract representations are formed and utilized will provide important insights into our understanding of working memory and higher-level cognition more generally.

References

Atkins AS, Berman MG, Reuter-Lorenz PA, Lewis RL, Jonides J (2011) Resolving semantic and proactive interference in memory over the short-term. Mem Cogn 39:806–817

Atkinson RC, Shiffrin RM (1968) Human memory: a proposed system and its control processes. Psychol Learn Motiv 2:89–195

Azuar C, Reyes P, Slachevsky A, Volle E, Kinkingnehun S, Kouneiher F, Bravo E, Dubois B, Koechlin E, Levy R (2014) Testing the model of caudo-rostral organization of cognitive control in the human with frontal lesions. Neuroimage 84:1053–1060

Baddeley AD (1986) Working memory. Oxford University Press, Oxford

Baddeley A (2012) Working memory: theories, models, and controversies. Annu Rev Psychol 63:1–29

Baddeley AD, Hitch GJ (1974) Working memory. In: Bower GH (ed) The psychology of learning and motivation. Academic Press, New York, pp 47–89

Badre D (2008) Cognitive control, hierarchy, and the rostro-caudal organization of the frontal lobes. Trends Cogn Sci 12:193–200

Badre D, D'Esposito M (2007) Functional magnetic resonance imaging evidence for a hierarchical organization of the prefrontal cortex. J Cogn Neurosci 19:2082–2099

Badre D, D'Esposito M (2009) Is the rostro-caudal axis of the frontal lobe hierarchical? Nat Rev Neurosci 10:659–669

Badre D, Hoffman J, Cooney JW, D'Esposito M (2009) Hierarchical cognitive control deficits following damage to the human frontal lobe. Nat Neurosci 12:515–522

Badre D, Kayser AS, D'Esposito M (2010) Frontal cortex and the discovery of abstract action rules. Neuron 66:315–326

Bays PM (2014) Noise in neural populations accounts for errors in working memory. J Neurosci Official J Soc Neurosci 34:3632–3645

Carpenter PA, Just MA, Shell P (1990) What one intelligence test measures: a theoretical account of the processing in the raven progressive matrices test. Psychol Rev 97:404–431

Chafee MV, Goldman-Rakic PS (1998) Matching patterns of activity in primate prefrontal area 8a and parietal area 7ip neurons during a spatial working memory task. J Neurophysiol 79:2919–2940

Chen AJ, Britton M, Turner GR, Vytlacil J, Thompson TW, D'Esposito M (2012) Goal-directed attention alters the tuning of object-based representations in extrastriate cortex. Front Hum Neurosci 6:187

Christophel TB, Hebart MN, Haynes JD (2012) Decoding the contents of visual short-term memory from human visual and parietal cortex. J Neurosci Official J Soc Neurosci 32:12983–12989

Clapp WC, Rubens MT, Gazzaley A (2010) Mechanisms of working memory disruption by external interference. Cereb Cortex 20:859–872

Cohen JD, Perlstein WM, Braver TS, Nystrom LE, Noll DC, Jonides J, Smith EE (1997) Temporal dynamics of brain activation during a working memory task. Nature 386:604–608

Constantinidis C, Steinmetz MA (1996) Neuronal activity in posterior parietal area 7a during the delay periods of a spatial memory task. J Neurophysiol 76:1352–1355

Courtney SM, Petit L, Maisog JM, Ungerleider LG, Haxby JV (1998) An area specialized for spatial working memory in human frontal cortex. Science 279:1347–1351

Curtis CE, D'Esposito M (2003) Persistent activity in the prefrontal cortex during working memory. Trends Cogn Sci 7:415–423

Curtis CE, Rao VY, D'Esposito M (2004) Maintenance of spatial and motor codes during oculomotor delayed response tasks. J Neurosci Official J Soc Neurosci 24:3944–3952

Daneman M, Carpenter PA (1980) Individual differences in working memory and reading. J Verbal Learn Verbal Behav 19:450–466

Daneman M, Merikle PM (1996) Working memory and language comprehension: a meta-analysis. Psychon Bull Rev 3:422–433

D'Esposito M (2007) From cognitive to neural models of working memory. Philos Trans R Soc Lond B Biol Sci 362:761–772

Druzgal TJ, D'Esposito M (2001) Activity in fusiform face area modulated as a function of working memory load. Brain Res Cogn Brain Res 10:355–364

Druzgal TJ, D'Esposito M (2003) Dissecting contributions of prefrontal cortex and fusiform face area to face working memory. J Cogn Neurosci 15:771–784

Duncan J, Owen AM (2000) Common regions of the human frontal lobe recruited by diverse cognitive demands. Trends Neurosci 23:475–483

Emrich SM, Riggall AC, Larocque JJ, Postle BR (2013) Distributed patterns of activity in sensory cortex reflect the precision of multiple items maintained in visual short-term memory. J Neurosci Official J Soc Neurosci 33:6516–6523

Erickson MA, Maramara LA, Lisman J (2010) A single brief burst induces GluR1-dependent associative short-term potentiation: a potential mechanism for short-term memory. J Cogn Neurosci 22:2530–2540

Ester EF, Anderson DE, Serences JT, Awh E (2013) A neural measure of precision in visual working memory. J Cogn Neurosci 25:754–761

Freedman DJ, Riesenhuber M, Poggio T, Miller EK (2003) A comparison of primate prefrontal and inferior temporal cortices during visual categorization. J Neurosci Official J Soc Neurosci 23:5235–5246

Fukuda K, Vogel E, Mayr U, Awh E (2010) Quantity, not quality: the relationship between fluid intelligence and working memory capacity. Psychon Bull Rev 17:673–679
Funahashi S, Bruce CJ, Goldman-Rakic PS (1989) Mnemonic coding of visual space in the monkey's dorsolateral prefrontal cortex. J Neurophysiol 61:331–349
Funahashi S, Bruce CJ, Goldman-Rakic PS (1993) Dorsolateral prefrontal lesions and oculomotor delayed-response performance: evidence for mnemonic "scotomas". J Neurosci Official J Soc Neurosci 13:1479–1497
Fuster JM, Alexander GE (1971) Neuron activity related to short-term memory. Science 173:652–654
Fuster JM, Jervey JP (1982) Neuronal firing in the inferotemporal cortex of the monkey in a visual memory task. J Neurosci Official J Soc Neurosci 2:361–375
Garcia JO, Srinivasan R, Serences JT (2013) Near-real-time feature-selective modulations in human cortex. Curr Biol CB 23:515–522
Goldman PS, Nauta WJ (1977) Columnar distribution of cortico-cortical fibers in the frontal association, limbic, and motor cortex of the developing rhesus monkey. Brain Res 122:393–413
Goldman-Rakic PS (1987) Circuitry of the primate prefrontal cortex and the regulation of behavior by representational memory. In: Plum F, Mountcastle VB (eds) Handbook of physiology. American Physiological Society, Bethseda, pp 373–417
Goldman-Rakic PS (1995) Cellular basis of working memory. Neuron 14:477–485
Harrison SA, Tong F (2009) Decoding reveals the contents of visual working memory in early visual areas. Nature 458:632–635
Jerde TA, Merriam EP, Riggall AC, Hedges JH, Curtis CE (2012) Prioritized maps of space in human frontoparietal cortex. J Neurosci Official J Soc Neurosci 32:17382–17390
Jha AP, Fabian SA, Aguirre GK (2004) The role of prefrontal cortex in resolving distractor interference. Cogn Affect Behav Neurosci 4:517–527
Jonides J, Lacey S, Nee D (2005) Processes of working memory in mind and brain. Curr Dir Psychol Sci 14:2–5
Jonides J, Lewis RL, Nee DE, Lustig CA, Berman MG, Moore KS (2008) The mind and brain of short-term memory. Annu Rev Psychol 59:193–224
Just MA, Carpenter PA (1992) A capacity theory of comprehension—individual-differences in working memory. Psychol Rev 99:122–149
Kanwisher N, McDermott J, Chun MM (1997) The fusiform face area: a module in human extrastriate cortex specialized for face perception. J Neurosci 17:4302–4311
Koechlin E, Ody C, Kouneiher F (2003) The architecture of cognitive control in the human prefrontal cortex. Science 302:1181–1185
Kubota K, Niki H (1971) Prefrontal cortical unit activity and delayed alternation performance in monkeys. J Neurophysiol 34:337–347
Lara AH, Wallis JD (2014) Executive control processes underlying multi-item working memory. Nat Neurosci 17:876–883
LaRocque JJ, Lewis-Peacock JA, Drysdale AT, Oberauer K, Postle BR (2013) Decoding attended information in short-term memory: an EEG study. J Cogn Neurosci 25:127–142
Larocque JJ, Lewis-Peacock JA, Postle BR (2014) Multiple neural states of representation in short-term memory? It's a matter of attention. Front Hum Neurosci 8:5
Lee TG, D'Esposito M (2012) The dynamic nature of top-down signals originating from prefrontal cortex: a combined fMRI-TMS study. J Neurosci Official J Soc Neurosci 32:15458–15466
Lee SH, Kravitz DJ, Baker CI (2013) Goal-dependent dissociation of visual and prefrontal cortices during working memory. Nat Neurosci 16:997–999
Lepsien J, Nobre AC (2007) Attentional modulation of object representations in working memory. Cereb Cortex 17:2072–2083
Levy R, Goldman-Rakic PS (2000) Segregation of working memory functions within the dorsolateral prefrontal cortex. Exp Brain Res 133:23–32
Lewis-Peacock JA, Postle BR (2008) Temporary activation of long-term memory supports working memory. J Neurosci Official J Soc Neurosci 28:8765–8771

Lewis-Peacock JA, Drysdale AT, Oberauer K, Postle BR (2012) Neural evidence for a distinction between short-term memory and the focus of attention. J Cogn Neurosci 24:61–79
Lewis-Peacock JA, Drysdale AT, Postle BR (2014) Neural evidence for the flexible control of mental representations. Cereb Cortex
Linden DE, Oosterhof NN, Klein C, Downing PE (2012) Mapping brain activation and information during category-specific visual working memory. J Neurophysiol 107:628–639
Luck SJ, Vogel EK (2013) Visual working memory capacity: from psychophysics and neurobiology to individual differences. Trends Cogn Sci 17:391–400
Ma WJ, Husain M, Bays PM (2014) Changing concepts of working memory. Nat Neurosci 17:347–356
Machizawa MG, Driver J (2011) Principal component analysis of behavioural individual differences suggests that particular aspects of visual working memory may relate to specific aspects of attention. Neuropsychologia 49:1518–1526
Mendoza-Halliday D, Torres S, Martinez-Trujillo JC (2014) Sharp emergence of feature-selective sustained activity along the dorsal visual pathway. Nat Neurosci 17:1255–1262
Meyers EM, Freedman DJ, Kreiman G, Miller EK, Poggio T (2008) Dynamic population coding of category information in inferior temporal and prefrontal cortex. J Neurophysiol 100:1407–1419
Miller EK, Cohen JD (2001) An integrative theory of prefrontal cortex function. Annu Rev Neurosci 24:167–202
Miller EK, Erickson CA, Desimone R (1996) Neural mechanisms of visual working memory in prefrontal cortex of the macaque. J Neurosci 16:5154–5167
Miller BT, Vytlacil J, Fegen D, Pradhan S, D'Esposito M (2011) The prefrontal cortex modulates category selectivity in human extrastriate cortex. J Cogn Neurosci 23:1–10
Miyashita Y, Chang HS (1988) Neuronal correlate of pictorial short-term memory in the primate temporal cortex. Nature 331:68–70
Mongillo G, Barak O, Tsodyks M (2008) Synaptic theory of working memory. Science 319:1543–1546
Nee DE, Brown JW (2012) Rostral-caudal gradients of abstraction revealed by multi-variate pattern analysis of working memory. Neuroimage 63:1285–1294
Nee DE, Brown JW (2013) Dissociable frontal-striatal and frontal-parietal networks involved in updating hierarchical contexts in working memory. Cereb Cortex 23:2146–2158
Nee DE, Jonides J (2011) Dissociable contributions of prefrontal cortex and the hippocampus to short-term memory: evidence for a 3-state model of memory. Neuroimage 54:1540–1548
Nee DE, Jonides J (2013a) Trisecting representational states in short-term memory. Front Hum Neurosci 7:796
Nee DE, Jonides J (2013b) Neural evidence for a 3-state model of visual short-term memory. Neuroimage 74:1–11
Nee DE, Brown JW, Askren MK, Berman MG, Demiralp E, Krawitz A, Jonides J (2013) A meta-analysis of executive components of working memory. Cereb Cortex 23:264–282
Nee DE, Jahn A, Brown JW (2014) Prefrontal cortex organization: dissociating effects of temporal abstraction, relational abstraction, and integration with FMRI. Cereb Cortex 24:2377–2387
Niendam TA, Laird AR, Ray KL, Dean YM, Glahn DC, Carter CS (2012) Meta-analytic evidence for a superordinate cognitive control network subserving diverse executive functions. Cogn Affect Behav Neurosci 12:241–268
Norman KA, Polyn SM, Detre GJ, Haxby JV (2006) Beyond mind-reading: multi-voxel pattern analysis of fMRI data. Trends Cogn Sci 10:424–430
Owen AM, McMillan KM, Laird AR, Bullmore E (2005) N-back working memory paradigm: a meta-analysis of normative functional neuroimaging studies. Hum Brain Mapp 25:46–59
Phillips WA, Christie DFM (1977) Components of visual memory. Q J Exp Psychol 29:117–133
Postle BR (2006) Working memory as an emergent property of the mind and brain. Neuroscience 139:23–38
Postle BR, Berger JS, Taich AM, D'Esposito M (2000) Activity in human frontal cortex associated with spatial working memory and saccadic behavior. J Cogn Neurosci 12(Suppl 2):2–14

Postle BR, Druzgal TJ, D'Esposito M (2003) Seeking the neural substrates of visual working memory storage. Cortex J Devoted Study Nerv Syst Behav 39:927–946
Puce A, Allison T, Gore JC, McCarthy G (1995) Face-sensitive regions in human extrastriate cortex studied by functional MRI. J Neurophysiol 74:1192–1199
Ranganath C, DeGutis J, D'Esposito M (2004a) Category-specific modulation of inferior temporal activity during working memory encoding and maintenance. Brain Res Cogn Brain Res 20:37–45
Ranganath C, Cohen MX, Dam C, D'Esposito M (2004b) Inferior temporal, prefrontal, and hippocampal contributions to visual working memory maintenance and associative memory retrieval. J Neurosci Official J Soc Neurosci 24:3917–3925
Riggall AC, Postle BR (2012) The relationship between working memory storage and elevated activity as measured with functional magnetic resonance imaging. J Neurosci Official J Soc Neurosci 32:12990–12998
Rottschy C, Langner R, Dogan I, Reetz K, Laird AR, Schulz JB, Fox PT, Eickhoff SB (2012) Modelling neural correlates of working memory: a coordinate-based meta-analysis. Neuroimage 60:830–846
Serences JT, Ester EF, Vogel EK, Awh E (2009) Stimulus-specific delay activity in human primary visual cortex. Psychol Sci 20:207–214
Sreenivasan KK, Vytlacil J, D'Esposito M (2014) Distributed and dynamic storage of working memory stimulus information in extrastriate cortex. J Cogn Neurosci 26:1141–1153
Stokes MG, Kusunoki M, Sigala N, Nili H, Gaffan D, Duncan J (2013) Dynamic coding for cognitive control in prefrontal cortex. Neuron 78:364–375
Sugase-Miyamoto Y, Liu Z, Wiener MC, Optican LM, Richmond BJ (2008) Short-term memory trace in rapidly adapting synapses of inferior temporal cortex. PLoS Comp Biol 4:e1000073
Super H, Spekreijse H, Lamme VA (2001) A neural correlate of working memory in the monkey primary visual cortex. Science 293:120–124
Tamber-Rosenau BJ, Esterman M, Chiu YC, Yantis S (2011) Cortical mechanisms of cognitive control for shifting attention in vision and working memory. J Cogn Neurosci 23:2905–2919
Todd MT, Nystrom LE, Cohen JD (2013) Confounds in multivariate pattern analysis: theory and rule representation case study. Neuroimage 77:157–165
van den Berg R, Awh E, Ma WJ (2014) Factorial comparison of working memory models. Psychol Rev 121:124–149
Wager TD, Smith EE (2003) Neuroimaging studies of working memory: a meta-analysis. Cogn Affect Behav Neurosci 3:255–274
Wilken P, Ma WJ (2004) A detection theory account of change detection. J Vis 4:1120–1135
Yoon JH, Curtis CE, D'Esposito M (2006) Differential effects of distraction during working memory on delay-period activity in the prefrontal cortex and the visual association cortex. Neuroimage 29:1117–1126
Zhang W, Luck SJ (2008) Discrete fixed-resolution representations in visual working memory. Nature 453:233–235

The Neuroscience of Human Decision-Making Through the Lens of Learning and Memory

Lesley K. Fellows

Abstract We are called upon to make decisions, large and small, many times a day. Whether in the voting booth, the stock exchange, or the cafeteria line, we identify potential options, estimate and compare their subjective values, and make a choice. Decision-making has only recently become a focus for cognitive neuroscience. The last two decades have seen rapid progress in our understanding of the brain basis of at least some aspects of this rather complex aspect of cognition. This work has provided fresh perspectives on poorly understood brain regions, such as orbitofrontal cortex and ventral striatum. It has led to interesting interdisciplinary exchanges with diverse fields, notably economics, but also ecology and political science, among others. The novel perspectives arising from these exchanges have begun to be related to better understood aspects of cognition. In particular, it is increasingly clear that decision-making is tightly interlinked with learning and memory. Key early insights in decision neuroscience came from what were essentially reinforcement learning tasks. Recent work has made similar links to aspects of declarative memory. Indeed, decision-making can be seen as the link between memory of the past and future actions. This chapter reviews selected topics in decision neuroscience, with a particular focus on the links to learning and memory, and a particular emphasis on regions within prefrontal cortex.

Keywords Reward · Reinforcement learning · Choice · Frontal lobes · Orbitofrontal cortex · Heuristics

Contents

1 Introduction 232
1.1 Overview of Decision-Making Processes 232
1.2 Neural Basis of Decision-Making 233

L.K. Fellows (✉)
Department of Neurology and Neurosurgery, Montreal Neurological Institute, McGill University, 3801 University Street, Montreal, QC H3A 2B4, Canada
e-mail: lesley.fellows@mcgill.ca

Curr Topics Behav Neurosci (2018) 37: 231–251
DOI 10.1007/7854_2016_468
Published Online: 18 February 2017

2 Decision-Making and Associative Learning 235
2.1 Learning from Reward Under Dynamic Conditions 235
3 Decision Contexts: Current Workspace and Future Time 241
3.1 Models and Schemas 241
3.2 Multi-attribute Decision-Making 243
3.3 Heuristics 243
3.4 Social Decision-Making and Decisions About Social Options 244
3.5 Time and Memory in Decision-Making 245
4 Conclusions 247
References 247

1 Introduction

1.1 Overview of Decision-Making Processes

Decision-making is a fundamental behavior. It is likely that the brain has developed to optimize decision-making: to make choices in the physical and social environment that achieve the general goals of avoiding harm and obtaining rewards while minimizing the costs of time, effort and missed opportunities. The centrality of these goals to survival presumably explains the multiplicity of mechanisms that organisms have developed to achieve them.

While plants turning their leaves toward the sun, or bacteria migrating up a sucrose gradient are in a sense "seeking" reward (Murray et al. 2011), it seems a stretch to label these phenomena as decision-making. Here, we focus on the repertoire of flexible behaviors that allow higher organisms, notably humans, to pursue motivated aims. Such aims may be motivated directly, yielding immediate rewards of obvious relevance to survival, such as food or a mating opportunity. Aims may instead be more distantly linked to such outcomes, for example, yielding currency (e.g., social, monetary) that might be eventually exchanged for primary rewards, or protection from direct threats to survival. Humans, at least, may pursue aims that are highly abstract and distant from personal survival, orienting their behavior toward socially constructed and codified values such as heroism or charity. Here, we consider decision-making as encompassing choices that can be analyzed in terms of the relative, context-sensitive subjective value, or utility, of their outcomes, a definition that encompasses this whole gamut of motivated behaviors.

Decision-making is, by its very nature, future-oriented. It involves selecting behaviors on the basis of prospects, possibilities, predictions. One may wish to undo decisions, or wistfully contemplate the road not taken, but these are not obligatory aspects of choice, which, fundamentally, concerns what to do next. Counterintuitively, it is this future orientation of decision-making that links it to the topic of memory. Optimal decision-making benefits enormously from learning and memory processes, particularly for organisms that must navigate complex, changing environments. Indeed, one could argue that memory of the past carries survival advantages only to the extent that it allows reliable predictions about the future, and that these predictions can be harnessed effectively to guide behavior. The relationship between memory and decision-making is circular and iterative: we learn about what we choose to experience, whether directly or vicariously, and our memory representations are adjusted through repeated experience, capturing the central tendency and the variance of outcomes in a given context, to improve subsequent predictions.

This prediction-choice-outcome loop is thought to be central to associative learning, in particular. This idea has been captured by formal models such as the influential temporal difference model, in which the difference between predicted and experienced reward drives learning of cue-outcome or action-outcome associations (Sutton and Barto 1998; Rescorla and Wagner 1972). In principle, the predictions that are central to learning can be co-opted to guide decision-making, even in settings where the outcome is not experienced (e.g., hypothetical choices). That is, a prediction about the rewardingness of each of two options could be generated, based, perhaps, on prior experience with each option, and then compared to allow the more rewarding option to be selected. This basic idea has been very influential in decision neuroscience to date (Padoa-Schioppa 2011; Glimcher 2002; Fellows 2004).

While predicted reward value is a plausible core element of decision-making, it is insufficient. What "cues" drive such predictions? How do they come to be linked to value? What factors are combined to generate a value estimate, and how might these vary with the decision context? Is value represented as a common currency? How does the context come to be defined, and how does it constrain the decision options that are considered? How general are the mechanisms that allow cue-value associations to be built up: do these support only the direct link between a cue and an immediate reward experience, or can they extend across time, and apply to more abstract outcomes?

1.2 Neural Basis of Decision-Making

This chapter reviews our current understanding of these issues, from a brain-based perspective. Decision-making behaviors have been extensively studied within several disciplines, including animal learning, ecology, human psychology, and behavioral economics. However, efforts to understand the neurobiological basis of many of these behaviors, particularly in humans, are less advanced; decision neuroscience has arguably only been a serious area of inquiry for 15 years or so.

There are two broad themes that can be discerned in this work, in part reflecting different behavioral starting points. One builds on our understanding of the brain basis of primary reward and punishment processing, particularly in associative learning, studied in animal models (Waelti et al. 2001; Schultz 2004). The other begins with human behavior, particularly as assessed in economics (Loewenstein et al. 2001) and decision psychology (Kahneman and Tversky 1979), applying cognitive neuroscience methods to define the neural correlates of classic behavioral constructs, such as subjective utility (or value) (Bartra et al. 2013), temporal discounting (Kable and Glimcher 2007), and risk (Hsu et al. 2005; Tom et al. 2007). There has also been much fruitful interaction between these two broad approaches, and across models and methods, as it has become clear that the constructs under study in reward learning and "economic" paradigms are in many cases identical or at least share important characteristics.

In broad terms, current views hold that information related to the subjective value of decision options is present in signals within a restricted set of brain regions: subcortical areas, including thalamus and various sectors of the striatum, and cortical areas, notably orbitofrontal (OFC) and ventromedial prefrontal cortex (vmPFC). More dorsal regions within the medial frontal and parietal lobes are also engaged in relation to subjective value. Damage to OFC-vmPFC disrupts value-based decision-making, even in very simple preference judgment paradigms (Henri-Bhargava et al. 2012; Fellows and Farah 2007), and damage to striatum disrupts some aspects of reward learning (Vo et al. 2014), as does striatal dysfunction related to Parkinson's disease and its treatment (Frank and Seeberger 2004; Cools and Robbins 2004).

Lateral PFC does not seem to be required for choices guided by value in their most basic form, but this region is activated in many decision paradigms, suggesting its engagement under certain conditions (McClure et al. 2004; Daw et al. 2005; Hutcherson et al. 2012). Finally, the frontal pole has been linked specifically to exploratory behavior. Monkey electrophysiology research has focused on prefrontal regions, including OFC, but has also found evidence for value-related signals in lateral parietal lobe (Glimcher and Rustichini 2004).

The OFC and vmPFC have received particular attention, in part, because decision-making provided a novel and useful perspective on the potential functions of these heretofore little studied and poorly understood regions (Bechara et al. 1997). Building on lesion work in rats and monkeys, and on a view of the function of these regions as "emotional," the initial emphasis in relation to decision-making was on the interactions with amygdala, and ventral striatum, reflecting a limbic network related to intuitive decisions and "gut feelings" (Bechara et al. 1997). More recently, there has been interest in the interplay between hippocampus and vmPFC, as the field has been reminded that there are strong projections connecting these regions, and that future predictions can involve explicit recall of past experiences under similar conditions. In addition, directly or indirectly through other frontal regions or the thalamus, OFC-vmPFC may influence visual-attentional processes in more posterior cortical regions (Chaumon et al. 2013; Rudebeck and Murray 2014).

Neuromodulatory systems are also thought to be important in aspects of reward processing, learning and decision-making. The role of dopamine has been a particular focus, spurred by the idea that phasic dopamine signals encode the difference between expected and actual reward outcome in simple associative learning tasks, i.e., could provide a biological implementation of the signal that is central to temporal difference learning models (Schultz 2004; Rangel et al. 2008; Dagher and Robbins 2009). The monoamines more generally are believed to be important in various aspects of motivated behavior (Aston-Jones and Cohen 2005; Robbins and Arnsten 2009). Because drugs of abuse are thought to act through dopaminergic mechanisms, this neuromodulatory perspective has been particularly linked to impulsivity and addiction; this will not be further addressed here.

Instead, this chapter will primarily review prefrontal contributions to decision-making. We will focus first on evidence that has come from reward learning paradigms. This work shows that different regions of prefrontal cortex make specific contributions to learning from reward, and accessing value representations to inform choice. A recurring theme in this literature is that prefrontal cortex becomes critical for learning under complex conditions: when contingencies are changing, or when there is ambiguity about the predictive features of the environment, for example. This is often summarized as "context-dependent" learning or choice. While thinking about context, ideas about time, semantic knowledge, task set, and decision strategies become relevant. This requires a broader perspective on decision-making, which will be addressed in the second half of the chapter.

2 Decision-Making and Associative Learning

Even very simple organisms can seek reward and avoid punishment. However, more complex organisms, with more complex nervous systems can vastly expand this basic repertoire to identify the most reward-predictive features of the environment, link these to specific outcomes, assess the current, relative reward value of those outcomes, organize behavior to acquire the predicted reward (potentially over multiple steps and long delays), and update these predictive links after the outcome is experienced. Prefrontal cortex, in interaction with several subcortical systems, makes critical contributions to this set of higher order, motivated behaviors.

2.1 Learning from Reward Under Dynamic Conditions

2.1.1 Reversal Learning

Reversal learning is a classic test of flexible reward learning. It involves first learning through trial and error to associate one of two cues with reward, in a two-choice discrimination task. Once a learning criterion is met, the reward contingencies are switched: the previously unrewarded cue becomes rewarding, the previously rewarded cue no longer leads to reward. This simple manipulation requires flexibly updating cue-reward associations. After the new associations have been learned, further reversals occur, requiring ongoing flexibility in responding based on the currently optimal strategy. The task has been implemented in rats, monkeys, and humans, and, despite important differences in the details (such as the extent of training required), has provided remarkably consistent findings in all three species. Lesion studies in all three species have shown that performance relies critically on intact orbitofrontal cortex (Butter 1968; Fellows and Farah 2003; Dias et al. 1996; Hornak et al. 2004; Ghods-Sharifi et al. 2008; McAlonan and Brown 2003). When the cue-reward associations are fully deterministic, these lesions do not disrupt acquisition of the initial discrimination, emphasizing that simply orienting behavior to the most rewarding feature of the environment is distinct from the problem the organism faces when contingencies reverse. Optimally solving the latter problem requires OFC.

Functional neuroimaging studies in humans suggest that reversal learning engages a larger set of regions, including (at least) lateral and dorsomedial prefrontal cortex, amygdala, and ventral striatum (Hampton and O'Doherty 2007; Budhani et al. 2007). Lesion studies combined with measurement of reward-related activity in OFC in non-human primates and humans have confirmed that amygdala contributes to this signal, but is not the sole, nor even the main input (Hampton et al. 2007; Rudebeck et al. 2013). On the other hand, amygdala lesions, or amygdala-orbitofrontal cortex disconnections in monkeys do disrupt reversal learning, among other value and emotion-related behaviors (Murray and Wise 2010).

Human studies have tended to make reversal learning tasks more difficult by introducing probabilistic cue-reward associations. Discriminating the better option now requires integrating feedback information across multiple trials, and filtering irrelevant negative feedback (that will inevitably occur despite choosing the "best" option) from relevant negative feedback (i.e., signaling the occurrence of a reversal). From a frontal lobe perspective, this variation of the task continues to depend uniquely on orbitofrontal cortex: a large study of patients with focal prefrontal damage showed that only medial and lateral orbitofrontal cortex damage (and, notably, not dorsomedial PFC damage) was reliably predictive of impaired performance (Tsuchida et al. 2010). That study also found that even the initial discrimination was impaired in these patients, supporting a more general role for OFC in optimal choice based on cue-reward associations under probabilistic feedback conditions. This idea is also supported by impairments in tracking dynamic reward feedback in three-choice tasks observed after OFC lesions in macaques (Noonan et al. 2012; Walton et al. 2011). That work suggests distinct contributions of medial and lateral OFC, with the former critical for tracking the value of potential options, and the latter engaged in linking specific feedback to specific cues in a stream of information.

One of the earliest findings suggesting that orbitofrontal and ventromedial PFC were important in decision-making came from research using what is now termed the Iowa Gambling Task (Bechara et al. 1997). This task taps several decision processes, including the ability to reverse initial tendencies as contingencies shift, and clearly relies critically on the ventral frontal lobe (Glascher et al. 2012). There is evidence that the impairments on this task in patients with vmPFC and orbitofrontal damage stem from this reversal learning requirement (Fellows and Farah 2005).

A recent excitotoxic OFC lesion study in macaques failed to replicate the reversal learning deficits seen in prior work (Rudebeck et al. 2013). Prior studies used aspiration lesion methods that may disrupt fibers of passage as well as the intended cortical targets. One interesting possibility is that reversal learning deficit after aspiration lesions is due to disruption of OFC—medial temporal lobe connections (Rudebeck and Murray 2014). Practically, human lesion studies cannot control the extent of lesions at this level of resolution. Such studies include patients with evident white matter, as well as cortical, damage in most cases. Further work will be needed to more firmly establish the specific anatomical substrates of these effects. However, whether through disconnection or direct local damage, across species, methods, and task variations it seems that OFC plays a role in flexibly linking cues to reward outcomes, particularly under conditions where that link must be updated dynamically.

2.1.2 Selective Satiety

Reversal learning requires flexibly updating cue-reward associations, but the cues themselves are unambiguous, as is the reward. A very different paradigm, also widely used to study flexible reward-related learning and decision-making, leaves the cues unchanged, but manipulates the subjective value of the associated reward instead. Subjects first show a preference for one of two cues, with each cue reliably associated with a particular reward (e.g., cue A predicts bananas, cue B peanuts). Next, they are provided with ad libitum access to the preferred reward, which they eat to satiety. They are then offered choices between the same two cues, but now in extinction: they do not receive and experience the rewards in the sated condition. This should lead to a reduction in the subjective value of the cue associated with the outcome that has been consumed to satiety. OFC lesions in macaques reliably impair this effect: these animals continue to choose the cue associated with the now-sated outcome, failing to update their value prediction in light of changes to their (internal) state (Rudebeck et al. 2013). Selective satiety manipulations have also been carried out in humans studied with fMRI, with signal in both medial and lateral OFC tracking the subjective value of cues as this value-dropped post-satiation (Valentin et al. 2007).

This paradigm illustrates a key concept in value-based decision-making, which is the subjective and changeable nature of "value." The organism must not only learn what outcome is predicted, but assay the current value of that predicted outcome in relation to internal state (hunger, etc.) and to external factors (e.g., what else is available).

2.1.3 Probability, Uncertainty, and Risk

The likelihood of a particular outcome somehow must enter into assessments of value. Risky decisions are of particular interest in economic research. When a risky choice is to be made repeatedly, if the probability and amount of the potential reward are known, a normative ("rational") solution can be calculated. This is called expected value (EV = probability × amount). Mathematically, a 100% chance of 5 $ is equivalent to a 50% chance of 10$. However, humans generally do not treat these choices as equivalent, preferring the sure thing over the risky option when expected values are equal, and requiring a substantial premium to accept the risk, at least in the domain of gains. The opposite pattern is seen in the domain of losses, where people are more willing to choose the risky option to avoid a loss (Try it for yourself: Do you prefer a certain 5$ loss, or a coin toss where heads means a loss of 10$, but tails means you lose nothing?) (Kahneman and Tversky 1979).

The brain basis of risky decisions of this kind has also been heavily studied. Functional MRI has demonstrated that regions sensitive to reward under conditions of certainty, such as striatum, vmPFC, and DMF also show activation that scales with the expected value of risky gambles, increasing for higher chances of larger rewards, and decreasing for higher chances of larger losses (Tom et al. 2007). Risky choices can be characterized in terms of the variance (uncertainty), as well as the probability, of the outcomes, and uncertainty also modulates activity in reward-sensitive brain regions (Schultz et al. 2011).

Patients with focal damage to various brain regions have been assessed with tasks requiring formal trade-offs between probability and magnitude of monetary outcomes, testing the necessary contributions of these regions to risky decision-making. One study reported that vmPFC damage was associated with less risk aversion, compared to healthy people. That is, these patients bet more money across the board. However, they remained sensitive to the odds, scaling their bets to the probability of winning. The same study found that patients with insula damage also made larger bets overall, but were less inclined to adjust their bets to the odds, suggesting an insensitivity to risk in general, rather than to loss aversion in particular (Clark et al. 2008). Earlier work with the same task also found an effect of amygdala damage on risky choice, similar to that seen after ventromedial frontal damage. More recent work with a different task found that VMF damage was associated with fewer gambles in a gain condition, and more in a "loss" condition compared to healthy controls (Pujara et al. 2015). One general difficulty with interpreting changes in risk-taking in patient populations is that healthy controls (like most people) are typically quite risk averse in the domain of gains, and risk-taking to avoid losses. Thus, non-specific impairments in the task (perhaps due to general difficulties estimating value) are likely to be reflected in greater risk-taking (for gains), and less risk-taking (for losses) because there is more scope for behavior to deviate in those directions.

Although financial and medical risks sometimes can be expressed in formal mathematical terms, decisions more often require an "intuitive" estimate of risk based on experience. Indeed, the clash between this source of information and formal knowledge of risk is the source of well-known decision biases. For example, the gambler's fallacy, in which people treat a sequence of chance events (repeated flips of a coin, for example) as though the outcome of one trial depended on the outcomes of prior trials. This well-known distortion likely arises from judging probability based on trial-by-trial experience: the brain is tuned to detect patterns across trials, and tends to find them even when they are spurious since, unlike with coin tosses, in many of the "foraging" situations so important to survival these patterns do exist, i.e., a "lucky" fishing spot really may have a higher probability of yielding more fish, cast after cast. Insula damage (and not damage to amygdala or ventral frontal lobe) disrupts at least some of these risk-related biases (Clark et al. 2014).

The experiential assessment of risk through repeated sampling makes evident contact with the probabilistic reward and reversal learning paradigms discussed above. Thus, progress in understanding the brain basis of dynamic learning from reward, an enterprise firmly rooted in animal learning theory, may well shed light on some "irrational" aspects of human economic decision-making under risk.

2.1.4 Foraging

Successful foraging depends on the ability to identify cues that predict reward, to gauge the rate at which rewards are, on average, available in the environment, compare that to the rate at which they are being harvested in the current state, and estimate the relative costs of obtaining the current rewards or switching to others. Costs might include delay, physical (or cognitive) effort, or exposure to dangers, such as predation, for example. Considering motivated behavior within a foraging framework again emphasizes the need for flexibility: decisions typically pit exploiting a current opportunity against searching for something better (exploration). This perspective has yielded some interesting insights into the brain regions engaged in decision-making. In contrast to the economic and reward learning approaches, foraging has emphasized potential contributions from dorsomedial PFC, notably dorsal anterior cingulate cortex (dACC). Single unit recordings from dACC in macaques performing a simple foraging task found evidence for a signal encoding the relative value of exploiting a current (dwindling) reward "patch" or exploring other options (Hayden et al. 2011). A human fMRI study also found evidence for signals in dACC that represented the average reward available in the task, the current and potential rewards within that setting, and the costs of switching away from the current option (Kolling et al. 2012).

Reward learning experiments have provided a different perspective on DMF/dACC: As mentioned above, flexibly learning the reward value of stimuli, as in probabilistic or reversal learning, does not rely critically on this part of the brain (Tsuchida et al. 2010). However, damage to DMF in humans and specifically to dACC in macaques has been shown to disrupt learning the value of distinct actions (Camille et al. 2011a; Rudebeck et al. 2008). Both those experiments showed dissociations, with preserved stimulus-reward learning in the DMF-damaged groups. One view of these results is that DMF may represent the predicted value of an action (an ongoing action, and perhaps, in foraging contexts, also the predicted value of alternative actions). This prediction could guide choice (i.e., behavioral shifts), and be drawn upon as a component of the signal to drive learning, i.e., to calculate the mismatch between predicted and experienced outcomes (Quilodran et al. 2008). We recently found evidence that another, more dorsal region within DMF [pre-supplementary motor area (preSMA)] is necessary for maintaining the value of the unattended option in a simple, two-choice value-based preference task (that did not involve learning) (Vaidya and Fellows 2015a).

The dACC in particular, and adjacent regions within DMF, including supplementary motor complex and preSMA, have, of course, been implicated in many other aspects of controlled behavior, notably conflict, error monitoring, and pain processing (Botvinick and Cohen 2014; Modirrousta and Fellows 2008; Holroyd et al. 2004; Lieberman and Eisenberger 2015). There has been some effort to provide a unifying view, proposing that dACC is involved in evaluating the "cost" of exerting cognitive control (Shenhav et al. 2013). It is safe to say that the jury is still out on reconciling these accounts of the processes supported by this much-debated set of brain regions.

2.1.5 Attention and Learning

In experimental tests of reward learning, the stimuli are typically simple (often just a single feature, such as cards distinguished by a color or symbol) and unambiguously different. However, in real-world settings, stimuli are defined by multiple features, and occur in a crowd of other potentially relevant environmental features. Filtering reward-predictive stimuli (or stimulus features) from non-predictive (distracting) features is likely a critical step in optimal learning. This important interplay between attention and reward has been recognized for many decades in the animal learning literature (Mackintosh 1975), but has only recently become a focus in decision neuroscience.

A first important question is when (and how) does a recent history of reward influence ongoing attentional allocation? Behavioral studies have demonstrated that reward can influence feature-based and spatial visual attention, even when the reward is incidental to the task (Anderson et al. 2011; Della Libera and Chelazzi 2006; Shomstein and Johnson 2013). Some work has addressed the neural basis of these effects: First, many studies have shown that the potential availability of reward modulates activity in early visual processing areas as well as parietal regions (notably lateral intraparietal sulcus) (Maunsell 2004; Glimcher and Fehr 2014). There is also at least preliminary evidence that OFC or ventromedial PFC is functionally connected to higher order visual areas, and that the strength of this connection can be modulated by the subjective rewardingness of the stimulus being viewed (Lim et al. 2013). We tested whether ventromedial frontal lobe was necessary for (incidental) reward modulation of feature-based attention, and found that damage to that region (and not to other frontal regions) disrupted trial-by-trial reward priming of attention evident in healthy subjects (Vaidya and Fellows 2015b). More work is needed to better define the characteristics of the interplay between reward and attention, and to more clearly establish the neural substrates of these phenomena. This seems a useful direction for further work, because it provides a potential mechanism by which OFC or vmPFC, for example, could influence decision-making and stimulus-reward learning, i.e., by biasing sensory processing toward reward-predictive aspects of the environment.

This putative ventral frontal-attentional mechanism adds a motivational piece to the more general view of the frontal lobes as important in goal-directed behavior. The contributions of lateral PFC to attention have been more intensively studied in this regard, with good evidence for critical contributions of these regions to shifting attentional sets, i.e., attending to a new feature of a task in response to an instruction or new task demand (Dias et al. 1996; Tsuchida and Fellows 2013; Aron et al. 2004). Attentional shifting (where the subject must learn to attend to a new feature of a compound stimulus to successfully perform the task) has been doubly dissociated from affective shifting (where the subject must learn the new value of an otherwise unchanged stimulus feature, as in reversal learning) after lateral and ventral frontal lobe damage, respectively (Dias et al. 1996). Lateral PFC may thus be important in "top-down" biasing of attention to specific features of complex stimuli in decision contexts that involve pursuing an explicit goal that is not the pre-potent response to these stimuli. For example, caloric density and predicted tastiness may dominate spontaneous assessment of food options (Tang and Fellows 2014), while considerations of healthiness may require more effortful control to guide attention to those aspects. The latter may particularly engage lateral PFC, while the former likely does not (Hare et al. 2011).

3 Decision Contexts: Current Workspace and Future Time

3.1 Models and Schemas

Computational models have been brought to bear in efforts to formalize the mechanisms that might underlie reward learning. Among other insights, the work has drawn distinctions between so-called "model-free" and "model-based" learning. Model-free learning produces stimulus-response associations, a process that can be described by temporal difference models (Barto 1994). This form of learning, also termed "habit learning," is robust and produces rapid, automatic stimulus-response tendencies, but requires many trials with consistent feedback, and is not readily generalized to new contexts. In contrast, model-based (goal-directed) learning is thought to involve more than the accumulated experience of a reward following a stimulus-response sequence. Model-based learning involves, for example, knowledge of the broader task structure, and reasoning about potential future outcomes based on generalization of past experience. That knowledge could include prioritization of currently relevant task dimensions, identified either explicitly (e.g., through instructions) or through learning. There is evidence that model-free and model-based reinforcement learning rely on different neural substrates, with the former reliant on the basal ganglia, the latter more heavily engaging various frontal regions (Doll et al. 2015). OFC in particular has been proposed as important in model-based learning, particularly required when value must be "inferred," i.e., updated based on specifics of the current context, rather than simply read out from associative memory. This region could hold information about context, both in terms of relevant (reward-predictive) environmental features, and expectations about the timing of reward (Stalnaker et al. 2015). This has been referred to as a flexible cognitive map or task-space representation (Wilson et al. 2014).

The concept of parallel neural systems for value-based learning is consistent with the extensive rodent literature on multiple memory systems, which proposes that a network centered on the basal ganglia is critical for habit learning, while a hippocampus-related system is important for stimulus-outcome learning (McDonald and White 1993). These forms of learning likely occur in parallel, although each may be more or less engaged, depending on the details of the task. Thus, lesion studies that disrupt one memory system can "unmask" the behavioral influence of other systems. For example, selective OFC lesions in macaques performing a challenging three-choice probabilistic stimulus-reward learning task found that medial and lateral OFC lesions had distinct effects on trial-by-trial learning. Lateral OFC lesions led to impairment in so-called "credit assignment," with the reward history of recent trials more diffusely associated with the outcome of a particular trial (Walton et al. 2011). Thus, these animals are still learning from reward, but with a system that has less fidelity in linking a specific choice to a specific outcome, in a multi-trial context.

A different perspective on the role of vmPFC/OFC comes from the observation that damage to vmPFC, typically after anterior communicating artery aneurysm rupture, occasionally results in a striking confabulatory amnesia (Gilboa et al. 2006; Moscovitch and Moscovitch 1989). Patients produce improbable, if not outright bizarre, memories, and strongly believe these false recollections are correct. This is not a consistent feature of vmPFC damage, and tends not to persist, although milder phenotypes that may not be clinically evident can be elicited in more patients with such damage with specific testing (Ghosh et al. 2014). One idea about the basis of this phenomenon is that it arises from impairment in strategic retrieval, with vmPFC damage disrupting the usual scaffolding of memory retrieval and monitoring by context (the term "schema" is often used in this literature) (Gilboa et al. 2006). Several studies have shown deficits in schema-facilitated processes after vmPFC damage: For example, there is evidence that vmPFC damage (when associated with a history of confabulation) disrupts the ability to rate words as related to a particular scenario (e.g., a visit to the doctor) (Ghosh et al. 2014), and that such damage reduces the declarative memory advantage provided by learning words or objects in a congruent (compared to an incongruent) context (Spalding et al. 2015). One wonders if this literature is identifying the same vmPFC-reliant process as is tapped by the context-sensitive reinforcement learning tasks described above: are these all instances where information must be situated in a specific context to be effectively learned, recalled, or brought to bear for a decision? The many differences between these paradigms make it difficult to go beyond loose speculation in this regard;

more work is needed to explore the connections between these different ways of thinking about learning and memory in relation to vmPFC.

3.2 Multi-attribute Decision-Making

Model-based learning of course requires a model, which raises questions about the brain mechanisms that allow us to prioritize information for decision-making. As discussed above, attentional mechanisms and reward associations can each contribute, separately or perhaps in some integrative way, perhaps optimizing a basic schema built up from semantic knowledge in many cases. One way of defining such models or "state spaces" is to examine the process of information acquisition in multi-attribute decision tasks (Payne 1992). We asked participants to discover the information they wanted in order to decide between several different apartments, providing them with a table of potential information about, for instance, rent, location, and size. Participants clicked on the cells of this table to reveal the information, and selected their preferred apartment whenever they felt ready to decide. In this particular task, healthy participants and patients with frontal damage sparing vmPFC-OFC generally acquired information in a way that allowed them to compare options across attributes, i.e., first they compared the rent across several options, then they compared the sizes. In contrast, patients with vmPFC-OFC damage collected information about a given option (i.e., the size, rent, and location of a particular apartment) before moving to the next option. These differences in information acquisition led to different choices, and suggest the vmPFC-OFC is somehow required for attribute-based comparisons (Fellows 2006). These findings also highlight the many routes to making a decision. Much as there seem to be multiple memory systems, there also appear to be multiple decision strategies relying on distinct neural mechanisms.

3.3 Heuristics

Heuristics refer to one set of decision strategies that have been extensively studied in decision psychology: A full, "rational" analysis of many decisions is beyond the cognitive ability of most of us, most of the time. Instead, we apply rules of thumb or simplifying strategies to make decisions that are generally good enough, saving time and cognitive resources (Marewski et al. 2010). For example, the "recognition heuristic" may be used to make decisions in which full knowledge of the options is not available, or if available, too onerous to acquire and synthesize. For example, if buying stock based only on the name of a company, you are likely better off with a company you recognize, than one you do not. This "shortcut" based on mere recognition works because you are more likely to know about large, profitable companies than the many smaller companies which, as a group, are less profitable on average. This particular heuristic obviously relies on recognition memory, which brings us back to ideas about context or schemas (Goldstein and Gigerenzer 2002).

These ideas also highlight that there are multiple routes to making a decision, i.e., that we must, in some sense, decide how to decide. This can be an explicit undertaking, as in choosing to carefully scrutinize a restaurant menu or to instead follow some predefined heuristic such as ordering whatever the waiter recommends, or whatever is ordered by your dinner companion. The decision strategy may also be imposed by the circumstances, with heuristics favored under extreme time pressure or in very information-poor contexts. Further, there is some evidence that tendencies toward particular decision strategies are personality traits. For example, individuals vary in the tendency to follow a maximizing strategy (searching exhaustively for the best available option) or a satisfying strategy (taking the first option that meets some pre-set criterion for acceptability, i.e., is "good enough") (Schwartz et al. 2002; Misuraca et al. 2015). The neural basis of these heuristic and "meta-decision" behaviors have been little studied, as yet, but are likely to be quite important in providing a full understanding of the brain basis of decision-making.

3.4 Social Decision-Making and Decisions About Social Options

Most of the work in decision neuroscience has focused on "economic" decisions related to primary or secondary rewards such as food or money, undertaken by an individual decision-maker, in simplified and highly structured tasks. There has also been interest in economic decision-making that involves interaction or negotiation between people. This has been most heavily studied in the context of economic games, such as the Dictator or Ultimatum games, examples of highly structured economic exchanges between two people. The Ultimatum game, for example, involves one person offering to split a sum of money with another person, who decides whether to accept the offer (both parties get the portion of the money as offered) or to reject it (neither party gets any money). Respondents often reject "unfair" offers (less than 50:50 splits), although this behavior is economically irrational since any amount of money is better than nothing.

These games bring additional complexity to the basic challenge of value assessment, with considerations of perspective-taking, intentionality, and social norms like fairness. A clear view of the neural substrates of these rather complex additional considerations has not yet emerged. The focus to date has largely been on insula, vmPFC, and dorsal ACC, regions that are implicated in more basic forms of decision-making and reinforcement learning, as described above. There is little consensus on the roles of these regions in these economic games in particular. For example, different studies have reported increased (Koenigs and Tranel 2007) or decreased (Gu et al. 2015) likelihood of rejecting unfair offers after vmPFC damage, tested in variations of the Ultimatum game. The extant functional MRI literature using this task has recently been reviewed (Gabay et al. 2014).

Other work has examined what might be considered decisions about the "value" of social stimuli, i.e., people; for example, selecting between political candidates or job applicants. There is some evidence that the same value-related regions identified in economic tasks are also involved in these social choices (Bartra et al. 2013; Ruff and Fehr 2014), although there may be important differences between different types of social choices (Krastev et al. 2016). Like other naturalistic decision options, even quite basic social stimuli (for example, black and white photos of faces) are characterized by multiple attributes which must somehow be weighted, traded-off or integrated to inform a value-based choice. The capacity to integrate and compare distinct stimulus attribute values may rely on OFC (Hunt et al. 2014): we found that damage to lateral OFC in particular disrupted political choice in a simulated "election" paradigm. The voting choices of control participants could be predicted by both the difference in subjective attractiveness and apparent competence between any two candidates. Although lateral OFC damage did not affect the ability to rate either attractiveness or competence, based on photos of the candidates, how these patients voted was only consistently related to attractiveness differences (Xia et al. 2015). As in the economic multi-attribute task discussed above, this may reflect a simplifying strategy, or perhaps impaired integration of attribute values.

3.5 *Time and Memory in Decision-Making*

This chapter began with the assertion that decision-making is inherently future-oriented, and then reviewed work on dynamic reward-related learning that showed how past reward experiences are integrated and linked to stimuli, and, separately, to actions, and then influence current choices. Time comes into decision-making in other ways as well. The future is a big place: how do decision-making processes in the brain represent the future, in general, and the value of future outcomes in particular? The second question has been considered at length, influenced by behavioral economics. Behavioral studies in pigeons, rats, monkeys, and humans show that delay systematically shrinks the subjective value of reward. You can see for yourself that 10$ right now is more appealing than 12$ in six months. Some people might even find 10$ now more appealing than 40 or 50 $ in 6 months. This phenomenon is called temporal discounting. It is a nonlinear phenomenon that is typically well-described by a hyperbolic discounting function, with value dropping quickly over short delays, and eventually flattening out, leading to irrational flip-flops in preference, i.e., you might strongly prefer 10$ now over 15$ in six months, but chances are you would prefer 15$ in 18 months over 10 $ in 12 months, even though both choices involve waiting 6 months for an extra 5 dollars (Ainslie 2001). In a given context, individuals tend to be rather consistent in their temporal discounting rate, but there is wide variability across individuals. This variability has been linked to real life impulsive behavior. For example, drug addicts tend to have steeper discounting than non-addicts (Monterosso and Ainslie 2007). Steep discounters are sometimes described as having "myopia for future rewards."

The neural basis of temporal discounting is not settled. FMRI studies have supported conflicting models: one proposes a struggle for decision control between a limbic, vmPFC-centered system representing immediate rewards, and a "rational," cognitive system related to lateral PFC (McClure et al. 2004). However, other work has shown that vmPFC carries a signal that scales with the individual subjective value of immediate and delay-discounted rewards, suggesting this region may provide an integrated "read-out" of the current value of options whether they are immediate or more distant (Kable and Glimcher 2007). Lesion studies are well-placed to adjudicate between these two accounts. That literature is not entirely concordant, with one study of a relatively large sample finding no effects of ventromedial frontal damage on temporal discounting rate (Fellows and Farah 2005), and a second study of a similar, but somewhat smaller, sample reporting steeper discounting after such damage (Sellitto et al. 2010). Neither of these results agrees with the two-system model, which would predict shallower discounting after ventromedial frontal damage. The common currency view would not predict a systematic effect of such damage on temporal discounting, except perhaps increased inconsistency in choices across trials, with delay simply adding an element of difficulty to what is basically a simple value-based preference judgment. We have observed such inconsistencies in patients with ventromedial frontal damage performing other kinds of preference judgments (Henri-Bhargava et al. 2012; Camille et al. 2011b).

Temporal discounting has been extensively studied, but there are also other important aspects of time to consider in relation to decision-making. Value may shrink as it is pushed into the future, but the very extent of "the future" itself is also an important consideration (Atance and O'Neill 2001). Indeed, perhaps ideas about decision "state space" (Wilson et al. 2014) should be expanded to include "state space and time." The window of time that spontaneously comes to mind when an individual is asked to contemplate the future varies. We found that one measure of this so-called "future time perspective" was specifically reduced in patients with ventromedial frontal damage, compared to patients with other frontal lobe damage, and to those with non-frontal brain injury (Fellows and Farah 2005). These same patients were not systematically steeper temporal discounters, suggesting that they were not "myopic" for future rewards, but rather were short-sighted in how long they considered as the future. The hippocampus has also been implicated in future time perspective, or more generally in the ability to imagine future events in detail (Andelman et al. 2010); interestingly, as for those with ventromedial frontal damage, a recent study found that temporal discounting is also intact after medial temporal lobe damage sufficient to disrupt episodic future thinking (Kwan et al. 2013).

4 Conclusions

This chapter has provided an overview of several of the key topics addressed in the emerging field of decision neuroscience, from the perspective of the component processes that contribute to decisions, and the neuroanatomical substrates of those processes. The emphasis on prefrontal subregions reflects the biases of the author, and also to some degree the initial focus of the field. I have shown that decision-making is closely intertwined with learning and memory, in ways that are only partly understood. Reward leaves its mark on the brain in several ways, and it is likely that learning is brought to bear for future decisions in several ways as well. There has been considerable progress in understanding some aspects of reinforcement learning in relation to decision-making, but much remains to be done. There is an emerging interest in better understanding how decision options are identified, their attributes prioritized, and how decision strategies are selected. At least some of these questions seem to make contact with similar concerns in relation to how declarative memories are organized and recalled. Looking forward, decision neuroscience has much to gain from the more mature fields of learning and memory, and hopefully something to offer in return as we bring together ideas about how the brain represents the past to chart a better course into the future.

References

Ainslie G (2001) Breakdown of will. Cambridge University Press, Cambridge

Andelman F, Hoofien D, Goldberg I, Aizenstein O, Neufeld MY (2010) Bilateral hippocampal lesion and a selective impairment of the ability for mental time travel. Neurocase Case Stud Neuropsychol Neuropsychiatry Behav Neurol 16(5):426–435

Anderson BA, Laurent PA, Yantis S (2011) Value-driven attentional capture. Proc Natl Acad Sci USA. 108(25):10367–10371

Aron AR, Monsell S, Sahakian BJ, Robbins TW (2004) A componential analysis of task-switching deficits associated with lesions of left and right frontal cortex. Brain 127(7):1561–1573

Aston-Jones G, Cohen JD (2005) An integrative theory of locus coeruleus-norepinephrine function: adaptive gain and optimal performance. Annu Rev Neurosci 28:403–450

Atance CM, O'Neill DK (2001) Episodic future thinking. Trends Cogn Sci 5(12):533–539

Barto AG (1994) Adaptive Critics and the Basal Ganglia. Model Inf Process Basal Ganglia 394

Bartra O, McGuire JT, Kable JW (2013) The valuation system: a coordinate-based meta-analysis of BOLD fMRI experiments examining neural correlates of subjective value. Neuroimage. 76:412–427

Bechara A, Damasio H, Tranel D, Damasio AR (1997) Deciding advantageously before knowing the advantageous strategy. Science 275:1293–1295

Botvinick MM, Cohen JD (2014) The computational and neural basis of cognitive control: charted territory and new frontiers. Cogn Sci 38(6):1249–1285

Budhani S, Marsh AA, Pine DS, Blair RJR (2007) Neural correlates of response reversal: considering acquisition. Neuroimage 34(4):1754–1765

Butter C (1968) Perseveration in extinction and in discrimination reversal tasks following selective frontal ablations in Macaca mulatta. Physiol Behav 4:163–171

Camille N, Tsuchida A, Fellows LK (2011a) Double dissociation of stimulus-value and action-value learning in humans with orbitofrontal or anterior cingulate cortex damage. J Neurosci 31(42):15048–15052

Camille N, Griffiths CA, Vo K, Fellows LK, Kable JW (2011b) Ventromedial frontal lobe damage disrupts value maximization in humans. J Neurosci 31(20):7527–7532

Chaumon M, Kveraga K, Barrett LF, Bar M (2013) Visual predictions in the orbitofrontal cortex rely on associative content. Cereb Cortex 1–9

Clark L, Bechara A, Damasio H, Aitken MRF, Sahakian BJ, Robbins TW (2008) Differential effects of insular and ventromedial prefrontal cortex lesions on risky decision-making. Brain 131:1311–1322

Clark L, Studer B, Bruss J, Tranel D, Bechara A (2014) Damage to insula abolishes cognitive distortions during simulated gambling. Proc Natl Acad Sci USA 111(16):6098–6103

Cools R, Robbins TW (2004) Chemistry of the adaptive mind. Philos Trans A Math Phys Eng Sci. 362(1825):2871–2888

Dagher A, Robbins TW (2009) Personality, addiction, dopamine: insights from Parkinson's disease. Neuron 61(4):502–510

Daw ND, Niv Y, Dayan P (2005) Uncertainty-based competition between prefrontal and dorsolateral striatal systems for behavioral control. Nat Neurosci 8(12):1704–1711

Della Libera C, Chelazzi L (2006) Visual selective attention and the effects of monetary rewards. Psychol Sci 17(3):222–227

Dias R, Robbins TW, Roberts AC (1996) Dissociation in prefrontal cortex of affective and attentional shifts. Nature 380(6569):69–72

Doll BB, Duncan KD, Simon DA, Shohamy D, Daw ND (2015) Model-based choices involve prospective neural activity. Nat Neurosci 18(5):767–772

Fellows LK (2004) The cognitive neuroscience of human decision making: a review and conceptual framework. Behav Cogn Neurosci Rev 3:159–172

Fellows LK (2006) Deciding how to decide: ventromedial frontal lobe damage affects information acquisition in multi-attribute decision making. Brain. 129:944–952

Fellows LK, Farah MJ (2003) Ventromedial frontal cortex mediates affective shifting in humans: evidence from a reversal learning paradigm. Brain. 126:1830–1837

Fellows LK, Farah MJ (2005a) Different underlying impairments in decision-making following ventromedial and dorsolateral frontal lobe damage in humans. Cereb Cortex 15:58–63

Fellows LK, Farah MJ (2005b) Dissociable elements of human foresight: a role for the ventromedial frontal lobes in framing the future, but not in discounting future rewards. Neuropsychologia 43:1214–1221

Fellows LK, Farah MJ (2007) The role of ventromedial prefrontal cortex in decision making: Judgment under uncertainty or judgment per se? Cereb Cortex 17:2669–2674

Frank MJ, Seeberger LC (2004) O'reilly RC. By carrot or by stick: cognitive reinforcement learning in parkinsonism. Science 306(5703):1940–1943

Gabay AS, Radua J, Kempton MJ, Mehta MA (2014) The Ultimatum Game and the brain: a meta-analysis of neuroimaging studies. Neurosci Biobehav Rev 47:549–558

Ghods-Sharifi S, Haluk DM, Floresco SB (2008) Differential effects of inactivation of the orbitofrontal cortex on strategy set-shifting and reversal learning. Neurobiol Learn Mem 89 (4):567–573

Ghosh VE, Moscovitch M, Melo Colella B, Gilboa A (2014) Schema representation in patients with ventromedial PFC lesions. J Neurosci 34(36):12057–12070

Gilboa A, Alain C, Stuss DT, Melo B, Miller S, Moscovitch M (2006) Mechanisms of spontaneous confabulations: a strategic retrieval account. Brain. 129(6):1399–1414

Glascher J, Adolphs R, Damasio H, Bechara A, Rudrauf D, Calamia M et al (2012) Lesion mapping of cognitive control and value-based decision making in the prefrontal cortex. Proc Natl Acad Sci 109(36):14681–14686

Glimcher P (2002) Decisions, decisions, decisions: choosing a biological science of choice. Neuron 36(2):323–32

Glimcher PW, Fehr E (2014) Neuroeconomics: decision making and the brain, 2nd edn. Elsevier/Academic Press, Amsterdam/Boston
Glimcher PW, Rustichini A (2004) Neuroeconomics: the consilience of brain and decision. Science 306:447–452
Goldstein DG, Gigerenzer G (2002) Models of ecological rationality: the recognition heuristic. Psychol Rev 109(1):75–90
Gu X, Wang X, Hula A, Wang S, Xu S, Lohrenz TM et al (2015) Necessary, yet dissociable contributions of the insular and ventromedial prefrontal cortices to norm adaptation: computational and lesion evidence in humans. J Neurosci 35(2):467–473
Hampton AN, O'Doherty JP (2007) Decoding the neural substrates of reward-related decision making with functional MRI. Proc Natl Acad Sci USA. 104(4):1377–1382
Hampton AN, Adolphs R, Tyszka MJ, O'Doherty JP (2007) Contributions of the amygdala to reward expectancy and choice signals in human prefrontal cortex. Neuron 55(4):545–555
Hare TA, Malmaud J, Rangel A (2011) Focusing attention on the health aspects of foods changes value signals in vmPFC and improves dietary choice. J Neurosci 31(30):11077–11087
Hayden BY, Pearson JM, Platt ML (2011) Neuronal basis of sequential foraging decisions in a patchy environment. Nat Neurosci 14:933–939
Henri-Bhargava A, Simioni A, Fellows LK (2012) Ventromedial frontal lobe damage disrupts the accuracy, but not the speed, of value-based preference judgments. Neuropsychologia 50:1536–1542
Holroyd CB, Nieuwenhuis S, Yeung N, Nystrom L, Mars RB, Coles MGH et al (2004) Dorsal anterior cingulate cortex shows fMRI response to internal and external error signals. Nat Neurosci 7(5):497–498
Hornak J, O'Doherty J, Bramham J, Rolls ET, Morris RG, Bullock PR et al (2004) Reward-related reversal learning after surgical excisions in orbito-frontal or dorsolateral prefrontal cortex in humans. J Cogn Neurosci 16:463–478
Hsu M, Bhatt M, Adolphs R, Tranel D, Camerer CF (2005) Neural systems responding to degrees of uncertainty in human decision-making. Science 310:1680–1683
Hunt LT, Dolan RJ, Behrens TEJ (2014) Hierarchical competitions subserving multi-attribute choice. Nat Neurosci 17(11):1613–1622
Hutcherson CA, Plassmann H, Gross JJ, Rangel A (2012) Cognitive regulation during decision making shifts behavioral control between ventromedial and dorsolateral prefrontal value systems. J Neurosci 32(39):13543–13554
Kable JW, Glimcher PW (2007) The neural correlates of subjective value during intertemporal choice. Nat Neurosci 10(12):1625–1633
Kahneman D, Tversky A (1979) Prospect theory: an analysis of decision under risk. Econom J Econom Soc. 47:263–292
Koenigs M, Tranel D (2007) Irrational economic decision-making after ventromedial prefrontal damage: evidence from the Ultimatum Game. J Neurosci 27(4):951–956
Kolling N, Behrens TEJ, Mars RB, Rushworth MFS (2012) Neural mechanisms of foraging. Science 336(6077):95–98
Krastev S, McGuire, JT, McNeney D, Kable JW, Stolle D, Gidengil E et al (2016) Do political and economic choices rely on common neural substrates? A systematic review of the emerging neuropolitics literature. Front Psychol 25(7):264
Kwan D, Craver CF, Green L, Myerson J, Rosenbaum RS (2013) Dissociations in future thinking following hippocampal damage: evidence from discounting and time perspective in episodic amnesia. J Exp Psychol Gen 142(4):1355–1369
Lieberman M, Eisenberger N (2015) The dorsal anterior cingulate cortex is selective for pain: results from large-scale reverse inference. Proc Natl Acad Sci USA 18:pii: 201515083
Lim S-L, O'Doherty JP, Rangel A (2013) Stimulus value signals in ventromedial PFC reflect the integration of attribute value signals computed in fusiform gyrus and posterior superior temporal gyrus. J Neurosci 33(20):8729–8741
Loewenstein GF, Weber EU, Hsee CK, Welch N (2001) Risk as feelings. Psychol Bull 127:267–286

Mackintosh NJ (1975) A theory of attention: variations in the associability of stimuli with reinforcement. Psychol Rev 82(4):276–298
Marewski JN, Gaissmaier W, Gigerenzer G (2010) Good judgments do not require complex cognition. Cogn Process 11(2):103–121
Maunsell JHR (2004) Neuronal representations of cognitive state: reward or attention? Trends Cogn Sci 8(6):261–265
McAlonan K, Brown VJ (2003) Orbital prefrontal cortex mediates reversal learning and not attentional set shifting in the rat. Behav Brain Res 146(1–2):97–103
McClure SM, Laibson DI, Loewenstein G, Cohen JD (2004) Separate neural systems value immediate and delayed monetary rewards. Science 306(5695):503–507
McDonald RJ, White NM (1993) A triple dissociation of memory-systems—hippocampus, amygdala, and dorsal striatum. Behav Neurosci (Internet) 107(1):3–22
Misuraca R, Faraci P, Gangemi A, Carmeci FA, Miceli S (2015) The decision making tendency inventory: a new measure to assess maximizing, satisficing, and minimizing. Pers Individ Dif 85:111–116
Modirrousta M, Fellows LK (2008) Dorsal medial prefrontal cortex plays a necessary role in rapid error prediction in humans. J Neurosci 28(51):14000–14005
Monterosso J, Ainslie G (2007) The behavioral economics of will in recovery from addiction. Drug Alcohol Depend 90(Suppl 1):S100–S111
Moscovitch M, Moscovitch M (1989) Confabulation and the Frontal Systems: Strategic versus Associative Retrieval in Neuropsychological Theories of Memory. Varieties of memory and consciousness: essays in honour of Endel Tulving, pp 133–61
Murray EA, Wise SP (2010) Interactions between orbital prefrontal cortex and amygdala: advanced cognition, learned responses and instinctive behaviors. Curr Opin Neurobiol 20 (2):212–220
Murray EA, Wise SP, Rhodes SEV (2011) What can different brains do with reward? [Internet]. CRC Press/Taylor & Francis
Noonan MP, Kolling N, Walton ME, Rushworth MFS (2012) Re-evaluating the role of the orbitofrontal cortex in reward and reinforcement. Eur J Neurosci 35(7):997–1010
Padoa-Schioppa C (2011) Neurobiology of economic choice: a good-based model. Annu Rev Neurosci [Internet]. Annual Reviews 34(1):333–359
Payne J (1992) Behavioral decision research: a constructive processing perspective. Annu Rev Psychol 43:87–131
Pujara MS, Wolf RC, Baskaya MK, Koenigs M (2015) Ventromedial prefrontal cortex damage alters relative risk tolerance for prospective gains and losses. Neuropsychologia 79(Pt A):70–75
Quilodran R, Rothé M, Procyk E (2008) Behavioral shifts and action valuation in the anterior cingulate cortex. Neuron 57(2):314–325
Rangel A, Camerer C, Montague PR (2008) A framework for studying the neurobiology of value-based decision making. Nat Rev Neurosci 9(7):545–556
Rescorla RA, Wagner AR (1972) A theory of Pavlovian conditioning: variations in the effectiveness of reinforcement and nonreinforcement. Class Cond II Curr Res Theory 21 (6):64–99
Robbins TW, Arnsten AFT (2009) The neuropsychopharmacology of fronto-executive function: monoaminergic modulation. Annu Rev Neurosci 32:267–287
Rudebeck PH, Murray EA (2014) Review The orbitofrontal oracle: cortical mechanisms for the prediction and evaluation of specific behavioral outcomes. Neuron 84(6):1143–1156
Rudebeck PH, Behrens TE, Kennerley SW, Baxter MG, Buckley MJ, Walton ME et al (2008) Frontal cortex subregions play distinct roles in choices between actions and stimuli. J Neurosci 28:13775–13785
Rudebeck PH, Mitz AR, Chacko RV, Murray EA (2013a) Effects of amygdala lesions on reward-value coding in orbital and medial prefrontal cortex. Neuron 80(6):1519–1531
Rudebeck PH, Saunders RC, Prescott AT, Chau LS, Murray EA (2013b) Prefrontal mechanisms of behavioral flexibility, emotion regulation and value updating. Nat Neurosci 16:1140–1145

Ruff CC, Fehr E (2014) The neurobiology of rewards and values in social decision making. Nat Rev Neurosci 15(8):549–562
Schultz W (2004) Neural coding of basic reward terms of animal learning theory, game theory, microeconomics and behavioural ecology. Curr Opin Neurobiol 14(2):139–147
Schultz W, O'Neill M, Tobler PN, Kobayashi S (2011) Neuronal signals for reward risk in frontal cortex. Ann N Y Acad Sci 1239(1):109–117
Schwartz B, Ward A, Monterosso J, Lyubomirsky S, White K, Lehman DR (2002) Maximizing versus satisficing: happiness is a matter of choice. J Pers Soc Psychol 83(5):1178–1197
Sellitto M, Ciaramelli E, di Pellegrino G (2010) Myopic discounting of future rewards after medial orbitofrontal damage in humans. J Neurosci 30(49):16429–16436
Shenhav A, Botvinick MM, Cohen JD (2013) The expected value of control: an integrative theory of anterior cingulate cortex function. Neuron 79(2):217–240
Shomstein S, Johnson J (2013) Shaping attention with reward: effects of reward on space- and object-based selection. Psychol Sci 24(12):2369–2378
Spalding KN, Jones SH, Duff MC, Tranel D, Warren DE (2015) Investigating the neural correlates of schemas: ventromedial prefrontal cortex is necessary for normal schematic influence on memory. J Neurosci 35(47):15746–15751
Stalnaker T, Cooch N, Schoenbaum G (2015) What the orbitofrontal cortex does not do. Nat Neurosci 18:620–627
Sutton R, Barto A (1998) Reinforcement Learning: an Introduction. MIT Press, Cambridge, Massachusetts
Tang DW, Fellows LK, Dagher A (2014) Behavioral and neural valuation of foods is driven by implicit knowledge of caloric content. Psychol Sci 25:2168–2176
Tom SM, Fox CR, Trepel C, Poldrack RA (2007) The neural basis of loss aversion in decision-making under risk. Science 315(5811):515–518
Tsuchida A, Fellows LK (2013) Are core component processes of executive function dissociable within the frontal lobes? Evidence from humans with focal prefrontal damage. Cortex. 49:1790–1800
Tsuchida A, Doll BB, Fellows LK (2010) Beyond reversal: a critical role for human orbitofrontal cortex in flexible learning from probabilistic feedback. J Neurosci 30:16868–16875
Vaidya A, Fellows L (2015a) Critical prefrontal contributions to fixation-based value assessment and updating. Nat Commun. 14(6):10120
Vaidya AR, Fellows LK (2015b) Ventromedial frontal damage in humans reduces attentional priming of rewarded visual features. J Neurosci 35:12813–12823
Valentin VV, Dickinson A, O'Doherty JP (2007) Determining the neural substrates of goal-directed learning in the human brain. J Neurosci 27(15):4019–4026
Vo K, Rutledge RB, Chatterjee A, Kable JW (2014) Dorsal striatum is necessary for stimulus-value but not action-value learning in humans. Brain 1–7
Waelti P, Dickinson A, Schultz W (2001) Dopamine responses comply with basic assumptions of formal learning theory. Nature 412(6842):43–48
Walton ME, Behrens TEJ, Noonan MP, Rushworth MFS (2011) Giving credit where credit is due: orbitofrontal cortex and valuation in an uncertain world. Ann N Y Acad Sci 1239:14–24
Wilson RC, Takahashi YK, Schoenbaum G, Niv Y (2014) Orbitofrontal cortex as a cognitive map of task space. Neuron 81(2):267–278
Xia C, Stolle D, Gidengil E, Fellows LK (2015) Lateral orbitofrontal cortex links social impressions to political choices. J Neurosci 35:8507–8514

Basal Forebrain Cholinergic System and Memory

M.G. Blake and M.M. Boccia

Abstract Basal forebrain cholinergic neurons constitute a way station for many ascending and descending pathways. These cholinergic neurons have a role in eliciting cortical activation and arousal. It is well established that they are mainly involved in cognitive processes requiring increased levels of arousal, attentive states and/or cortical activation with desynchronized activity in the EEG. These cholinergic neurons are modulated by several afferents of different neurotransmitter systems. Of particular importance within the cortical targets of basal forebrain neurons is the hippocampal cortex. The septohippocampal pathway is a bidirectional pathway constituting the main septal efferent system, which is widely known to be implicated in every memory process investigated. The present work aims to review the main neurotransmitter systems involved in modulating cognitive processes related to learning and memory through modulation of basal forebrain neurons.

Keywords Acetylcholine · Learning · Modulation · Consolidation · GABA · Glutamate · Noradrenaline · Hypocretin · Orexin · Vasopressin · Oxytocin · Substance P

Blake and Boccia—Both authors contributed equally.

M.G. Blake
Facultad de Medicina UBA, CONICET. Paraguay, Instituto de Fisiología y Biofísica Bernardo Houssay, 7th Floor (C1121ABG), 2155, Buenos Aires, Argentina

M.M. Boccia (✉)
Laboratorio de Neurofarmacología de la Memoria, Cátedra de Farmacología, Facultad de Farmacia y Bioquímica, UBA. Junín, 5th Floor (C1113AAD), 956, Buenos Aires, Argentina
e-mail: mboccia@ffyb.uba.ar

Curr Topics Behav Neurosci (2018) 37: 253–273
DOI 10.1007/7854_2016_467
Published Online: 18 February 2017

Contents

1 Cholinergic System: Historical Perspective 254
2 Acetylcholine Receptors and Distribution 255
3 Cholinergic Pathways 256
4 Acetylcholine and Cognitive Functions 257
5 Modulation of Cortical Projecting Cholinergic Neurons 259
6 GABA 259
7 Glutamate 260
8 Catecholamines 261
9 Acetylcholine 262
10 Hypocretin/Orexin and Adenosine 262
11 Vasopressin and Oxytocin 263
12 Substance P 263
13 Neuropeptide-Y 264
14 Galanin 264
15 Concluding Remarks 266
References 266

1 Cholinergic System: Historical Perspective

Acetylcholine (ACh) is known as the main neurotransmitter of the parasympathetic branch (PNS) of the Autonomous Nervous System (ANS). Probably, ACh signaling evolved long before the development of a nervous system since it is well conserved in nature (bacteria, fungi, protozoa and plants), along with the machinery to synthesize and degrade it (Semba 2004). Moreover, as Gaddum has demonstrated, ACh is present in several nonneural tissues, such as the placenta, suggesting that ACh not only acts as a neurotransmitter but also serves other cellular functions (Chang and Gaddum 1933). ACh was the first neurotransmitter to be identified in 1906 by Hunt and Taveou (1906). Soon after, in the 1910s, Henry Dale demonstrated that ACh mimics the effects of parasympathetic nerve stimulation and he also established the distinction between the muscarinic and nicotinic actions of ACh (Dale et al. 1910). It was not known until the 1920s, by the famous, classical, simple and elegant experiments performed by Otto Loewi, using two frog hearts, that ACh was released upon nerve stimulation (Loewi 1921). Loewi's experiments renewed interest in the scientific community of the possibility of chemical involvement in neuronal communication. At that time, ACh had not been found to occur naturally in the body; however, Dale was excited and intrigued by the "vagustoff" substance described by Loewi and commented to a friend:

> We are still struggling with the ACh problem, which I mentioned to you when I saw you in the autumn. I am more and more convinced that the thing is there to be found, if only we can overcome the technical difficulties. (Letter, Dale to Richards, 22 March 1929, Archives of the National Institute for Medical Research, File 647; quoted in (Tansey 1991).
>
> Later on, Dudley and Dale finally demonstrated that ACh was a natural constituent of mammalian bodies (Dale and Dudley 1929).

A seminal step in the study of ACh as a mediator of different effects was the introduction of physostigmine or eserine, a drug that inhibits the activity of the enzyme acetylcholinesterase. Heman George Führer introduced, for the first time, physostigmine to an organ bath in which a leech muscle was suspended. With this pharmacological tool he was able to tackle the transiency of the effects of ACh and the muscle became extremely sensitive to it (Führer 1918). For many years the eserinised leech muscle assay remained the most reproducible, sensitive method of identifying and quantifying ACh. It was not until the 1960s that more sophisticated chemical methods for detecting and measuring ACh were developed. And it was toward the end of that decade that a collaborative project between staff of the University of California and the Karolinska Institute in Stockholm, using a gas chromatography/mass spectrometry technique, was able to determine ACh in submicrogram amounts (Hammar et al. 1968).

The first evidence that ACh is released from cerebral cortex came much later. MacIntosh and Obering were the first to demonstrate an outflow of ACh from the cerebral cortex (MacIntosh and Oboring 1955). They collected the superfusate from the cortex of anesthetized cat using a "surface cup" applied to the cortical surface using an anticholinesterase to prevent breakdown of ACh. ACh was then determined by bioassay. This technique was the forerunner of the actual microdialysis technique which makes it possible to study ACh, and also other substances, released in behaving animals.

2 Acetylcholine Receptors and Distribution

Acetylcholine is recognized by two different families of receptors, nicotinic receptors (nAchRs, ligand-gated ion channels) (Albuquerque et al. 2009) and muscarinic receptors (mAchRs, seven transmembrane G-protein-coupled receptors) (Bubser et al. 2012). So far, five different mAchRs are known (M1–M5). Each subtype has a unique distribution in central and peripheral nervous system, being expressed both pre- and post-synaptically (Levey 1993). While M1 mAchRs are predominantly expressed in all major areas of the forebrain, including the hippocampus, cerebral cortex, corpus striatum and thalamus, the M2 mAchRs are highly abundant in non-cholinergic neurons that project throughout the brain, including the hippocampus and neocortex. M3 mAchRs are widely distributed in the CNS, being found in the hypothalamus and in many other regions including the hippocampus. M4 mAchRs are mainly expressed in the corpus striatum in the CNS and on various prejunctional nerve terminals in the periphery. Finally, M5 mAchRs are distributed in the pars compacta of the substantia nigra and in the ventral tegmental area (VTA; Bubser et al. 2012).

Nicotinic ACh receptors (nAchRs) participate in a variety of physiological functions, including the regulation of neurotransmitter release and neuronal excitability (Levin 1992; Albuquerque et al. 2009). The nAchRs are well distributed in the CNS and also in the peripheral nervous system, immune system, and peripheral tissues (Albuquerque et al. 2009). To date, nine different nAchRs subunits

(α_2–α_7 and β_2–β_4) are known to exist in the mammalian brain, which combine as either homo- or heteromeric complexes into diverse functional pentameric structures (Albuquerque et al. 2009; Fasoli and Gotti 2015). The main subtypes functionally expressed in the brain are the α_7* subunit-containing receptors (homo- or heteromeric) and those including both α and β subunits, named $\alpha_4\beta_2$* and $\alpha_3\beta_4$* (* indicates that nAchRs can contain other α and β subunits). 90% of the high nicotine binding sites in the brain correspond to the $\alpha_4\beta_2$* receptor subtype; the $\alpha_3\beta_4$* nAchR is found primarily in the parasympathetic ganglia and also expressed in a variety of brain areas (interpeduncular nucleus and medial habenula, VTA, BLA, among others) (Albuquerque et al. 2009; Yakel 2014). The α_7* nAchRs are expressed on a variety of cells on the periphery, as well as in several brain regions involved in learning and memory. In the brain, the α_7* receptors are expressed on neurons and non-neuronal cells (such as astrocytes, oligodendrocyte precursor cells, endothelial cells, and microglia) (Yakel 2014). Expression in these non-neuronal cells suggests a potential role in brain neuroprotection, inflammation, and brain immunity (Di Cesare Mannelli et al. 2015; Liu et al. 2015). Neuronal nAchRs are relatively expressed in low density in the human brain compared to mAchRs. Their distribution pattern is homogeneous and not restricted to well-defined brain cholinergic pathways. Binding studies using high affinity radioactive ligands against the nAchR help us to characterize its neuroanatomical distribution. In general terms, nAchRs are present in a variety of brain structures, with high density in the thalamus, caudate nucleus, substantia nigra, nucleus basalis of Meynert (NBM), and some areas with relatively lower levels such as the hippocampus, caudate, putamen and cortex, among others (Albuquerque et al. 2009; Ragozzino and Brown 2013).

3 Cholinergic Pathways

The central cholinergic system was mainly characterized using antibodies against the main enzyme responsible for synthesizing ACh, choline acetyltransferase (ChAT). Ten subpopulations of cholinergic neurons were identified and named Ch1–Ch10. Most of them are projecting neurons and subpopulations of interneurons. The main and most studied cholinergic neurons are those found in the basal forebrain because of its undoubted degeneration in Alzheimer's disease (AD). The basal forebrain contains four overlapping cholinergic and noncholinergic cell groups tangled among each other. The Ch1–Ch4 nomenclature designates the cholinergic neurons within these four cell groups (Table a) (Mesulam 2004b).

Nomenclature	Cholinergic neurons associated with
Ch1	Medial septal nucleus
Ch2	Vertical nucleus of the diagonal band
Ch3	Horizontal limb of the diagonal band nucleus
Ch4	Nucleus basalis of Meynert

Experiments performed with tracers in several animal species have shown that Ch1 and Ch2 mainly innervate the hippocampal formation; Ch3, the olfactory bulb; and Ch4, the rest of the cortex and the amygdala. The Ch4 group contains, in the brain of primates, a component of the NBM. According to Mesulam, the term "nucleus basalis" designates "the cholinergic, as well as noncholinergic, components in the nucleus, whereas Ch4 designation is reserved for its cholinergic neurons" (Mesulam et al. 1983; Mesulam 2004a).

The main cortical inputs to the nucleus basalis are mostly glutamatergic, but can also be GABAergic, and they come from cortical projections from limbic and paralimbic structures, including the amygdala (Zaborszky et al. 1997). In rodents, there are projections from the VTA (dopamine), raphe nucleus (5-HT), and locus coeruleus (noradrenaline) areas reaching the Ch4 spot (Jones and Cuello 1989; Smiley and Mesulam 1999). Moreover, the nucleus basalis has receptors which recognize all these neurotransmitters. Saper (1984), proposed that the cholinergic basal forebrain system "diffusely" innervates the cortex. However, later, Zaborsky et al. demonstrated that the cholinergic and non-cholinergic projections to the neocortex are not diffuse (Zaborszky et al. 1997, 2015). They are organized into overlapping and segregated groups of neurons that may transmit information which originates in the basal forebrain to cortical areas, themselves interconnected. The results described suggest that basal forebrain—cortex projections patterns are very similar to those described for the cortico—striatal pathway (Zaborszky et al. 2015).

4 Acetylcholine and Cognitive Functions

In AD there are several signs of brain dysfunction, mainly memory loss. This loss of memory function was attributed to the profound degeneration of Ch4 neurons and loss of cortical cholinergic innervations (Mesulam 2004a). Moreover, before that, it was already known that the active ingredients of *Atropa belladonna* (Solanaceous family) in moderate doses have a great impact on memory and orientation. However, at that time, it was unknown which were the active ingredients contained in those plants. One of the first studies regarding the brain pharmacodynamics of atropine and related drugs was performed by Macht 1923. However, he did not propose a mechanism of action for their effects—the authors were aware of the parasympathetic activity of these drugs. Later on, several experiments were performed in both animals and humans, which have shown learning and memory impairments after treatment with anticholinergic drugs. Altogether, it has led to "the cholinergic hypothesis of geriatric memory dysfunction," by Bartus et al. (1982) However, in the last decades, the role of Ach in memory has been debated (Blokland 1995; Gold 2003), mainly due to conflicting results obtained with cholinergic lesions (Easton et al. 2012).

At this point it is worth mentioning that memory is not a unitary process. It requires and relies on multiple cognitive functions supported by different brain systems and, moreover, memory requires multiple processes to be formed (encoding, consolidation, retrieval) (Hasselmo 2006; Hasselmo and Sarter 2011). The idea that cholinergic projections from BF to the cortex are involved in synaptic plasticity leading to learning and memory is supported by several observations (reviewed in Zaborszky et al. 1999). When a subject needs to perform and learn a specific task, the enhancement of a sensory representation in the task-relevant sensory cortical area depends on activation of a specific sensory cortex-prefrontal-BF-sensory cortex loop signaling the behavioral significance of the situation.

Cholinergic neurons constitute a way station for many ascending and descending pathways, thus receiving multiple inputs. The BF Ch neurons have a role in eliciting cortical activation and arousal, and it is well established that they are mainly involved in cognitive processes requiring increased level of arousal, an attentive state and/or cortical activation with desynchronized activity in the EEG. Cholinergic neurons present in the medial septum/horizontal diagonal band project to the hippocampus, and the neurons present in the horizontal limb of the diagonal band, substantia innominata, and the peripallidal regions, project to neocortical areas, constituting the nucleus basalis magnocellularis. The different populations of BF neurons appear to be innervated by different combinations of afferents (Zaborszky et al. 1991). Therefore, the BF populations seem to be compartmentalized, each compartment being involved in different cognitive operations (Everitt and Robbins 1997)—that is, the different BF circuits may be involved in modality-specific attention.

Of particular importance within the cortical targets of BF neurons is the hippocampal cortex. The septohippocampal pathway is the main septal efferent system, which is widely known to be implicated in cognitive processes. It is a bidirectional pathway composed of three separate components: cholinergic, GABAergic, and glutamatergic fibers (Goldbach et al. 1998; Pascual et al. 2004, Farr et al. 1999). The GABAergic fibers end on hippocampal interneurons, while cholinergic projections have a wider distribution, ending on many hippocampal cell types (Pascual et al. 2004).

Recently, Dannenberg et al. (2015), described that the septohippocampal system comprises two components: a direct cholinergic projection causing increased firing of hippocampal inhibitory interneurons with concomitantly decreased firing of principal cells, and an indirect pathway involved in hippocampal theta synchronization, comprising noncholinergic neurons within the MSDB that are recruited by cholinergic neurons. Activation of both pathways causes a reduction in pyramidal neuron firing and a more precise coupling to the theta oscillatory phase. These two anatomically and functionally distinct pathways are likely relevant for cholinergic control of encoding versus retrieval modes in the hippocampus (Dannenberg et al. 2015).

The septohippocampal pathway has been implicated in every memory process investigated. A suggestive finding concerning the importance of the septo-hippocampal pathway is the fact that food-storing birds have enlarged hippocampal region (dorso-medial cortex), relative to brain and body size, when compared with the non-storers, and the volume of one of the major afferent-efferent pathways (the septohippocampal pathway) is also greater in food storing species (Krebs 1990).

The septohippocampal pathway is less active in AD (Krügel et al. 2001). In a mouse model of AD, it was shown that tau pathology presents early in the hippocampus and basal forebrain, and several morphological and functional alterations in the septohippocampal pathway suggest that there is a disconnection between both structures in AD and that tau pathology may have a role in cholinergic neurons' degeneration (Belarbi et al. 2009).

Intracerebroventricular injections of the immunotoxin 192 IgG-saporin lesion cholinergic neurons, including those in the basal forebrain (Leanza et al. 1995; Torres et al. 1994). Rats receiving 192 IgG-saporin displayed a significant delay-dependent decline in performance, indicating that this pathway is implicated in short-term memory processing. Administration of the mAChR blocker scopolamine (0.5 mg/kg, i.p.) produced more pronounced impairment in the performance of the normal control rats across all delays, and also induced further impairment in animals with 192 IgG-saporin lesions (Winters and Dunnett 2004).

After a dorsal septohippocampal pathway lesion, BF grafts allow successful cholinergic reinnervation of hippocampal neurons, but do not enhance cognitive functions in rats, and may also have adverse effects after partial septohippocampal system lesions (Dalrymple-Alford 1994). This might indicate that although these grafts reinnervate hippocampus, they probably fail to incorporate into circuits, suggesting that appropriate modulation of BF neurons by afferents appears to be critical for their activity.

5 Modulation of Cortical Projecting Cholinergic Neurons

The cholinergic neurons of the BF providing the major source of the diffuse cortical innervation are the Ch4 subgroup. Some cortical innervation (to medial prefrontal cortex, for example) also appears to be provided by Ch1–3, arising from the medial septum and the diagonal band (Eckenstein et al. 1988; Lamour et al. 1984). The stimulation of Ch4 neurons increases cortical release of ACh, thus causing desynchronization of the electroencephalogram. These cholinergic neurons are modulated by several afferents of BF, corresponding to different neurotransmitter systems.

6 GABA

The administration of GABAA agonists decreases cortical ACh release (Scatton and Bartholini 1982) by inhibiting the firing of cholinergic neurons (Khateb et al. 1998). The intra-NBM administration of the GABAA agonist muscimol impairs performance in the reference and working memory components of rats' performance in a Y-maze task (Smith et al. 1994). GABAA antagonists produce opposite effects (Bertorelli et al. 1991). Cognitive effects observed after the administration of GABAA-modulating drugs correlate with the expected behavior of this pathway. For example, the administration of FG7142 (N-methyl-betacarboline- 3-carboxamide, a partial agonist of the BZD binding site of the GABAA receptor) in the

NBM, increases ACh release to the cortex (Fadel et al. 1996; Moore et al. 1995), and enhances performance on the working memory components of a double Y-maze task (Smith et al. 1994).

However, the administration of the GABAA receptor agonist muscimol and the antagonist bicuculline impairs long-term memory consolidation, but not acquisition, of an inhibitory avoidance memory in rats (Morón et al. 2002). The administration of the cognitive enhancer NS-105 ((+)-5-oxo-D-prolinepiperidinamide monohydrate) reverses the amnestic actions caused by cholinergic dysfunction in a variety of animal models (Ogasawara et al. 1999), enhancing cholinergic neuronal activity by the suppression of GABAB receptor-mediated responses.

GABAergic modulation of cholinergic BF neurons, and also GABAergic projecting neurons from BF, is involved not only in cognitive processes, but also in the sleep-wake cycle.

7 Glutamate

Glutamatergic agonists administered in the NBM increase cortical ACh release (Kurosawa et al. 1989), via stimulation of cholinergic neurons, apparently through AMPA receptors (Page et al. 1993; Weiss et al. 1994; but see also Rasmusson et al. 1996). Again, these pharmacological effects appear to be correlated with behavior. In rats subjected to a task in which darkness is associated with the opportunity to consume a sweetened pellet, the intra-NBM administration of NMDA following exposure to darkness/cereal stimulus potentiated both the magnitude and duration of stimulated cortical ACh release (Fadel et al. 2001). Moreover, kynurenate, a normal product of the metabolism of the amino acid L-tryptophan that blocks ionotropic glutamate receptors (Elmslie and Yoshikami 1985), reduced ACh release in the prefrontal cortex in rats subjected to this task (Fadel et al. 2001).

Glutamatergic modulation of BF cholinergic neurons activity appears to be critical for AD progression. In this sense, Aβ 1-42 increases glutamate levels in the synaptic cleft by inhibiting the astroglial glutamate transporter, thus increasing intracellular Ca^{2+} levels (Harkany et al. 2000), through enhancement of NMDA receptor activity (Molnár et al. 2004), although other possible mechanisms have been proposed. Such glutamate dysregulation has a profound impact on the selective degeneration of BF cholinergic neurons in the early stage of AD, and constitutes the rationale for the clinical use of memantine, a moderate affinity uncompetitive NMDA receptor antagonist that does not interfere with normal NMDA receptor function (Chen and Lipton 2006). It was also shown that activation of metabotropic glutamate receptor mGluR7 protected BF neurons from NMDA-induced excitotoxicity. This protective effect of mGluR7 activation on BF cholinergic neurons is selectively impaired by Aβ accumulation. Hence, this fact suggests an additional potential basis for the Aβ-induced disruption of calcium homeostasis (Gu et al. 2014).

8 Catecholamines

Noradrenergic inputs to the BF seems to arise from collateral connections to cholinergic neurons branching from an ascending visceral projection system (Knox et al. 2004) which cause depolarization and spike-discharges in cholinergic neurons through α_1 adrenergic receptors (Fort et al. 1995). The majority of noradrenergic input to medial septal area, the medial preoptic area, and the substantia innominata is ipsilateral and provided by the locus coeruleus (España and Berridge 2006), degeneration of locus coeruleus neurons has been linked to age-related dementia (Chan-Palay 1991; Palmer and DeKosky 1993) and to AD progression (German et al. 1992). It was shown that this degeneration of locus coeruleus neurons correlates with amyloid plaque formation and neurofibrillary tangles in the areas receiving projections from the locus coeruleus, and also correlates with the severity of dementia (Bondareff et al. 1987), thus suggesting a protective role for noradrenergic innervation of these areas (Heneka et al. 2006). Indirect evidence of a behavioral correlation with this innervation arose from experiments performed with the neurotoxins 6-OH-DA and DSP-4 in an inhibitory avoidance task (Cornwell-Jones et al. 1989), in a radial arm maze task (Heneka et al. 2006), and in a social partner recognition test (Heneka et al. 2006). DSP-4 causes hippocampal and cortical depletion of monoamines, but leaves BF monoaminergic neurotransmission less affected (Fritschy and Grzanna 1989), allowing adrenergic modulation of cholinergic neurons at this level. Catecholaminergic enhancing effects on arousal are exerted by modulation of BF cholinergic neurons (Lelkes et al. 2013). This modulation of BF cholinergic neurons is also involved in non-REM sleep suppression, although these neurons seem not to be implicated in the REM-sleep supressing effect of noradrenaline (Lelkes et al. 2013).

Specifically regarding the septohippocampal pathway, septal neurons are modulated by catecholamines. Dopamine exerts a negative modulatory effect on cholinergic septal neurons. Lesions of dopaminergic afferents by 6-hydroxydopamine lead to a disinhibition of cholinergic neurons, causing a specific increase in cholinergic hippocampal activity in mice (Galey et al. 1989), enhancing spatial discrimination performance related to improved working memory (but not reference memory) in an 8-arm radial maze. On the contrary, norepinephrin modulates these neurons positively through alpha receptors. In this sense, the intra-septal administration of phenoxybenzamine (an alpha-noradrenergic blocker), impairs the cholinergic activation induced by retrieval, causing a selective working memory deficit in the 8-arm radial maze (Marighetto et al. 1989).

9 Acetylcholine

Cholinergic innervation of the basal forebrain neurons comes from the pedunculepontine tegmentum (PPT), the Ch5 group of Mesulam et al. (1983, 1993). These efferents exert widespread control over neocortical EEG activity and aid the maintenance of high-frequency EEG activation during waking and REM sleep. Stimulation of Ch5 neurons with 100 Hz for 2 s, causes EEG desynchronization, an effect markedly reduced by blockers of neural firing (tetrodotoxin, procaine, lidocaine) and also by blockers of synaptic transmission (calcium-free solution plus magnesium or cobalt) on BF cholinergic neurons (Rasmusson et al. 1994; Dringenberg and Olmstead 2003). Although this desynchronizing effect of Ch5 group stimulation involves relays not only in the BF but also in the central thalamus, the BF is the predominant structure (Dringenberg and Olmstead 2003). Infusion of the glutamate antagonist kynurenate within the NBM also reduces the EEG desynchronization elicited by PPT stimulation, suggesting that the major excitatory input to the cholinergic neurons of the NBM from the PPT is exerted via modulation of glutamatergic synapses (Rasmusson et al. 1994), and it has been proposed that M2 presynaptic receptors of glutamatergic neurons of the BF might be involved (Bertorelli et al. 1991; Sim and Griffith 1996).

10 Hypocretin/Orexin and Adenosine

Another major modulatory system of BF involves the orexin neurons, which project widely throughout the CNS. Orexin system projections in the BF modulate cortical ACh release, a fact that could be important for the cognitive components of motivated behavior such as the behavioral adaptation to food deprivation by promoting the detection and selection of stimuli related to physiological needs (Fadel and Frederick-Duus 2008). Orexin/hypocretin also modulates hippocampal function by direct inputs to hippocampal neurons, and a robust innervation of medial septum cholinergic neurons, a pathway that appears to play a critical role in attention (Fadel and Burk 2010). Additionally, orexin/hypocretin release in BF or directly in the hippocampus caused increased GABA and glutamate release from hippocampal neurons. Aging is accompanied by a significant reduction in orexin fiber innervation of GABAergic neurons in the BF, while the direct hippocampal innervation remains unaltered (Stanley and Fadel 2011). This alteration in innervation of BF might contribute to age-related cognitive dysfunctions.

This neuromodulatory system interacts with the neuroregulator adenosine. Adenosine is considered a sleep/wake cycle homeostatic regulator, because a significant increase in extracellular adenosine levels is observed after prolonged wakefulness in specific areas of the brain, particularly the BF and cortex (Porkka-Heiskanen et al. 2000; Basheer et al. 2001, 2007; Kalinchuk et al. 2011). Adenosine promotes sleep by an A1 receptor-mediated inhibition of glutamatergic inputs to cortically projecting cholinergic and GABA/PV neurons. Conversely, blockade of A1 receptors in the BF promotes attentive wakefulness by promoting the high-frequency oscillations in the cortex required for attention and cognition (Yang et al. 2013).

11 Vasopressin and Oxytocin

The two neurohypophyseal hormones have also been proposed to modulate the firing of BF cholinergic neurons. In the human forebrain, the distribution of receptors differs between vasopressin and oxytocin, but they overlap in the brainstem. Vasopressin receptors are present in the dorsal part of the lateral septal nucleus, in midline nuclei and adjacent intralaminar nuclei of the thalamus, in the hilus of the dentate gyrus, and the dorsolateral part of the basal amygdaloid nucleus (Loup et al. 1991). Oxytocin receptors are observed in the NBM, the nucleus of the vertical limb of the diagonal band of Broca, the ventral part of the lateral septal nucleus, the preoptic/anterior hypothalamic area, the posterior hypothalamic area, and variably in the globus pallidus and ventral pallidum (Loup et al. 1991). The presence of oxytocin and vasopressin binding sites in all these areas suggests they play a neuromodulatory role in the central nervous system, probably by modulating cholinergic transmission in the BF. In this sense, the administration of oxytocin to Swiss mice impairs consolidation of avoidance memory, and this impairment is prevented by the anticholinesterase drug physostigmine. Additionally, blockade of oxytocin receptors enhances memory consolidation in mice, an effect blocked by antagonists of central cholinergic receptors (muscarinic and nicotinic), suggesting that oxytocin modulates ACh release in downstream neurons related to consolidation of avoidance memory (Boccia and Baratti 2000). Following the same line, arginine-vasopressin enhances memory consolidation of an avoidance memory through V1 receptors, and an antagonist of these receptors impairs memory (Boccia et al. 1998; Tanabe et al. 1999). These effects of vasopressin seem to be exerted by positively modulating ACh release in cholinergic neurons, since vasopressin reversed the memory deficit caused by administration of scopolamine either in an inhibitory avoidance in mice (Tanabe et al. 1999), and in an eight-arm radial maze in rats (Mishima et al. 2001).

12 Substance P

Substance P is a neurokinin widely distributed in the brain, particularly important in the hippocampus, diencephalon, BF cholinergic neurons, and in the amygdala (Ribeiro-da-Silva and Hökfelt 2000). Substance P administered in the NBM (Huston and Hasenöhrl 1995) or in the medial septal nucleus (Stäubli and Huston 1980) exerts memory enhancing effects. The related modulator neuropeptide K also enhances memory retention when injected into the rostral and caudal portions of the hippocampus and the amygdala, but exerts no effect when injected into the septum (Flood et al. 1990).

13 Neuropeptide-Y

Neuropeptide-Y (NPY) is present both in local neurons as well as in fibers in the BF, particularly in the more caudal areas, and has been proposed to have a role in locomotor activity, eating behavior, stress responses, memory processing, blood pressure, some neuroendocrine functions, and also in the integration of sleep and behavioral stages via the BF (Tóth et al. 2007; Wettstein et al. 1995). Neuropeptide Y was proven to be affected in AD, and there is also clinical evidence for its implication in depression, schizophrenia and anorexia nervosa, among others (Wettstein et al. 1995; Eaton et al. 2007). NPY-containing neurons have wide axonal arborizations, and they are believed to modulate the GABA/Ach interactions in the BF. NPY enhanced avoidance memory in a T-maze in mice when administered into the rostral portion of the hippocampus and septum, but impaired retention when injected in the caudal portion of the hippocampus and amygdala, and its interaction with the central cholinergic system is evidenced by its ability to reverse scopolamine-induced amnesia (Flood et al. 1987, 1989). These effects of NPY on memory retention appear to be mediated through presynaptic (Y2) NPY receptors (Flood and Morley 1989). It was also reported that septal cholinergic efferents in the dentate gyrus exert a powerful modulation of NPY-containing interneurons (Dougherty and Milner 1999). In a colchicine-induced AD-like condition in rats (a rat model of AD) intracerebroventricular administration of nicotine induces reversal of amnesia in a Morris water maze (Rangani et al. 2012). The administration of NPY mimics nicotine effects, whereas an NPY-Y1 receptor antagonist impairs the nicotine-induced reversal of amnesia (Rangani et al. 2012), showing that the memory enhancing effects of ACh through nicotinic cholinergic receptors may be mediated, at least in part, by NPY interneurons. Regarding the possible involvement of NPY signaling on AD, it was found that intracerebroventricular administration of aggregated Aβ(1–40) induced depressive-like behavior and spatial memory impairment, but these effects were prevented by pretreatment with NPY, and these effects are probably mediated by NPY-Y2 receptors (dos Santos et al. 2103). Thus, this peptide has a relevant physiological role as a modulator of memory processing within the BF and the hippocampus.

14 Galanin

Galanin is widely distributed within the central and peripheral nervous system, exerting behavioral and non-behavioral actions. In the BF neurons of the nucleus basalis of Meynert/medial septum/diagonal band innervating the cerebral cortex and hippocampus, galanin is an inhibitory modulator of cholinergic transmission (Crawley 1996). Galanin does not colocate with ACh in the BF of humans and great apes, but it does in monkeys, suggesting that an evolutionary change occurred in galanin-ACh coexistence within the primate BF, at the branch point between monkeys and apes (Benzing et al. 1993). The degree of cognitive decline observed in elderly patients is highly correlated with the magnitude of the reduction in central cholinergic activity, independent of age (Bartus et al. 1982; Francis et al. 1999), and the same applies for AD patients (Bierer et al. 1995; Neugroschl and Wang 2011; Querfurth and LaFerla 2010). Among all of them, the most consistent marker of neuronal loss in AD is the decline in number of cholinergic neurons of the NBM (Coyle et al. 1983). Dramatic reductions in ChAT and AChsterase are routinely seen in postmortem samples of BF and cortical samples from AD patients, as compared to age-matched controls (Bierer et al. 1995). Also, cholinergic nicotinic receptors were found to be reduced in 30–40%, mainly due to a reduction of the $\alpha_4\beta_2$ subtype, with relative preservation of the α_7-nicotinic receptors (Court et al. 2001; Perry et al. 1995). However, cholinergic dysfunction does not provide a complete account of age-related cognitive deficits, and age-related changes in cholinergic function typically occur within the context of changes in several other neuromodulatory systems (Decker and McGaugh 1991). Remarkably, in post-mortem studies performed in the BF of AD patients, galanin concentrations are much higher in the NBM than in age matched controls (Beal et al. 1990). Histological analyses show that galanin hyperinnervates cholinergic neurons in the BF in AD patients, and this hypertrophic network widens with the severity of AD symptoms (Chan-Palay 1988; Mufson et al. 1993). Galanin injected in the medial septum/diagonal band, inhibits ACh release in the hippocampus, in vivo and in vitro (Fisone et al. 1987), and this effect is blocked by galanin receptor blockers (Bartfai et al. 1991). It appears that this inhibitory effect of galanin on cholinergic projecting neurons is exerted by interaction with presynaptic receptors (Dutar et al. 1989), and occurs physiologically only at high discharge frequency (Hökfelt et al. 1987). That is, endogenous galanin inhibits ACh release only when the neuron is firing at high rates, and this is hypothesized to happen during the progressive loss of the large majority of BF cholinergic neurons in AD (Hökfelt et al. 1987).

Behavioral effects observed after administration of galanin are in accordance with these observations. In this sense, centrally administered galanin has inhibitory actions in several rodent learning tasks. Galanin was reported to impair learning in a Morris water maze (Sundström et al. 1988), in a one-trial discriminative reward learning task with a starburst five-arm radial maze (Malin et al. 1992), in a delayed nonmatching-to-sample task (Robinson and Crowley 1993a, b), in the latter case aggravating a deficit caused by a muscarinic cholinergic receptor antagonist, and in a step-down inhibitory avoidance task (Ukai et al. 1995). In the latter case, the impairing effect of galanin was reversed by improving cholinergic system signaling. Intra-septal administration of galanin also impairs performance in a spontaneous alternation task in rats, an effect that can be ameliorated by co-administration of glucose (Stefani and Gold 1998). Galanin antagonists were reported to enhance learning in a Morris water maze (Ogren et al. 1992).

Galanin administered in the medial septum decreases choice accuracy in a working memory task in a T-maze, and also decreases hippocampal theta activity recorded from the dentate hilus in a dose-dependent manner (Givens et al. 1992).

Galanin overexpressing transgenic mice exhibit cognitive and neurochemical deficits similar to those occurring in AD: they show learning and memory deficits, impaired long-term potentiation, reduced hippocampal excitability, lower evoked glutamate release, and reduced numbers of cholinergic neurons in the horizontal limb of the diagonal band compared to wild type mice (Crawley et al. 2002).

It is undoubtedly true that galanin exacerbates cognitive impairment in AD. However, galanin hyperinnervation promotes BF cholinergic neuronal function and survival (Ding et al. 2006; Elliott-Hunt et al. 2004). Galanin also exerts neuroprotective effects in rodent models of neurotoxicity, supporting the idea that galanin may delay the onset of symptoms of AD (Counts et al. 2010).

There is also evidence for the existence of interactions between galanin receptors and NPY receptors in the nucleus of the solitarii tract, hypothalamus and dorsal raphe nucleus, and it has been suggested that these interactions might exist in other areas, serving to equilibrate the physiological actions of the two receptors (Díaz-Cabiale et al. 2014).

15 Concluding Remarks

Modulation of basal forebrain cholinergic neurons by neurotransmitter systems critically modifies cognitive functions, including learning and memory processes. Although loss of these cholinergic neurons occurs during the progression of AD, neurons also degenerate in several nuclei projecting to the basal forebrain, making it hard to distinguish causes from consequences. Increasing knowledge of these modulatory systems allows the discovery of pharmacological targets, important for ameliorating AD cognitive symptoms.

References

Albuquerque EX, Pereira EFR, Alkondon M, Rogers SW (2009) Mammalian nicotinic acetylcholine receptors: from structure to function. Physiol Rev 89:73–120

Bartfai T, Bedecs K, Land T, Langel U, Bertorelli R, Girotti P, Consolo S, Xu XJ, Wiesenfeld-Hallin Z, Nilsson S et al (1991) M-15: high-affinity chimeric peptide that blocks the neuronal actions of galanin in the hippocampus, locus coeruleus, and spinal cord. Proc Natl Acad Sci USA 88(23):10961–5

Bartus RT, Dean RL, Beer B, Lippa AS (1982) The cholinergic hypothesis of geriatric memory dysfunction. Sci 217:408–414

Basheer R, Bauer A, Elmenhorst D, Ramesh V, McCarley RW (2007) Sleep deprivation upregulates A1 adenosine receptors in the rat basal forebrain. NeuroReport 18:1895–1899

Basheer R, Halldner L, Alanko L, McCarley RW, Fredholm BB, Porkka-Heiskanen T (2001) Opposite changes in adenosine A1 and A2A receptor mRNA in the rat following sleep deprivation. Neuroreport 12(8):1577–1580

Beal MF, MacGarvey U, Swartz KJ (1990) Galanin immunoreactivity is increased in the nucleus basalis of Meynert in Alzheimer's disease. Ann Neurol 28(2):157–161

Belarbi K, Schindowski K, Burnouf S, Caillierez R, Grosjean ME, Demeyer D, Hamdane M, Sergeant N, Blum D, Buée L (2009) Early Tau pathology involving the septo-hippocampal pathway in a Tau transgenic model: relevance to Alzheimer's disease. Curr Alzheimer Res 6 (2):152–157

Benzing WC, Kordower JH, Mufson EJ (1993) Galanin immunoreactivity within the primate basal forebrain: evolutionary change between monkeys and apes. J Comp Neurol 336(1):31–39

Bertorelli R, Forloni G, Consolo S (1991) Modulation of cortical in vivo acetylcholine release by the basal nuclear complex: role of the pontomesencephalic tegmental area. Brain Res 563:353–356

Bierer LM, Haroutunian V, Gabriel S, Knott PJ, Carlin LS, Purohit DP, Perl DP, Schmeidler J, Kanof P, Davis KL (1995) Neurochemical correlates of dementia severity in Alzheimer's disease: relative importance of the cholinergic deficits. J Neurochem 64(2):749–760

Blokland A (1995) Acetylcholine: a neurotransmitter for learning and memory? Brain Res Rev 21:285–300

Boccia MM, Baratti CM (2000) Involvement of central cholinergic mechanisms in the effects of oxytocin and an oxytocin receptor antagonist on retention performance in mice. Neurobiol Learn Mem 74(3):217–228

Boccia MM, Kopf SR, Baratti CM (1998) Effects of a single administration of oxytocin or vasopressin and their interactions with two selective receptor antagonists on memory storage in mice. Neurobiol Learn Mem 69(2):136–146

Bondareff W, Mountjoy CQ, Roth M, Rossor MN, Iversen LL, Reynolds GP, Hauser DL (1987) Neuronal degeneration in locus ceruleus and cortical correlates of Alzheimer disease. Alzheimer Dis Assoc Disord 1:256–262

Bubser M, Byun N, Wood MR, Jones CK (2012) Muscarinic receptor pharmacology and circuitry for the modulation of cognition. Handb Exp Pharmacol 208:121–166

Chang HC, Gaddum JH (1933) Choline esters in tissue extracts. J Physiol 79:255–285

Chan-Palay V (1988) Galanin hyperinnervates surviving neurons of the human basal nucleus of Meynert in dementias of Alzheimer's and Parkinson's disease: a hypothesis for the role of galanin in accentuating cholinergic dysfunction in dementia. J Comp Neurol 273(4): 543–557

Chan-Palay V (1991) Alterations in the locus coeruleus in dementias of Alzheimer's and Parkinson's disease. Prog Brain Res 88:625–630

Chen HS, Lipton SA (2006) The chemical biology of clinically tolerated NMDA receptor antagonist. J Neurochem 97:1611–1626

Cornwell-Jones CA, Decker MW, Chang JW, Cole B, Goltz KM, Tran T et al (1989) Neonatal 6-hydroxydopa, but not DSP-4, elevates brainstem monoamines and impairs inhibitory avoidance learning in developing rats. Brain Res 493:258–268

Counts SE, Perez SE, Ginsberg SD, Mufson EJ (2010) Neuroprotective role for galanin in Alzheimer's disease. Experientia Suppl 102:143–162

Court J, Martin-Ruiz C, Piggott M, Spurden D, Griffiths M, Perry E (2001) Nicotinic receptor abnormalities in Alzheimer's disease. Biol Psychiatry 49(3):175–184

Coyle JT, Price DL, DeLong MR (1983) Alzheimer's disease: a disorder of cortical cholinergic innervation. Sci 219(4589):1184–1190

Crawley JN (1996) Galanin-acetylcholine interactions: relevance to memory and Alzheimer's disease. Life Sci 58(24):2185–2199

Crawley JN, Mufson EJ, Hohmann JG, Teklemichael D, Steiner RA, Holmberg K, Xu ZQ, Blakeman KH, Xu XJ, Wiesenfeld-Hallin Z, Bartfai T, Hökfelt T (2002) Galanin overexpressing transgenic mice. Neuropeptides 36(2–3):145–156

Dale HH, Dudley HW (1929) The presence of histamine and acetylcholine in the spleen of the ox and the horse. J Physiol (Lond) 68:97–123

Dale HH, Laidlaw PP, Symons CT (1910) A reversed action of the vagus on the mammalian heart. J Physiol (Lond) 41:1–18
Dalrymple-Alford JC (1994) Behavioral effects of basal forebrain grafts after dorsal septohippocampal pathway lesions. Brain Res 661(1–2):243–258
Dannenberg H, Pabst M, Braganza O, Schoch S, Niediek J, Bayraktar M, Mormann F, Beck H (2015) Synergy of direct and indirect cholinergic septo-hippocampal pathways coordinates firing in hippocampal networks. J Neuroscience 35(22):8394–8410
Decker MW, McGaugh JL (1991) The role of interactions between the cholinergic system and other neuromodulatory systems in learning and memory. Synapse 7(2):151–168
Di Cesare Mannelli L, Tenci B, Zanardelli M et al (2015) α_7 Nicotinic Receptor Promotes the Neuroprotective Functions of Astrocytes against Oxaliplatin Neurotoxicity. Neural Plast 2015:396908
Díaz-Cabiale Z, Flores-Burgess A, Parrado C, Narváez M, Millón C, Puigcerver A, Coveñas R, Fuxe K, Narváez JA (2014) Galanin receptor/neuropeptide y receptor interactions in the central nervous system. Curr Protein Pept Sci 15(7):666–672
Ding X, MacTavish D, Kar S, Jhamandas JH (2006) Galanin attenuates beta-amyloid (Abeta) toxicity in rat cholinergic basal forebrain neurons. Neurobiol Dis 21(2):413–420
dos Santos VV, Santos DB, Lach G, Rodrigues AL, Farina M, De Lima TC, Prediger RD (2013) Neuropeptide Y (NPY) prevents depressive-like behavior, spatial memory deficits and oxidative stress following amyloid-β (Aβ(1–40)) administration in mice. Behav Brain Res 244:107–115
Dougherty KD, Milner TA (1999) Cholinergic septal afferent terminals preferentially contact neuropeptide Y-containing interneurons compared to parvalbumin-containing interneurons in the rat dentate gyrus. J Neuroscience 19(22):10140–10152
Dringenberg HC, Olmstead MC (2003) Integrated contributions of basal forebrain and thalamus to neocortical activation elicited by pedunculopontine tegmental stimulation in urethane-anesthetized rats. Neuroscience 119(3):839–853
Dutar P, Lamour Y, Nicoll RA (1989) Galanin blocks the slow cholinergic EPSP in CA1 pyramidal neurons from ventral hippocampus. Eur J Pharmacol 164(2):355–360
Easton A, Douchamps V, Eacott M, Lever C (2012) A specific role for septohippocampal acetylcholine in memory? Neuropsychologia 50:3156–3168
Eaton K, Sallee FR, Sah R (2007) Relevance of neuropeptide Y (NPY) in psychiatry. Curr Top Med Chem 7(17):1645–1659
Eckenstein FP, Baughman RW, Quinn J (1988) An anatomical study of cholinergic innervation in rat cerebral cortex. Neuroscience 25:457–474
Elliott-Hunt CR, Marsh B, Bacon A, Pope R, Vanderplank P, Wynick D (2004) Galanin acts as a neuroprotective factor to the hippocampus. Proc Natl Acad Sci USA 101(14):5105–5110
Elmslie KS, Yoshikami D (1985) Effects of kynurenate on root potentials evoked by synaptic activity and amino acids in the frog spinal cord. Brain Res 330(2):265–272
España RA, Berridge CW (2006) Organization of noradrenergic efferents to arousal-related basal forebrain structures. J Comp Neurol 496(5):668–683
Everitt BJ, Robbins TW (1997) Central cholinergic systems and cognition. Annu Rev Psychol 48:649–684
Fadel J, Burk JA (2010) Orexin/hypocretin modulation of the basal forebrain cholinergic system: Role in attention. Brain Res 1314:112–123
Fadel J, Frederick-Duus D (2008) Orexin/hypocretin modulation of the basal forebrain cholinergic system: insights from in vivo microdialysis studies. Pharmacol Biochem Behav 90(2):156–162
Fadel J, Moore H, Sarter M, Bruno JP (1996) Trans-synaptic stimulation of cortical acetylcholine release after partial 192 IgG-saporin-induced loss of cortical cholinergic afferents. J Neuroscience 16:6592–6600
Fadel J, Sarter M, Bruno JP (2001) Basal forebrain glutamatergic modulation of cortical acetylcholine release. Synapse 39:201–212
Farr SA, Uezu K, Flood JF, Morley JE (1999) Septo-hippocampal drug interactions in post-trial memory processing. Brain Res 847(2):221–230

Fisone G, Wu CF, Consolo S, Nordström O, Brynne N, Bartfai T, Melander T, Hökfelt T (1987) Galanin inhibits acetylcholine release in the ventral hippocampus of the rat: histochemical, autoradiographic, in vivo, and in vitro studies. Proc Natl Acad Sci USA 84(20):7339–7343

Flood JF, Baker ML, Hernandez EN, Morley JE (1989) Modulation of memory processing by neuropeptide Y varies with brain injection site. Brain Res 503(1):73–82

Flood JF, Baker ML, Hernandez EN, Morley JE (1990) Modulation of memory retention by neuropeptide K. Brain Res 520(1–2):284–290

Flood JF, Hernandez EN, Morley JE (1987) Modulation of memory processing by neuropeptide Y. Brain Res 421(1–2):280–290

Flood JF, Morley JE (1989) Dissociation of the effects of neuropeptide Y on feeding and memory: evidence for pre- and postsynaptic mediation. Peptides 10(5):963–966

Fort P, Khateb A, Pegna A, Muhlethaler M, Jones BE (1995) Noradrenergic modulation of cholinergic nucleus basalis neurons demonstrated by in vitro pharmacological and immunohistochemical evidence in the guinea-pig brain. Eur J Neuroscience 7:1502–1511

Francis PT, Palmer AM, Snape M, Wilcock GK (1999) The cholinergic hypothesis of Alzheimer's disease: A review of progress. J Neurol Neurosurg Psychiatry 66(2):137–147

Fritschy JM, Grzanna R (1989) Immunohistochemical analysis of the neurotoxic effects of DSP-4 identifies two populations of noradrenergic axon terminals. Neuroscience 30:181–197

Führer H (1918) Untersuchunge über den Synergismus von Giften. Arch f Exper Pathol u Pharmacol 82:51–80

Galey D, Toumane A, Durkin T, Jaffard R (1989) In vivo modulation of septo-hippocampal cholinergic activity in mice: relationships with spatial reference and working memory performance. Behav Brain Res 32(2):163–172

German DC, Manaye KF, White CL III, Woodward DJ, McIntire DD, Smith WK, Kalaria RN, Mann DM (1992) Disease-specific patterns of locus coeruleus cell loss. Ann Neurol 32:667–676

Givens BS, Olton DS, Crawley JN (1992) Galanin in the medial septal area impairs working memory. Brain Res 582(1):71–77

Gold PE (2003) Acetylcholine modulation of neural systems involved in learning and memory. Neurobiol Learn Mem 80:194–210

Goldbach R, Allgaier C, Heimrich B, Jackisch R (1998) Postnatal development of muscarinic autoreceptors modulating acetylcholine release in the septohippocampal cholinergic system. I. Axon terminal region: hippocampus. Brain Res Dev Brain Res 108(1–2):23–30

Gu Z, Cheng J, Zhong P, Qin L, Liu W, Yan Z (2014) Aβ selectively impairs mGluR7 modulation of NMDA signaling in basal forebrain cholinergic neurons: implication in Alzheimer's disease. J Neuroscience 34(41):13614–13628

Hammar CG, Hanin I, Holmstedt B et al (1968) Identification of acetylcholine in fresh rat brain by combined gas chromatography-mass spectrometry. Nature 220:915–917

Harkany T, Abrahám I, Timmerman W, Laskay G, Tóth B, Sasvári M et al (2000) Beta-amyloid neurotoxicity is mediated by a glutamate-triggered excitotoxic cascade in rat nucleus basalis. Eur J Neuroscience 12:2735–2745

Hasselmo ME (2006) The role of acetylcholine in learning and memory. Curr Opin Neurobiol 16:710–715

Hasselmo ME, Sarter M (2011) Modes and models of forebrain cholinergic neuromodulation of cognition. Neuropsychopharmacology 36:52–73

Heneka MT, Ramanathan M, Jacobs AH, Dumitrescu-Ozimek L, Bilkei-Gorzo A, Debeir T, Sastre M, Galldiks N, Zimmer A, Hoehn M, Heiss WD, Klockgether T, Staufenbiel M (2006) Locus ceruleus degeneration promotes Alzheimer pathogenesis in amyloid precursor protein transgenic mice. J Neuroscience 26(5):1343–1354

Hökfelt T, Millhorn D, Seroogy K, Tsuruo Y, Ceccatelli S, Lindh B, Meister B, Melander T, Schalling M, Bartfai T et al (1987) Coexistence of peptides with classical neurotransmitters. Experientia 43(7):768–780

Hunt R, Taveau R de M (1906) On the physiological action of cetain cholin derivatives and new methods for detecting cholin. Br Med J 3:1788–1791

Huston JP, Hasenöhrl RU (1995) The role of neuropeptides in learning: focus on the neurokinin substance P. Behav Brain Res 66:117–127

Jones BE, Cuello AC (1989) Afferents to the basal forebrain cholinergic cell area from pontomesencephalic–catecholamine, serotonin, and acetylcholine–neurons. Neuroscience 31:37–61

Kalinchuk AV, McCarley RW, Porkka-Heiskanen T, Basheer R (2011) The time course of adenosine, nitric oxide (NO) and inducible NO synthase changes in the brain with sleep loss and their role in the non-rapid eye movement sleep homeostatic cascade. J Neurochem 116:260–272

Khateb A, Fort P, Williams S, Serafin M, Muhlethaler M, Jones BE (1998) GABAergic input to cholinergic nucleus basalis neurons. Neuroscience 86:937–947

Knox D, Sarter M, Berntson GG (2004) Visceral afferent bias on cortical processing: role of adrenergic afferents to the basal forebrain cholinergic system. Behav Neuroscience 118:1455–1459

Krebs JR (1990) Food-storing birds: adaptive specialization in brain and behaviour? Philos Trans R Soc Lond B Biol Sci 329(1253):153–160

Krügel U, Bigl V, Eschrich K, Bigl M (2001) Deafferentation of the septo-hippocampal pathway in rats as a model of the metabolic events in Alzheimer's disease. Int J Dev Neuroscience 19 (3):263–277

Kurosawa M, Sato A, Sato Y (1989) Stimulation of the nucleus basalis of Meynert increases acetylcholine release in the cerebral cortex in rats. Neurosci Lett 98:45–50

Lamour Y, Dutar P, Jobert A (1984) Cortical projections of the nucleus of the diagonal band of Broca and of the substantia innominata in the rat: an anatomical study using the anterograde transport of a conjugate of wheat germ agglutinin and horseradish peroxidase. Neuroscience 12:395–408

Leanza G, Nilsson OG, Wiley RG, Björklund A (1995) Selective lesioning of the basal forebrain cholinergic system by intraventricular 192 IgG-saporin: behavioural, biochemical and stereological studies in the rat. Eur J Neuroscience 7(2):329–343

Lelkes Z, Porkka-Heiskanen T, Stenberg D (2013) Cholinergic basal forebrain structures are involved in the mediation of the arousal effect of noradrenaline. J Sleep Res 22(6):721–726

Levey AI (1993) Immunological localization of m1-m5 muscarinic acetylcholine receptors in peripheral tissues and brain. Life Sci 52:441–448

Levin ED (1992) Nicotinic systems and cognitive function. Psychopharmacology 108:417–431

Liu Y, Zeng X, Hui Y et al (2015) Activation of α_7 nicotinic acetylcholine receptors protects astrocytes against oxidative stress-induced apoptosis: implications for Parkinson's disease. Neuropharmacology 91:87–96

Loewi O (1921) Über humorale Übertragbarkeit der Herznervenwirkung. Pflügers Arch Ges Physiol 189:239–242

Loup F, Tribollet E, Dubois-Dauphin M, Dreifuss JJ (1991) Localization of high-affinity binding sites for oxytocin and vasopressin in the human brain. An autoradiographic study. Brain Res 555(2):220–232

Macht DI (1923) A pharmacodynamic analysis of the cerebral effects of atropin, homatropin, scopolamin and related drugs. J Pharmacol Exp Ther 22:35–48

MacIntosh F.C., Oboring P. E. (1955) Release of acetylcholine from the intact cerebral cortex. Abstr 19th Int Physiol Congr 580–581

Malin DH, Novy BJ, Lett-Brown AE, Plotner RE, May BT, Radulescu SJ, Crothers MK, Osgood LD, Lake JR (1992) Galanin attenuates retention of one-trial reward learning. Life Sci 50(13):939–944

Marighetto A, Durkin T, Toumane A, Lebrun C, Jaffard R (1989) Septal alpha-noradrenergic antagonism in vivo blocks the testing-induced activation of septo-hippocampal cholinergic neurones and produces a concomitant deficit in working memory performance of mice. Pharmacol Biochem Behav 34(3):553–558

Mesulam M (2004a) The cholinergic lesion of Alzheimer's Disease: pivotal factor or side show? Learn Mem 11:43–49

Mesulam MM (2004b) The cholinergic innervation of the human cerebral cortex. Prog Brain Res 145:67–78

Mesulam MM, Mufson EJ, Wainer BH, Levey AI (1983) Central cholinergic pathways in the rat: an overview based on an alternative nomenclature (Ch1-Ch6). Neuroscience 10:1185–1201

Mishima K, Tsukikawa H, Inada K, Fujii M, Iwasaki K, Matsumoto Y, Abe K, Egawa T, Fujiwara M (2001) Ameliorative effect of vasopressin-(4–9) through vasopressin V(1A) receptor on scopolamine-induced impairments of rat spatial memory in the eight-arm radial maze. Eur J Pharmacol 427(1):43–52

Molnár Z, Soós K, Lengyel I, Penke B, Szegedi V, Budai D (2004) Enhancement of NMDA responses by beta-amyloid peptides in the hippocampus in vivo. NeuroReport 15:1649–1652

Moore H, Stuckman S, Sarter M, Bruno JP (1995) Stimulation of cortical acetylcholine efflux by FG 7142 measured with repeated microdialysis sampling. Synapse 21:324–331

Morón I, Ramírez-Lugo L, Ballesteros MA, Gutiérrez R, Miranda MI, Gallo M, Bermúdez-Rattoni F (2002) Differential effects of bicuculline and muscimol microinjections into the nucleus basalis magnocellularis in taste and place aversive memory formation. Behav Brain Res 134(1–2):425–431

Mufson EJ, Cochran E, Benzing W, Kordower JH (1993) Galaninergic innervation of the cholinergic vertical limb of the diagonal band (Ch2) and bed nucleus of the stria terminalis in aging, Alzheimer's disease and Down's syndrome. Dementia 4(5):237–50

Neugroschl J, Wang S (2011) Alzheimer's disease: diagnosis and treatment across the spectrum of disease severity. Mt Sinai J Med 78(4):596–612

Ogasawara T, Itoh Y, Tamura M, Mushiroi T, Ukai Y, Kise M, Kimura K (1999) Involvement of cholinergic and GABAergic systems in the reversal of memory disruption by NS-105, a cognition enhancer. Pharmacol Biochem Behav 64(1):41–52

Ogren SO, Hökfelt T, Kask K, Langel U, Bartfai T (1992) Evidence for a role of the neuropeptide galanin in spatial learning. Neuroscience 51(1):1–5

Page KJ, Saha A, Everitt BJ (1993) Differential activation and survival of basal forebrain neurons following infusions of excitatory amino acids: studies with the immediate early gene c-fos. Exp Brain Res 93:412–422

Palmer AM, DeKosky ST (1993) Monoamine neurons in aging and Alzheimer's disease. J Neural Transm Gen Sect 91:135–159

Pascual M, Pérez-Sust P, Soriano E (2004) The GABAergic septohippocampal pathway in control and reeler mice: target specificity and termination onto Reelin-expressing interneurons. Mol Cell Neurosci 25(4):679–691

Perry EK, Morris CM, Court JA, Cheng A, Fairbairn AF, McKeith IG et al (1995) Alteration in nicotine binding sites in Parkinson's disease, Lewy body dementia and Alzheimer's disease: possible index of early neuropathology. Neuroscience 64(2):385–395

Porkka-Heiskanen T, Strecker RE, McCarley RW (2000) Brain site-specificity of extracellular adenosine concentration changes during sleep deprivation and spontaneous sleep: an in vivo microdialysis study. Neuroscience 99:507–517

Querfurth HW, LaFerla FM (2010) Alzheimer's disease. N Engl J Med 362(4):329–344

Ragozzino ME, Brown HD (2013) Muscarinic Cholinergic Receptor Agonists and Antagonists. Encyclopedia of Psychopharmacology. Springer, Berlin Heidelberg, pp 1–6

Rangani RJ, Upadhya MA, Nakhate KT, Kokare DM, Subhedar NK (2012) Nicotine evoked improvement in learning and memory is mediated through NPY Y1 receptors in rat model of Alzheimer's disease. Peptides 33(2):317–328

Rasmusson DD, Clow K, Szerb JC (1994) Modification of neocortical acetylcholine release and electroencephalogram desynchronization due to brainstem stimulation by drugs applied to the basal forebrain. Neuroscience 60(3):665–677

Rasmusson DD, Szerb IC, Jordan JL (1996) Differential effects of alpha-amino-3-hydroxy-5-methyl-4-isoxazole propionic acid and N-methyl-d-aspartate receptor antagonists applied to the basal forebrain on cortical acetylcholine release and electroencephalogram desynchronization. Neuroscience 72:419–427

Ribeiro-da-Silva A, Hökfelt T (2000) Neuroanatomical localisation of Substance P in the CNS and sensory neurons. Neuropeptides 34:256–271
Robinson JK, Crawley JN (1993a) Intraseptal galanin potentiates scopolamine impairment of delayed nonmatching to sample. J Neurosci 13(12):5119–5125
Robinson JK, Crawley JN (1993b) Intraventricular galanin impairs delayed nonmatching-to sample performance in rats. Behav Neurosci 107(3):458–467
Saper CB (1984) Organization of cerebral cortical afferent systems in the rat. II. Magnocellular basal nucleus. J Comp Neurol 222:313–342
Scatton B, Bartholini G (1982) Gamma-Aminobutyric acid (GABA) receptor stimulation. IV. Effect of progabide (SL 76002) and other GABAergic agents on acetylcholine turnover in rat brain areas. J Pharmacol Exp Ther 220:689–695
Semba K (2004) Phylogenetic and ontogenetic aspects of the basal forebrain cholinergic neurons and their innervation of the cerebral cortex. Prog Brain Res 145:3–43
Sim JA, Griffith WH (1996) Muscarinic inhibition of glutamatergic transmissions onto rat magnocellular basal forebrain neurons in a thin-slice preparation. Eur J Neurosci 8:880–891
Smiley JF, Mesulam MM (1999) Cholinergic neurons of the nucleus basalis of Meynert receive cholinergic, catecholaminergic and GABAergic synapses: an electron microscopic investigation in the monkey. Neuroscience 88:241–255
Smith CG, Beninger RJ, Mallet PE, Jhamandas K, Boegman RJ (1994) Basal forebrain injections of the benzodiazepine partial inverse agonist FG 7142 enhance memory of rats in the double Y-maze. Brain Res 666:61–67
Stanley EM, Fadel JR (2011) Aging-related alterations in orexin/hypocretin modulation of septohippocampal amino acid neurotransmission. Neuroscience 195:70–79
Stäubli U, Huston JP (1980) Facilitation of learning by post-trial injection of substance P into the medial septal nucleus. Behav Brain Res 1:245–255
Stefani MR, Gold PE (1998) Intra-septal injections of glucose and glibenclamide attenuate galanin-induced spontaneous alternation performance deficits in the rat. Brain Res 813(1):50–56
Sundström E, Archer T, Melander T, Hökfelt T (1988) Galanin impairs acquisition but not retrieval of spatial memory in rats studied in the Morris swim maze. Neurosci Lett 88(3):331–335
Tanabe S, Shishido Y, Nakayama Y, Furushiro M, Hashimoto S, Terasaki T, Tsujimoto G, Yokokura T (1999) Effects of arginine-vasopressin fragment 4–9 on rodent cholinergic systems. Pharmacol Biochem Behav 63(4):549–553
Tansey EM (1991) Chemical neurotransmission in the autonomic nervous system: Sir Henry Dale and acetylcholine. Clin Auton Res 1:63–72
Torres EM, Perry TA, Blockland A, Wilkinson LS, Wiley RG, Lappi DA, Dunnet SB (1994) Behavioural, histochemical and biochemical consequences of selective immunolesions in discrete regions of the basal forebrain cholinergic system. Neuroscience 63(1):95–122
Tóth A, Hajnik T, Záborszky L, Détári L (2007) Effect of basal forebrain neuropeptide Y administration on sleep and spontaneous behavior in freely moving rats. Brain Res Bull 72(4–6):293–301
Ukai M, Miura M, Kameyama T (1995) Effects of galanin on passive avoidance response, elevated plus-maze learning, and spontaneous alternation performance in mice. Peptides 16(7):1283–1286
Weiss JH, Yin HZ, Choi DW (1994) Basal forebrain cholinergic neurons are selectively vulnerable to AMPA/kainate receptor-mediated neurotoxicity. Neuroscience 60:659–664
Wettstein JG, Earley B, Junien JL (1995) Central nervous system pharmacology of neuropeptide Y. Pharmacol Ther 65(3):397–414
Winters BD, Dunnett SB (2004) Selective lesioning of the cholinergic septo-hippocampal pathway does not disrupt spatial short-term memory: a comparison with the effects of fimbria-fornix lesions. Behav Neurosci 118(3):546–562
Yakel JL (2014) Functional distribution and regulation of neuronal nicotinic ach receptors in the mammalian brain. In: Lester RAJ (ed) Nicotinic Receptors. Springer, New York, pp 93–114

Yang C, Franciosi S, Brown RE (2013) Adenosine inhibits the excitatory synaptic inputs to Basal forebrain cholinergic, GABAergic, and parvalbumin neurons in mice. Front Neurol 4:77
Záborszky L, Cullinan WE, Braun A (1991) Afferents to basal forebrain cholinergic projection neurons: an update. Adv Exp Med Biol 295:43–100
Zaborszky L, Gaykema RP, Swanson DJ, Cullinan WE (1997) Cortical input to the basal forebrain. Neuroscience 79:1051–1078
Záborszky L, Pang K, Somogyi J, Nadasdy Z, Kallo I (1999) The basal forebrain corticopetal system revisited. Ann N Y Acad Sci 877:339–367
Zaborszky L., Duque A., Gielow M., et al (2015) Organization of the basal forebrain cholinergic projection system: specific or diffuse?. Rat Nerv. Syst. (Fourth Edition)—Chapter 19

Habit Formation and the Striatum

Barbara J. Knowlton and Tara K. Patterson

Abstract Data from experimental animals and human subjects has provided convergent evidence for the key role of the striatum in the formation of stimulus-response habits. Habits can be distinguished from associative memories that support goal-directed actions based on their insensitivity to reward devaluation and contingency degradation. Behavior on many instrumental learning tasks can be supported by both declarative knowledge and habits, and these contributions shift with the amount of training. This shift appears to be accompanied by the involvement of different cortico-striatal loops in controlling behavior. Factors that encourage the shift toward and maintenance of habits include learning under conditions of stress, distraction, and interval or probabilistic schedules of reinforcement.

Keywords Flexibility · Nondeclarative memory · Probabilistic classification · Reinforcement · Spatial learning · Stress

Contents

1 Introduction 276
2 The Habit Memory System 276
  2.1 Maze Navigation Tasks 276
  2.2 Instrumental Learning Tasks 281
3 Conditions Favoring the Formation of Habits 285
  3.1 Concurrent Discriminations 285
  3.2 Probabilistic Associations 286
  3.3 Distraction 287
  3.4 Stress 287
4 Implications for Understanding Compulsions and Addictions 290
5 Conclusion 291
References 292

B.J. Knowlton (✉) · T.K. Patterson
UCLA Department of Psychology, University of California, Los Angeles, USA
e-mail: knowlton@psych.ucla.edu

Curr Topics Behav Neurosci (2018) 37: 275–295
DOI 10.1007/7854_2016_451
Published Online: 28 September 2016

1 Introduction

The idea of multiple memory systems largely arose from the observation that patients with lesions to the hippocampal formation located in the medial temporal lobe have profound declarative memory deficits but have intact ability to perform tasks that rely on other types of memory (e.g., Milner 1962). Habits, which are memories based on stimulus–response (S-R) associations, are one of the important forms of memory that are spared by damage to the medial temporal lobe (Knowlton 2002; Packard and Knowlton 2002; Packard 2009b). Some of the features of habit memory are that unlike hippocampus-dependent declarative memory representations, habit memory representations are relatively inflexible, and are acquired gradually over many trials through feedback (Knowlton and Moody 2008; Foerde and Shohamy 2011).

Multiple lines of evidence have supported the idea that the basal ganglia, specifically the dorsal striatum, mediate the acquisition and expression of S-R habits. A great deal of evidence for the dissociation between habits and declarative memories comes from studies using navigation tasks. The first section of this chapter will discuss findings from these studies, in which one can measure the extent to which performance is based on "place" learning (also called "spatial" or "cognitive" learning) as compared to associations between a specific stimulus and a specific behavior (i.e., "S-R" learning, also called "response," "nonspatial," or "habit" learning). The second section will focus on the other commonly used assay for habit memory, in which animals are trained to perform an instrumental response to obtain a desired outcome, and the degree of habit formation is assessed with a test of goal-directed action selection. While these procedures have primarily been used in rodent subjects, there have been some attempts to compare habit vs. goal-directed memory in human subjects. The third section of this chapter will discuss manipulations used in humans and nonhuman primates to disentangle habit vs. declarative memories in associative learning tasks. Finally, the effects of stress and Pavlovian cues on S-R habit behavior will be discussed along with the implications for understanding compulsions and drug addiction.

2 The Habit Memory System

2.1 *Maze Navigation Tasks*

The strongest evidence for a striatal-based habit system comes from experiments in maze learning in rats. The spatial navigation assay for dissociating place-based memory from S-R habit memory has its roots in a task developed by Tolman and colleagues to resolve a debate between S-R learning theorists (e.g., Hull 1943) and cognitive learning theorists (e.g., Tolman 1932). In 1946, Tolman and colleagues published a paper in which they asked, "what is it that [rats trained to find food in a

T-maze] have learned?" (Tolman et al. 1946). According to S-R learning theory, the rat has learned to respond in a certain way to a stimulus (e.g., to turn right at the choice point), and the role of the food is to strengthen the association between stimulus and response. In contrast, according to cognitive learning theory, the rat has instead learned where the food is located, and orients itself toward that place to solve the maze. To distinguish between these two accounts, the experimenters trained two groups of rats on a plus-shaped maze with two opposing food arms and two opposing start arms. Animals in the place learning group always received food in the same arm and had to learn to run to this arm regardless of their starting point. Animals in the response learning group always received food in the arm located to the right of their starting position. The rats that were assigned to place learning acquired their task more rapidly, indicated by a steeper drop in arm choice errors. The authors concluded that the learning done by the place group is "simpler and more primitive" than the learning done by the response group, posing a challenge for S-R theory (Tolman et al. 1946).

Several decades later, three experiments using spatial navigation tasks were conducted to provide evidence for the existence of multiple anatomically distinct memory systems in the mammalian brain (Packard et al. 1989; Packard and McGaugh 1992; McDonald and White 1994). In the first of these studies, Packard et al. (1989) compared the behavior of animals performing two radial arm maze tasks, the win-shift task and the win-stay task. In the win-shift task, all eight arms of the maze were baited with food, and performance was measured by counting the number of incorrect (i.e., repeated) arm entries. Thus, the animal was required to demonstrate knowledge of where it had been during the trial. In contrast, the win-stay task required animals to retrieve food from four lit arms and avoid the four unlit arms of the maze. Successful performance of this task therefore required the development of an association between the stimulus (i.e., the light), and the correct response (i.e., running down the arm). Within the two task groups there were three lesion groups: fimbria-fornix, striatum, and control. Rats with fimbria-fornix lesions were impaired at acquiring the win-shift task, but were unimpaired at acquiring the win-stay task (and in fact outperformed controls). The opposite was true of rats with striatal lesions: these animals were impaired at acquiring the win-stay task, but performed as well as controls on the win-shift task. Thus, this experiment constituted an important first step in elucidating the role of the striatum in learning and memory and distinguishing it from the role of the hippocampal system.

Packard and McGaugh (1992) replicated this double dissociation using two water maze tasks, the spatial discrimination water maze and the visual discrimination water maze. Both mazes contained two visible platforms, one that provided escape and one that did not. In the spatial discrimination water maze, the escape platform was always located in the same quadrant but its visual appearance changed. Successful performance of this task thus required learning where the escape platform was located using extra-maze cues. In contrast, in the visual discrimination water maze task, the escape platform had a consistent visual appearance, but its spatial location varied from trial to trial. Successful performance of this task required learning to swim to the platform that had the visual appearance paired with escape. The results showed that rats with lesions to the fimbria-fornix had impaired acquisition of the spatial discrimination task, but were unimpaired on acquisition of the visual discrimination task. In contrast, rats with lesions to the striatum showed the opposite pattern. Thus, taken together, these two experiments indicated that behaviors driven by place knowledge (performance of the win-shift radial maze and the spatial discrimination water maze) are supported by the hippocampal system, while behaviors driven by S-R associations (performance of the win-stay radial maze and the visual discrimination water maze) are supported by the striatum.

A third experiment that gave support to the idea of a double dissociation between hippocampus-dependent learning and striatum-dependent learning was conducted by McDonald and White (1994). Animals with fimbria-fornix lesions, striatum lesions, and control animals were trained to swim to a visible escape platform that was in a consistent spatial location and had a consistent visual appearance. Interspersed among these visible platform trials were several hidden platform trials, where the visual cue was removed, but escape was still available at the same spatial location as in the visible platform trials. Consistent with the results of Packard et al. (1989) and Packard and McGaugh (1992), McDonald and White (1994) found that rats with fimbria-fornix lesions were impaired on the hidden platform trials compared to rats with striatum lesions and controls. S-R learning was measured after the last acquisition trial and was investigated with a "transfer test" in which the visible platform was moved to a new spatial location. On the transfer test, rats with fimbria-fornix lesions had latencies equivalent to those of control animals, while rats with striatum lesions were impaired. The majority of the rats in the striatum group swam first to the place where the escape platform had been previously located, while the rats in the fimbria-fornix group all swam directly toward the visible platform in its new location.

These initial experiments employed irreversible lesions done before training occurred to study how they affected subsequent learning. A second set of experiments took advantage of reversible lesion techniques to inactivate specific brain regions after learning, which allowed the experimenters to investigate the expression and consolidation of place and response memory (Packard and McGaugh 1996; Schroeder et al. 2002). In the first such study, Packard and McGaugh (1996) used a dual solution version of the plus-maze task. Rats were trained with a consistent start box and reward box, and were tested twice, once early in training and once late in training. On test trials, the rat was placed in the arm opposite the usual start arm and experimenters recorded whether the rat went to the arm that had been rewarded previously (exhibiting place learning) or performed the turning behavior that had been rewarded previously (exhibiting response learning). Control animals predominantly expressed place learning on the early test trial and response learning on the late test trial, consistent with a shift in control of behavior from the hippocampus to the striatum with task experience. The two lesion groups were given intracerebral injections immediately prior to the two test trials. Inactivation of the hippocampus, but not the striatum, blocked expression of place learning on the early test trial, whereas inactivation of the striatum, but not the hippocampus, blocked the expression of response learning on the late test trial. In fact, the animals that received striatum inactivation expressed significant place learning on the late test trial, indicating that the hippocampus-dependent learning no longer expressed by control animals could be "unmasked" by inactivation of the dominant habit system. Additionally, because the lesions were administered after learning had occurred, the authors concluded that the same regions that had been shown to mediate the acquisition of place and response learning also mediate their expression.

Schroeder et al. (2002) similarly used reversible lesions administered post-training to investigate place versus response learning in the plus-maze, but in their experiment the administration of the post-training lesions was timed to disrupt consolidation of learning rather than interfere with expression of learning. Rats were trained on either a place water plus-maze or a response water plus-maze. Post-training inactivation of the hippocampus both attenuated performance on the place maze and enhanced performance on the response maze; these effects were not present when inactivation was delayed. The observed enhancement was consistent with the enhancement reported by Packard et al. (1989), and provided strong support for the idea that interactions between multiple memory systems can be competitive. Similarly, a subsequent study using the plus-maze also demonstrated that inactivation of the hippocampus before training impaired acquisition of a place plus-maze and enhanced acquisition of a response plus-maze (Chang and Gold 2003; for a review of literature on competition between multiple memory systems see Poldrack and Packard 2003).

Work on the roles of the medial temporal lobe and the striatum in the human navigational system has been generally consistent with the animal literature reviewed above. A human analog of the radial maze task used in animals is the four-on-eight virtual maze task (Iaria et al. 2003; Bohbot et al. 2004, 2007). In this task, participants are required to visit four accessible arms of an eight-arm radial maze, and then in the second half of the trial, visit the four arms that were not visited previously. The same arm visitation patterns are repeated across trials. Two alternative strategies support performance of this task: participants can use the extra-maze landmarks in the virtual environment or they can use a series of turns relative to a single starting point. Verbal report of strategy and performance on probe trials in which the extra-maze landmarks are concealed are used to classify participants as either "spatial" or "nonspatial/response" learners.

Iaria et al. (2003) used fMRI to investigate the brain regions activated in these two types of learners and found that the hippocampus supports the spatial strategy, while the striatum, specifically the caudate nucleus, supports the response strategy. Consistent with this finding, a subsequent study of patients with medial temporal lobe lesions revealed that patients who used a spatial strategy were impaired on the four-on-eight virtual maze relative to patients who used a response strategy and control subjects (Bohbot et al. 2004). Finally, using voxel-based morphometry, Bohbot et al. (2007) found that across healthy participants, the number of errors on landmark-absent probe trials was positively correlated with hippocampal gray matter density and negatively correlated gray matter density in the caudate.

Despite the conceptual similarities between work in human and nonhuman animals in distinguishing hippocampus-dependent learning from striatum-dependent learning, there has been divergence in terms of the role of specific regions within the striatum. Packard and Knowlton (2002) suggested on the basis of its anatomical connections and a review of supporting literature that the dorsolateral striatum (corresponding to the putamen in humans) may mediate the learning of S-R habits, while the dorsomedial striatum (corresponding to the caudate nucleus in humans) may mediate a type of memory more similar to that typically associated with the hippocampus. A subsequent study by Yin and Knowlton (2004) provided empirical support for this hypothesis. Employing a procedure similar to the one used by Packard and McGaugh (1996), Yin and Knowlton trained rats in a T-maze from a consistent start box to run to a consistent reward box. The rats were tested twice during acquisition by placing the start arm in the opposite position, and the experimenters recorded whether animals entered the arm where the reward was previously located (demonstrating place learning) or the opposite arm (demonstrating response learning). Pre-training lesions of the dorsomedial striatum resulted in greater expression of response learning on the early test probe compared to controls, similar to the effects of hippocampus and fimbria-fornix lesions observed in previous studies. On the late test probe, animals with dorsomedial striatum lesions behaved like control animals, consistent with previous work showing a shift toward greater use of response learning in normal animals with extended training (e.g., Packard and McGaugh 1996).

These results highlighted a need for revision of models where the hippocampus subserves place learning and the striatum subserves response learning; instead, Yin and Knowlton (2004) proposed that the hippocampus and dorsomedial striatum were both parts of a larger corticostriatal system that mediates flexible, goal-directed behavior. Support for the idea that different striatal subregions are parts of different corticostriatal circuits was initially reviewed by Alexander et al. (1986). In their model, each circuit contains information that flows from cortex to the basal ganglia to thalamus and back to cortex. Based on this model, Yin and Knowlton (2006) described a framework for understanding habit formation in which the dorsolateral striatum is a node in a "sensorimotor" corticostriatal circuit subserving behavior driven by S-R associations, and the dorsomedial striatum is a node in a separate "associative" corticostriatal circuit subserving behavior driven by action–outcome associations. A shift in control from the associative to the sensorimotor network corresponds to behavior becoming more habitual (Fig. 1).

One point where the human literature diverges somewhat from the animal literature is in demonstrating functional heterogeneity within the striatum. Although work in rats using maze navigation tasks indicates that the dorsolateral striatum in particular supports S-R habit behavior (Yin and Knowlton 2004), the region most commonly associated with human habit behavior in these types of tasks is the caudate nucleus, which corresponds to the rodent dorsomedial striatum. Whether this inconsistency represents an actual difference between the human brain and the rodent brain remains to be seen. Due to improvements in image resolution, subregions of the human striatum can now be defined with greater precision than in the

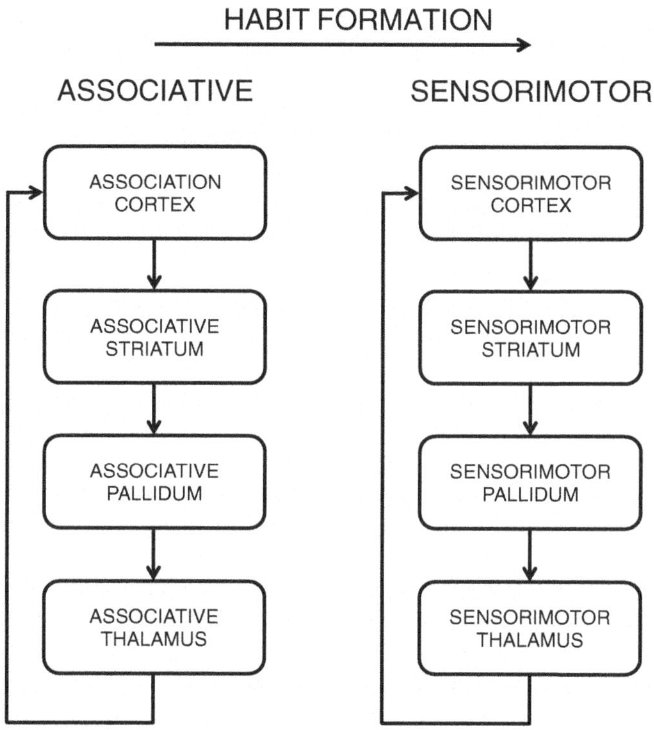

Fig. 1 Corticostriatal networks underlying habit formation. In the associative network, targets in prefrontal and parietal association cortex project to the associative striatum (dorsomedial striatum in the rodent, caudate nucleus in the human). In the sensorimotor network, targets in sensorimotor cortex project to the sensorimotor striatum (dorsolateral striatum in the rodent, putamen in the human). According to one framework for understanding habit formation (Yin and Knowlton 2006), a shift from control of behavior by action–outcome associations to stimulus–response associations corresponds to a shift from activation of the associative network to the sensorimotor network

past. Moreover, tractography methods now allow researchers the ability to define striatal subregions on the basis of structural or functional connectivity. These tools may afford researchers the ability to test models of striatal function that were previously only testable in rodents in the human brain.

2.2 *Instrumental Learning Tasks*

Instrumental learning relies on associations that are formed between stimuli, responses, and outcomes. Behaviors that are governed by S-R associations are called "habit" behaviors, and behaviors that are governed by response–outcome associations are called "goal-directed" behaviors or "actions" (Dickinson 1985).

The response made in instrumental learning tasks is generally similar whether behavior is based on a habit or a goal-directed action. Thus, it is necessary to use a probe test to assess the nature of the underlying association. The first such test, which is commonly considered to be the "gold standard" for determining whether a behavior has become habitual, is the outcome devaluation test.

As with the maze navigation tasks reviewed above, the use of outcome devaluation to study habit behavior also has its origin in a conflict between S-R learning theory (e.g., Hull 1943) and cognitive learning theory (e.g., Tolman 1932), because these theories make opposing predictions about how an experimental subject should respond when an instrumental outcome is devalued. According to S-R learning theory, outcomes are not associated with performance of the instrumental response but rather only serve to reinforce the S-R association. Therefore, devaluation of the outcome should have no effect on an animal's performance of a learned behavior, because after learning has occurred, the trained response should be elicited automatically by the contextual stimuli present during training. In contrast, according to cognitive learning theory, an animal should stop performing an instrumental behavior when the reward it previously received for doing so is no longer valued.

These predictions were tested in an experiment conducted by Adams and Dickinson (1981). Rats were trained to press a lever to receive a food reward, and then a conditioned taste aversion to the food that had been used to reinforce the lever pressing behavior during training was induced in half of the animals. Next, the animals were tested in extinction to determine whether they would continue to press the lever or cease to do so. The extinction responding of the rats that received reward devaluation was diminished compared to the extinction responding of control rats, indicating that they were sensitive to the value of the outcome of their behavior. Adams (1982) replicated this finding in rats that received limited acquisition training, but additionally showed that rats that received extended acquisition training became insensitive to reward devaluation (i.e., they maintained extinction responding levels equivalent to controls). This finding is consistent with the idea that with extended training, behavior shifts from being controlled by response–outcome associations to being controlled by S-R associations. Dickinson (1985) argued for this account, advocating the term "action" to refer to a behavior that is purposeful and goal-directed, in order to distinguish such behaviors from simple "responses" that are automatically elicited by stimuli and relatively insensitive to devaluation. Outcome devaluation procedures can be implemented within subjects (Colwill and Rescorla 1985) and can be performed using the original conditioned taste aversion procedure or a selective satiation procedure (Colwill and Rescorla 1985; Balleine and Dickinson 1998b). In the satiation version of the procedure, a food reward is devalued by giving the subject access to it until it is no longer consumed. Devaluation is shown by a decrease in performance of the behavior associated with the sated food relative to performance of the behavior associated with a non-sated food.

After the establishment of the outcome devaluation test as a means to determine whether instrumental behavior was goal-directed or habit-based, a number of lesion studies were conducted using outcome devaluation to determine the brain regions involved in these two types of responding. The first such study demonstrated that in rats, the prelimbic region of medial prefrontal cortex supports goal-directed action, as animals with prelimbic lesions were found to be insensitive to reward devaluation, while control animals were not (Balleine and Dickinson 1998a). This result was replicated by Killcross and Coutureau (2003) in a study that included two training levels, similar to the behavioral study conducted by Adams (1982). Rats with prelimbic lesions showed insensitivity to devaluation after both limited and extended training, while control rats were only insensitive to devaluation after extended training. A pair of studies focused on dissociations within the rat striatum showed that, like prelimbic lesions, lesions to the dorsomedial striatum also produce outcome insensitivity under conditions where control animals show goal-directed behavior, and that conversely, lesions to the dorsolateral striatum disrupt habit-based responding (Yin et al. 2004, 2005).

The outcome devaluation procedure has also been used to assess goal-directedness of behavior in a maze navigation task used to dissociate striatal-dependent and hippocampal-dependent learning. Sage and Knowlton (2000) demonstrated that animals overtrained on a win-stay radial maze were insensitive to outcome devaluation, while animals performing the win-stay radial maze with minimal training and animals performing the win-shift radial maze showed sensitivity to devaluation. Although devaluation did not affect arm choice accuracy in any group, the under-trained win-stay animals and the win-shift animals had greater trial latencies after devaluation compared to controls. These results support the view that in the hippocampus-dependent win-shift task, behavior is controlled by a representation of the outcome; in the striatum-dependent win-stay task, behavior is initially also controlled by a representation of the outcome, but becomes independent of this representation with extended training.

Another powerful method for testing goal-directed action selection is through measuring sensitivity to degradation of the action–outcome contingency. Unlike habit learning, goal-directed learning involves learning about the value of the outcome of an action and the contingency between the action and the outcome (Balleine and Dickinson 1998a). In the contingency degradation procedure, rats are trained to make an instrumental response and given a degradation session in which they are presented with rewards unpaired with performance of the response. Animals are later tested under extinction conditions and the impact of the degradation session is assessed. Habit-based responding would not be affected by the degradation session because the contingency between the response and the outcome is not part of the S-R association (Balleine and Dickinson 1998a). In an extreme form of contingency degradation, the contingency between the response and the outcome are reversed. After training, animals are exposed to an omission contingency, in which free rewards are given and are actually delayed when the trained response is made. Again, habit is inferred when responding is not affected by the omission session. Yin et al. (2006) showed that in rats that had learned an S-R habit, deactivation of the dorsolateral striatum during the omission session resulted in sensitivity to the changes in the action–outcome contingency. These results indicate that the dorsolateral striatum prevents the updating of contingency information in habit representations.

In human subjects, instrumental behavior can be habitual, persisting when the outcome is devalued. This was first demonstrated experimentally in a study by Tricomi et al. (2009), which showed that people overtrained to make instrumental responses for food rewards had equivalent response rates for the valued food and the devalued food in extinction, while the extinction behavior of a group of people who received minimal training indicated that they were sensitive to outcome value. Indeed, although it may operate below the level of conscious awareness, habitual responding likely contributes to a great deal of human behavior in daily life. For example, people can be habitual popcorn eaters when at the movies, in that their popcorn consumption is not sensitive to hunger level or whether or not the popcorn is stale (Neal et al. 2011). Consistent with this behavior being based on an S-R association, popcorn eating in these subjects can become less habitual if stimuli associated with the cinema context are removed, or if the nondominant hand is used for eating (a different response).

Considerable effort has been put into adapting the instrumental learning tasks used in nonhuman animals for use with humans in order to study the transition between goal-directed behavior and habit behavior in the human brain. Findings from this line of work have been quite consistent with the animal lesion literature. First, Tricomi et al. (2004) found that activity in the human caudate was modulated by the perceived contingency between instrumental actions and their outcomes, indicating that this striatal subregion may play a role in mediating goal-directed behavior. In another study, Valentin et al. (2007) found that activity in ventromedial prefrontal cortex was higher when participants performed an instrumental behavior associated with a valued food or an instrumental behavior that avoided a devalued food, compared to when participants performed an instrumental behavior associated with a devalued food or an instrumental behavior that avoided a valued food. Taken together, these studies indicate that the caudate and ventromedial prefrontal cortex support goal-directed behavior in humans, which is consistent with previous findings showing that the dorsomedial striatum and medial prefrontal cortex are required for goal-directed behavior in rats (Balleine and Dickinson 1998a; Killcross and Coutureau 2003; Yin et al. 2005).

Consistent with the findings of Valentin et al. (2007), Tricomi et al. (2009) found that in humans responding for food rewards, ventromedial prefrontal cortex was more active as anticipation of upcoming rewards increased. In the same experiment, increased activity in the striatum (putamen extending into globus pallidus) was found in response to stimulus onset after extended training. This effect is consistent with animal work showing the involvement of the dorsolateral striatum in habitual behavior (Yin et al. 2004). Unlike the effect found in the striatum, the effect in ventromedial prefrontal cortex was significant early in training. Thus, it appears that the role of the ventromedial prefrontal cortex in goal-directed behavior may be in representing the value of an outcome, while different subregions of the striatum appear to be involved at different points in the learning process, with the caudate supporting behavior during goal-directed action and the putamen supporting behavior as the human brain shifts to a more habitual mode of responding.

3 Conditions Favoring the Formation of Habits

Several factors present during learning have been described that result in a greater tendency for learning to result in habits. Overtraining is known to result in more habitual behavior, suggesting that habits are learned more gradually than goal-directed actions. Habit learning is also likely to occur when training with interval schedules of reinforcement, in which the first action after an interval earns a reward, such as in checking for mail. The amount of reward received is not proportional to the number of responses made. In contrast, ratio schedules, in which the reward earned is proportional to the rate of responding, tends to result in goal-directed learning. This difference is thought to result from the reduced contingency between responses and rewards under interval schedules (Dickinson et al. 1983). Other factors that have been shown to increase use of the habit system in humans and nonhuman primates can be viewed as factors that reduce or challenge the use of declarative memory; these factors include the widely spaced learning intervals present in concurrent discrimination studies, probabilistic associations between stimuli and reinforcement, distraction with a secondary task, and stress, which are described in detail in the following sections.

3.1 Concurrent Discriminations

In human and nonhuman primates, associative learning in laboratory situations is generally declarative, perhaps because of the highly developed primate medial temporal lobe memory system. Thus, habit learning tends to occur in situations in which the declarative system is thwarted. For example, discrimination learning is generally disrupted by hippocampal lesions in monkeys, except when only one trial per day is given (Teng et al. 2000). Introducing a long delay may make it difficult for monkeys to memorize these associations, and thus learning can only be supported by the more robust habit system. Interestingly, lesions of the striatum in monkeys abolishes discrimination learning when pairs are presented once per day, but not when multiple trials are given per day, which may enable the hippocampal system to support learning (Fernandez-Ruiz et al. 2001). Consistent with the idea that concurrent discrimination can be learned as a habit, Bayley et al. (2005) showed that the profoundly amnesic patients EP and GP were able to learn to discriminate between pairs of items despite having no memory of the training episode. Learning in the amnesic patients was much more gradual, and their knowledge of the rewarded element of the pair was less flexible than it was in controls. For example, they could not pick out which items were rewarded in an array of training items, but could only indicate their learning in the identical context to training, as would be expected if performance was based on S-R associations. Moody et al. (2010) also showed that nearly half of healthy control subjects exhibited concurrent discrimination learning that was supported by inflexible associations when learning conditions were not conducive to declarative learning (e.g., a large number of stimuli that were difficult to verbalize and a response deadline). Although they were able to learn to select the rewarded element of the pair, they were not able to indicate which items were rewarded when they were presented in a different context. In contrast, patients with Parkinson's disease, a disorder affecting striatal function, did not show any evidence of learning inflexible associations, suggesting that this type of learning was dependent on the striatum.

3.2 *Probabilistic Associations*

Another method for investigating human habit learning is by assessing participants' ability to perform probabilistic classification (Knowlton et al. 1994, 1996). In this type of experiment, memory for individual trials is not as important as information gleaned gradually across trials. Thus, there is opportunity for the associations between stimuli and correct responses to be learned as habits. In these studies, participants are shown a set of cue stimuli and have to make a prediction based on the cue; e.g., in the weather prediction task, participants are shown cards with geometric shapes and are required to predict whether the weather will be sunny or rainy based on the cards. After making each prediction, participants are given feedback. Over many trials, participants' ability to make correct predictions gradually improves. Knowlton et al. (1994) demonstrated that amnesic patients are capable of learning to do the task despite lacking explicit knowledge about the testing session, indicating the task does not require declarative memory. A subsequent study by Knowlton et al. (1996) showed that probabilistic classification ability is impaired in patients with Parkinson's disease, implicating striatal dependence. Knowlton et al. (1996) also found that the performance of Parkinson's patients eventually improved after extended training, suggesting that probabilistic classification performance can be supported by declarative memory when the habit memory system is impaired.

Investigations into the neural structures supporting probabilistic classification performance continued in a series of experiments employing fMRI. Consistent with the evidence of impaired performance in patients with Parkinson's disease, two initial studies reported that healthy human subjects performing the weather prediction task activate the caudate nucleus (Poldrack et al. 1999, 2001). Another fMRI study conducted by Moody et al. (2004) reported that during performance of the weather prediction task, healthy control subjects showed greater activation in the basal ganglia (caudate, putamen, and globus pallidus) compared to patients with

Parkinson's disease. In the medial temporal lobe, the Parkinson's group showed significant task-evoked activation, consistent with a compensatory reliance on declarative memory.

3.3 Distraction

Taken together, these neuropsychological and neuroimaging studies of probabilistic classification learning suggested a clear contribution of striatal-dependent habit memory to probabilistic classification performance. What was less clear, however, was the extent to which declarative memory contributes to performance of these types of tasks; although studies with amnesic patients demonstrated that declarative memory is not required, it could still contribute to performance in the intact system. Two additional studies looking at the effects of distraction on probabilistic classification helped clarify this contribution (Foerde et al. 2006, 2007). In both studies, healthy young adults were required to do a concurrent tone-counting task during acquisition of the classification task. Both studies provided evidence that classification learning proceeds similarly under single and dual task conditions, but that having a secondary task impairs acquisition of flexible knowledge of cue–outcome associations, a product of declarative memory. Furthermore, across subjects, classification performance under single task conditions correlated with activity in the hippocampus, while classification under dual task conditions correlated with activity in the putamen (Foerde et al. 2006). This suggests that attentional demands modulate the extent to which the declarative memory system and habit memory system are recruited, and that healthy young adults likely use a combination of declarative memory and habit learning to perform probabilistic classification tasks in the absence of distraction.

3.4 Stress

Recent work has shown that stress increases the learning and performance of habit behaviors. The first experiment in this line of research was conducted by Kim et al. (2001), and demonstrated that rats exposed to acute stress show impaired spatial learning and enhanced S-R learning when searching for the escape platform in a water maze. This study also provided evidence in support of the idea that the observed spatial impairment was related to reduced hippocampal long-term potentiation, and that the amygdala mediates this effect. Schwabe et al. (2007) subsequently replicated this finding in humans, showing that acute stress increased subjects' likelihood of using an S-R strategy over a spatial strategy to locate a goal object. Similarly, it has been demonstrated that acute stress before or after instrumental learning produces insensitivity to outcome devaluation (Schwabe and Wolf 2009, 2010).

Experiments on the effects of chronic stress have shown that, like acute stress, chronic stress also induces a shift toward greater habit responding. Schwabe et al. (2008) found that both people and mice exposed to chronic stress were more likely to use an S-R strategy than a spatial strategy. Importantly, in the mouse study, the behavioral changes induced by chronic stress were evident after a week-long stress-free recovery period. That a stress-induced bias toward habit responding persists far beyond the duration of the immediate stress response indicates that chronic stress may cause structural changes in brain regions that control the degree to which habit behavior is expressed. And indeed, studies that have looked directly at neural changes associated with stress-induced shifts toward habit responding have found evidence consistent with this idea. In nonhuman animals, chronic stress leading to increased habit behavior has been associated with decreased hippocampal neurogenesis, associative loop atrophy, and sensorimotor loop hypertrophy (Dias-Ferreira et al. 2009; Ferragud et al. 2010). In a study with humans, chronic stress lasting on the order of months was found to result in diminished sensitivity to outcome devaluation, diminished associative loop activation and volume, and increased sensorimotor loop activation and volume (Soares et al. 2012). These changes largely reversed after a six-week stress-free recovery period, indicating a high level of plasticity in the neural circuits required for performing S-R habits and goal-directed actions.

The majority of studies investigating the effects of stress on S-R habit behavior have looked at the effects of stress occurring in adulthood, leaving open the question of how stress that occurs during development may affect the brain regions involved in supporting and overriding habits. Stressful experiences that occur during sensitive periods can have a heightened impact on neural trajectories and behavior later in life (Lupien et al. 2009; Tottenham 2014). For example, in humans, prolonged institutional rearing has been associated with atypical development of amygdala-prefrontal circuitry (Tottenham et al. 2010; Gee et al. 2013). This neural embedding of stress may extend to systems for learning and memory, biasing how readily an organism forms and expresses habits.

Two lines of research support the hypothesis that early-life stress persistently affects the neural substrates of S-R habit learning. First, stress can have deleterious effects on the structure and function of brain regions that support goal-directed behavior. Specifically, stress is associated with dendritic atrophy in prefrontal cortex, and with dendritic atrophy, reduced long-term potentiation, and reduced neurogenesis in the hippocampus (McEwen 2000; Joels et al. 2007). Underactivation of these regions during a sensitive period, therefore, may result in a compensatory dominance of habit responding, and concomitant overactivation of sensorimotor loop structures. The second line of evidence in support of the hypothesis that early-life stress persistently affects neural circuits for instrumental behavior comes from studies of how early-life stress affects S-R habits during navigation tasks later in life. First, male rats exposed to maternal separation during the first two weeks of life are more likely to use an S-R navigation strategy in early adolescence (Grissom et al. 2012). Second, people exposed to stress prenatally are more likely to use an S-R navigation strategy in adulthood (Schwabe et al. 2012).

Although this area of research is still very much in its infancy, the results are consistent with a neural embedding hypothesis.

Given the evidence for persistent effects of early-life stress on navigation, one direction for future research is to establish whether the observed behavioral effects of early-life stress extend to non-navigational tasks. We recently addressed this question by investigating instrumental learning in young adults with a history of early-life stress (Patterson et al. 2013). Specifically, we were interested in whether sensitivity to reinforcement history, indexed by the partial reinforcement extinction effect (PREE), differed as a function of early-life stress history. The PREE is the well-known phenomenon whereby behaviors learned under a schedule of partial reinforcement are more resistant to extinction than behaviors learned under continuous reinforcement (e.g., Mowrer and Jones 1945; Vogel-Sprott 1967). A leading explanation for this effect is that it is based on memory for events during training (Capaldi 1966, 1967). In the partial reinforcement condition, because of the similarity between the extinction trials and the memory for the training trials, responding persists for an interval in extinction at a similar rate to training. In contrast, if every trial was reinforced during training, extinction trials are very different than memory for the training trials, and response rate sharply decreases.

Based on this hypothesis, the PREE should be reduced in individuals with a history of early-life stress to the extent to which responding is based on habit versus declarative memory We reported results from two experiments. In Experiment 1, we found that the PREE was reduced in people with a history of early-life stress. There was a large effect of reinforcement schedule, with participants trained on a continuous reinforcement schedule responding at a much lower rate during extinction. There was no difference between the groups after continuous reinforcement training, but for participants trained under partial reinforcement, there was a significant reduction in responding in the early-life stress group.

In Experiment 2, we demonstrated dosage effects of early-life stress, with the reduced PREE effect largely driven by participants who reported three or more adverse childhood experiences compared to those who reported one or two. We also included a condition in which participants learned the instrumental response while performing a concurrent tone-counting task similar to the one used by Foerde et al. (2006), (2007). This concurrent task served as a declarative memory challenge, and thus may result in more habitual learning in vulnerable individuals. Indeed, we found that while the concurrent task had no effect on the PREE in participants who did not report any adverse childhood experiences, it significantly reduced the PREE in the group that reported a moderate level of early-life stress. Thus, even in a resilient sample of university students, dose-dependent effects of early-life stress appear to be present in instrumental learning. Furthermore, the reduced PREE observed in the moderate early-life stress group under the declarative memory challenge is consistent with our framework of the PREE acting as an assessment of the influence of declarative memory on instrumental behavior. An approach combining probes that test sensitivity to reinforcement history with the more traditional tests of sensitivity to outcome devaluation and contingency degradation would be useful in determining how far these observed early-life stress effects extend, and whether sensitivity to reinforcement history is a reliable measure of goal-directed action selection.

4 Implications for Understanding Compulsions and Addictions

An overreliance on habit learning may substantially contribute to disorders such as obsessive-compulsive disorder (OCD) and drug addiction. In patients with OCD, behaviors are present that are automatically elicited by stimuli in the environment, even when the patient does not feel that the behaviors are useful. Patients with OCD showed less sensitivity to reinforcer devaluation in a discrimination task than were control subjects, suggesting that these patients have a greater tendency to learn habits (Gillan et al. 2014, 2015). Interestingly, the patients showing evidence of habit learning showed greater activation of the caudate nucleus during the devaluation test. Because this region is generally thought to support goal-directed learning based on lesion studies, it may be that this activation represents dysfunction in this circuit, impairing the ability to direct behavior using goal-directed actions that compete with habits (Gillan et al. 2015).

Drug addiction, like OCD, can also be understood in terms of increased habitual responding. Addiction can be viewed as the endpoint of a transition from control of behavior by action–outcome associations to control by S-R associations (Everitt and Robbins 2005; Belin et al. 2009). In the brain, this is thought to occur through neural reorganization both within the striatum and in the looping connections between the striatum and other brain regions. Toxic effects of drugs cause further neural dysfunction, such as damage to prefrontal executive control structures, resulting in even greater use of the habit system.

Several papers on the stress-induced shift to S-R habit behavior have suggested that this line of research may also have implications for the understanding and treatment of drug addiction (e.g., Packard 2009a; Packard and Goodman 2012); indeed, some have speculated that beta blockers or glucocorticoid antagonists could be useful in the prevention of drug relapse (Schwabe et al. 2011; Schwabe and Wolf 2011). Future research should target the role that the mechanisms underlying the stress-induced shift to habit behavior may play in addiction. Schwabe et al. (2011) proposed a two-part model in which acute and chronic stress may have separate effects on addiction processes. In the acute stress model, stress occurring during a drug-free period in the addicted organism activates the habit system and creates a vulnerability to relapse through increased responsivity to drug-related cues. In the chronic stress model, stress may accelerate the transition from goal-directed to habitual control of behavior that characterizes addiction.

Another domain in which stress could affect habit formation is through Pavlovian–Instrumental Transfer. Instrumental conditioning differs from Pavlovian conditioning in that a response must be made to receive reinforcement. While these two types of conditioning are fundamentally different, there are situations in which the presence of a Pavlovian conditional stimulus can affect instrumental conditioning (for review, see Rescorla and Solomon 1967; Balleine 2001). For example, presenting an appetitive conditional stimulus will energize pressing a bar for food (Lovibond 1983; Hall et al. 2001). Similarly, in the presence of alcohol-associated cues, alcoholics show a higher level of instrumental responding for alcohol rewards (Ludwig et al. 1974). It has long been known that that the presence of Pavlovian conditional stimuli can enhance instrumental learning (Rescorla and Solomon 1967). As discussed above, as practice on an instrumental task increases, goal-directed responding decreases and habitual responding increases. Thus, if Pavlovian conditional stimuli can accelerate instrumental conditioning, they may also facilitate the transition from goal-directed to habit learning.

In animal models, unconditional stressors (footshock or predator odors) have been shown to bias behavior toward habit responding (Packard 2009a). It is possible that Pavlovian fear cues would have a similar effect, and this effect may be potentiated in individuals with amygdala dysfunction. For example, people with a history of early-life stress have been shown to have amygdala hypertrophy and enhanced fear conditioning (for review, seeMcEwen 2003; Tottenham and Sheridan 2009); therefore, it is likely that Pavlovian fear cues are particularly salient to these individuals and could bias their behavior toward habit responding. This would be an example of Pavlovian-to-instrumental transfer, in that it would mean an aversive conditional stimulus alters the nature of instrumental responding. Thus, individuals with a history of early-life stress may become even more driven by habit responding when reminded of past aversive events, which may help to explain the association between stress during early life and increased rates of substance abuse and addictive behaviors in adulthood (Dube et al. 2003a, b; Macleod et al. 2013).

5 Conclusion

Work in nonhuman animals and humans has been consistent in demonstrating that the striatum is crucial for the development of habitual behavior, and that habits represent a type of learning that occurs gradually, is relatively inflexible, and does not rely on the medial temporal lobe structures that underlie declarative memory. Many developments in this area are very recent, and there are still many open questions. Future directions of interest include developmental and long-lasting chronic stress effects, and what precise mechanisms facilitate the shift in control of behavior from a goal-directed system to a habit system. The development of experimental assays to complement the current work in spatial navigation, instrumental learning, concurrent discrimination, and probabilistic classification could also help clarify mechanisms behind this transition, and might inform more generally on the characteristics of the habit memory system. Research on these topics has widespread implications for addiction and other public health issues.

References

Adams CD (1982) Variations in the sensitivity of instrumental responding to reinforcer devaluation. Q J Exp Psychol Sect B 34B(2):77–98

Adams CD, Dickinson A (1981) Instrumental responding following reinforcer devaluation. Q J Exp Psychol Sect B 33B(2):109–121

Alexander GE, DeLong MR, Strick PL (1986) Parallel organization of functionally segregated circuits linking basal ganglia and cortex. Annu Rev Neurosci 9:357–381

Balleine BW (2001) Incentive processes in instrumental conditioning. In: Handbook of contemporary learning theories, Psychology Press, USA, pp 307–366

Balleine BW, Dickinson A (1998a) Goal-directed instrumental action: contingency and incentive learning and their cortical substrates. Neuropharmacology 37(4–5):407–419

Balleine BW, Dickinson A (1998b) The role of incentive learning in instrumental outcome revaluation by sensory-specific satiety. Learn Behav 26(1):46–59

Bayley PJ, Frascino JC, Squire LR (2005) Robust habit learning in the absence of awareness and independent of the medial temporal lobe. Nature 436(7050):550–553

Belin D, Jonkman S, Dickinson A, Robbins TW, Everitt BJ (2009) Parallel and interactive learning processes within the basal ganglia: Relevance for the understanding of addiction. Behav Brain Res 199(1):89–102

Bohbot VD, Iaria G, Petrides M (2004) Hippocampal function and spatial memory: evidence from functional neuroimaging in healthy participants and performance of patients with medial temporal lobe resections. Neuropsychology 18(3):418–425

Bohbot VD, Lerch J, Thorndycraft B, Iaria G, Zijdenbos AP (2007) Gray matter differences correlate with spontaneous strategies in a human virtual navigation task. J Neurosci 27 (38):10078–10083

Capaldi EJ (1966) Partial reinforcement: a hypothesis of sequential effects. Psychol Rev 73 (5):459–477

Capaldi EJ (1967) A sequential hypothesis of instrumental learning. In: Spence KW, Spence JT (eds) The psychology of learning and motivation: advances in research and theory, vol 1. Academic Press, New York, pp 67–156

Chang Q, Gold PE (2003) Intra-hippocampal lidocaine injections impair acquisition of a place task and facilitate acquisition of a response task in rats. Behav Brain Res 144(1–2):19–24

Colwill RM, Rescorla RA (1985) Postconditioning devaluation of a reinforcer affects instrumental responding. J Exp Psychol: Anim Behav Processes 11(1):120–132

Dias-Ferreira E, Sousa JC, Melo I, Morgado P, Mesquita AR, Cerqueira JJ, Costa RM, Sousa N (2009) Chronic stress causes frontostriatal reorganization and affects decision-making. Science 325(5940):621–625

Dickinson A (1985) Actions and habits: the development of behavioural autonomy. Philos Trans R Soc Lond Ser B: Biol Sci 308(1135):67–78

Dickinson A, Nicholas DJ, Adams CD (1983) The effect of the instrumental training contingency on susceptibility to reinforcer devaluation. Q J Exp Psychol Sect B-Comp Physiol Psychol 35 (FEB):35–51

Dube SR, Felitti VJ, Dong M, Chapman DP, Giles WH, Anda RF (2003a) Childhood abuse, neglect, and household dysfunction and the risk of illicit drug use: the adverse childhood experiences study. Pediatrics 111(3):564–572

Dube SR, Felitti VJ, Dong M, Giles WH, Anda RF (2003b) The impact of adverse childhood experiences on health problems: evidence from four birth cohorts dating back to 1900. Prev Med 37(3):268–277

Everitt BJ, Robbins TW (2005) Neural systems of reinforcement for drug addiction: From actions to habits to compulsion. Nat Neurosci 8(11):1481–1489

Fernandez-Ruiz J, Wang J, Aigner TG, Mishkin N (2001) Visual habit formation in monkeys with neurotoxic lesions of the ventrocaudal neostriatum. Proc Natl Acad Sci USA 98(7):4196–4201

Ferragud A, Haro A, Sylvain A, Velazquez-Sanchez C, Hernandez-Rabaza V, Canales JJ (2010) Enhanced habit-based learning and decreased neurogenesis in the adult hippocampus in a murine model of chronic social stress. Behav Brain Res 210(1):134–139

Foerde K, Shohamy D (2011) The role of the basal ganglia in learning and memory: Insight from parkinson's disease. Neurobiol Learn Mem 96(4):624–636

Foerde K, Knowlton BJ, Poldrack RA (2006) Modulation of competing memory systems by distraction. Proc Natl Acad Sci USA 103(31):11778–11783

Foerde K, Poldrack RA, Knowlton BJ (2007) Secondary-task effects on classification learning. Mem Cognition 35(5):864–874

Gee DG, Gabard-Durnam LJ, Flannery J, Goff B, Humphreys KL, Telzer EH, Hare TA, Bookheimer SY, Tottenham N (2013) Early developmental emergence of human amygdala-prefrontal connectivity after maternal deprivation. Proc Natl Acad Sci USA 110 (39):15638–15643

Gillan CM, Morein-Zamir S, Urcelay GP, Sule A, Voon V, Apergis-Schoute AM, Fineberg NA, Sahakian BJ, Robbins TW (2014) Enhanced avoidance habits in obsessive-compulsive disorder. Biol Psychiatry 75(8):631–638

Gillan CM, Apergis-Schoute AM, Morein-Zamir S, Urcelay GP, Sule A, Fineberg NA, Sahakian BJ, Robbins TW (2015) Functional neuroimaging of avoidance habits in obsessive-compulsive disorder. Am J Psychiatry 172(3):284–293

Grissom EM, Hawley WR, Bromley-Dulfano SS, Marino SE, Stathopoulos NG, Dohanich GP (2012) Learning strategy is influenced by trait anxiety and early rearing conditions in prepubertal male, but not prepubertal female rats. Neurobiol Learn Mem 98(2):174–181

Hall J, Parkinson JA, Connor TM, Dickinson A, Everitt BJ (2001) Involvement of the central nucleus of the amygdala and nucleus accumbens core in mediating pavlovian influences on instrumental behaviour. Eur J Neurosci 13(10):1984–1992

Hull CL (1943) Principles of behavior. Appleton-Century-Crofts, New York

Iaria G, Petrides M, Dagher A, Pike B, Bohbot VD (2003) Cognitive strategies dependent on the hippocampus and caudate nucleus in human navigation: Variability and change with practice. J Neurosci 23(13):5945–5952

Joels M, Karst H, Krugers HJ, Lucassen PJ (2007) Chronic stress: implications for neuronal morphology, function and neurogenesis. Front Neuroendocrin 28(2–3):72–96

Killcross S, Coutureau E (2003) Coordination of actions and habits in the medial prefrontal cortex of rats. Cereb Cortex 13(4):400–408

Kim JJ, Lee HJ, Han J-S, Packard MG (2001) Amygdala is critical for stress-induced modulation of hippocampal long-term potentiation and learning. J Neurosci 21(14):5222–5238

Knowlton BJ (2002) The role of the basal ganglia in learning and memory. In: Squire LR, Schacter DL (eds) Neuropsychology of memory, 3rd edn. Guilford Press, New York, pp 143–153

Knowlton BJ, Moody TD (2008) Procedural learning in humans. In: Byrne JH (ed) Learning and memory: a comprehensive reference: Vol 3 memory systems. Academic Press, Oxford, pp 321–340

Knowlton BJ, Squire LR, Gluck MA (1994) Probabilistic classification learning in amnesia. Learn Memory 1(2):106–120

Knowlton BJ, Mangels JA, Squire LR (1996) A neostriatal habit learning system in humans. Science 273(5280):1399–1402

Lovibond PF (1983) Facilitation of instrumental behavior by a pavlovian appetitive conditioned-stimulus. J Exp Psychol-Anim Behav Process 9(3):225–247

Ludwig AM, Wikler A, Stark LH (1974) The first drink: Psychobiological aspects of craving. Arch Gen Psychiatry 30(4):539–547

Lupien SJ, McEwen BS, Gunnar MR, Heim C (2009) Effects of stress throughout the lifespan on the brain, behaviour and cognition. Nat Rev Neurosci 10(6):434–445

Macleod J, Hickman M, Jones HE, Copeland L, McKenzie J, De Angelis D, Kimber J, Robertson JR (2013) Early life influences on the risk of injecting drug use: case control study based on the edinburgh addiction cohort. Addiction 108(4):743–750

McDonald RJ, White NM (1994) Parallel information processing in the water maze: evidence for independent memory systems involving dorsal striatum and hippocampus. Behav Neural Biol 61(3):260–270
McEwen BS (2000) Effects of adverse experiences for brain structure and function. Biol Psychiatry 48(8):721–731
McEwen BS (2003) Early life influences on life-long patterns of behavior and health. Mental Retard Dev Disabil Res Rev 9(3):149–154
Milner B (1962) Les troubles de la memoire accompagnant les lesions hippocampiques bilaterales. In: Passouant P (ed) Physiologie de l'hippocampe. Centre National de la Recherche Scientifique, Paris, pp 257–272
Moody TD, Bookheimer SY, Vanek Z, Knowlton BJ (2004) An implicit learning task activates medial temporal lobe in patients with parkinson's disease. Behav Neurosci 118(2):438–442
Moody TD, Chang GY, Vanek ZF, Knowlton BJ (2010) Concurrent discrimination learning in parkinson's disease. Behav Neurosci 124(1):1–8
Mowrer OH, Jones H (1945) Habit strength as a function of the pattern of reinforcement. J Exp Psychol 35(4):293–311
Neal DT, Wood W, Wu M, Kurlander D (2011) The pull of the past: when do habits persist despite conflict with motives? Pers Soc Psychol Bull 37(11):1428–1437
Packard MG (2009a) Anxiety, cognition, and habit: a multiple memory systems perspective. Brain Res 1293:121–128
Packard MG (2009b) Exhumed from thought: Basal ganglia and response learning in the plus-maze. Behav Brain Res 199(1):24–31
Packard MG, Goodman J (2012) Emotional arousal and multiple memory systems in the mammalian brain. Front Behav Neurosci 6:14
Packard MG, Knowlton BJ (2002) Learning and memory functions of the basal ganglia. Annu Rev Neurosci 25(1):563–593
Packard MG, McGaugh JL (1992) Double dissociation of fornix and caudate-nucleus lesions on acquisition of 2 water maze tasks: Further evidence for multiple memory systems. Behav Neurosci 106(3):439–446
Packard MG, McGaugh JL (1996) Inactivation of hippocampus or caudate nucleus with lidocaine differentially affects expression of place and response learning. Neurobiol Learn Mem 65 (1):65–72
Packard MG, Hirsh R, White NM (1989) Differential effects of fornix and caudate nucleus lesions on two radial maze tasks: Evidence for multiple memory systems. J Neurosci 9(5):1465–1472
Patterson TK, Craske MG, Knowlton BJ (2013) The effect of early-life stress on memory systems supporting instrumental behavior. Hippocampus 23(11):1025–1034
Poldrack RA, Packard MG (2003) Competition among multiple memory systems: converging evidence from animal and human brain studies. Neuropsychologia 41(3):245–251
Poldrack RA, Prabhakaran V, Seger CA, Gabrieli JDE (1999) Striatal activation during acquisition of a cognitive skill. Neuropsychology 13(4):564–574
Poldrack RA, Clark J, Pare-Blagoev EJ, Shohamy D, Creso Moyano J, Myers C, Gluck MA (2001) Interactive memory systems in the human brain. Nature 414(6863):546–550
Rescorla RA, Solomon RL (1967) 2-process learning theory—relationships between pavlovian conditioning and instrumental learning. Psychol Rev 74(3):151
Sage JR, Knowlton BJ (2000) Effects of us devaluation on win-stay and win-shift radial maze performance in rats. Behav Neurosci 114(2):295–306
Schroeder JP, Wingard JC, Packard MG (2002) Post-training reversible inactivation of hippocampus reveals interference between memory systems. Hippocampus 12(2):280–284
Schwabe L, Wolf OT (2009) Stress prompts habit behavior in humans. J Neurosci 29(22):7191–7198
Schwabe L, Wolf OT (2010) Socially evaluated cold pressor stress after instrumental learning favors habits over goal-directed action. Psychoneuroendocrinology 35(7):977–986
Schwabe L, Wolf OT (2011) Stress-induced modulation of instrumental behavior: From goal-directed to habitual control of action. Behav Brain Res 219(2):321–328

Schwabe L, Oitzl MS, Philippsen C, Richter S, Bohringer A, Wippich W, Schachinger H (2007) Stress modulates the use of spatial versus stimulus-response learning strategies in humans. Learn Mem 14(1):109–116

Schwabe L, Dalm S, Schachinger H, Oitzl MS (2008) Chronic stress modulates the use of spatial and stimulus-response learning strategies in mice and man. Neurobiol Learn Mem 90(3): 495–503

Schwabe L, Dickinson A, Wolf OT (2011) Stress, habits, and drug addiction: a psychoneuroendocrinological perspective. Exp Clin Psychopharmacol 19(1):53–63

Schwabe L, Bohbot VD, Wolf OT (2012) Prenatal stress changes learning strategies in adulthood. Hippocampus 22(11):2136–2143

Soares JM, Sampaio A, Ferreira LM, Santos NC, Marques F, Palha JA, Cerqueira JJ, Sousa N (2012) Stress-induced changes in human decision-making are reversible. Transl Psychiatry 2: e131

Teng E, Stefanacci L, Squire LR, Zola SM (2000) Contrasting effects on discrimination learning after hippocampal lesions and conjoint hippocampal-caudate lesions in monkeys. J Neurosci 20 (10):3853–3863

Tolman EC (1932) Purposive behavior in animals and men. Century, New York

Tolman EC, Ritchie BF, Kalish D (1946) Studies in spatial learning: 2. Place learning versus response learning. J Exp Psychol 36(3):221–229

Tottenham N (2014) The importance of early experiences for neuro-affective development. In: Andersen SL, Pine DS (eds) Current topics in behavioral neurosciences: Vol 16 the neurobiology of childhood. Springer, New York, pp 109–129

Tottenham N, Sheridan MA (2009) A review of adversity, the amygdala and the hippocampus: A consideration of developmental timing. Front Hum Neurosci 3:68

Tottenham N, Hare TA, Quinn BT, McCarry TW, Nurse M, Gilhooly T, Millner A, Galvan A, Davidson MC, Eigsti IM et al (2010) Prolonged institutional rearing is associated with atypically large amygdala volume and difficulties in emotion regulation. Dev Sci 13(1):46–61

Tricomi EM, Delgado MR, Fiez JA (2004) Modulation of caudate activity by action contingency. Neuron 41(2):281–292

Tricomi E, Balleine BW, O'Doherty JP (2009) A specific role for posterior dorsolateral striatum in human habit learning. Eur J Neurosci 29(11):2225–2232

Valentin VV, Dickinson A, O'Doherty JP (2007) Determining the neural substrates of goal-directed learning in the human brain. J Neurosci 27(15):4019–4026

Vogel-Sprott M (1967) Partial-reward training for resistance to punishment and to subsequent extinction. J Exp Psychol 75(1):138–140

Yin HH, Knowlton BJ (2004) Contributions of striatal subregions to place and response learning. Learn Memory 11(4):459–463

Yin HH, Knowlton BJ (2006) The role of the basal ganglia in habit formation. Nat Rev Neurosci 7 (6):464–476

Yin HH, Knowlton BJ, Balleine BW (2004) Lesions of dorsolateral striatum preserve outcome expectancy but disrupt habit formation in instrumental learning. Eur J Neurosci 19(1):181–189

Yin HH, Ostlund SB, Knowlton BJ, Balleine BW (2005) The role of the dorsomedial striatum in instrumental conditioning. Eur J Neurosci 22(2):513–523

Yin HH, Knowlton BJ, Balleine BW (2006) Inactivation of dorsolateral striatum enhances sensitivity to changes in the action–outcome contingency in instrumental conditioning. Behav Brain Res 166(2):189–196

The Anatomy and Physiology of Eyeblink Classical Conditioning

Kaori Takehara-Nishiuchi

Abstract This chapter reviews the past research toward identifying the brain circuit and its computation underlying the associative memory in eyeblink classical conditioning. In the standard delay eyeblink conditioning paradigm, the conditioned stimulus (CS) and eyeblink-eliciting unconditioned stimulus (US) converge in the cerebellar cortex and interpositus nucleus (IPN) through the pontine nuclei and inferior olivary nucleus. Repeated pairings of CS and US modify synaptic weights in the cerebellar cortex and IPN, enabling IPN neurons to activate the red nucleus and generate the conditioned response (CR). In a variant of the standard paradigm, trace eyeblink conditioning, the CS and US are separated by a brief stimulus-free trace interval. Acquisition in trace eyeblink conditioning depends on several forebrain regions, including the hippocampus and medial prefrontal cortex as well as the cerebellar–brainstem circuit. Details of computations taking place in these regions remain unclear; however, recent evidence supports a view that the forebrain encodes a temporal sequence of the CS, trace interval, and US in a specific environmental context and signals the cerebellar–brainstem circuit to execute the CR when the US is likely to occur. Together, delay eyeblink conditioning represents one of the most successful cases of understanding the neural substrates of long-term memory in mammals, while trace eyeblink conditioning demonstrates its utility for uncovering detailed computations in the whole brain network underlying long-term memory.

Keywords Associative memory · Nictitating membrane · Cerebellum · Hippocampus

---

K. Takehara-Nishiuchi (✉)
Department of Psychology, Cell and Systems Biology, Neuroscience Program, University of Toronto, Toronto M5S 3G3, Canada
e-mail: takehara@psych.utoronto.ca

Curr Topics Behav Neurosci (2018) 37: 297–323
DOI 10.1007/7854_2016_455
Published Online: 27 December 2016

Contents

1 Introduction 298
2 Delay Eyeblink Conditioning 299
2.1 CS and US Pathways 299
2.2 Converging Sites of the CS and US Information 301
2.3 Circuit Mechanism of CR Acquisition 302
2.4 Synaptic Plasticity Underlying CR Acquisition 303
3 Trace Eyeblink Conditioning 304
3.1 Key Components of the Network 304
3.2 Putting It All Together: A Theory of the Forebrain–Cerebellar Interaction During Trace Eyeblink Conditioning 308
4 Common Features Among Associative Memory in Trace Eyeblink Conditioning and Other Hippocampus-Dependent Memories 310
4.1 Reorganization of the Cortical Network During Systems Consolidation 310
4.2 Conscious Awareness of Memory Content 312
5 Conclusions and Future Directions 313
References 314

1 Introduction

Eyeblink classical conditioning is one of the best characterized models for associative memory in mammals. What animals are *expected* to learn in eyeblink classical conditioning is simple: the association between a neutral conditioned stimulus (CS) and an aversive, eyeblink-eliciting unconditioned stimulus (US). An auditory, visual, or somatosensory stimulus is typically used as the CS, and an air puff to the cornea or electric shock near the eye is used as the US. Initially, naïve animals show only a reflexive eyelid response (eyeblinks or extension of nictitating membrane) upon the presentation of US as an unconditioned response (UR). With repeated pairings of the CS and US, the animals come to exhibit a defensive eyelid response upon the presentation of CS as a conditioned response (CR). The CR differs from the UR and spontaneous blinks in terms of kinematics and the underlying circuitry (Gruart et al. 1995, 2000; Schade Powers et al. 2010). The CR is detected by mechanically measuring the extension of the nictitating membrane, recording the electromyogram in the eyelid or monitoring eyelid movement with a high-speed camera or infrared reflective sensors. The strength of the acquired CS–US association is quantified as the percentage of CS presentations that elicit the CR.

Each trial in eyeblink classical conditioning is very brief. The duration of CS is typically less than one second, and the duration of US is around one hundred milliseconds. In a standard paradigm, called delay eyeblink conditioning, the CS precedes the US and terminates with the US. In a frequently used variant, trace eyeblink conditioning, the CS, and US are separated by a stimulus-free interval (typically up to 500 ms in rabbits and rodents, and up to one second in humans). Daily conditioning includes the pairing of CS and US thirty to a few hundred times. CR acquisition in delay eyeblink conditioning is possible in a minimum of one conditioning day, whereas trace eyeblink conditioning requires more days of conditioning to establish reliable CRs.

In the past few decades, studies in eyeblink classical conditioning have uncovered profound details concerning the neuronal bases of associative memory. The goal of this chapter was to outline these achievements, with special focus on the outcome of two types of investigations. The first type of investigation is a behavioral study that examined behavioral changes induced by targeted manipulations to a specific brain region. The second type is a neurophysiological study that monitors changes of neuronal activity while animals acquire and express CRs. The first section of this chapter describes specific anatomical connections and synaptic plasticity within the cerebellum and brainstem that underlie memory acquisition in delay eyeblink conditioning (Sect. 2). The subsequent section first reviews evidence suggesting that trace eyeblink conditioning depends on an integrated network that includes the cerebellar–brainstem circuit and several forebrain regions, and then discusses a hypothesis concerning the computations that take place within each region and through the inter-region interaction (Sect. 3). The hypothesis posits that the primary function of forebrain is to encode the temporal sequence of CS, trace interval, and US in a specific context. This hints a possibility to link the forebrain function in trace eyeblink conditioning to a broader theoretical framework of forebrain mnemonic function. To strengthen this point, this chapter subsequently introduces two additional features of memory acquired in trace eyeblink conditioning that are common in other types of hippocampus-dependent memory (Sect. 4). The chapter ends with a conclusion and future directions (Sect. 5).

2 Delay Eyeblink Conditioning

The neuronal mechanisms underlying delay eyeblink conditioning have been extensively characterized. The seminal work by Richard Thompson and his colleagues identified that the neuronal plasticity underlying the acquisition of the CR exists in the cerebellum (Lincoln et al. 1982; McCormick et al. 1982). The entire essential circuit has been mapped (Sects. 2.1 and 2.2) and the mechanism of how the circuit generates the CR has been uncovered (Sects. 2.3 and 2.4).

2.1 CS and US Pathways

The pontine nuclei and dorsal accessory inferior olive (DAO) provide the cerebellum with the CS and US information, respectively (Fig. 1a). The pontine nuclei receive the projection of the brainstem and subcortical sources, which process auditory, visual, and somatosensory stimuli (Steinmetz and Sengelaub 1992; Mihailoff 1993). It in turn sends projections through the middle cerebellar peduncle to the deep cerebellar nuclei and granule cells in the cerebellar cortex. In parallel,

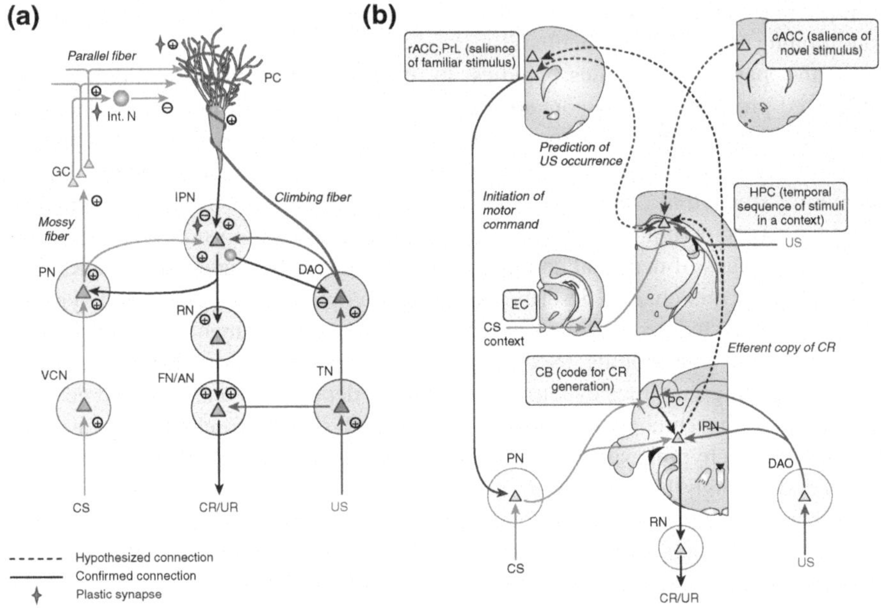

Fig. 1 Neural circuit underlying delay and trace eyeblink conditioning with an auditory-conditioned stimulus (CS). **a** In delay eyeblink conditioning, the CS reaches the pontine nuclei (PN) via the vestibulocochlear nucleus (VCN). The PN projects to the interpositus nucleus (IPN) and granule cells (GC) in the cerebellar cortex. GC projects to inhibitory interneurons (Int. N) as well as Purkinje cells (PC). The unconditioned stimulus (US) reaches the dorsal accessory inferior olive (DAO) via the trigeminal nucleus (TN). The DAO projects to the IPN and PC. The PC sends inhibitory projections to the IPN, which are combined with the excitatory inputs from the PN and DAO. The outcome of IPN reaches the RN, which in turn activates the facial nucleus (FN) and abducens nuclei (AN) to generate the conditioned response (CR). In parallel, the TN projects to the FN and AN, which elicits the unconditioned response (UR). **b** Most parts of neural circuit in trace eyeblink conditioning remain undefined; however, the figure represents a hypothesis based on available experimental evidence. The CS and contextual information is first processed in the entorhinal cortex (EC) and then sent to the hippocampus (HPC). The US information may proceed to the HPC via pathways that are yet to be defined. The HPC combines these streams of information and forms a coherent representation of the temporal sequence of CS, trace interval and, US in a specific context, which is used to generate a moment-to-moment estimation of the probability of US occurrence. This process may be modulated by the activity of the caudal and rostral anterior cingulate cortex (cACC, rACC) as well as the prelimbic cortex (PrL), which signal the salience of novel and familiar stimuli. The output of HPC is sent back to the rACC and PrL, which send the signal to the PN to initiate a motor command to express the CR. In parallel, the IPN sends, via multi-synaptic connections, the efferent copy of the CR back to the HPC and rACC, which may be used to refine their ability to generate the prediction of the timing of the US

axons of neurons in the DAO form climbing fibers and project to the interpositus nucleus as well as Purkinje cells in the cerebellar cortex via the inferior cerebellar peduncle (Matsushita and Ikeda 1970; Swenson and Castro 1983). Damage to either pontine nuclei or DAO completely prevents CR acquisition (McCormick et al.

1985; Steinmetz et al. 1987; Mintz et al. 1994; Welsh and Harvey 1998; Bao et al. 2000). Moreover, reliable CRs were acquired when the microstimulation of pontine nuclei and DAO was used as the CS and US, respectively (Mauk et al. 1986; Steinmetz et al. 1986, 1989; Hesslow et al. 1999).

The activity of neurons in the pontine nuclei and DAO shows learning-related changes during delay eyeblink conditioning. In naïve rabbits, neurons in the pontine nuclei respond to the CS with a very short latency, and the response terminates at the offset of CS (McCormick et al. 1983). When the conditioning progresses, the CS-evoked response becomes stronger toward the onset of US (McCormick et al. 1983; Clark and Gohl 1997; Bao et al. 2000). This "conditioned" neuronal response is abolished by damage to the interpositus nucleus, suggesting that it is mediated by feedback excitation from the interpositus nucleus, but not by local synaptic plasticity (Cartford et al. 1997; Clark and Gohl 1997; Bao et al. 2000). Neurons in the DAO, on the other hand, increase the firing rate upon the US presentation, but more importantly they reduce the US-induced firing as animals acquire the CR (Sears and Steinmetz 1991; Hesslow and Ivarsson 1996). This property is consistent with the prediction of the Rescorla–Wagner model (Rescorla and Wagner 1972). The reduction of the US-induced response is likely mediated by the feedback inhibition from the interpositus nucleus to the DAO (Fredette and Mugnaini 1991; Nicholson and Freeman 2003; Svensson et al. 2006) and/or by feedback from the red nucleus to the trigeminal/dorsal column nuclei (Weiss et al. 1990).

2.2 Converging Sites of the CS and US Information

The CS and US information converges at two sites within the cerebellum: the cerebellar cortex and the interpositus nucleus (Fig. 1a). Dysfunction of either structure significantly impairs the acquisition and expression of CR (McCormick et al. 1982; Lavond et al. 1985, 1987; Yeo et al. 1985a, b; Clark et al. 1992; Steinmetz et al. 1992; Krupa et al. 1993; Krupa and Thompson 1997; Perrett et al. 1993; Chen et al. 1996, 1999).

The cerebellar cortex consists of five cell types which are orderly arranged in three layers (Palay and Chan-Palay 1974). Among them, granule cells receive mossy fibers (axons of pontine neurons) and send excitatory projections to Purkinje cells as parallel fibers. In addition, each Purkinje cell is wrapped up in an axon of an olivary neuron. The two excitatory inputs induce two distinct patterns of spiking response in Purkinje cells. One is a simple spike, which is a regular action potential induced by parallel fiber inputs. The other is a complex spike, which is a burst firing induced by the climbing fiber input (Eccles et al. 1966a, b).

Purkinje cells are the sole output of cerebellar cortex and make inhibitory synapses on neurons in deep cerebellar nuclei, including the interpositus nucleus. In parallel, the interpositus nucleus receives the CS information from the pontine nuclei and the US information from the DAO. The output of the interpositus nucleus reaches the red nucleus, which controls eyelid movement and eyeball retraction, via the accessory facial nucleus, as well as a nictitating membrane, via the accessory abducens nucleus (Disterhoft et al. 1985; Fig. 1a). As expected based on the anatomical pattern, electric stimulation of the restricted portion of interpositus nuclei elicits an eyeblink response in both trained and naive animals (McCormick and Thompson 1984a), suggesting that the interpositus nucleus converts the cerebellar output to the eyelid movement.

2.3 Circuit Mechanism of CR Acquisition

In naive animals, the CS does not induce eyeblink responses because the excitation of the interpositus nucleus by the pontine nuclei is canceled out by the inhibition by Purkinje cells (Fig. 1a). The US, on the other hand, induces the UR through the excitation of the accessory facial and abducens nuclei by the trigeminal nucleus. The CR is developed as the repeated CS–US pairings change the CS-induced firing response of Purkinje cells and nuclear neurons, which modulates or adds onto the UR. With learning, Purkinje cells come to pause the simple spike firing during the CS (Berthier and Moore 1986; Hesslow and Ivarsson 1994; Green and Steinmetz 2005; Kotani et al. 2006; Jirenhed et al. 2007). In contrast, neurons in the interpositus nucleus increase the firing rate upon CS presentation (McCormick and Thompson 1984a, b; Berthier and Moore 1990; Gould and Steinmetz 1996; Freeman and Nicholson 2000; Rogers et al. 2001). Importantly, both the conditioned neuronal responses demonstrate a pattern that precedes and predicts the temporal form of the behavioral CR. The changes in response to Purkinje cells are likely mediated by the synaptic plasticity in the cerebellar cortex (Sect. 2.4), which in turn drive the changes of nuclear neurons' response. The reduced inputs from the Purkinje cells result in the disinhibition of nuclear neurons, thereby enabling them to respond to the CS input from the pontine nuclei. The output of nuclear neurons, in turn, activates the red nucleus, which executes the CR. Therefore, the conditioned pause in Purkinje cells' spiking and the associated increase in nuclear neurons' firing are the neuronal bases of behavioral CR.

It is important to note a potential functional dissociation between the cerebellar cortex and interpositus nucleus in CR generation. With extensive training, the acquisition of CR is possible without the cerebellar cortex, but it does not occur without the interpositus nucleus (Lavond and Steinmetz 1989). Moreover, several studies used pharmacological manipulations to disconnect the cerebellar cortex from the interpositus nucleus (Garcia and Mauk 1998; Bao et al. 2002; Kalmbach et al. 2010) and showed that when the input from Purkinje cells is eliminated (via microinfusion of $GABA_A$ antagonist into the interpositus nucleus), the latency of CR becomes shorter, and the response amplitude does not increase toward US onset. In contrast, when the interpositus nucleus is inhibited (via the microinfusion of $GABA_A$ agonist into the interpositus nuclei), the CR is abolished. These results suggest that the activity of the interpositus nucleus drives the expression of CR while the cerebellar cortex shapes the adaptive temporal pattern of CR.

2.4 Synaptic Plasticity Underlying CR Acquisition

Several forms of synaptic plasticity have been implicated as a mechanism of the learning-induced pause of simple spikes in Purkinje cells. Among them, the most studied is the long-term depression (LTD) at synaptic connections between parallel fibers and Purkinje cells. In slice preparations, the conjunctive activation of parallel fibers and climbing fibers results in persistent depression in the excitatory postsynaptic potentials evoked by the stimulation of parallel fibers (Ito and Kano 1982; Linden and Connor 1995; Aiba et al. 1994; De Zeeuw et al. 1998; Feil et al. 2003). Importantly, the LTD was induced most effectively when the parallel fiber stimulation preceded the climbing fiber stimulation by 250 ms (Chen and Thompson 1995), which corresponds to a typical interval between the CS and US in delay eyeblink conditioning. In addition, in several lines of transgenic mice, the impairment in LTD at parallel fiber-Purkinje cell synapses accompanies the impaired acquisition in delay eyeblink conditioning (Aiba et al. 1994; Chen et al. 1995; Shibuki et al. 1996; Kishimoto et al. 2001a, b; Miyata et al. 2001; Koekkoek et al. 2003; but see Welsh et al. 2005; Schonewille et al. 2011).

The LTD in parallel fiber-Purkinje cell synapses is one of many forms of synaptic plasticity that exist in the cerebellar cortex (for a review, see Gao et al. 2012), and it is possible that the other forms of plasticity may also play a role in the conditioned pause of Purkinje cell firing. As is evident in the neuronal recording studies (Berthier and Moore 1986; Hesslow and Ivarsson 1994; Green and Steinmetz 2005; Kotani et al. 2006; Jirenhed et al. 2007), Purkinje cells show tonic simple spike firings in the absence of any overt stimuli, and several lines of evidence suggested that the spontaneous firing of Purkinje cells is due to their intrinsic membrane properties (Gahwiler 1975; Hausser and Clark 1997; Raman and Bean 1997; Cerminara and Rawson 2004). This questions whether the reduction of responsiveness to excitatory inputs from parallel fibers (i.e., LTD) is sufficient to generate the pause of Purkinje firing. Rather, the strengthening of inhibitory inputs to Purkinje cells is required to actively suppress their firing (Jorntell et al. 2010; Schonewille et al. 2011; Gao et al. 2012). One possible source of this inhibition is interneurons (basket and stellate cells) in the molecular layer of the cerebellar cortex. These interneurons receive excitatory inputs from parallel fibers and climbing fibers and form inhibitory synapses onto Purkinje cells (Palay and Chan-Palay 1974). Importantly, the synaptic transmission between parallel fibers and the interneurons is enhanced by the coactivation of climbing fiber and parallel fiber inputs (Jorntell and Ekerot 2002, 2003; Liu and Cull-Candy 2005; Soler-Llavina and Sabatini 2006). This would increase the inhibition onto Purkinje cells during the CS, thereby inducing the CS-locked pause of the spiking of Purkinje cells (Jorntell et al. 2010; Schonewille et al. 2011; Gao et al. 2012).

In parallel with the synaptic plasticity in the cerebellar cortex, synapses between the mossy fibers and neurons in the interpositus nucleus are plastic. In slice preparation, long-term potentiation (LTP) is induced at these synapses when high-frequency trains of synaptic excitation precede a period of inhibition and disinhibition of nuclear neurons (Pugh and Raman 2006, 2008). This corresponds to a situation during delay eyeblink conditioning in which the excitatory input of pontine nuclei precedes the period of inhibition and disinhibition by Purkinje cells. Moreover, the number of excitatory synapses between the mossy fibers and nuclear neurons is increased with delay eyeblink conditioning (Kleim et al. 2002; Weeks et al. 2007), which may serve as a structural basis of LTP at mossy fiber-nuclear neuron synapses. The LTP would enhance the response of nuclear neurons to the excitatory inputs from mossy fibers, which enhances the nuclear neurons' ability to activate the red nucleus.

3 Trace Eyeblink Conditioning

Trace eyeblink conditioning differs from delay eyeblink conditioning in that a short interval is inserted between the offset of the CS and the onset of the US. Despite the small modification, acquisition in trace eyeblink conditioning depends on several regions in the forebrain in addition to the above-reviewed circuit in the cerebellum and brainstem (Fig. 1b). These forebrain regions include the hippocampus (Solomon and Vander Schaaf 1986; Moyer et al. 1990; Weiss et al. 1999; Beylin et al. 2001; Tseng et al. 2004; Walker and Steinmetz 2008), medial prefrontal cortex (Kronforst-Collins and Disterhoft 1998; Weible et al. 2000, McLaughlin et al. 2002; Takehara-Nishiuchi et al. 2005; Oswald et al. 2006), rhinal cortex (Ryou et al. 2001; Suter et al. 2013; Tanninen et al. 2015), sensory cortex (Galvez et al. 2007; Steinmetz et al. 2013), caudate nucleus (Flores and Disterhoft 2009, 2013), and thalamic nuclei (Powell and Churchwell 2002; Oswald et al. 2007). The following section first discusses computations that may be taking place in these regions (Sect. 3.1) and then offers a theory on how these regions cooperate during trace eyeblink conditioning (Sect. 3.2).

3.1 Key Components of the Network

3.1.1 The Cerebellum and Brainstem

It is generally agreed that trace eyeblink conditioning depends on the same circuit in the cerebellum and brainstem as delay eyeblink conditioning (Weiss and Disterhoft 1996; Woodruff-Pak and Disterhoft 2008); however, whether they share the same computation remains a matter of debate. Previous studies agreed that the acquisition of the CR critically depends on an intact interpositus nucleus (Woodruff-Pak et al. 1985; Pakaprot et al. 2009; Hu et al. 2010). In contrast, two lines of evidence conflict on how the cerebellar cortex is involved in trace eyeblink conditioning. Several transgenic mice with the disrupted cerebellar cortex are impaired in delay eyeblink conditioning, but they acquire CRs in trace eyeblink conditioning at a rate comparable to that of controls (Kishimoto et al. 2001a, b; Woodruff-Pak et al. 2006; Brown et al. 2010). These findings suggest that the cerebellar cortex may not be involved in CR acquisition in trace eyeblink conditioning or that dysfunction of the cerebellar cortex can be compensated by other regions (likely the forebrain). In contrast, pharmacological disconnection of the cerebellar cortex from the interpositus nucleus does not affect CR expression per se, but it disrupts the temporal pattern of CR in both trace and delay eyeblink conditioning (Kalmbach et al. 2010; Hu et al. 2010). These findings suggest that the cerebellar cortex shapes the timing of CR in both delay conditioning and trace eyeblink conditioning.

3.1.2 Hippocampus

In the 1970s, the hippocampus was presumed to contain a neuronal substrate of delay eyeblink conditioning because hippocampal neurons developed the response that modeled the amplitude and time course of CRs (Berger et al. 1976; Patterson et al. 1979). The argument was soon challenged by the observation that rabbits with bilateral ablations of the hippocampus show intact CR acquisition (Schmaltz and Theios 1972; Solomon and Moore 1975; Solomon et al. 1983). These findings resulted in the current consensus that the hippocampus is not necessary for the acquisition of CR in delay eyeblink conditioning. Subsequently, it was demonstrated that damage to the hippocampus impairs several variants of eyeblink conditioning, including trace eyeblink conditioning (Solomon and Vander Schaaf 1986; Moyer et al. 1990; Weiss et al. 1999; Beylin et al. 2001; Tseng et al. 2004; Walker and Steinmetz 2008), long delay conditioning (Beylin et al. 2001), latent inhibition (Solomon and Moore 1975; Schmajuk et al. 1994), discrimination reversal (Berger and Orr 1983; Weikart and Berger 1986), and extinction (Schmaltz and Theios 1972; Akase et al. 1989; Moyer et al. 1990). These observations suggest that the hippocampus may be involved in computations whose outcome cannot be detected as CR expression unless it is explicitly tested.

The first two variants, such as trace eyeblink and long delay conditioning, challenge an animal's ability to temporally integrate stimulus information. Long delay conditioning involves a longer interval between the onsets of the CS and US than standard delay conditioning, but the CS still coterminates with the US. CR acquisition in the long delay conditioning is impaired in rats with hippocampal damage; however, the impairment is not as severe as that in trace eyeblink conditioning (Beylin et al. 2001). In trace eyeblink conditioning, the degree of impairment depends on the length of the interval between the *offset* of the CS and the onset of the US (trace interval). Impairment following hippocampal damage is more severe with longer compared with shorter trace intervals (Tseng et al. 2004; Walker and Steinmetz 2008). Together, these findings suggest that the hippocampus may become necessary to associate the CS and US when their temporal contiguity is weakened by long inter-stimulus or trace intervals.

Another common feature of some hippocampus-dependent variants is the dependence on contextual information. Contextual information here refers to information that is present during the conditioning, but not explicitly reinforced, such as the location of conditioning chamber, the visual appearance of the conditioning chamber, and the sequence of procedures to place animals in the conditioning chamber. Although none of these pieces of "background" information is absolutely necessary to form the CR, they are nonetheless automatically captured and integrated with the CS and US information. For example, changing the conditioning chamber markedly attenuates the expression of previously acquired CR (Penick and Solomom 1991). In addition, the pre-exposure of a CS in one conditioning chamber later impairs animals' ability to associate the CS with the US in the chamber (i.e., latent inhibition; Solomon and Moore 1975; Schmajuk et al. 1994). Importantly, both these features are eliminated by hippocampal damage (Penick and Solomom 1991; Solomon and Moore 1975; Schmajuk et al. 1994; but see Shohamy et al. 2000). Together, these results suggest that the hippocampus may play a role in encoding an associative representation of foreground stimuli (i.e., the CS and US) and background contextual information.

The two features discussed above, temporal integration and associative encoding, are also evident in firing patterns of single neurons in the hippocampus during trace eyeblink conditioning. Neurons in the CA1 region of the hippocampus show a variety of conditioning-related responses, including increased or decreased firing upon the presentation of CS, US, or both (Weiss et al. 1996; McEchron and Disterhoft 1997; McEchron et al. 2001; Munera et al. 2001; Weible et al. 2006). In addition, some hippocampal neurons change their firing rate during the trace interval (McEchron and Disterhoft 1997). Repeated pairings of a CS and US, therefore, form a sequential firing pattern of hippocampal neurons: CS-responding neurons, trace-interval-responding neurons, and US-responding neurons (Modi et al. 2014). This sequential activity may also be locked to representations of the local environment: Conditioning-related neuron responses appear to depend on the conjunction between environmental location and CS (Moita et al. 2003). Thus, hippocampal representations may combine temporal sequence information with information about spatial location.

How does the hippocampus use the context-specific representation of CS-trace interval-US sequence to support CR acquisition and later expression in trace eyeblink conditioning? One possibility is that the hippocampus may use the representation to generate a prediction on whether the US would occur at a given moment. The prediction can be computed based on how well the present incoming inputs activate the neuronal activity state related to the US. When animals are placed in a conditioning chamber used during the previous conditioning session, the conditioning chamber should activate the contextual representation that was formed during the previous conditioning session (Vazdarjanova and Guzowski 2004; Lee et al. 2004; Leutgeb et al. 2004). When the CS is presented in the conditioning chamber, it should activate CS-responding neurons bound to the contextual representation (Moita et al. 2003). Importantly, during the previous conditioning session, the CS-responding neurons were associated with the trace-interval-responding neurons and US-responding neurons because these neurons repeatedly fired in a specific temporal sequence (McEchron and Disterhoft 1997; Modi et al. 2014).

Thus, this acquired representation of stimulus sequence would complete the neural activity state of US presentation from the activity of CS-responding neurons before the US is actually presented. This internally generated US representation signals the upcoming US which would facilitate the process in other brain regions that generate the behavioral CR (a computational model on a similar idea, see Moustafa et al. 2013).

3.1.3 The Medial Prefrontal Cortex

The medial prefrontal cortex consists of several subregions, including the anterior cingulate, prelimbic, and infralimbic cortices, and each of them has distinct patterns of afferent and efferent projections (Uyling et al. 2003). Accumulating evidence suggests that trace eyeblink conditioning depends on the anterior cingulate and prelimbic cortices; however, they appear to be involved in different stages of memory processes. CR acquisition is impaired by damage that includes the caudal anterior cingulate cortex (Kronforst-Collins and Disterhoft 1998; Weible et al. 2000, but see McLaughlin et al. 2002). In contrast, manipulations to the rostral part of anterior cingulate or prelimbic cortex have either no effect (Kronforst-Collins and Disterhoft 1998) or only moderately retard acquisition (McLaughlin et al. 2002; Takehara-Nishiuchi et al. 2005) but impair expression of CRs performed after acquisition (Takehara-Nishiuchi et al. 2006; Oswald et al. 2008, 2010). These findings suggest that the caudal anterior cingulate cortex plays a critical role in the early stage of acquisition, and that the rostral anterior cingulate and prelimbic cortices are involved in later expression.

The stage-dependent involvement is also supported by activity patterns of single neurons. Neurons in the caudal anterior cingulate cortex respond to the CS from the first acquisition session, and these responses weaken as animals acquire the CR (Weible et al. 2003; Hattori et al. 2014). On the other hand, there are neurons in the rostral anterior cingulate and prelimbic cortices that respond to the CS and maintain activity during the trace interval (Takehara-Nishiuchi and McNaughton 2008; Siegel et al. 2012; Siegel and Mauk 2013; Hattori et al. 2014). This persistent activity becomes stronger as animals acquire the CR (Takehara-Nishiuchi and McNaughton 2008; Siegel et al. 2012; Siegel and Mauk 2013; Hattori et al. 2014). Despite the difference in the time course of development, an important shared feature of these activity patterns is that the selectivity emerges with training. Neurons initially respond to the CS regardless of CS–US contingency (i.e., the paired or unpaired presentation), but with learning, they become capable of selectively responding to the CS only when it predicts the US (Weible et al. 2003; Takehara-Nishiuchi and McNaughton 2008; Siegel et al. 2012; Siegel and Mauk 2013; Hattori et al. 2014). This pattern suggests that prefrontal neurons signal the behavioral relevance of stimuli, as estimated by the past experience.

As reviewed above, the caudal part of the medial prefrontal cortex is involved in memory acquisition, and the rostral part is involved in later expression. This suggests that the selective neuron activity in the caudal and rostral prefrontal cortex may have different functional significance. Specifically, the neuronal activity in the caudal anterior cingulate cortex may signal the relevance of novel stimuli during the early stage of learning, which would facilitate their incorporation into the associative encoding of the present situation in the hippocampus (discussed in Sect. 3.1.2). The neuronal activity in the rostral anterior cingulate and prelimbic cortices, on the other hand, may signal the relevance of familiar stimuli that have already been incorporated into the associative code in the hippocampus. The relevance signal can facilitate the pattern completion of the associative code in the hippocampus, thereby allowing for the generation of the prediction of the upcoming US (discussed in Sect. 3.1.2).

The hippocampal prediction signal may return back to the rostral medial prefrontal cortex, in which it could be converted into the execution signal of CR by sending the final outcome of forebrain processing to the cerebellar–brainstem circuit. This idea was supported by several pieces of experimental evidence: The rostral medial prefrontal cortex sends direct projections to the pontine nuclei (Wiesendanger and Wiesendanger 1982; Buchanan et al. 1994; Schmahmann and Pandya 1997; Weible et al. 2007; Siegel et al. 2012; Moya et al. 2014). In the part of the pontine nuclei that receives the prefrontal projection, some neurons show a similar persistent activity to that observed in the rostral anterior cingulate cortex (Siegel et al. 2012). The persistent activity of the pontine neuron is causally related to CR acquisition in trace eyeblink conditioning because CR acquisition becomes possible only when the cerebellum receives two inputs, one that mimics the persistent activity in the rostral medial prefrontal cortex and the other that mimics CS-elicited activity in the pontine nuclei (Kalmbach et al. 2009, 2010).

3.2 *Putting It All Together: A Theory of the Forebrain–Cerebellar Interaction During Trace Eyeblink Conditioning*

This section proposes a hypothesis of how forebrain and cerebellar–brainstem circuit cooperate during trace eyeblink conditioning (Fig. 1b). It posits that the function of forebrain is to (1) learn a sequence of stimuli (CS, trace interval, and US) in a specific context (the conditioning environment) and (2) generate a prediction of whether and when the US occurs by comparing the incoming inputs to the learned temporal sequence. The cerebellar–brainstem circuit, on the other hand, learns to generate a motor command of CR, which allows an animal to adaptively respond to the upcoming US. The output of the forebrain informs the cerebellar–brainstem circuit of when to execute the CR, and the cerebellar circuit feeds back an efferent copy of the CR to the forebrain to enhance the accuracy of prediction of US. The bidirectional interaction between the forebrain and cerebellum likely takes place in both delay conditioning and trace eyeblink conditioning; however, in delay eyeblink conditioning, the outcome of forebrain does not have any impact on the computation in the cerebellar–brainstem circuit because the cerebellar circuit can be activated directly by the CS input. The following section outlines the anatomical and neurophysiological substrates that may underlie the proposed computations.

In the forebrain, neutral sensory stimuli, including the CS and any stimuli available in conditioning environment, are first processed in the sensory cortex (Galvez et al. 2007; Steinmetz et al. 2013). The stimulus information is likely to be transferred from the sensory cortex to the hippocampus via the rhinal cortex because damage to the rhinal cortex abolishes both CS-induced firing rate changes (Ryou et al. 2001) and neuron firing selective for the environment (Brun et al. 2008; Van Cauter et al. 2008; Lu et al. 2013) in the hippocampus. It is less clear which brain region conveys US information to the hippocampus in which neurons robustly respond to the US (Berger et al. 1976; Weiss et al. 1996). Upon receiving these two streams of information, the hippocampal neurons form a coherent representation of the CS, trace interval, US, and their temporal sequence in a specific context (Weiss et al. 1996; McEchron and Disterhoft 1997; McEchron et al. 2001; Munera et al. 2001; Moita et al. 2003; Weible et al. 2006). This process may be facilitated by additional inputs from the caudal part of the medial prefrontal cortex, which signals the relevance of CS during the early stage of learning (Weible et al. 2003; Hattori et al. 2014). Once formed, the coherent representation enables the hippocampus to calculate the moment-to-moment similarity between the acquired temporal sequence of stimuli and current incoming stimulus inputs, with which it generates the prediction of whether and when the US is likely to occur. The hippocampal prediction is sent to the rostral part of medial prefrontal cortex through the monosynaptic connections (Jay and Witter 1991) and/or the rhinal cortex (Ryou et al. 2001; Suter et al. 2013; Tanninen et al. 2015). The rostral medial prefrontal cortex in turn sends outputs to the cerebellar–brainstem circuit to initiate the execution of the appropriate behavior assigned to the situation (i.e., CR expression).

In the cerebellar–brainstem circuit, the pontine nuclei receive the output of the rostral medial prefrontal cortex and provide sustained activation to the interpositus nuclei as well as granule cells in the cerebellar cortex. Currently available experimental evidence is not sufficient to propose the exact computation taking place within the cerebellar circuit during trace eyeblink conditioning; however, the final outcome is the activation of the interpositus nucleus (Woodruff-Pak et al. 1985; Pakaport et al. 2009; Hu et al. 2010), which generates the CR. In parallel, the activity of the interpositus nucleus is also sent back to the forebrain, thereby potentially serving as an efferent copy of CR: Lesions of the interpositus nucleus abolish the hippocampal neuronal response that models the CR (Clark et al. 1984; Sears and Steinmetz 1990) as well as the prefrontal neuronal response to the trace interval (Siegel and Mauk 2013). This suggests that the hippocampus and medial prefrontal cortex may update their code based on whether their output results in an appropriate behavior (i.e., CR). The anatomical pathway that transmits the signal from the interpositus nucleus to the hippocampus and medial prefrontal cortex has not been identified, but the ventrolateral and mediodorsal thalamic nuclei (Sears et al. 1996; Powell and Churchwell 2002; Oswald et al. 2007) and/or the cerebello-thalamo-striatal loop (Flores and Disterhoft 2009, 2013; Bostan et al. 2013) may play a role.

The model proposed here helps incorporate findings from trace eyeblink conditioning into a broader theoretical framework of the hippocampal and prefrontal roles in long-term memory. Temporal integration and associative encoding are the two commonly proposed functions of the hippocampus (Hirsh 1974; O'Keefe and Nadel 1978; Wickelgren 1979; Teyler and DiScenna 1986; Sutherland and Rudy 1989; Squire and Zola-Morgan 1991). Although these functions are often linked to the unique contribution of the hippocampus to encoding contextual details of daily experiences as episodic memory, other theories highlight the hippocampal contribution of making predictions of future events using the acquired representation of past episodes (Vinogradova 2001; Kumaran and Maguire 2006; Lisman and Redish 2009; Davachi and DuBrow 2015). In parallel, the selective prefrontal neuron activity for behaviorally salient stimuli has also been reported in other behavioral paradigms, which are assigned to various labels, including the code for the prediction of the aversive stimulus (Baeg et al. 2001; Gilmartin and McEchron 2005), motivational significance (Pratt and Mizumori 2001), reward expectation (Burton et al. 2009), and value (Insel and Barnes 2014).

4 Common Features Among Associative Memory in Trace Eyeblink Conditioning and Other Hippocampus-Dependent Memories

The previous section suggested that the role of the forebrain in trace eyeblink conditioning may fit into a common framework of forebrain function in long-term memory. This section further expands this idea by introducing two features of trace eyeblink conditioning that are shared with other hippocampus-dependent memories. One is network reorganization during systems consolidation (Sect. 4.1), and the other is conscious awareness of memory content (Sect. 4.2).

4.1 Reorganization of the Cortical Network During Systems Consolidation

Following learning, hippocampus-dependent memory undergoes two types of memory consolidation. The first, called cellular consolidation, stabilizes modified synaptic connections during learning by the synthesis of new proteins, which last for a few hours immediately after learning (Squire and Kandel 2009). The second, called systems consolidation, is presumed to involve modifications of synaptic connections between neurons in different brain regions (Squire and Kandel 2009; Frankland and Bontempi 2005). This results in the gradual reorganization of brain regions supporting memory over weeks. Some models assume that over the course of systems consolidation, synaptic connections between the neocortical regions are gradually strengthened (Squire and Zola-Morgan 1991; Squire and Alvarez 1995; McClelland et al. 1995), whereas others assume that the original neocortical activity pattern is bound to additional synaptic connections in the hippocampus and surrounding cortical regions (Nadel and Moscovitch 1997). Research in the past decade significantly refined these models (for a review, see Frankland and Bontempi 2005; Insel and Takehara-Nishiuchi 2013), and studies using trace eyeblink conditioning have made a significant contribution to this endeavor.

Like certain types of hippocampus-dependent memory (Squire and Zola-Morgan 1991; Winocur et al. 2010), memory expression in trace eyeblink conditioning is initially impaired by damage to the hippocampus (whether the dorsal part only or both dorsal and ventral parts), but it remains intact when the damage was made one month after learning (Kim et al. 1995; Takehara et al. 2002, 2003). This time-limited involvement of the hippocampus sharply contrasts with the time course of involvement of other cortical regions. Specifically, damage to the medial prefrontal cortex, which includes the prelimbic, rostral, and caudal anterior cingulate cortex, impairs memory expression (Takehara et al. 2003; Simon et al. 2005). Notably, the impairment was more severe when the lesion was made one month after learning compared with one day after learning (Takehara et al. 2003). Subsequent studies with electric, neurochemical lesion, or pharmacological silencing to a specific prefrontal subregion showed that the prelimbic cortex is involved in memory expression at several weeks after learning (Takehara-Nishiuchi et al. 2006; Oswald et al. 2008, 2010), and the anterior cingulate cortex is necessary for memory expression immediately after learning (Oswald et al. 2010). In parallel, reversible inactivation of the lateral part of the entorhinal cortex significantly impairs memory expression one day as well as one month later, and the degree of impairment is comparable between the two time points (Morrissey et al. 2012; Tanninen et al. 2013). Together, these behavioral studies suggest that the memory in trace eyeblink conditioning initially depends on the integrated network of the hippocampus, lateral entorhinal cortex, and anterior cingulate cortex, but with time the network is reorganized to a form that includes only the lateral entorhinal cortex and prelimbic cortex.

Although similar network reorganization has been demonstrated in other types of hippocampus-dependent memory (for a review, see Frankland and Bontempi 2005), the studies using trace eyeblink conditioning are unique in that the behavioral findings are accompanied by the characterization of single neuron firings over the corresponding time period. When the activity of multiple single neurons was recorded in the prelimbic cortex for months, starting from the first acquisition session, prelimbic neurons develop robust selective activity for a memory over two weeks after the memory was acquired (Takehara-Nishiuchi and McNaughton 2008). The selective activity developed even without continued conditioning sessions (Takehara-Nishiuchi and McNaughton 2008; Hattori et al. 2014), suggesting that memory acquisition initiates gradual reinforcement processes that result in lasting changes in prefrontal memory representation. In contrast, the selectivity of neurons in the anterior cingulate cortex (Hattori et al. 2014) or the hippocampus (Hattori et al. 2015) did not change after a one-month period of consolidation. These studies provide direct support for continuous modifications of neocortical memory representation after learning (Squire and Zola-Morgan 1991; Squire and Alvarez 1995; McClelland et al. 1995) while challenging an account that the medial prefrontal cortex facilitates the retrieval of weak memory stored in other regions through their top-down, attentional control (Rudy et al. 2005).

4.2 Conscious Awareness of Memory Content

Eyeblink conditioning can be applied to both humans and non-human animals with very minor modifications. Many studies applied eyeblink conditioning to patients with focal brain damage or measured brain activation patterns in healthy subjects and have shown cross-species correspondence in the brain circuitry involved in delay and trace eyeblink conditioning (for a review, see Green and Woodruff-Pak 2000; Christian and Thompson 2003; Gerwig et al. 2007; Woodruff-Pak and Disterhoft 2008). This provides a unique opportunity to unify detailed circuit mechanisms uncovered in animal research and the higher cognitive process that can be examined only in human subjects.

As reviewed above, trace eyeblink conditioning shares many features of other memories dependent on the hippocampus, one of which is declarative memory in humans. This raises the question of which aspect of trace eyeblink conditioning requires the function of the hippocampus that subserves declarative memory. A series of experiments by Clark and Squire addressed this point by monitoring verbal reports of memory contents. They tested amnesic patients and healthy subjects on delay and trace eyeblink conditioning and assessed the extent to which the subjects became capable of verbally explaining the relational feature of the CS and US. The initial report (Clark and Squire 1998) used a differential paradigm, in which one CS was paired with the US and another CS was presented alone while a silent movie was played. This was followed by a list of questions that asked the subject about the content of the movie, the strength of the memory of presented stimuli, and their temporal relationship. The study found that knowledge of the stimulus contingency was not related to the performance during the delay differential paradigm. In contrast, for the trace differential paradigm, subjects were able to acquire the CR only when they were aware of the stimulus contingency. Subsequent studies strengthened the link between conscious awareness of stimulus contingency and trace eyeblink conditioning by demonstrating that the performance in trace but not delay differential conditioning is bi-directionally modulated by manipulations to the level of awareness (Clark and Squire 1999) and that the awareness of stimulus contingency is related to acquisition in standard, single CS-trace but not delay eyeblink conditioning (Manns et al. 2000a, 2001). Another follow-up study incorporated a self-report on stimulus contingency *during* the conditioning and demonstrated that the awareness of stimulus contingency occurs in parallel with the development of CR, but the former develops faster than the latter (Manns et al. 2000b). Together, these studies highlight the fundamental difference between delay conditioning and trace eyeblink conditioning: Trace eyeblink conditioning relies on declarative knowledge of the stimulus contingency, but delay eyeblink conditioning does not (but see Knuttinen et al. 2001).

5 Conclusions and Future Directions

What are the prerequisites for elucidating the complete neuronal mechanism of memory? Identifying the brain circuitry underlying the memory is the most important step because it serves as the basis to achieving other steps, such as uncovering representations of information and computations in all involved brain regions. Studies using eyeblink classical conditioning constitute only a few cases in which researchers have come very close to this ultimate goal of memory research.

In delay eyeblink conditioning, the pathways for input (i.e., CS and US) and output (i.e., CR and UR) have been mapped onto specific anatomical connections within the cerebellar–brainstem circuit. The CS and US information converges in the cerebellar cortex and interpositus nucleus, in which several types of synaptic plasticity have been identified as critical neuronal substrates of CR acquisition. These studies were on the cutting edge of memory research in the late twentieth century and preceded the recent focus on circuit-level mechanisms of behavior. By using the recently developed, cell-type specific, temporally precise manipulations with light-activated proteins (Nguyen-Vu et al. 2013; Heiney et al. 2014; Lee et al. 2015), future studies with delay eyeblink conditioning would further refine the circuit mechanisms underlying this specific form of associative memory as well as a model of cerebellar computation in general.

In parallel, several variants of eyeblink conditioning involve the interaction between the cerebellar–brainstem circuit and the forebrain. In a variant, trace eyeblink conditioning, many forebrain regions, including the hippocampus and medial prefrontal cortex, are involved in CR acquisition and later expression; however, evidence is still scarce on how these regions represent and compute the conditioning-related information. Moreover, it remains largely unknown how the local computations are coupled with one another to coherently operate as an integrated network. The above section proposed a hypothesis concerning the forebrain–cerebellum interaction during trace eyeblink conditioning, which generates a number of testable predictions for future behavioral and electrophysiological investigations. These future studies would not only uncover detailed operations of the whole brain network during temporal associative memory, but also contribute to forming a unified framework for forebrain function in long-term memory.

Together, this chapter highlights the significant contribution of eyeblink classical conditioning to uncovering the circuit mechanisms of associative memory. It also emphasizes great potentialities of this valuable paradigm for bringing us a step closer to the complete mechanistic understanding of long-term memory.

Acknowledgments The author thanks Drs. Craig Weiss and Nathan Insel for their helpful comments. This work was supported by NSERC Discovery Grant (KT).

References

Aiba A, Kano M, Chen C, Stanton ME, Fox GD, Herrup K, Zwingman TA, Tonegawa S (1994) Deficient cerebellar long-term depression and impaired motor learning in mGluR1 mutant mice. Cell 79(2):377–388

Akase E, Alkon D, Disterhoft JF (1989) Hippocampal lesions impair memory of short-delay conditioned eye blink in rabbits. Behav Neurosci 103(5):935–943

Baeg EH, Kim YB, Jang J, Kim HT, Mook-Jung I, Jung MW (2001) Fast spiking and regular spiking neural correlates of fear conditioning in the medial prefrontal cortex of the rat. Cereb Cortex 11(5):441–451

Bao S, Chen L, Thompson RF (2000) Learning- and cerebellum-dependent neuronal activity in the lateral pontine nucleus. Behav Neurosci 114(2):254–261

Bao S, Chen L, Kim JJ, Thompson RF (2002) Cerebellar cortical inhibition and classical eyeblink conditioning. Proc Natl Acad Sci USA 99(3):1592–1597

Berger TW, Orr WB (1983) Hippocampectomy selectively disrupts discrimination reversal conditioning of the rabbit nictitating membrane response. Behav Brain Res 8(1):49–68

Berger TW, Alger B, Thompson RF (1976) Neuronal substrate of classical conditioning in the hippocampus. Science 192(4238):483–485

Berthier NE, Moore JW (1986) Cerebellar Purkinje cell activity related to the classically conditioned nictitating membrane response. Exp Brain Res 63(2):341–350

Berthier NE, Moore JW (1990) Activity of deep cerebellar nuclear cells during classical conditioning of nictitating membrane extension in rabbits. Exp Brain Res 83(1):44–54

Beylin AV, Gandhi CC, Wood GE, Talk AC, Matzel LD, Shors TJ (2001) The role of the hippocampus in trace conditioning: temporal discontinuity or task difficulty? Neurobiol Learn Mem 76(3):447–461

Bostan AC, Dum RP, Strick PL (2013) Cerebellar networks with the cerebral cortex and basal ganglia. Trends Cogn Sci 17(5):241–254

Brown KL, Agelan A, Woodruff-Pak DS (2010) Unimpaired trace classical eyeblink conditioning in Purkinje cell degeneration (pcd) mutant mice. Neurobiol Learn Mem 93(3):303–311

Brun VH, Leutgeb S, Wu H-Q, Schwarcz R, Witter MP, Moser EI, Moser M-B (2008) Impaired spatial representation in CA1 after lesion of direct input from entorhinal cortex. Neuron 57 (2):290–302

Buchanan SL, Thompson RH, Maxwell BL, Powell DA (1994) Efferent connections of the medial prefrontal cortex in the rabbit. Exp Brain Res 100(3):469–483

Burton BG, Hok V, Save E, Poucet B (2009) Lesion of the ventral and intermediate hippocampus abolishes anticipatory activity in the medial prefrontal cortex of the rat. Behav Brain Res 199 (2):222–234

Cartford MC, Gohl EB, Singson M, Lavond DG (1997) The effects of reversible inactivation of the red nucleus on learning-related and auditory-evoked unit activity in the pontine nuclei of classically conditioned rabbits. Learn Mem 3(6):519–531

Cerminara NL, Rawson JA (2004) Evidence that climbing fibers control an intrinsic spike generator in cerebellar Purkinje cells. J Neurosci 24(19):4510–4517

Chen C, Thompson RF (1995) Temporal specificity of long-term depression in parallel-fiber Purkinje synapses in rat cerebellar slice. Learn Mem 2(1):185–198

Chen C, Kano M, Abeliovich A, Chen L, Bao S, Kim JJ, Hashimoto K, Thompson RF, Tonegawa S (1995) Impaired motor coordination correlates with persistent multiple climbing fiber innervation in PKC gamma mutant mice. Cell 83(7):1233–1242

Chen L, Bao S, Lockard JM, Kim JK, Thompson RF (1996) Impaired classical eyeblink conditioning in cerebellar-lesioned and Purkinje cell degeneration (pcd) mutant mice. J Neurosci 16(8):2829–2838

Chen L, Bao S, Thompson RF (1999) Bilateral lesions of the interpositus nucleus completely prevent eyeblink conditioning in Purkinje cell-degeneration mutant mice. Behav Neurosci 113 (1):204–210

Christian KM, Thompson RF (2003) Neural substrates of eyeblink conditioning: acquisition and retention. Learn Mem 10(6):427–455

Clark RE, Gohl E (1997) The learning-related activity that develops in the pontine nuclei during classical eye-blink conditioning is dependent on the interpositus nucleus. Learn Mem 3 (1):532–544

Clark RE, Squire LR (1998) Classical conditioning and brain systems: the role of awareness. Science 280(5360):77–81

Clark RE, Squire LR (1999) Human eyeblink classical conditioning: Effects of manipulating awareness of the stimulus contingencies. Psychol Sci 10(1):14–18

Clark GA, McCormick DA, Lavond DG, Thompson RF (1984) Effects of lesions of cerebellar nuclei on conditioned behavioral and hippocampal neuronal responses. Brain Res 291(1):125–136

Clark RE, Zhang AA, Lavond DG (1992) Reversible lesions of the cerebellar interpositus nucleus during acquisition and retention of a classically conditioned behavior. Behav Neurosci 106 (6):879–888

Davachi L, DuBrow S (2015) How the hippocampus preserves order: the role of prediction and context. Trends Cogn Sci 19(2):92–99

De Zeeuw CI, Hansel C, Bian F, Koekkoek SK, van Alphen AM, Linden DJ, Oberdick J (1998) Expression of a protein kinase C inhibitor in Purkinje cells blocks cerebellar LTD and adaptation of the vestibulo-ocular reflex. Neuron 20(3):495–508

Disterhoft JF, Quinn KJ, Weiss C, Shipley MT (1985) Accessory abducens nucleus and conditioned eye retraction/nictitating membrane extension in rabbit. J Neurosci 5(4):941–950

Eccles JC, Llinas R, Sasaki K (1966a) The excitatory synaptic action of climbing fibres on the Purkinje cells of the cerebellum. J Physiol 182(2):268–296

Eccles JC, Llinas R, Sasaki K (1966b) Parallel fibre stimulation and the responses induced thereby in the Purkinje cells of the cerebellum. Exp Brain Res 1(1):17–39

Feil R, Hartmann J, Luo C, Wolfsgruber W, Schilling K, Feil S, Barski JJ et al (2003) Impairment of LTD and cerebellar learning by Purkinje cell-specific ablation of cGMP-dependent protein kinase I. J Cell Biol 163(2):295–302

Flores LC, Disterhoft JF (2009) Caudate nucleus is critically involved in trace eyeblink conditioning. J Neurosci 29(46):14511–14520

Flores LC, Disterhoft JF (2013) Caudate nucleus in retrieval of trace eyeblink conditioning after consolidation. J Neurosci 33(7):2828–2836

Frankland PW, Bontempi B (2005) The organization of recent and remote memories. Nat Rev Neurosci 6(2):119–130

Fredette BJ, Mugnaini E (1991) The GABAergic cerebello-olivary projection in the rat. Anat Embryol (Berl) 184(3):225–243

Freeman JH Jr, Nicholson DA (2000) Developmental changes in eye-blink conditioning and neuronal activity in the cerebellar interpositus nucleus. J Neurosci 20(2):813–819

Gahwiler BH (1975) The effects of GABA, picrotoxin and bicuculline on the spontaneous bioelectric activity of cultured cerebellar Purkinje cells. Brain Res 99(1):85–95

Galvez R, Weible AP, Disterhoft JF (2007) Cortical barrel lesions impair whisker-CS trace eyeblink conditioning. Learn Mem 14(1–2):94–100

Gao Z, van Beugen BJ, De Zeeuw CI (2012) Distributed synergistic plasticity and cerebellar learning. Nat Rev Neurosci 13(9):619–635

Garcia KS, Mauk MD (1998) Pharmacological analysis of cerebellar contributions to the timing and expression of conditioned eyelid responses. Neuropharmacology 37(4–5):471–480
Gerwig M, Kolb FP, Timmann D (2007) The involvement of the human cerebellum in eyeblink conditioning. Cerebellum 6(1):38–57
Gilmartin MR, McEchron MD (2005) Single neurons in the medial prefrontal cortex of the rat exhibit tonic and phasic coding during trace fear conditioning. Behav Neurosci 119(6): 1496–1510
Gould TJ, Steinmetz JE (1996) Changes in rabbit cerebellar cortical and interpositus nucleus activity during acquisition, extinction, and backward classical eyelid conditioning. Neurobiol Learn Mem 65(1):17–34
Green JT, Steinmetz JE (2005) Purkinje cell activity in the cerebellar anterior lobe after rabbit eyeblink conditioning. Learn Mem 12(3):260–269
Green JT, Woodruff-Pak DS (2000) Eyeblink classical conditioning: hippocampal formation is for neutral stimulus associations as cerebellum is for association-response. Psychol Bull 126 (1):138–158
Gruart A, Blazquez P, Delgado-Garcia JM (1995) Kinematics of spontaneous, reflex, and conditioned eyelid movements in the alert cat. J Neurophysiol 74(1):226–248
Gruart A, Schreurs BG, del Toro ED, Delgado-Garcia JM (2000) Kinetic and frequency-domain properties of reflex and conditioned eyelid responses in the rabbit. J Neurophysiol 83(2): 836–852
Hattori S, Yoon T, Disterhoft JF, Weiss C (2014) Functional reorganization of a prefrontal cortical network mediating consolidation of trace eyeblink conditioning. J Neurosci 34(4):1432–1445
Hattori S, Chen L, Weiss C, Disterhoft JF (2015) Robust hippocampal responsivity during retrieval of consolidated associative memory: hippocampal involvement after memory consolidation. Hippocampus 25:655–669
Hausser M, Clark BA (1997) Tonic synaptic inhibition modulates neuronal output pattern and spatiotemporal synaptic integration. Neuron 19(3):665–678
Heiney SA, Kim J, Augustine GJ, Medina JF (2014) Precise control of movement kinematics by optogenetic inhibition of Purkinje cell activity. J Neurosci 34(6):2321–2330
Hesslow G, Ivarsson M (1994) Suppression of cerebellar Purkinje cells during conditioned responses in ferrets. NeuroReport 5(5):649–652
Hesslow G, Ivarsson M (1996) Inhibition of the inferior olive during conditioned responses in the decerebrate ferret. Exp Brain Res 110(1):36–46
Hesslow G, Svensson P, Ivarsson M (1999) Learned movements elicited by direct stimulation of cerebellar mossy fiber afferents. Neuron 24(1):179–185
Hirsh R (1974) The hippocampus and contextual retrieval of information from memory: a theory. Behav Biol 12(4):421–444
Hu B, Chen H, Feng H, Zeng Y, Yang L, Fan ZL, Wu YM, Sui JF (2010) Disrupted topography of the acquired trace-conditioned eyeblink responses in guinea pigs after suppression of cerebellar cortical inhibition to the interpositus nucleus. Brain Res 1337:41–55
Insel N, Barnes CA (2014) Differential activation of fast-spiking and regular-firing neuron populations during movement and reward in the dorsal medial frontal cortex. Cereb Cortex 25 (9):2631–2647
Insel N, Takehara-Nishiuchi K (2013) The cortical structure of consolidated memory: a hypothesis on the role of the cingulate-entorhinal cortical connection. Neurobiol Learn Mem 106:343–350
Ito M, Kano M (1982) Long-lasting depression of parallel fiber-Purkinje cell transmission induced by conjunctive stimulation of parallel fibers and climbing fibers in the cerebellar cortex. Neurosci Lett 33(3):253–258
Jay TM, Witter MP (1991) Distribution of hippocampal CA1 and subicular efferents in the prefrontal cortex of the rat studied by means of anterograde transport of *Phaseolus vulgaris-leucoagglutinin*. J Comp Neurol 313(4):574–586
Jirenhed DA, Bengtsson F, Hesslow G (2007) Acquisition, extinction, and reacquisition of a cerebellar cortical memory trace. J Neurosci 27(10):2493–2502

Jorntell H, Ekerot CF (2002) Reciprocal bidirectional plasticity of parallel fiber receptive fields in cerebellar Purkinje cells and their afferent interneurons. Neuron 34(5):797–806
Jorntell H, Ekerot CF (2003) Receptive field plasticity profoundly alters the cutaneous parallel fiber synaptic input to cerebellar interneurons in vivo. J Neurosci 23(29):9620–9631
Jorntell H, Bengtsson F, Schonewille M, De Zeeuw CI (2010) Cerebellar molecular layer interneurons—computational properties and roles in learning. Trends Neurosci 33(11): 524–532
Kalmbach BE, Ohyama T, Kreider JC, Riusech F, Mauk MD (2009) Interactions between prefrontal cortex and cerebellum revealed by trace eyelid conditioning. Learn Mem 16(1): 86–95
Kalmbach BE, Davis T, Ohyama T, Riusech F, Nores WL, Mauk MD (2010) Cerebellar cortex contributions to the expression and timing of conditioned eyelid responses. J Neurophysiol 103 (4):2039–2049
Kim JJ, Clark RE, Thompson RF (1995) Hippocampectomy impairs the memory of recently, but not remotely, acquired trace eyeblink conditioned responses. Behav Neurosci 109(2):195–203
Kishimoto Y, Hirono M, Sugiyama T, Kawahara S, Nakao K, Kishio M, Katsuki M, Yoshioka T, Kirino Y (2001a) Impaired delay but normal trace eyeblink conditioning in PLCbeta4 mutant mice. NeuroReport 12(13):2919–2922
Kishimoto Y, Kawahara S, Suzuki M, Mori H, Mishina M, Kirino Y (2001b) Classical eyeblink conditioning in glutamate receptor subunit delta 2 mutant mice is impaired in the delay paradigm but not in the trace paradigm. Eur J Neurosci 13(6):1249–1253
Kleim JA, Freeman JH Jr, Bruneau R, Nolan BC, Cooper NR, Zook A, Walters D (2002) Synapse formation is associated with memory storage in the cerebellum. Proc Natl Acad Sci U S A 99 (20):13228–13231
Knuttinen MG, Power JM, Preston AR, Disterhoft JF (2001) Awareness in classical differential eyeblink conditioning in young and aging humans. Behav Neurosci 115(4):747–757
Koekkoek SK, Hulscher HC, Dortland BR, Hensbroek RA, Elgersma Y, Ruigrok TJ, De Zeeuw CI (2003) Cerebellar LTD and learning-dependent timing of conditioned eyelid responses. Science 301(5640):1736–1739
Kotani S, Kawahara S, Kirino Y (2006) Purkinje cell activity during classical eyeblink conditioning in decerebrate guinea pigs. Brain Res 1068(1):70–81
Kronforst-Collins MA, Disterhoft JF (1998) Lesions of the caudal area of rabbit medial prefrontal cortex impair trace eyeblink conditioning. Neurobiol Learn Mem 69(2):147–162
Krupa DJ, Thompson RF (1997) Reversible inactivation of the cerebellar interpositus nucleus completely prevents acquisition of the classically conditioned eye-blink response. Learn Mem 3(6):545–556
Krupa DJ, Thompson JK, Thompson RF (1993) Localization of a memory trace in the mammalian brain. Science 260(5110):989–991
Kumaran D, Maguire EA (2006) An unexpected sequence of events: mismatch detection in the human hippocampus. PLoS Biol 4(12):e424
Lavond DG, Steinmetz JE (1989) Acquisition of classical conditioning without cerebellar cortex. Behav Brain Res 33(2):113–164
Lavond DG, Hembree TL, Thompson RF (1985) Effect of kainic acid lesions of the cerebellar interpositus nucleus on eyelid conditioning in the rabbit. Brain Res 326(1):179–182
Lavond DG, Steinmetz JE, Yokaitis MH, Thompson RF (1987) Reacquisition of classical conditioning after removal of cerebellar cortex. Exp Brain Res 67(3):569–593
Lee I, Yoganarasimha D, Rao G, Knierim JJ (2004) Comparison of population coherence of place cells in hippocampal subfields CA1 and CA3. Nature 430(6998):456–459
Lee KH, Mathews PJ, Reeves AM, Choe KY, Jami SA, Serrano RE, Otis TS (2015) Circuit mechanisms underlying motor memory formation in the cerebellum. Neuron 86(2):529–540
Leutgeb S, Leutgeb JK, Treves A, Moser MB, Moser EI (2004) Distinct ensemble codes in hippocampal areas CA3 and CA1. Science 305(5688):1295–1298

Lincoln JS, McCormick DA, Thompson RF (1982) Ipsilateral cerebellar lesions prevent learning of the classically conditioned nictitating membrane/eyelid response. Brain Res 242(1):190–193
Linden DJ, Connor JA (1995) Long-term synaptic depression. Annu Rev Neurosci 18(1):319–357
Lisman J, Redish AD (2009) Prediction, sequences and the hippocampus. Philos T R Soc B 364 (1521):1193–1201
Liu SJ, Cull-Candy SG (2005) Subunit interaction with PICK and GRIP controls Ca2+ permeability of AMPARs at cerebellar synapses. Nat Neurosci 8(6):768–775
Lu L, Leutgeb JK, Tsao A, Henriksen EJ, Leutgeb S, Barnes CA, Witter MP, Moser M-B, Moser EI (2013) Impaired hippocampal rate coding after lesions of the lateral entorhinal cortex. Nat Neurosci 16(8):1085–1093
Manns JR, Clark RE, Squire LR (2000a) Awareness predicts the magnitude of single-cue trace eyeblink conditioning. Hippocampus 10(2):181–186
Manns JR, Clark RE, Squire LR (2000b) Parallel acquisition of awareness and trace eyeblink classical conditioning. Learn Mem 7(5):267–272
Manns JR, Clark RE, Squire L (2001) Single-cue delay eyeblink conditioning is unrelated to awareness. Cogn Affect Behav Neurosci 1(2):192–198
Matsushita M, Ikeda M (1970) Olivary projections to the cerebellar nuclei in the cat. Exp Brain Res 10(5):488–500
Mauk MD, Steinmetz JE, Thompson RF (1986) Classical conditioning using stimulation of the inferior olive as the unconditioned stimulus. Proc Natl Acad Sci USA 83(14):5349–5353
McClelland JL, McNaughton BL, O'Reilly RC (1995) Why there are complementary learning systems in the hippocampus and neocortex: insights from the successes and failures of connectionist models of learning and memory. Psychol Rev 102(3):419–457
McCormick DA, Thompson RF (1984a) Cerebellum: essential involvement in the classically conditioned eyelid response. Science 223(4633):296–299
McCormick DA, Thompson RF (1984b) Neuronal responses of the rabbit cerebellum during acquisition and performance of a classically conditioned nictitating membrane-eyelid response. J Neurosci 4(11):2811–2822
McCormick DA, Clark GA, Lavond DG, Thompson RF (1982) Initial localization of the memory trace for a basic form of learning. Proc Natl Acad Sci USA 79(8):2731–2735
McCormick DA, Lavond DG, Thompson RF (1983) Neuronal responses of the rabbit brainstem during performance of the classically conditioned nictitating membrane (NM)/eyelid response. Brain Res 271(1):73–88
McCormick DA, Steinmetz JE, Thompson RF (1985) Lesions of the inferior olivary complex cause extinction of the classically conditioned eyeblink response. Brain Res 359(1–2):120–130
McEchron MD, Disterhoft JF (1997) Sequence of single neuron changes in CA1 hippocampus of rabbits during acquisition of trace eyeblink conditioned responses. J Neurophysiol 78(1):1030–1044
McEchron MD, Weible AP, Disterhoft JF (2001) Aging and learning-specific changes in single-neuron activity in CA1 hippocampus during rabbit trace eyeblink conditioning. J Neurophysiol 86(4):1839–1857
McLaughlin J, Skaggs H, Churchwell J, Powell DA (2002) Medial prefrontal cortex and pavlovian conditioning: trace versus delay conditioning. Behav Neurosci 116(1):37–47
Mihailoff GA (1993) Cerebellar nuclear projections from the basilar pontine nuclei and nucleus reticularis tegmenti pontis as demonstrated with PHA-L tracing in the rat. J Comp Neurol 330 (1):130–146
Mintz M, Lavond David G, Zhang AA, Yun Y, Thompson Richard F (1994) Unilateral inferior olive NMDA lesion leads to unilateral deficit in acquisition and retention of eyelid classical conditioning. Behav Neural Biol 61(1):218–224
Miyata M, Kim HT, Hashimoto K, Lee TK, Cho SY, Jiang H, Wu Y et al (2001) Deficient long-term synaptic depression in the rostral cerebellum correlated with impaired motor learning in phospholipase C beta4 mutant mice. Eur J Neurosci 13(10):1945–1954
Modi MN, Dhawale AK, Bhalla US (2014) CA1 cell activity sequences emerge after reorganization of network correlation structure during associative learning. Elife 3:e01982

Moita MA, Rosis S, Zhou Y, LeDoux JE, Blair HT (2003) Hippocampal place cells acquire location-specific responses to the conditioned stimulus during auditory fear conditioning. Neuron 37(3):485–497

Morrissey MD, Maal-Bared G, Brady S, Takehara-Nishiuchi K (2012) Functional dissociation within the entorhinal cortex for memory retrieval of an association between temporally discontiguous stimuli. J Neurosci 32(16):5356–5361

Moustafa AA, Wufong E, Servatius RJ, Pang KCH, Gluck MA, Myers CE (2013) Why trace and delay conditioning are sometimes (but not always) hippocampal dependent: a computational model. Brain Res 1493:48–67

Moya MV, Siegel JJ, McCord ED, Kalmbach BE, Dembrow N, Johnston D, Chitwood RA (2014) Species-specific differences in the medial prefrontal projections to the pons between rat and rabbit. J Comp Neurol 522(13):3052–3074

Moyer JR, Deyo RA, Disterhoft JF (1990) Hippocampectomy disrupts trace eye-blink conditioning in rabbits. Behav Neurosci 104(2):243–252

Munera A, Gruart A, Munoz MD, Fernandez-Mas R, Delgado-Garcia JM (2001) Hippocampal pyramidal cell activity encodes conditioned stimulus predictive value during classical conditioning in alert cats. J Neurophysiol 86(5):2571–2582

Nadel L, Moscovitch M (1997) Memory consolidation, retrograde amnesia and the hippocampal complex. Curr Opin Neurobiol 7(2):217–227

Nguyen-Vu TD, Kimpo RR, Rinaldi JM, Kohli A, Zeng H, Deisseroth K, Raymond JL (2013) Cerebellar Purkinje cell activity drives motor learning. Nat Neurosci 16(12):1734–1736

Nicholson DA, Freeman JH Jr (2003) Addition of inhibition in the olivocerebellar system and the ontogeny of a motor memory. Nat Neurosci 6(5):532–537

O'Keefe J, Nadel L (1978) The hippocampus as a cognitive map. Clarendon Press, Oxford

Oswald B, Knuckley B, Mahan K, Sanders C, Powell DA (2006) Prefrontal control of trace versus delay eyeblink conditioning: role of the unconditioned stimulus in rabbits (*Oryctolagus cuniculus*). Behav Neurosci 120(5):1033–1042

Oswald BB, Knuckley B, Maddox SA, Powell DA (2007) Ibotenic acid lesions to ventrolateral thalamic nuclei disrupts trace and delay eyeblink conditioning in rabbits. Behav Brain Res 179 (1):111–117

Oswald BB, Maddox SA, Powell DA (2008) Prefrontal control of trace eyeblink conditioning in rabbits: role in retrieval of the CR? Behav Neurosci 122(4):841–848

Oswald BB, Maddox SA, Tisdale N, Powell DA (2010) Encoding and retrieval are differentially processed by the anterior cingulate and prelimbic cortices: a study based on trace eyeblink conditioning in the rabbit. Neurobiol Learn Mem 93(1):37–45

Pakaprot N, Kim S, Thompson RF (2009) The role of the cerebellar interpositus nucleus in short and long term memory for trace eyeblink conditioning. Behav Neurosci 123(1):54–61

Palay SL, Chan-Palay V (1974) Cerebellar xortex. Cytology and organization. Springer, Berlin

Patterson MM, Berger TW, Thompson RF (1979) Neuronal plasticity recorded from cat hippocampus during classical conditioning. Brain Res 163(2):339–343

Penick S, Solomom PR (1991) Hippocampus, context, and conditioning. Behav Neurosci 105 (5):611–617

Perrett SP, Ruiz BP, Mauk MD (1993) Cerebellar cortex lesions disrupt learning-dependent timing of conditioned eyelid responses. J Neurosci 13(4):1708–1718

Powell DA, Churchwell J (2002) Mediodorsal thalamic lesions impair trace eyeblink conditioning in the rabbit. Learn Mem 9(1):10–17

Pratt WE, Mizumori SJY (2001) Neurons in rat medial prefrontal cortex show anticipatory rate changes to predictable differential rewards in a spatial memory task. Behav Brain Res 123 (2):165–183

Pugh JR, Raman IM (2006) Potentiation of mossy fiber EPSCs in the cerebellar nuclei by NMDA receptor activation followed by postinhibitory rebound current. Neuron 51(1):113–123

Pugh JR, Raman IM (2008) Mechanisms of potentiation of mossy fiber EPSCs in the cerebellar nuclei by coincident synaptic excitation and inhibition. J Neurosci 28(42):10549–10560

Raman IM, Bean BP (1997) Resurgent sodium current and action potential formation in dissociated cerebellar Purkinje neurons. J Neurosci 17(12):4517–4526

Rescorla RA, Wagner AR (1972) A theory of Pavlovian conditioning: variations in the effectiveness of reinforcement and nonreinforcement. In Black AH, Prokasy WF (eds) Classical conditioning II. Appleton-Century-Crofts, pp 64–99

Rogers RF, Britton GB, Steinmetz JE (2001) Learning-related interpositus activity is conserved across species as studied during eyeblink conditioning in the rat. Brain Res 905(1–2):171–177

Rudy JW, Biedenkapp JC, O'Reilly RC (2005) Prefrontal cortex and the organization of recent and remote memories: an alternative view. Learn Mem 12(5):445–446

Ryou JW, Cho SY, Kim HT (2001) Lesions of the entorhinal cortex impair acquisition of hippocampal-dependent trace conditioning. Neurobiol Learn Mem 75(2):121–127

Schade Powers A, Coburn-Litvak P, Evinger C (2010) Conditioned eyelid movement is not a blink. J Neurophysiol 103(2):641–647

Schmahmann JD, Pandya DN (1997) Anatomic organization of the basilar pontine projections from prefrontal cortices in rhesus monkey. J Neurosci 17(1):438–458

Schmajuk NA, Lam Y-W, Christiansen BA (1994) Latent inhibition of rat eyeblink response. Physiol Behav 55(3):597–601

Schmaltz LW, Theios J (1972) Acquisition and extinction of a classically conditioned response in hippocampectomized rabbits (*Oryctolagus cuniculus*). J Comp Physiol Psych 79(2):328–333

Schonewille M, Gao Z, Boele H-J, Vinueza Veloz MF, Amerika WE, Šimek AAM, De Jeu MT et al (2011) Reevaluating the role of LTD in cerebellar motor learning. Neuron 70(1):43–50

Sears LL, Steinmetz JE (1990) Acquisition of classically conditioned-related activity in the hippocampus is affected by lesions of the cerebellar interpositus nucleus. Behav Neurosci 104 (5):681–692

Sears LL, Steinmetz JE (1991) Dorsal accessory inferior olive activity diminishes during acquisition of the rabbit classically conditioned eyelid response. Brain Res 545(1–2):114–122

Sears LL, Logue SF, Steinmetz JE (1996) Involvement of the ventrolateral thalamic nucleus in rabbit classical eyeblink conditioning. Behav Brain Res 74(1–2):105–117

Shibuki K, Gomi H, Chen L, Bao S, Kim JJ, Wakatsuki H, Fujisaki T et al (1996) Deficient cerebellar long-term depression, impaired eyeblink conditioning, and normal motor coordination in GFAP mutant mice. Neuron 16(3):587–599

Shohamy D, Allen MT, Gluck MA (2000) Dissociating entorhinal and hippocampal involvement in latent inhibition. Behav Neurosci 114(5):867–874

Siegel JJ, Mauk MD (2013) Persistent activity in prefrontal cortex during trace eyelid conditioning: dissociating responses that reflect cerebellar output from those that do not. J Neurosci 33(38):15272–15284

Siegel JJ, Kalmbach B, Chitwood RA, Mauk MD (2012) Persistent activity in a cortical-to-subcortical circuit: bridging the temporal gap in trace eyelid conditioning. J Neurophysiol 107(1):50–64

Simon B, Knuckley B, Churchwell J, Powell DA (2005) Post-training lesions of the medial prefrontal cortex interfere with subsequent performance of trace eyeblink conditioning. J Neurosci 25(46):10740–10746

Soler-Llavina GJ, Sabatini BL (2006) Synapse-specific plasticity and compartmentalized signaling in cerebellar stellate cells. Nat Neurosci 9(6):798–806

Solomon PR, Moore JW (1975) Latent inhibition and stimulus generalization of the classically conditioned nictitating membrane response in rabbits (*Oryctolagus cuniculus*) following dorsal hippocampal ablation. J Comp Physiol Psychol 89(10):1192–1203

Solomon PR, Vander Schaaf ER (1986) Hippocampus and trace conditioning of the rabbit's classically conditioned nictitating membrane response. Behav Neurosci 100(5):729–744

Solomon PR, Solomon SD, Schaaf EV, Perry HE (1983) Altered activity in the hippocampus is more detrimental to classical conditioning than removing the structure. Science 220 (4594):329–331

Squire LR, Alvarez P (1995) Retrograde amnesia and memory consolidation: a neurobiological perspective. Curr Opin Neurobiol 5(2):169–177

Squire LR, Kandel ER (2009) Memory from mind to molecules. 2nd edn Aufl. Roberts & Company,Colorado

Squire LR, Zola-Morgan S (1991) The medial temporal lobe memory system. Science 253 (5026):1380–1386

Steinmetz JE, Sengelaub DR (1992) Possible conditioned stimulus pathway for classical eyelid conditioning in rabbits. I. Anatomical evidence for direct projections from the pontine nuclei to the cerebellar interpositus nucleus. Behav Neural Biol 57(2):103–115

Steinmetz JE, Rosen DJ, Chapman PF, Lavond DG, Thompson RF (1986) Classical conditioning of the rabbit eyelid response with a mossy-fiber stimulation CS: I. Pontine nuclei and middle cerebellar peduncle stimulation. Behav Neurosci 100(6):878–887

Steinmetz JE, Logan CG, Rosen DJ, Thompson JK, Lavond DG, Thompson RF (1987) Initial localization of the acoustic conditioned stimulus projection system to the cerebellum essential for classical eyelid conditioning. Proc Natl Acad Sci USA 84(10):3531–3535

Steinmetz JE, Lavond DG, Thompson RF (1989) Classical conditioning in rabbits using pontine nucleus stimulation as a conditioned stimulus and inferior olive stimulation as an unconditioned stimulus. Synapse 3(3):225–233

Steinmetz JE, Logue SF, Steinmetz SS (1992) Rabbit classically conditioned eyelid responses do not reappear after interpositus nucleus lesion and extensive post-lesion training. Behav Brain Res 51(1):103–114

Steinmetz AB, Harmon TC, Freeman JH (2013) Visual cortical contributions to associative cerebellar learning. Neurobiol Learn Mem 104:103–109

Suter EE, Weiss C, Disterhoft JF (2013) Perirhinal and postrhinal, but not lateral entorhinal, cortices are essential for acquisition of trace eyeblink conditioning. Learn Mem 20(2):80–84

Sutherland RJ, Rudy JW (1989) Configural association theory: the role of the hippocampal formation in learning, memory, and amnesia. Psychobiology 17(2):129–144

Svensson P, Bengtsson F, Hesslow G (2006) Cerebellar inhibition of inferior olivary transmission in the decerebrate ferret. Exp Brain Res 168(1–2):241–253

Swenson RS, Castro AJ (1983) The afferent connections of the inferior olivary complex in rats. An anterograde study using autoradiographic and axonal degeneration techniques. Neuroscience 8 (2):259–275

Takehara K, Kawahara S, Takatsuki K, Kirino Y (2002) Time-limited role of the hippocampus in the memory for trace eyeblink conditioning in mice. Brain Res 951(2):183–190

Takehara K, Kawahara S, Kirino Y (2003) Time-dependent reorganization of the brain components underlying memory retention in trace eyeblink conditioning. J Neurosci 23 (30):9897–9905

Takehara-Nishiuchi Kaori, McNaughton Bruce L (2008) Spontaneous changes of neocortical code for associative memory during consolidation. Science 322(5903):960–963

Takehara-Nishiuchi K, Kawahara S, Kirino Y (2005) NMDA receptor-dependent processes in the medial prefrontal cortex are important for acquisition and the early stage of consolidation during trace, but not delay eyeblink conditioning. Learn Mem 12(6):606–614

Takehara-Nishiuchi K, Nakao K, Kawahara S, Matsuki N, Kirino Y (2006) Systems consolidation requires postlearning activation of NMDA receptors in the medial prefrontal cortex in trace eyeblink conditioning. J Neurosci 26(19):5049–5058

Tanninen SE, Morrissey MD, Takehara-Nishiuchi K (2013) Unilateral lateral entorhinal inactivation impairs memory expression in trace eyeblink conditioning. PLoS ONE 8(12): e84543

Tanninen SE, Yu X, Giritharan T, Tran L, Bakir R, Volle J, Morrissey MD, Takehara-Nishiuchi K (2015) Cholinergic, but not NMDA, receptors in the lateral entorhinal cortex mediate acquisition in trace eyeblink conditioning. Hippocampus 25(11):36–45

Teyler TJ, DiScenna P (1986) The hippocampal memory indexing theory. Behav Neurosci 100 (2):147–154

Tseng W, Guan R, Disterhoft JF, Weiss C (2004) Trace eyeblink conditioning is hippocampally dependent in mice. Hippocampus 14(1):58–65

Uylings HB, Groenewegen HJ, Kolb B (2003) Do rats have a prefrontal cortex? Behav Brain Res 146(1–2):3–17
Van Cauter T, Poucet B, Save E (2008) Unstable CA1 place cell representation in rats with entorhinal cortex lesions. Eur J Neurosci 27(8):1933–1946
Vazdarjanova A, Guzowski JF (2004) Differences in hippocampal neuronal population responses to modifications of an environmental context: evidence for distinct, yet complementary, functions of CA3 and CA1 ensembles. J Neurosci 24(29):6489–6496
Vinogradova OS (2001) Hippocampus as comparator: role of the two input and two output systems of the hippocampus in selection and registration of information. Hippocampus 11 (5):578–598
Walker AG, Steinmetz JE (2008) Hippocampal lesions in rats differentially affect long- and short-trace eyeblink conditioning. Physiol Behav 93(3):570–578
Weeks AC, Connor S, Hinchcliff R, LeBoutillier JC, Thompson RF, Petit TL (2007) Eye-blink conditioning is associated with changes in synaptic ultrastructure in the rabbit interpositus nuclei. Learn Mem 14(6):385–389
Weible AP, McEchron MD, Disterhoft JF (2000) Cortical involvement in acquisition and extinction of trace eyeblink conditioning. Behav Neurosci 114(6):1058–1067
Weible AP, Weiss C, Disterhoft JF (2003) Activity profiles of single neurons in caudal anterior cingulate cortex during trace eyeblink conditioning in the rabbit. J Neurophysiol 90(2):599–612
Weible AP, O'Reilly JA, Weiss C, Disterhoft JF (2006) Comparisons of dorsal and ventral hippocampus cornu ammonis region 1 pyramidal neuron activity during trace eye-blink conditioning in the rabbit. Neuroscience 141(3):1123–1137
Weible AP, Weiss C, Disterhoft JF (2007) Connections of the caudal anterior cingulate cortex in rabbit: neural circuitry participating in the acquisition of trace eyeblink conditioning. Neuroscience 145(1):288–302
Weikart CL, Berger TW (1986) Hippocampal lesions disrupt classical learning of cross-modality reversal learning of the rabbit nictitating response membrane. Behav Brain Res 22(1):85–89
Weiss C, Disterhoft JF (1996) Eyeblink conditioning, motor control, and the analysis of limbic-cerebellar interactions. Behav Brain Sci 19(3):479–481
Weiss C, Houk JC, Gibson AR (1990) Inhibition of sensory responses of cat inferior olive neurons produced by stimulation of red nucleus. J Neurophysiol 64(4):1170–1185
Weiss C, Kronforst-Collins MA, Disterhoft JF (1996) Activity of hippocampal pyramidal neurons during trace eyeblink conditioning. Hippocampus 6(2):192–209
Weiss C, Bouwmeester H, Power JM, Disterhoft JF (1999) Hippocampal lesions prevent trace eyeblink conditioning in the freely moving rat. Behav Brain Res 99(2):123–132
Welsh JP, Harvey JA (1998) Acute inactivation of the inferior olive blocks associative learning. Eur J Neurosci 10(11):3321–3332
Welsh JP, Yamaguchi H, Zeng XH, Kojo M, Nakada Y, Takagi A, Sugimori M, Llinas RR (2005) Normal motor learning during pharmacological prevention of Purkinje cell long-term depression. Proc Natl Acad Sci USA 102(47):17166–17171
Wickelgren WA (1979) Chunking and consolidation: a theoretical synthesis of semantic networks, configuring in conditioning, S-R versus cognitive learning, normal forgetting, the amnesic syndrome, and the hippocampal arousal system. Psychol Rev 86(1):44–60
Wiesendanger R, Wiesendanger M (1982) The corticopontine system in the rat. II. The projection pattern. J Comp Neurol 208(3):227–238
Winocur G, Moscovitch M, Bontempi B (2010) Memory formation and long-term retention in humans and animals: convergence towards a transformation account of hippocampal-neocortical interactions. Neuropsychologia 48(8):2339–2356
Woodruff-Pak DS, Disterhoft JF (2008) Where is the trace in trace conditioning? Trends Neurosci 31(2):105–112
Woodruff-Pak DS, Lavond DG, Thompson RF (1985) Trace conditioning: abolished by cerebellar nuclear lesions but not lateral cerebellar cortex aspirations. Brain Res 348(2):249–260

Woodruff-Pak DS, Green JT, Levin SI, Meisler MH (2006) Inactivation of sodium channel Scn8A (Na-sub(v)1.6) in Purkinje neurons impairs learning in Morris water maze and delay but not trace eyeblink classical conditioning. Behav Neurosci 120(2):229–240
Yeo CH, Hardiman MJ, Glickstein M (1985a) Classical conditioning of the nictitating membrane response of the rabbit. I. Lesions of the cerebellar nuclei. Exp Brain Res 60(1):87–98
Yeo CH, Hardiman MJ, Glickstein M (1985b) Classical conditioning of the nictitating membrane response of the rabbit. II. Lesions of the cerebellar cortex. Exp Brain Res 60(1):99–113

Exploration of the Neurobiological Basis for a Three-System, Multiattribute Model of Memory

Raymond P. Kesner

Abstract The structure and utilization of memory is central to one's knowledge of the past, interpretation of the present, and prediction of the future. Therefore, the understanding of the structural and process components of memory systems at the psychological and neurobiological level is of paramount importance. There have been a number of attempts to divide learning and memory into multiple memory systems. Schacter and Tulving, Memory systems 1994. MIT Press, Cambridge (1994) have suggested that one needs to define memory systems in terms of the kind of information to be represented, the processes associated with the operation of each system, and the neurobiological substrates, including neural structures and mechanisms, that subserve each system. Furthermore, it is likely that within each system there are multiple forms or subsystems associated with each memory system and there are likely to be multiple processes that define the operation of each system. Finally, there are probably multiple neural structures that form the overall substrate of a memory system.

Keywords Event-based memory · Hippocampus · Dentate gyrus · CA3 · CA1 · Knowledge-based memory · Posterior parietal cortex · Rule-based memory · Prefrontal cortex

Contents

1 Event-Based Memory System 334
1.1 Dentate Gyrus and Conjunctive Encoding 335
1.2 Dentate Gyrus and Spatial Pattern Separation 336
1.3 CA3 and Arbitrary Associations 338
1.4 CA3 and Pattern Completion 341
1.5 CA1 and Temporal Pattern Separation 342
2 Knowledge-Based Memory System 344
3 Rule-Based Memory System 345
3.1 Precentral Cortex (PC) 346
3.2 Anterior Cingulate Cortex (AC) 346

R.P. Kesner (✉)
Department of Psychology, University of Utah, Salt Lake City, USA
e-mail: ray.kesner@psych.utah.edu

Curr Topics Behav Neurosci (2018) 37: 325–359
DOI 10.1007/7854_2016_454
Published Online: 20 September 2015

3.3 Prelimbic and Infralimbic Plus Medial Orbital Cortex (PL-IL/MO) ... 347
3.4 Agranular Insular and Lateral Orbital Cortex (AI/LO) ... 348
4 Interactions Between Event-Based Memory and Rule-Based Memory ... 349
5 Interactions Between Knowledge-Based Memory and Event-Based Memory ... 350
6 Interactions Between Knowledge-Based Memory and Rule-Based Memory ... 352
7 Conclusion ... 354
References ... 354

The first model of hippocampal function and the processing of spatial information was described by O'Keefe and Nadel (1978; see Nadel 1994 as well). They developed a memory model with a concentration on space as the critical attribute of specific memories. They further divided the spatial attribute into a locale system, which codes places in the environment into cognitive maps, and a taxon system, which codes motor responses in terms of specific orientations within a spatial environment. In terms of neural mediation of the locale versus taxon system, they propose that the hippocampus is important in mediating only one form of memory, namely spatial, within the locale system and other neural regions as important for subserving the taxon system. With respect to the operation of each system, it was assumed that learning within the locale system is based in part on consolidation processes and is (a) all-or-none, (b) sensitive to interference, (c) involved in separating traces, and (d) flexible, whereas learning in the taxon system is (a) incremental, (b) not sensitive to interference, (c) involved in combining traces, and (d) not flexible. Even though the hippocampus was assumed to be the mediator to the locale system, the neural circuit subserving spatial information does include a number of neural regions such as the entorhinal cortex, the retrosplenial cortex, the pre-, para-, and post-subiculum, the parietal cortex, and the pre- and infralimbic cortex. Nadel's focus on the hippocampus might be too limiting. The taxon system is large and needs to be differentiated. Furthermore, a genuine neurobiological system analysis requires the identification of neural regions that subserve the response component associated with the taxon system. However, there is no mention of a memory contribution of the prefrontal cortex (PFC) and there is no mention of other brain areas that support memory for other attributes (e.g., amygdala and affect attribute).

A second model of hippocampal function and the processing of spatial information was presented by Olton (1983). He proposed is a somewhat different system emphasizing more the importance of process. He suggested that within every learning task there are two types of memories that organize the critical information into two systems, labeled working memory and reference memory (Olton 1983). He suggested that the specific, personal, and temporal context of a situation is coded in working memory. This would translate into memory for events that occur on a specific trial in a task, biasing mnemonic coding toward the processing of incoming data. In contrast, information concerning rules and procedures (general knowledge) of specific situations is coded in reference memory. This would translate into memory for events that happen on all trials in a task, biasing mnemonic coding toward the processing of expectancies based on the organization of the extant memory. The working versus reference memory system emphasizes the role of the hippocampus and interconnected neural systems as the critical substrate of memory for a single process, namely working memory, and the neocortex as the critical neural substrate within reference memory for all forms or attributes of memory. It was assumed that the two memory systems are independent of each other. Different terms have been used to reflect the same distinction including episodic versus semantic memory (Tulving 1983).

The Olton model has some limits in that the emphasis is placed only on the hippocampus and interconnected neural circuits as the neural system subserving working memory for all information. However, it is clear that in the Olton model the hippocampus is limited to working memory for only spatial, temporal, and linguistic information. There is no mention of a memory contribution of the PFC and there is no mention of other brain areas that support memory for other attributes (e.g., caudate and the response attribute). Furthermore, the hippocampus is also involved in processes other than short term or working memory, such as pattern separation, consolidation, and retrieval of information (Kesner et al. 1996).

A third model and the most popular model of memory was presented by Squire (1994), Squire et al. (2004) and can be characterized as a dual memory system with an emphasis on the hippocampus and medial temporal lobe including perirhinal cortex, parahippocampal cortex, and entorhinal cortex for one component of the model and a composite of other brain structures as the other component. For example, they have suggested that memory can be divided into a medial temporal lobe-dependent declarative memory that provides for conscious recollection of facts and events, and a non-hippocampal-dependent nondeclarative memory that provides for memory without conscious access for skills and habits mediated by the caudate nucleus and interconnected systems. Furthermore, priming is mediated by the neocortex, simple classical conditioning of emotional responses by the amygdala, simple classical conditioning of skeletal musculature by the cerebellum, and nonassociative learning is mediated by reflex pathways. A limitation is that there is no mention of the PFC contribution to memory, in the context of declarative memory different attributes mediated by the amygdala or caudate do not play a role, and the emphasis is primarily on one single process, namely consolidation. Different models have used different terms to reflect the same type of distinction, including a hippocampal-dependent explicit memory versus a non-hippocampal-dependent implicit memory (Schacter 1987).

A fourth model was presented by Eichenbaum (Cohen and Eichenbaum 1993; Eichenbaum 1994, 2004). They proposed that the declarative memory system is dependent on the hippocampus and provides for a substrate for relational representation of all forms of memory as well as representational flexibility allowing for the retrieval of memories in novel situations. Relational processing is carried out by the hippocampus, but the processing of individual items resides in the perirhinal and parahippocampal cortex. In contrast, a nondeclarative system is independent of the hippocampus and is characterized by individual representations and inflexibility in retrieving memories in novel situations. The limitations include that there is no mention of a memory contribution of the PFC as part of the model (see Kesner and

Churchwell 2011), Also, there is not enough emphasis on different attributes of memory, and processes such as pattern separation and pattern completion are not incorporated in the model.

Because memory is complex and involves many neural systems in addition to the hippocampus, Kesner (2007) has proposed a three-system (event-based, knowledge-based, and rule-based) multiple attribute-based theoretical model of memory. The model is an extension of models presented above. For example, I have accepted Olton's working-reference memory and Tulving's episodic–semantic dual memory model distinctions and labeled them as event-based memory versus knowledge-based memory, but in addition I have added a third rule-based system subserved by mnemonic processes associated with the PFC. Also, I have adopted the attribute model described by Underwood (1969) and Spear (1976). They presented a good case that there are many different forms or attributes of memory such as *space, time, response, sensory-perception, reward value (affect), and language*. These attributes are processed by different neural regions and interconnected networks across all three (event-based, knowledge-based, and rule-based) memory systems. This is an enrichment of the previous mentioned memory models, which emphasize one or two attributes or do not differentiate among attributes. Finally, each memory system operates in processing mnemonic information based on a unique set of processes that involve more than just consolidation. The selection of some these processes have been influenced greatly by computational models of specific brain regions (see Rolls and Kesner 2006).

In the three-system multiattribute model of memory, one can characterize each system as composed of the same set of multiple attributes or forms of memory, characterized by a set of process-oriented operating characteristics, and mapped onto multiple neural regions and interconnected neural circuits (for more detail see Kesner 1998b, 2007).

On a psychological level (see Fig. 1), the event-based memory system provides for temporary representations of incoming data concerning the present, with an emphasis upon data and events that are usually personal or egocentric and that occur within specific external and internal contexts. The emphasis is upon the processing of new and current information. During initial learning great emphasis is placed on the event-based memory system, which will continue to be of importance even after initial learning in situations where unique or novel trial information needs to be remembered. This system is akin to episodic memory (Tulving 1983) and some aspects of declarative memory (Squire 1994).

The knowledge-based memory system (see Fig. 2) provides more permanent representations of previously stored information in long-term memory and can be thought of as one's general knowledge of the world. The knowledge-based memory system would tend to be of greater importance after a task has been learned given that the situation is invariant and familiar. The organization of these attributes within the knowledge-based memory system can take many forms and are organized as a set of attribute-dependent cognitive maps and their interactions that are unique for each memory. This system is akin to semantic memory (Tulving 1983).

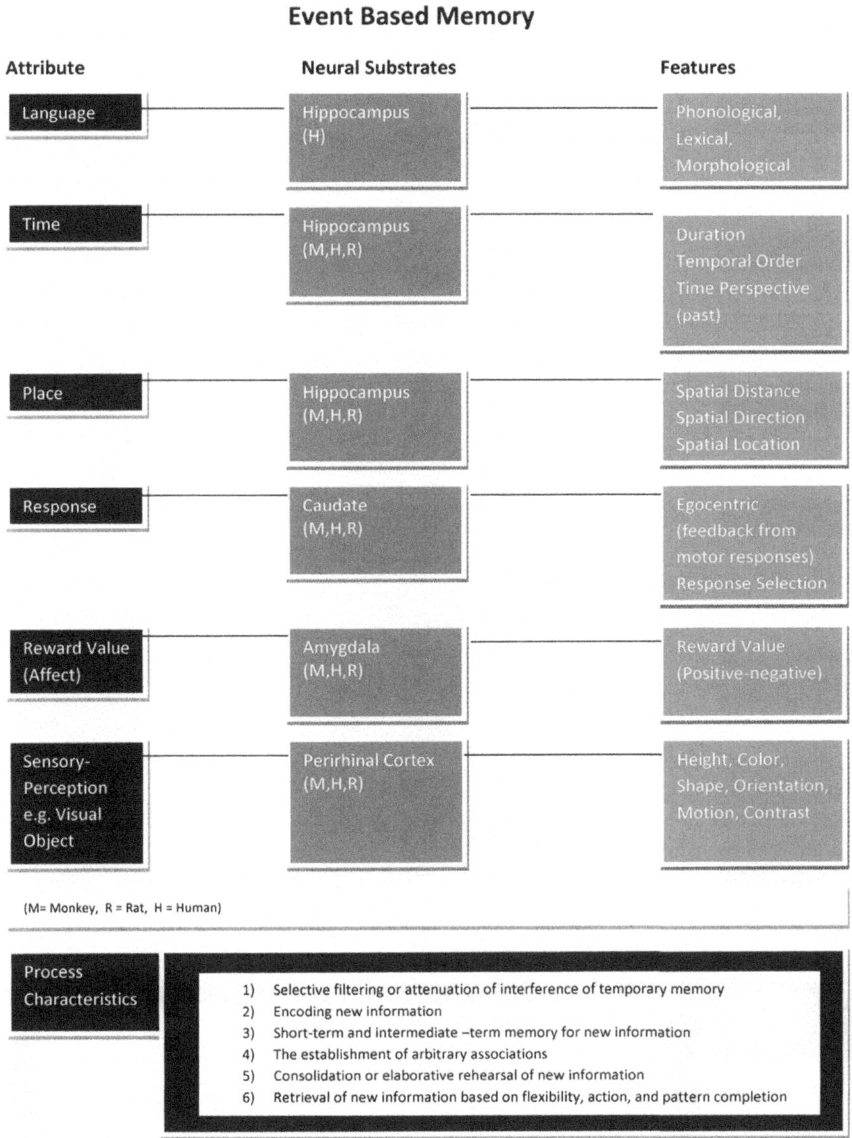

Fig. 1 Representation of the neural substrates, features, and process characteristics associated with the event-based memory system for the language, time, place, response, value (affect), and sensory- perception attributes

The rule-based memory system (see Fig. 3) receives information from the event-based and knowledge-based systems and integrates the information by applying rules and strategies and decisions for subsequent action. In every learning

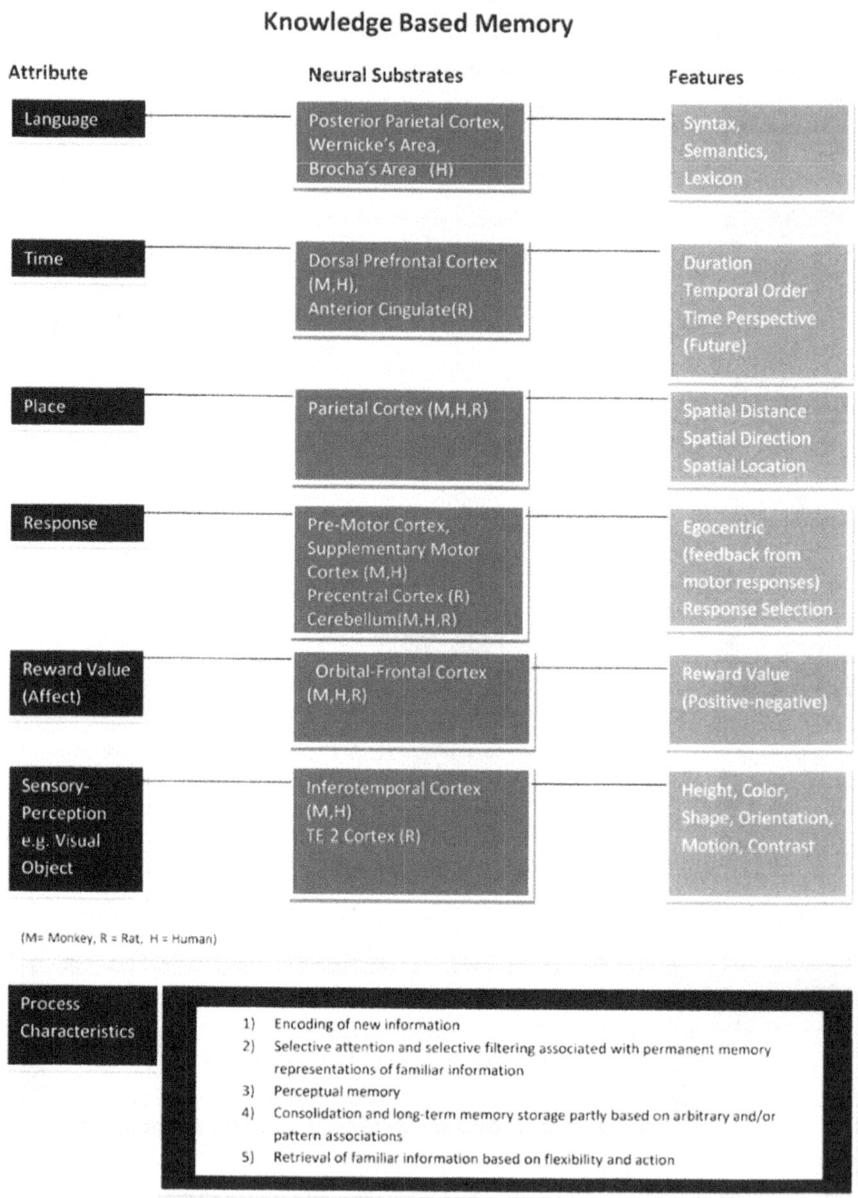

Fig. 2 Representation of the neural substrates, features, and process characteristics associated with the knowledge-based memory system for the language, time, place, response, value (affect), and sensory—perception attributes

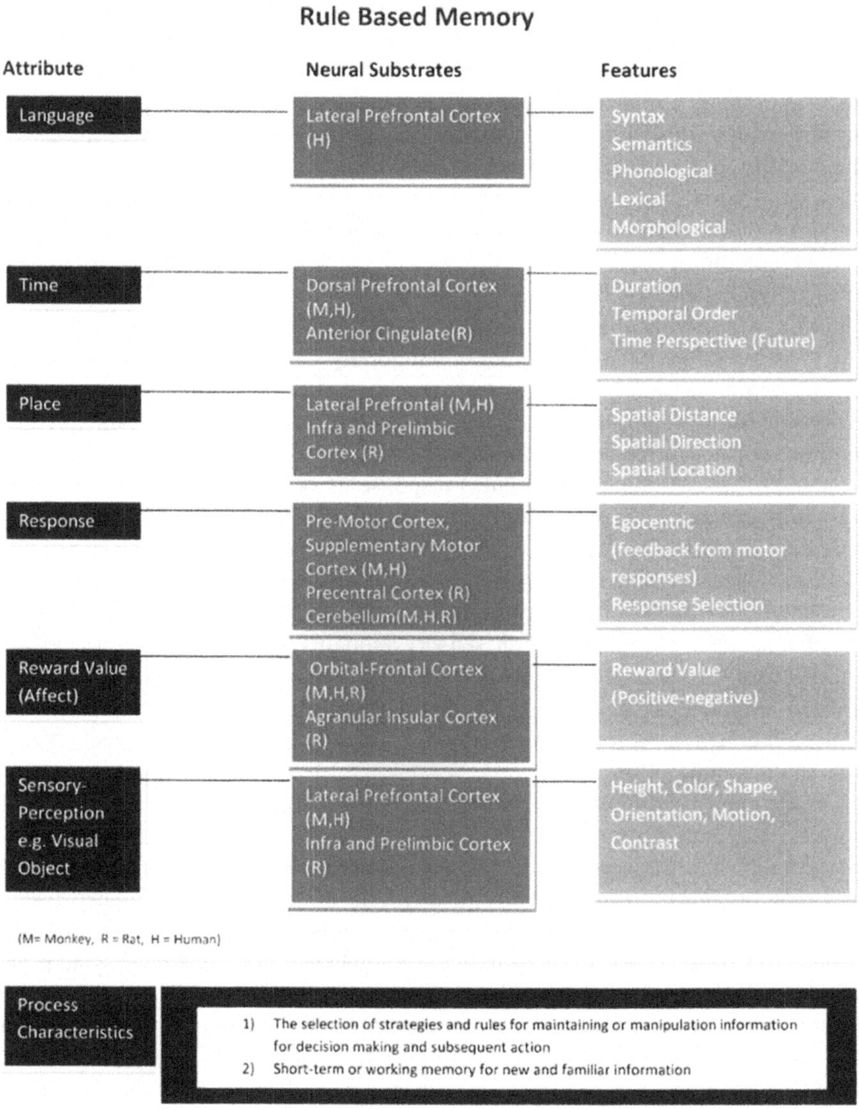

Fig. 3 Representation of the neural substrates, features, and process characteristics associated with the rule-based memory system for the language, time, place, response, value (affect), and sensory—perception attributes

and memory tasks, the subject has to select an appropriate strategy or set of rules to aid in memory consolidation of the task. The processes associated with rule-based memory are most likely mediated by the PFC. In most situations, however, one would expect a contribution of all three systems with a varying proportion of involvement of one relative to the other.

The three memory systems are composed of the same forms, domains, or attributes of memory. Even though there could be many attributes, the most important attributes include *space, time, response, sensory-perception, and reward value (affect)*. In humans, a *language* attribute is also added. A spatial (space) attribute within this framework involves memory representations of places or relationships between places. It is exemplified by the ability to encode and remember spatial maps and to localize stimuli in external space. Memory representations of the spatial attribute can be further subdivided into specific spatial features including allocentric spatial distance, egocentric spatial distance, allocentric direction, egocentric direction, and spatial location. A temporal (time) attribute within this framework involves memory representations of the duration of a stimulus, the succession or temporal order of temporally separated events or stimuli, and memory representations of the past. A response attribute within this framework involves memory representations based on feedback from motor responses (often based on proprioceptive and vestibular cues) that occur in specific situations as well as memory representations of stimulus-response associations. A reward value (affect) attribute within this framework involves memory representations of a hedonic continuum of positive and negative values and the associations between stimuli and rewards. A sensory-perceptual attribute within this framework involves memory representations of a set of sensory stimuli that are organized in the form of cues as part of a specific experience. Each sensory modality (olfaction, auditory, vision, somatosensory, and taste) can be considered as apart of the sensory-perceptual attribute component of memory. A language attribute within this framework involves memory representations of phonological, lexical, morphological, syntactical, and semantic information.

The attributes within each memory system can be organized in many different ways and are likely to interact extensively with each other, even though it can be demonstrated that these attributes do in many cases operate independent of each other. The organization of these attributes within the event-based memory system can take many forms and are probably organized hierarchically and in parallel. The organization of these attributes within the knowledge-based memory system can take many forms and are (assumed to be) organized as a set of cognitive maps or neural nets and their interactions that are unique for each memory. It is assumed that long-term representations within cognitive maps are more abstract and less dependent upon specific features. The organization of these attributes within the rule-based memory system can also take many forms; these are (assumed to be) organized to provide flexibility in executive function in developing rules and goals, as well as decision processes.

Within each system, attribute information is processed in different ways based on different operational characteristics. For the event-based memory system (see Fig. 1), specific processes involve (a) selective filtering or attenuation of interference of temporary memory representations of new information and this process is labeled pattern separation, (b) encoding of new information, (c) short-term and intermediate-term memory for new information, (d) the establishment of arbitrary associations, (e) consolidation or elaborative rehearsal of new information, and (f) retrieval of new information based on flexibility, action, and pattern completion.

For the knowledge-based memory system (see Fig. 2), specific processes include (a) encoding of new information, (b) selective attention and selective filtering associated with permanent memory representations of familiar information, (c) perceptual memory, (d) consolidation and long-term memory storage partly based on arbitrary and/or pattern associations, and (e) retrieval of familiar information based on pattern completion, flexibility, and action.

For the rule-based memory system (see Fig. 3), it is assumed that information is processed through the integration of information from the event-based and knowledge-based memory systems for the use of major processes that include the selection of strategies and rules for maintaining or manipulating information for subsequent decision making and action.

On a neurobiological level, each attribute maps onto a set of neural regions and their interconnected neural circuits (see Figs. 1, 2 and 3). For example, within the event-based memory system, it has been demonstrated that in animals and humans (a) the hippocampus supports memory for spatial, temporal, and language attribute information, (b) the caudate mediates memory for response attribute information, (c) the amygdala subserves memory for reward value (affect) attribute information, and (d) the perirhinal and extrastriate visual cortex support memory for visual object attribute information as an example of a sensory-perceptual attribute (for more detail see Kesner 1998b, 2007).

Within the knowledge-based memory system, it has been demonstrated that in animals and humans (a) the posterior parietal cortex (PPC) supports memory for spatial attributes, (b) the dorsal and dorsolateral PFC and/or anterior cingulate support memory for temporal attributes, (c) the premotor, supplementary motor, and cerebellum in monkeys and humans and precentral cortex (PC) and cerebellum in rats support memory for response attributes, (d) the orbital PFC supports memory for reward value (affect) attributes, (e) the inferotemporal cortex in monkeys and humans and TE2 cortex in rats subserves memory for sensory-perceptual attributes, e.g., visual objects, and (f) parietal cortex, Broca and Wernicke's areas subserve memory for the language attribute (for more details see Kesner 1998b, 2007).

Within the rule-based memory system, it can be shown that different subdivisions of the PFC support different attributes. For example, (a) the dorsolateral and ventrolateral PFC in humans support spatial, object, and language attributes and the infralimbic and prelimbic cortex in rats supports spatial and visual object attributes, (b) the premotor and supplementary motor cortex in monkeys and humans and PC in rats support response attributes, (c) the dorsal, dorsolateral, and mid-dorsolateral PFC in monkeys and humans and anterior cingulate in rats mediate primarily temporal attributes, and (d) the orbital PFC in monkeys and humans and agranular insular cortex in rats support affect attributes (for more detail see Kesner 2000a, 2007).

1 Event-Based Memory System

Given the complexity of memory representations in the brain, how is one to test the neurobiological basis of the attribute model of memory? In order to test whether different brain regions subserve the processing of different attributes within the event-based memory system, I selected the process of short term or working memory. The short-term memory task designed to test this consists of a study phase comprising one item (e.g., object, spatial location, motor response, or reward) and then following a delay there is a test phase consisting of two items with one item identical to the study phase and a new item leading to reinforcement for a match or mismatch with the study phase. After the task is learned, lesions of specific neural substrates are used. With this paradigm, it has been shown that there is a triple dissociation among the hippocampus (spatial location), caudate (response) and extra striate visual cortex (visual object) (Kesner et al. 1993), a double dissociation between the hippocampus (spatial location) and the amygdala (affect) (Gilbert and Kesner 2002a; Gilbert and Kesner 2006), a double dissociation between hippocampus (spatial location) and perirhinal cortex (visual object) (Gilbert and Kesner 2003a; Kesner 1999), as well as a double dissociation within the hippocampus in terms of spatial (dentate gyrus) versus temporal (CA1) processing of information (Gilbert et al. 2001). Thus it appears that within the event-based memory system, different neuroanatomical circuits are involved in the processing of different attributes in that they can operate independently of each other.

In subsequent research, I have concentrated on determining the importance of examining multiple processes associated with the event-based memory system, including (1) conjunctive encoding to create a spatial representation, (2) selective filtering or attenuation of interference with encoding of information labeled as pattern separation, especially for spatial location and spatial contextual information, (3) formation of arbitrary associations, (4) retrieval of familiar information based on pattern completion, (5) temporal processing of information including temporal pattern separation, (6) short-term and intermediate-term memory for new information, and (7) promotion of consolidation or elaborative rehearsal of new information. I will concentrate on the different subregions of the hippocampus and I will mention other brain areas that subserve the same function for a different attribute given the availability of empirical studies. First, I will examine the role of the dentate gyrus (DG) subregion of the hippocampus in supporting conjunctive encoding to create a spatial representation and selective filtering or attenuation of interference with encoding of information labeled pattern separation, especially spatial. Second, I will examine the role of the CA3 subregion of the hippocampus in supporting formation of arbitrary associations and retrieval of familiar information based on pattern completion. Third, I will examine the role of the CA1 subregion of the hippocampus in supporting temporal processing of information including temporal pattern separation. I will not discuss intermediate-term memory for new information and promotion of consolidation or elaborative rehearsal of new information because of space limitations.

1.1 Dentate Gyrus and Conjunctive Encoding

The DG has been shown to receive multiple sensory inputs, including vestibular, olfactory, visual, auditory, and somatosensory, from the perirhinal and lateral entorhinal cortex in conjunction with spatially organized grid cells from the medial entorhinal cortex (Hafting et al. 2005) to represent metric spatial representations. The perforant path input of the DG can be divided into medial and lateral components. The medial component processes spatial information and the lateral component processes nonspatial (e.g., objects, odors) information (Hargreaves et al. 2005; Witter et al. 1989). Based on the idea that the medial perforant path (MPP) input into the DG mediates spatial information via activation of NMDA receptors and the lateral perforant path (LPP) input into the DG mediates visual object information via activation of opioid receptors, the following experiment was conducted. Using a paradigm developed by Poucet (1989) rats were tested for detection of a novel spatial change and detection of a novel visual object change while under the influence of direct infusions of AP5 (an NMDA antagonist) or naloxone (a μ opiate antagonist) into the DG. The results are shown in Fig. 4 and indicate that naloxone infusions into the DG disrupted both novelty detection of a spatial location and a visual object, whereas AP5 infusions into the DG disrupted only detection of a novel spatial location, but had no effect on detection of a novel object (Hunsaker et al. 2007). These data suggest that the DG uses conjunctive encoding of visual object and spatial information to provide for a spatial representation that may be based on metric information.

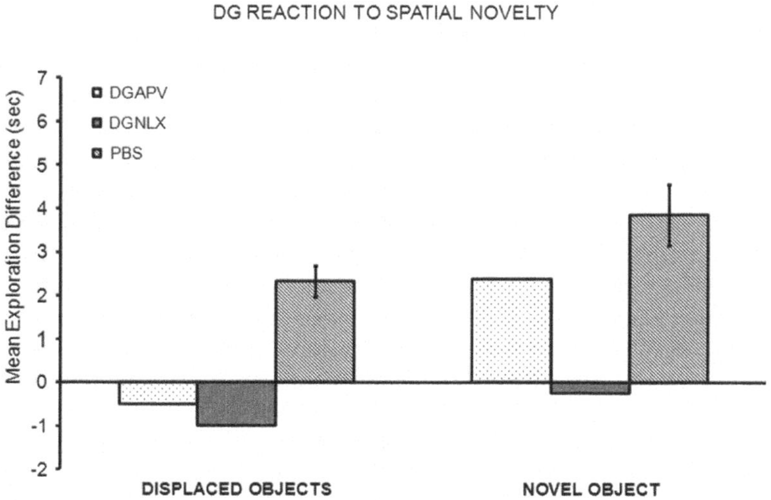

Fig. 4 The effects of naloxone (NLX), 2-amino-5-phosphonovaleric acid (APV) or Phsophate Burrered Saline (PBS) infusions within the dentate gyrus for spatial (bars on the *left*) and nonspatial (visual object; bars on the *right*) novelty detection

1.2 Dentate Gyrus and Spatial Pattern Separation

Pattern separation is defined as a process to remove redundancy from similar inputs so that events can be separated from each other and interference can be reduced, and in addition can produce a more orthogonal, sparse, and categorized set of outputs. Computational models have emphasized the importance of the hippocampus in mediating spatial pattern separation, which has been developed extensively by computational models of the subregions of the hippocampus with a special emphasis on the DG. Based on the empirical findings that all sensory inputs are processed by the DG subregion of the hippocampus, it has been suggested that a possible role for the hippocampus might be to provide for sensory markers to demarcate a spatial location, so that the hippocampus can more efficiently mediate spatial information. It is thus possible that one of the main process functions of the hippocampus is to encode and separate spatial locations from each other. This would ensure that new highly processed sensory information is organized within the hippocampus and enhances the possibility of remembering and temporarily storing one place as separate from another place. It is assumed that this is accomplished via pattern separation of spatial information, so that spatial locations can be separated from each other and spatial interference is reduced.

Rolls' (1996) model proposes that pattern separation is facilitated by sparse connections in the mossy fiber system, which connects DG granular cells to CA3 pyramidal neurons. Separation of patterns is accomplished based on the low probability that any two CA3 neurons will receive mossy fiber input synapses from a similar subset of DG cells. Mossy fiber inputs to CA3 from DG are suggested to be essential during learning and may influence that CA3 neurons fire based on the distributed activity within the DG. Cells of the DG are suggested to act as a competitive learning network with Hebb-like modifiability to reduce redundancy and produce sparse, orthogonal outputs. O'Reilly and McClelland (1994) and Shapiro and Olton (1994) also suggested that the mossy fiber connections between the DG and CA3 may support pattern separation.

To examine the contribution of the DG to spatial pattern separation, Gilbert et al. (2001) tested rats with DG lesions using a paradigm that measured short-term memory for spatial location information as a function of spatial similarity between locations. Specifically, the study was designed to examine the role of the DG subregion in discriminating spatial locations when rats were required to remember a spatial location based on distal environmental cues and to differentiate between the to-be-remembered location and a distractor location with different degrees of similarity or overlap among the distal cues. Rats were tested using a cheeseboard maze apparatus (the cheese board is similar to a dry-land water maze with 177 circular, recessed holes on a 119 cm diameter board) on a delayed-match-to-sample for spatial location task. Animals were trained to displace an object that was randomly positioned to cover a baited food well in 1 of 15 locations along a row of food wells. Following a short delay, the animals were required to choose between objects which were identical to the sample phase object: one object was in the same location as the sample phase object and the second object was in a different location along the row of food wells. Rats were rewarded for displacing the object in the same spatial location as the sample phase object (correct choice), but they received no reward for displacing the foil object (incorrect choice). Five spatial separations, from 15 cm to 105 cm, were used to separate the correct object and the foil object during the choice phase. The results are shown in Fig. 5 and indicate that rats with DG lesions were significantly impaired at short spatial separations; however, during the choice phase, performance of DG-lesioned animals increased as a function of greater spatial separation between the correct and foil objects. The performance of rats with DG lesions matched control rats at the largest spatial separation. The graded nature of the impairment and the significant linear improvement in performance as a function of increased separation illustrate a deficit in pattern separation. Based on these results, it was concluded that lesions of the DG decrease the efficiency of spatial pattern separation, which results in impairments on trials with increased spatial proximity and increased spatial similarity among working memory representations. Thus, the DG may function to encode and to separate events in space producing spatial pattern separation. Such a spatial pattern separation ensures that new highly processed sensory information is organized within the hippocampus, which in turn enhances the possibility of encoding and temporarily remembering one spatial location as separate from another.

In further support of the attribute model, it has been shown that lesions of the amygdala, but not hippocampus, disrupt memory-based pattern separation for affect information (Gilbert and Kesner 2002a), lesions of the caudate nucleus, but not hippocampus, disrupt memory-based pattern separation for motor responses (Kesner and Gilbert 2006), lesions of the perirhinal cortex, but not hippocampus, disrupt memory-based pattern separation for objects (Gilbert and Kesner 2003a), and ventral DG lesions disrupt memory-based pattern separation for odors (Weeden et al. 2014).

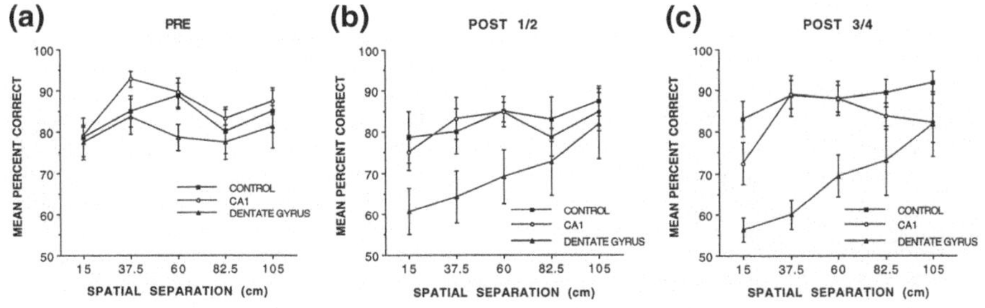

Fig. 5 **a** Mean percent correct performance as a function of spatial separation (number of intervening locations) for the control group, CA1 lesion group, and dentate gyrus lesion group on preoperative trials. **b**, **c** Mean percent correct performance as a function of spatial separation for the control group, CA1 lesion group, and dentate gyrus lesion group on two sets 1/2 and 3/4 of 30 postoperative trials. Note the inter-cue distance-dependent impairment in performance in the dentate gyrus-lesioned group, demonstrating the role of dentate gyrus in spatial pattern separation

1.3 CA3 and Arbitrary Associations

In the standard model (Marr 1971; McNaughton and Morris 1987; Levy 1996; Hasselmo and Wyble 1997; Rolls and Treves 1998; Rolls and Kesner 2006) the CA3 system acts as an auto-association system. This enables arbitrary (especially spatial in animals and likely language for humans as well) associations to be formed within the hippocampus. The CA3 recurrent collateral associative connections enable bidirectional associations to be formed between whatever stimuli are represented in the hippocampus, in that, for example, any place could be associated with any object, and in that the object could be recalled with a spatial recall cue, or the place with an object recall cue (Rolls and Treves 1998).

In the Kesner laboratory, a visual object recall for aspatial location task has been developed based on the Day, Langston and Morris (2003) experiment. In this task, after training to displace objects for food, rats in the study phase of each trial are placed in the start box (see Fig. 6 where each shape represents an object). When the door in front of the start box is opened the rats are allowed to displace one object in one location, and then return to the start box, after which the door is opened again and the rats are allowed to displace a second object in another location. Fifty possible objects and 48 locations were used to ensure that each trial was unique. In the test phase of each trial (see Fig. 6 where the open square represents spatial locations covered by a neutral block), the rat is shown one of the previously presented objects (first or second randomized) in the start box as a cue for which spatial location to choose, and then, after a 10 s delay, the door is opened and the rats must go to the correct location (choosing and displacing one of two identical neutral objects). The rats receive a reward for selecting the correct location that was

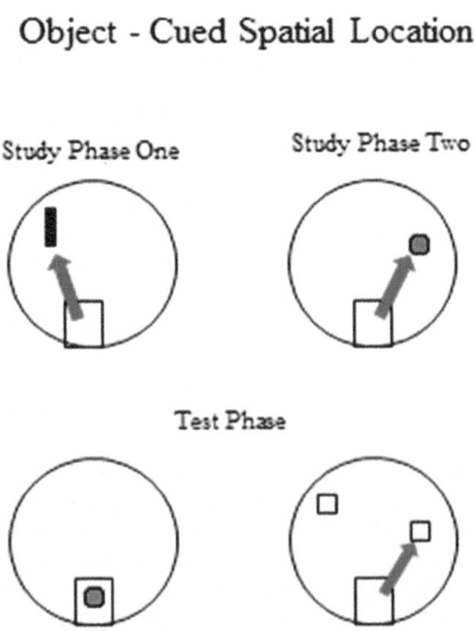

Fig. 6 Object-cued spatial location recall. Each shape represents a different object, and the *open squares* represent spatial locations covered by neutral blocks. Each trial consisted of two spatial location based study phases followed 15 s later by an object cue and 10 s later by a test between two previously experienced spatial locations

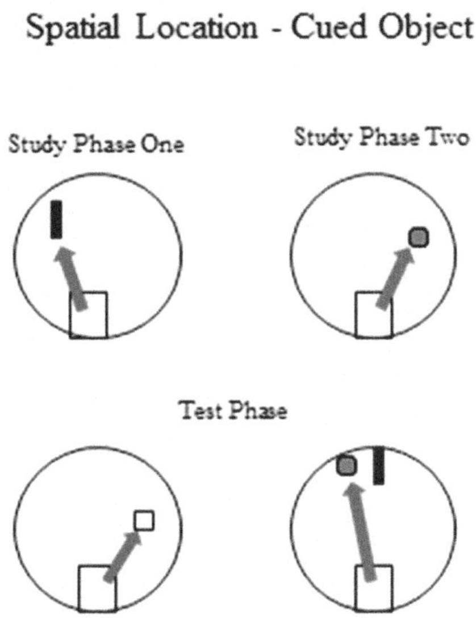

Fig. 7 Spatial location-cued object recall. Each shape represents a different object. The *open square* represents a neutral block placed on the correct spatial location as a cue. Each trial consisted of two object based study phases followed 15 s later by a spatial location cue, and 10 s later by a test between two previously experienced objects

associated with that specific object cue. A spatial location-cued recall for a visual object task has also been developed (see Fig. 7). For the spatial-cued recall for a visual object task, the study phase (See Fig. 7 where each shape represents a different object) is the same, but in this case in the test phase (See Fig. 7 where the open square represents the correct location that is covered by a neutral block given as a cue), when the door is opened the rat is allowed to displace a neutral object in one of the previous locations (first or second, randomized) on the maze as a location cue, return to the start box, and then, after a 10 s delay, the door is opened and the rats must select the correct object (choosing and displacing one of two visual objects placed in different locations than during the Study phases). The rats receive a reward for selecting the correct visual object that was associated with the location cue. Rats learn both tasks with 75 % or better accuracy.

Results are shown in Figs. 8 and 9 and indicate that CA3 lesions produce chance performance on both the object-cued place recall and the place-cued object recall task (Kesner et al. 2008).

The potential implications of such results are that indeed the CA3 supports arbitrary associations as well as episodic memory based on one-trial learning. A control fixed visual conditional to place task with the same delay was not impaired, showing that it is recall after one-trial (or rapid) learning that is impaired. Thus, some hippocampal neurons appear to reflect spatial recall given an object recall cue. These data are consistent with the prediction of the standard computational model that emphasizes the importance of CA3 in mediating the development of arbitrary associations.

There is anatomical support for CA3 involvement in support of the mediation of associative processes including arbitrary associations. The perforant path from the entorhinal cortex can be divided into a medial and lateral component. It has been

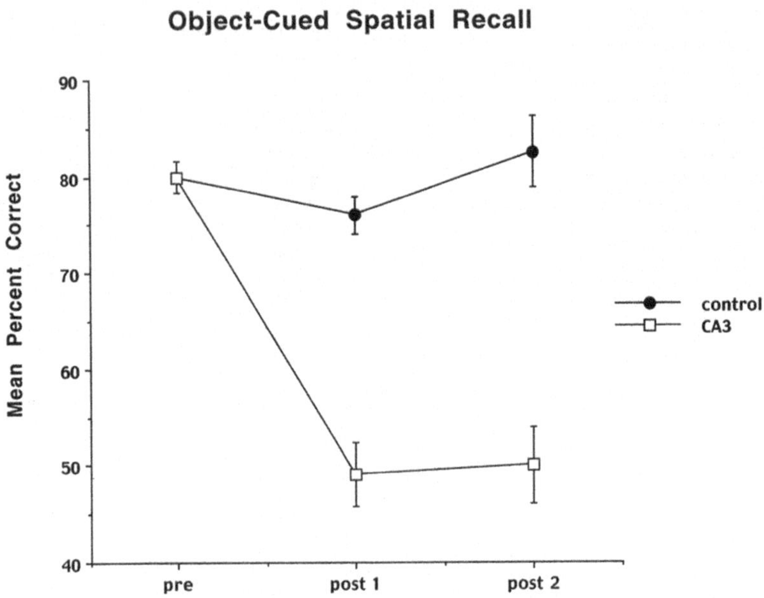

Fig. 8 Mean percent correct performance for the control and CA3-lesioned rats on the object-cued spatial location recall task before (pre) and after surgery (30 trials of post 1 and 30 trials of post 2). Note the profound CA3 lesion effect

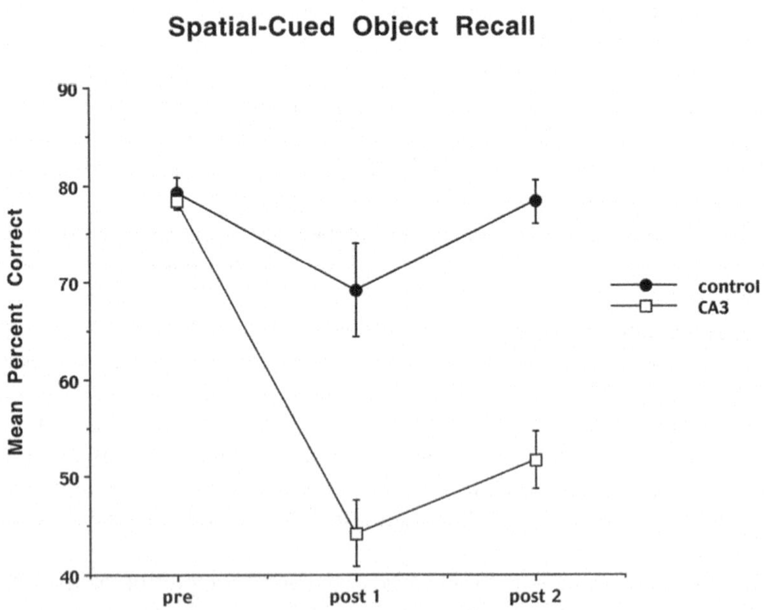

Fig. 9 Mean percent correct performance for the control and CA3-lesioned rats on the spatial location-cued object recall task before (pre) and after surgery (30 trials post 1 and 30 trials post 2). Note the profound CA3 lesion effect

suggested that the medial component processes spatial information and that the lateral component processes nonspatial (e.g., object, odor) information (Witter et al. 1989; Hargreaves et al. 2005). In one study Ferbinteanu et al. (1999) showed that lesions of the MPP disrupted water maze learning, whereas LPP lesions had no effect. In a more recent study based on the idea that the MPP input into the CA3 mediates spatial information via activation of NMDA receptors and the LPP input into the CA3 mediates visual object information via activation of opioid receptors, the following experiment was conducted using the same paradigm described in the dentate and conjunctive encoding section except that direct infusions of AP5 (an NMDA antagonist) or naloxone (a μ opiate antagonist) into CA3 were administered. The results indicated that naloxone or AP5 infusions into the CA3 disrupted both novelty detection of a spatial location and a visual object (Hunsaker et al. 2007).

1.4 CA3 and Pattern Completion

Marr (1971) suggested that hippocampal recurrent collaterals should play a significant role during the retrieval of previously stored information patterns in the face of partial inputs to the hippocampus ("collateral effect" or pattern completion). According to McNaughton and Morris (1987) and Rolls and Treves (1998), an auto-associative network within CA3 should be able to support pattern completion. Experimental efforts to find evidence of pattern completion within the CA3 region have been successful in recent years. For example, Gold and Kesner (2005) trained rats on a delayed matching to sample for a spatial location task to study spatial pattern completion. Animals were tested on the cheese board task, which was surrounded by a black curtain with four extra-maze cues. In the sample phase of the task, rats were trained to move a small black block covering a food well that could appear in one of five possible spatial locations that were in front of four extra-maze cues (i.e., the rat could see all four cues when approaching the spatial location as they were within the 180 degrees visible immediately upon leaving the start box). During the choice phase of the task, rats were required to find the same food well, with the block removed in order to receive a food reward. After reaching stable performance, rats were randomly assigned to receive bilateral intracranial neurotoxic infusions or vehicle control infusions into the CA3 subregion of the hippocampus. Following recovery from surgery, each animal was retested on the delayed matching to sample task. During the sample phase, the animal was presented with all four extra-maze cues; however, the number of available cues (zero, one, two, three, or four cues) varied during the choice phase. The results are shown in Fig. 10 and indicate that control rats performed well on the task regardless of the availability of one, two, three, or four cues, suggesting intact spatial pattern completion. Following the CA3 lesion, however, there were impairments in accuracy compared to the controls especially when only one or two cues were available,

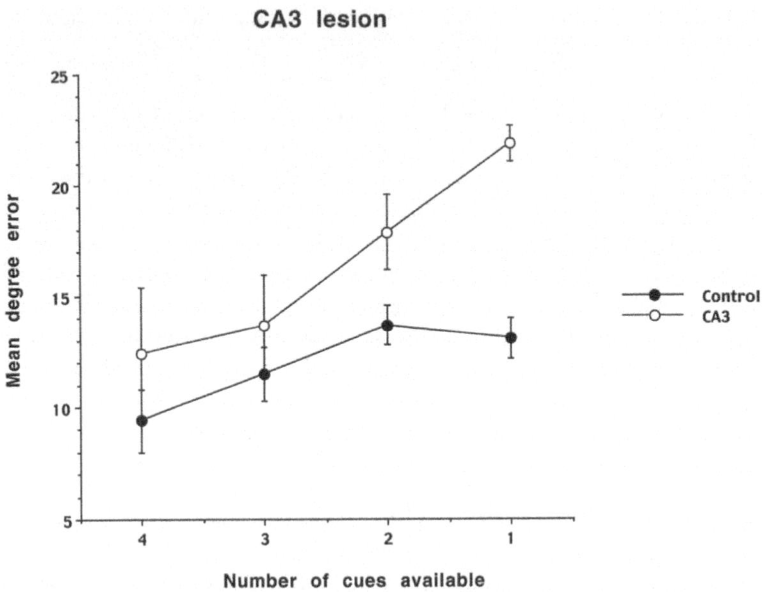

Fig. 10 Pattern completion impairment produced by CA3 lesions. The mean (with SEM) degree of error in finding the correct place on the cheeseboard task when rats were tested with 1, 2, 3 or 4 of the extra-maze cues available. A graded impairment in the CA3 lesion group as a function of the number of cues available was found. Prior to surgery the task was learned in the study phase with the 4 cues present. The performance of the control group is also shown

suggesting impairment in spatial pattern completion in CA3-lesioned rats (Gold and Kesner 2005). Similar results were observed for naloxone (μ opioid receptor antagonist) infusions into CA3 (Kesner and Warthen 2010).

1.5 CA1 and Temporal Pattern Separation

Estes (1986) summarized data demonstrating that, in human memory, there are fewer errors for distinguishing items (by specifying the order in which they occurred) that are far apart in a sequence than those that are temporally adjacent. This phenomenon is referred to as a temporal distance effect [sometimes referred to as a temporal pattern separation effect (Kesner et al. 2004)]. The temporal distance effect is assumed to occur because there is more interference for temporally proximal events than for temporally distant events. Based on these findings, Gilbert et al. (2001) tested memory for the temporal order of items in a one-trial sequence learning paradigm in rodents. In the task, each rat was given one daily trial consisting of a sample phase followed by a choice phase. During the sample phase, the animal visited each arm of an 8-arm radial maze once in a randomly predetermined order and was given a reward at the end of each arm. The choice phase began immediately following the presentation of the final arm in the sequence. In the choice phase, two arms were opened simultaneously and the animal was allowed to choose between the arms. To obtain a food reward, the animal had to enter the arm that occurred earlier in the sequence that it had just followed. Temporal separations of 0, 2, 4, and 6 were randomly selected for each choice phase. These values represented the number of arms in the sample phase that intervened between the arms that were used in the test phase. After reaching criterion, rats received CA1, DG, or control lesions. The results are shown in Fig. 11 and indicate that control and DG-lesioned rats matched their preoperative performance across all temporal separations. In contrast, rats with CA1 lesions performed at chance across 0, 2, 4, and 6 temporal separations.

The results suggest that the CA1 subregion is involved in memory for spatial location as a function of temporal separation of spatial locations. Thus, lesions of the CA1 decrease efficiency in temporal pattern separation. CA1 lesioned rats cannot separate events across time, perhaps due to an inability to inhibit interference that may be associated with sequentially occurring events. The increase in temporal interference impairs the rat's ability to remember the order of specific events. For additional functions of CA1, see Hunsaker et al. (2008).

In summary, the hippocampus was used to detail the multiple operations that characterize the overall activity of this brain region within the event-based memory system. The processes that were discussed include DG mediation of conjunctive encoding and spatial pattern separation, CA3 mediation of arbitrary associations and pattern completion, and CA1 mediation of temporal pattern separation. It should be noted that that there are parallel brain-function relationships between the rodent data and the human data. With the use of similar behavioral paradigms with humans, it can be shown that there is extensive support for the attribute-based

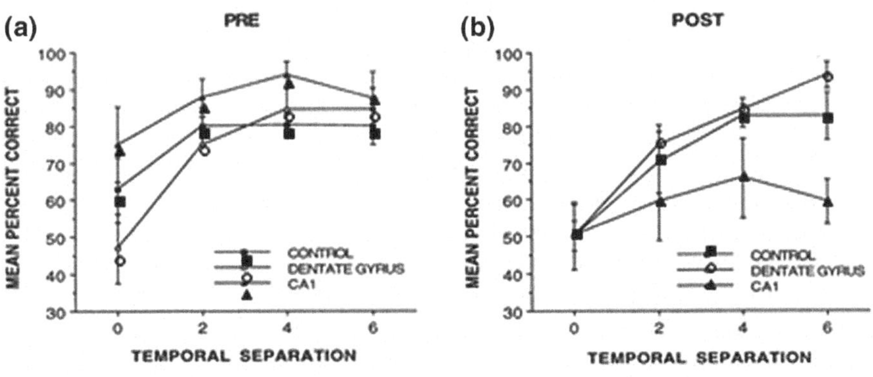

Fig. 11 **a** Mean percent correct performance as a function of temporal pattern separation for the control group, dentate gyrus lesion group, and CA1 lesion group on preoperative trials. **b** Mean percent correct performance as a function of temporal separation for the control group, dentate gyrus lesion group, and CA1 lesion group on postoperative trials

theoretical model of memory that is organized into event-based, knowledge-based, and rule-based memory systems. For review of the hippocampus see Kesner and Hopkins (2006) and Kesner and Warthen (2010), for a review of parietal cortex see Kesner and Creem-Regehr (2013), and for a review of PFC, see Kesner and Churchwell (2011).

2 Knowledge-Based Memory System

The model suggests that different brain regions subserve the processing of different attributes within the knowledge-based memory system. To illustrate this, I selected processes that mediate perceptual memory associated within long-term memory including repetition priming and object recognition. The emphasis will be on visual and spatial perceptual processing and object recognition within the knowledge-based system. I will concentrate on temporal cortex (TE2) and make comparisons with the PPC in this section. In order to study one process associated with the knowledge- based system, a positive priming task was selected. Rats were then trained on tasks that resulted in a positive priming effect as indexed by facilitation of responding following a repetition of a spatial location or a visual object. TE2 lesions produced a deficit in processing positive priming for features of visual objects (a component of the knowledge-based memory system), but the rats performed well in positive priming for spatial location (Kesner, in preparation), whereas PPC lesions produced a deficit in processing positive priming for spatial locations (a component of the knowledge-based memory system), but performed well in positive priming for visual objects (in preparation). Thus, there is a double dissociation between TE2 and PPC for visual object versus spatial location priming. In a somewhat different study, a continuous recognition procedure was used to train rats on a 12-arm radial maze. Each rat was allowed to visit a sequence of 12 arms per day in an order predetermined for that trial. Of the 12 arms visited, either 3 or 4 of the arms were repeated within the running sequence. The arms selected for repetition varied according to lag (0–6), or the number of arms that occurred between the first visit to an arm and its repetition. In order to gain access to each arm, the animal was required to orient to a cue on the Plexiglas door at the entrance of the arm. Once the animal oriented to the cue, the door was lowered and the latency for the animal to reach the end of the arm was measured. Three groups of rats were trained on the knowledge-based perceptual memory training procedure. The perceptual/implicit memory group received reinforcement at the end of each arm regardless of whether the arm was a novel arm or a repeated arm. This group showed decreased latencies when visiting repeated arms displaying a repetition priming effect. After training, rats received PPC, sham-operated or cortical control lesions. After retesting, the results indicated that relative to the sham-operated and control groups control, the PPC-lesioned rats were impaired in the knowledge-based perceptual memory condition (Chiba et al. 2002).

Using a visual object-place recognition task, TE2-lesioned rats failed to detect a visual object change, whereas PPC-lesioned rats failed to detect a spatial location (Tees 1999) suggesting that the two cortical areas play a distinctive role in perceptual processing of visual versus spatial location information. Similar results were reported by Ho et al. (2011) who showed that rats with TE2 lesions had object recognition problems at 20 min, but not at 5 min delays. Lesions of the rat PPC disrupted retention of a spatial navigation task using either the water maze or dry-land version of the water maze task (DiMattia and Kesner 1988; Kesner et al. 1991; Save and Moghaddam 1996). Furthermore, in a multiple object scene task, PPC lesions disrupted retention of a previously learned discrimination in which rats had to detect a change in the location of the object in a scene, but had no effect in a previously learned discrimination in which the rat had to detect a change in one of the objects (DeCoteau and Kesner 1998). Finally, rats with PPC lesions do not react to a change consisting of removing a stimulus requiring a retrieval dependent pattern completion process (Save et al. 1992).

Other examples of a role for PPC in storing spatial information into long-term memory include a study by Kesner et al. (1987), who had shown that in an 8-arm maze task PPC lesions placed in rats after training on 4 unbaited and 4 baited arms resulted in a deficit in retrieval from knowledge-based memory, but not from event-based memory. If one assumes that the presentation of unbaited arms reflects the operation of long-term memory and that the presentation of baited arms reflects the operation of event-based memory, then lesions of the PPC only disrupted long-term memory, but not event-based memory.

Finally, there is evidence to suggest that the parietal cortex may be a site for long-term representation of complex spatial information. Cho and Kesner (1996), Cho et al. 1995) have shown that rats with parietal cortex lesions have a non-graded retrograde amnesia for four, but not two previously learned spatial discriminations prior to surgery, suggesting that the deficit cannot be due to a performance or anterograde amnesia problem, but rather appears to be a function of the number or complexity of the spatial information to be stored and to be remembered.

In summary, within the knowledge-based memory system different brain regions process different attributes in support of perceptual processes. Data are presented to support this assertion by demonstrating that the PPC mediates the spatial attribute for spatial perceptual information and spatial recognition, whereas the TE2 cortex mediates the sensory-perceptual attribute for visual object information and visual object recognition.

3 Rule-Based Memory System

The model assumes that different brain regions subserve the processing of different attributes within the rule-based memory system. I selected a variety of tasks to illustrate this, because processing of mnemonic information is likely to incorporate rules and strategies and is associated with the emphasis on PFC function. Wise et al. (1996) suggested that the subregions of the PFC can be divided on the basis of rules and strategies. Furthermore, they proposed a hierarchy in terms of the complexity of the rules required, which they labeled lower order, higher order, and highest order. I have proposed that the PFC in the rat can be fractionated in terms of functions associated with a slightly revised rule model that incorporates the rule-based memory system component of the attribute model (Kesner 2000a).

One can organize the subregions of the prefrontal cortex (PFC) in the rat according to the schema proposed by Uylings and van Eden (1990). These subregions include the medial PFC which can be subdivided into a dorsal medial region including the PC, the dorsal and ventral anterior cingulate cortices (AC), and a ventral medial region including the prelimbic and infralimbic as well as medial orbital cortices (PL-IL/MO), the lateral PFC which includes the dorsal and ventral agranular insular and the lateral orbital cortices (AI/LO), and the ventral PFC which includes the ventral orbital and ventrolateral orbital cortices (VLO/VO).

3.1 Precentral Cortex (PC)

The PC cortex appears to play an important role in working memory for motor responses requiring temporal processing of information, and paired-associate learning. Supporting evidence is based on the findings that lesions of the AC and PC cortex that spare the PL-IL/MO cortex produce a deficit in working memory for motor response information such as working memory for a motor (right-left turn) response (Kesner et al. 1996), acquisition of an egocentric turn response (Kesner et al. 1989), and acquisition of visual-motor associative conditional discriminations (Passingham et al. 1988; Winocur 1991; Winocur and Eskes 1998).

3.2 Anterior Cingulate Cortex (AC)

The AC cortex appears to play an important role in memory requiring temporal processing of information, and paired-associate learning. These lesions disrupt performance associated with processing of information in complex tasks, such as memory for temporal order of spatial information (Chiba et al. 1994, 1997; Kesner and Holbrook 1987; Kesner 1998a), memory for frequency information (Kesner 1990), use of a prospective code in a spatial 12-arm working memory task (Kesner 1989), and working memory for a list of five spatial locations (Kesner and Holbrook 1987).

The AC and PC cortex lesions, however, do not disrupt acquisition of visual, spatial or olfactory discrimination (Eichenbaum et al. 1983; Harrison and Mair 1996;

Ragozzino et al. 1999a), spatial discrimination reversal, cross-modal switching from visual cue to place or place to visual cue, switching between win-stay and win-shift rules or switching from a delayed nonmatching to sample to a delayed matching to sample rule (Harrison and Mair 1996; Neave et al. 1994; Ragozzino et al. 1998), spatial location navigation (deBruin et al. 1997; Kesner et al. 1989; King and Corwin 1992), working memory for visual object (Ennaceur et al. 1997; Kesner et al. 1996; Shaw and Aggleton 1993), duration (Jackson et al. 1998), or affect information (Decoteau et al. 1997). There are also no deficits, with a few exceptions, in working memory for spatial information using delayed nonmatching to position, delayed spatial alternation or nonmatching to sample in a T-maze, 8-arm maze, or continuous spatial recognition memory procedure (Ennaceur et al. 1997; Kesner et al. 1996; Ragozzino et al. 1998). Thus, the data suggest that the AC and PC cortex process rule-dependent working memory for motor response information, conditioned learning with response association as an important component to be learned, and/or higher order cognitive processes, but do not process rule-dependent working memory for visual object, spatial, affect (taste), or time as duration information as well as intra-modal or cross-modal shifting of set and acquisition of spatial location navigation.

3.3 Prelimbic and Infralimbic Plus Medial Orbital Cortex (PL-IL/MO)

The PL-IL/MO cortex appears to play an important role in working memory for visual object and spatial location information as well as rules associated with cross-modal set switching. Supporting evidence is based on the findings that lesions of the PL-IL/MO cortex produced deficits in working memory for spatial information (Delatour and Gisquet-Verrier 1996; Ragozzino et al. 1998; Seamans et al. 1995), working memory for visual object information (Kesner et al. 1996), and cross-modal switching between place and visual cue or visual cue and place as well as motor response and place and place and motor response (Ragozzino et al. 1999a, b). These lesions, however, do not affect the acquisition of spatial, motor response and visual discriminations, or visual, motor response and spatial intra-modal (reversal) learning (Bussey et al. 1997; Ragozzino et al. 1998; 1999b) or learning of spatial location navigation (Maaswinkel et al. 1996), working memory for affect or motor response (DeCoteau et al. 1997; Ragozzino and Kesner 1998), and no deficit in a visual-response conditional associative task (Bussey et al. 1996). Thus, the data suggest that the PL-IL/MO cortex mediates working memory for spatial and visual object information as well as cross-modal switching involving spatial locations and visual objects as well as spatial locations and motor responses, but is not involved in motor response working memory, visual-response conditional processing, or intra-modal switching.

3.4 Agranular Insular and Lateral Orbital Cortex (AI/LO)

Based on anatomical and behavioral data, the AI/LO cortex appears to play an important role in working memory for affect information usually involving odor and taste. Supporting evidence is based on the findings that lesions of the AI/LO cortex produce deficits in working memory for affect based on taste or odor information (DeCoteau et al. 1997; Otto and Eichenbaum 1992; Ragozzino and Kesner 1999). There is also some evidence that this region plays a role in mediation of cross-modal associations in that many neurons within the AI/LO region fire differentially for a cross-modal association between odors and locations (Lipton et al. 1999). There are also deficits in acquisition and retention of a tactile-odor configuration task (Whishaw et al. 1992). However, there are mild or no significant deficits in odor discrimination or taste preferences (DeCoteau et al. 1997; Eichenbaum et al. 1983; Whishaw et al. 1992), in spatial working memory (Eichenbaum et al. 1983; Ragozzino and Kesner 1998) and in learning a spatial location navigation task (Corwin et al. 1994). Also, there are no deficits in spatial discrimination or its reversal (Harrison and Mair 1996). Analysis of single cell recording from the agranular insular, lateral orbital and ventrolateral orbital cortices revealed that there are cells that respond primarily when the animal makes a reliable shift to perform in a go no-go olfactory discrimination task. A few cells reverse their firing selectivity during reversal training, but the exact location of these cells within the agranular insular, lateral orbital and ventrolateral orbital cortex was not specified (Schoenbaum et al. 1999). Thus, the data suggest that the AI/LO cortex mediates working memory for odor and taste information as well as cross-modal associations with odor and other sensory modalities, but is not involved in spatial processing of information. There is not much data available for the contribution of the ventral orbital and ventrolateral orbital cortices, but lesions in this area in conjunction with lateral orbital cortex contribute to reversal learning (Kim and Ragozzino 2005; McAlonan and Brown 2003).

In summary, based on the Wise et al. (1996) rule model, the PC cortex supports higher order rules for motor responses, the AC cortex supports the highest rules for temporal ordering, paired-associate learning, list learning and planning that include the use of temporal and prospective strategies, the PL-IL/MO cortex supports higher order rules for spatial and visual object information, the AI/LO cortex supports higher order rules for odor and taste information, and the VLO/VO cortex supports lower order rules.

Also there are dissociations based on different attributes characterizing the contribution of (a) the response memory attribute mediated by the PC, but not the AC, PL-IL/MO or AI/LO cortical regions, (b) the temporal memory attribute mediated by AC, but not PC, PL-IL/MO, or AI/LO cortical regions, (c) the object and spatial memory attributes mediated by the PL-IL/MO region, but not the PC, AC, or AI/LO cortical regions, and (d) the affect memory attribute mediated by AI/LO but not PL-IL/MO cortical region. There is also a clear correspondence between rats and humans in terms of mediation of the above-mentioned attributes. For more details see Kesner (2000a) and Kesner and Churchwell (2011).

4 Interactions Between Event-Based Memory and Rule-Based Memory

Are there interactions between the event-based memory system (e.g., hippocampus) and rule-based memory system (e.g., IL/PL)? I present two examples based on temporal processing of information. In the first study, Lee and Kesner (2003) examined the dynamic interactions between the PFC and hippocampus by training and testing rats on a delayed nonmatching-to-place task on an 8-arm radial maze. Rats had to remember a single spatial location following short-term delays (i.e., 10 s or 5 min). The results showed that inactivating both regions at the same time resulted in a severe impairment of short-term and intermediate memory for spatial information suggesting that one of the structures needs to function properly for intact processing of short- or intermediate-term spatial memory. Thus, the two regions interact with each other to ensure the processing of spatial information across a dynamic temporal range including both short- and intermediate-term memory. These results provide compelling evidence indicating that a mnemonic time-window is a critical factor in dissociating the function of the hippocampal system from that of the medial PFC in a delayed choice task. That is, the dorsal hippocampus and medial PFC appear to process spatial memory in parallel within a short-term range, whereas the dorsal hippocampal function becomes more essential once the critical time-window requires spatial memory for a time period exceeding that range. In the second study, rats were also trained on a spatial delayed non-match to-sample working memory task using short (10 s) and long (5 min) time delays to evaluate the hypothesis that the intermediate CA1 region of the HPC (iCA1) and prelimbic cortex (PL) interact and operate in parallel under different temporal working memory constraints. In order to assess the functional role of these structures, an inactivation strategy was used in which each subject received bilateral chronic cannula implantation of the iCA1 and PL, allowing one to perform bilateral, contralateral, ipsilateral, and combined bilateral inactivation of structures and structure pairs within each subject. Compared to saline infusions, rats receiving contralateral infusions of muscimol into PL and iCA1 displayed impairment for the 5 min delay, but not the10 s delay. In contrast, rats receiving ipsilateral infusions of muscimol into PL and iCA1 displayed no impairment at either delay. These results suggest that there is an interaction in terms of temporal processing of information between the PL and iCA. However, bilateral infusions of muscimol into both PL and iCA1 resulted in a deficit at both the 5 min and 10 s delay, suggesting that either structure may independently represent spatial information sufficient to successfully complete the task (Churchwell and Kesner 2011). This result is similar to what was reported by Lee and Kesner (2003). The findings of these studies suggest that there are interactions and parallel processing of temporal information between the event-based and rule-based memory systems. From an anatomical point of view, there is a direct one way connection from iCA to PL region (Jay and Witter 1991) and information from PL can reach the hippocampus either via nucleus reuniens or entorhinal cortex (Vertes 2006). This circuit could subserve the functions described above.

5 Interactions Between Knowledge-Based Memory and Event-Based Memory

Are there interactions and dissociations between different attributes within the knowledge-based memory system (e.g., PPC) and event-based memory system (e.g., hippocampus)? I present two experiments dissociating knowledge-based perceptual memory versus event-based memory processing of information as well as experiments examining binding of objects and places.

In the first experiment, two spatial continuous recognition training procedures designed to query knowledge-based perceptual memory and event-based episodic memory were employed. A continuous recognition procedure was used to train rats on a 12-arm radial maze. The details of the experimental protocol can be found in the knowledge-based memory section. After training, rats received PPC, hippocampus, or sham-operated and cortical control lesions. After retesting, the results indicated that relative to control and pretraining performance, the PPC-lesioned rats were impaired in the knowledge-based perceptual memory condition, but showed no deficits in the event-based episodic memory condition. In contrast, the hippocampal-lesioned rats were impaired in the event-based episodic memory condition, but showed no deficits in the knowledge-based perceptual memory condition (Chiba et al. 2002).

In order to have an even better measure of knowledge-based perceptual memory, a new experiment was generated to measure positive as well as negative repetition priming for spatial locations in rats similar to paradigms used with humans. Based on 48 repetition trials, all rats in the positive priming condition ran more quickly to the repeated spatial location. In the negative priming condition, it was assumed that rats not only actively attend to the positive stimulus, but also actively inhibit responding to the negative stimulus (Neill and Mathis 1995). Based on 48 repetition trials, all rats in the negative priming condition ran more slowly to the repeated spatial location, because the correct location had resulted in some inhibition on the previous trial. After training, rats received PPC lesions and then were retested. The results indicate that PPC-lesioned rats are impaired for both positive and negative priming (Kesner 2000b). In the positive priming paradigm, different rats received lesions of the hippocampus (Kesner 2000b). The results indicate that rats with hippocampal lesions showed normal positive priming. Thus, it appears that the PPC, but not the hippocampus, is directly involved in knowledge-based perceptual memory for spatial location information. The observation that the PPC does not mediate event-based episodic memory is supported by the observations that PPC lesions do not disrupt performance in a five-choice serial reaction time task (Muir et al. 1996). The data of both experiments suggest that there is a double dissociation between the two systems indicating that the two systems can operate independent of each other. Thus, a double dissociation appears to exist between PPC and hippocampus for knowledge-based perceptual memory versus event-based episodic memory operations, suggesting that the two neural circuits mediated by the hippocampus and PPC can operate independent of each other. This functional independence would require that spatial information reach the hippocampus and PPC via separate neural pathways. Indeed spatial information that reaches the dorsal lateral thalamus in the rat can be directed to the hippocampus via connections with the pre and parasubiculum and medial entorhinal cortex and the PPC via direct connections. In the rat there are no direct connections between the PPC and hippocampus. The parietal cortex and the hippocampus can interact via the entorhinal cortex or retrosplenial cortex and pre- and parasubiculum (Kohler 1985; Van Groen and Wyss 1990; Witter et al. 1989).

A second possible role for the rodent parietal cortex could be to bind across modalities to maintain the association between landmark and spatial location information. In other words, the parietal cortex may not be involved in memory for a single landmark or a single spatial location, but rather in the processing that assigns a specific landmark to a specific spatial location. To test this hypothesis, rats with small lesions of the parietal cortex were tested in an object/spatial location paired-associate task that required concurrent memory for both object and spatial location information. In addition, memory for landmark only or spatial location only information was also assessed. A deficit in the paired-associate task (which requires memory for both landmark and spatial location information), in the absence of deficits in either the landmark or the spatial location only memory, would support the idea that the PPC is involved in the memory for the combination of landmark and spatial location information. The results indicated that small lesions of the PPC as defined by Reep, Chandler, King, and Corwin (1994) and larger PPC lesions disrupted learning of the object-place paired-associate task, but did not disrupt the learning of a spatial or object discrimination (Long and Kesner 1998). Furthermore, lesions of the hippocampus and especially the CA3 subregion of the hippocampus disrupted object-place paired-associate learning (Gilbert and Kesner 2002b, 2003b; Long et al. 1998), although it should be noted that a disruption only occurs when one component of the paired-associate is a spatial location. In a subsequent study unilateral lesions were made to the dorsal hippocampus or posterior PC contralaterally or ipsilaterally. It was hypothesized that if the hippocampus and PC interact, then contralateral lesioned animals should be markedly impaired compared to ipsilateral lesions. The results indicate that contralateral lesioned animals were significantly more impaired than animals with ipsilateral lesions during object-place paired-associate learning; however, both groups readily learned single discriminations (i.e., objects or places) (Rogers and Kesner 2007). These results suggest that in this case there is an interaction between the PPC and hippocampus.

It appears that both parallel and interactive processing of information characterize the relationships between the PPC (a component of the knowledge-based memory system) and hippocampus (a component of the event-based memory system).

6 Interactions Between Knowledge-Based Memory and Rule-Based Memory

Are there interactions between different attributes within the knowledge-based memory system (e.g., PPC) and rule-based memory system (e.g., PFC)? I selected egocentric versus allocentric spatial processing to illustrate possible interactions between the knowledge-based and rule-based memory systems. Rats with medial PFC or parietal cortex lesions and sham-operated and nonoperated controls were tested for the acquisition of an adjacent arm task where the rats were placed at the end of a randomly selected arm in an 8-arm radial maze and trained to run to the adjacent right or left arm to receive a reinforcement. This task accentuated the importance of egocentric spatial localization. In a second task, a cheese board spatial navigation task that accentuated the importance of allocentric spatial localization was used. Results indicated that relative to controls, animals with medial prefrontal cortex lesions were impaired on the adjacent arm task but displayed facilitation on the cheese board task. In contrast, relative to controls, rats with parietal cortex lesions were impaired on the cheese board task but showed no impairment on the adjacent arm task. (Kesner et al. 1989; King and Corwin 1992). The data suggest a double dissociation of function between medial prefrontal cortex and parietal cortex in terms of coding of egocentric versus allocentric spatial information. However, there are data to suggest that in a less structured task such as the water maze, that the PPC can also mediate egocentric spatial information. For example, Save and Poucet (2000) showed that in the Morris water maze PPC-lesioned rats were impaired in finding a hidden platform when three salient cues were located in the pool close to the correct location (proximal cues), but they were not impaired when only room cues (distal cues) were available to find the platform. Kolb and Walkey (1987) showed that PPC-lesioned rats were impaired in finding a platform location in a landmark task in which the rats had to associate a visual cue with a site that was spatially discontiguous and where the relevant cue moved relative to the rest of the extra-maze cues. This impairment manifested itself in the adoption of a looping strategy to locate a hidden platform. Foreman et al. (1992) found that the trajectories of rats turning and running between familiar visible targets at opposite ends of an area were less accurate in PPC-lesioned rats than in controls.

It appears that both parallel and potential interactive processing of information characterize the relationships between the PPC (a component of the knowledge-based memory system) and PFC (a component of the rule-based memory system).

Even though the event-based, knowledge-based, and rule-based memory systems are supported by neural substrates and different operating characteristics, the systems can operate independent of each other and there are also important interactions between the three systems. Clearly for each attribute, there is a neural circuit that encompasses all three memory systems in representing specific attribute information. I will present one example depicting the neural substrates and their

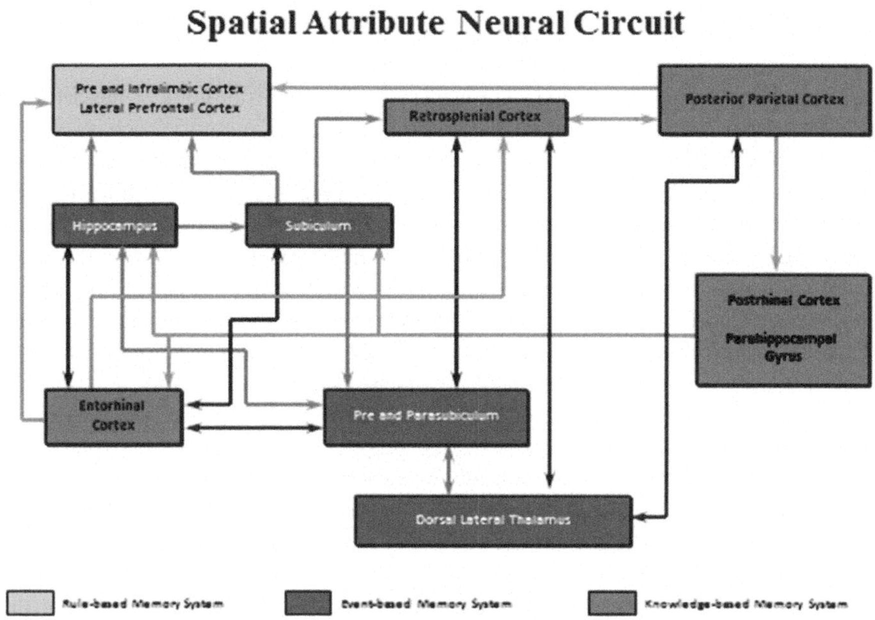

Fig. 12 A representation of the spatial attribute neural circuit incorporating neural regions that mediate event-based, knowledge-based, and rule-based memory

interconnections associated with the spatial (place) attribute across all three memory systems (see Fig. 12). Note that the dorsal lateral thalamus, pre- and parasubiculum, hippocampus, and subiculum represent neural substrates that support the event-based memory system, the entorhinal cortex, parahippocampal gyrus or postrhinal cortex, PPC, and retrosplenial cortex support the knowledge-based memory system, and the lateral PFC or pre- and infralimbic cortex support the rule-base memory system. This circuit provides anatomical support for a possible independence in the operation of the hippocampus as part of the event-based memory system and PPC as part of the knowledge-based memory system in that spatial information that is processed via the dorsal lateral thalamus can activate both the hippocampus and the PPC in parallel. Also, information can reach the lateral PFC or pre- and infralimbic cortex as part of the rule-based memory system via direct connections from the PPC part of the knowledge-based memory system and hippocampus as part of the event-based memory system. Finally, spatial information can interact with other specific attributes via a series of direct connections including, for example, an interaction with reward value attribute information via hippocampus–amygdala connection or lateral prefrontal cortex-orbital frontal cortex connections and an interaction with response attribute information via hippocampus–caudate or lateral prefrontal–premotor and supplementary motor connections.

7 Conclusion

In this chapter, I have presented data in support of a neurobiological basis for an attribute model based on different forms or attributes of memory such as *space, time, response, sensory-perception, reward value (affect)* and in humans a *language* attribute is also added. These attributes are processed by different neural regions and interconnected networks across all three (event-based, knowledge-based, and rule-based) memory systems. The model is a major extension of previous mentioned brain-based memory models (Nadel 1994; Olton 1983; Tulving 1983; Squire 1994; Cohen and Eichenbaum 1993). Each memory system operates the processing of mnemonic information based on a unique set of processes. The selection of some of these processes has been influenced greatly by computational models of specific brain regions. For each brain area, there are a large number of processes that define the operation of each memory system. The hippocampus is used extensively, but not exclusively, to detail the multiple operations that characterize the overall activity of this brain region within the event-based memory system. The processes that are discussed for the event-based memory system include conjunctive encoding, spatial pattern separation, formation of arbitrary associations, pattern completion, and temporal pattern separation. The processes that are discussed for the knowledge-based memory system include perceptual memory and repetition priming. For the rule-based memory system, the process of working memory is presented. Furthermore, based on brain-behavior experiments, there are interactions and parallel processing operations between the event-based memory system and the knowledge-based systems, between the event-based and rule-based memory systems, and between the rule-based and knowledge-based systems.

References

Bussey TJ, Muir JL, Everitt BJ, Robbins TW (1996) Dissociable effects of anterior and posterior cingulate cortex lesions on the acquisition of a conditional visual discrimination: facilitation of early learning vs. impairment of late learning. Behav Brain Res 82:45–56

Bussey TJ, Muir JL, Everitt BJ, Robbins TW (1997) Triple dissociation of anterior cingulate, posterior cingulate, and medial frontal cortices on visual discrimination tasks using a touchscreen testing procedure for the rat. Behav Neurosci 111:920–936

Chiba AA, Kesner RP, Reynolds AM (1994) Memory for spatial location as a function of temporal lag in rats: role of hippocampus and medial prefrontal cortex. Behav Neural Biol 61:123–131

Chiba AA, Kesner RP, Gibson CJ (1997) Memory for temporal order of new and familiar spatial location sequences: Role of the medial prefrontal cortex. Learning and Memory 4:311–317

Chiba AA, Kesner RP, Jackson P (2002) Two forms of spatial memory: a double dissociation between the parietal cortex and the hippocampus in the rat. Behav Neurosci 116:874–883

Cho YH, Kesner RP (1996) Involvement of entorhinal cortex or parietal cortex in long-term spatial discrimination memory in rats: retrograde amnesia. Behav Neurosci 110:436–442

Cho YH, Kesner RP, Brodale S (1995) Retrograde and anterograde amnesia for spatial discrimination in rats: role of hippocampus, entorhinal cortex and parietal cortex. Psychobiology 23:185–194

Churchwell JC, Kesner RP (2011) Hippocampal-prefrontal dynamics in spatial working memory: interactions and independent parallel processing. Behav Brain Res 225:389–395
Cohen NJ, Eichenbaum HB (1993) Memory, amnesia, and hippocampal function. MIT Press, Cambridge
Corwin JV, Fussinger M, Meyer RC, King VR, Reep RL (1994) Bilateral destruction of the ventrolateral orbital cortex produces allocentric but not egocentric spatial deficits in rats. Behav Brain Res 61:79–86
Day M, Langston R, Morris RG (2003) Glutamate-receptor-mediated encoding and retrieval of paired-associate learning. Nature 424:205–209
deBruin JPC, Swinkels WAM, deBrabander JM (1997) Response learning of rats in a Morris water maze: Involvement of the medial prefrontal cortex. Behav Brain Res 85:47–55
DeCoteau WE, Kesner RP (1998) Effects of hippocampal and parietal cortex lesions on the processing of multiple object scenes. Behav Neurosci 112:68–82
DeCoteau WE, Kesner RP, Williams JM (1997) Short-term memory for food reward magnitude: the role of the prefrontal cortex. Behav Brain Res 88:239–249
Delatour B, Gisquet-Verrier P (1996) Prelimbic cortex specific lesions disrupt delayed variable response tasks in the rat. Behav Neurosci 110:1282–1298
DiMattia BV, Kesner RP (1988) Spatial cognitive maps: Differential role of parietal cortex and hippocampal formation. Behav Neurosci 102:471–480
Eichenbaum H (1994) The hippocampal system and declarative memory in humans and animals: experimental analysis and historical origins. In Schacter DL, Tulving E (eds) Memory systems 1994. MIT Press, Cambridge, pp 39–63
Eichenbaum H (2004) Hippocampus: cognitive processes and neural representation that underlie declarative memory. Neuron 44:109–120
Eichenbaum H, Clegg RA, Feeley A (1983) Reexamination of functional subdivisions of the rodent prefrontal cortex. Exp Neurol 79:434–451
Ennaceur A, Neave N, Aggleton JP (1997) Spontaneous object recognition and object location memory in rats: the effects of lesions in the cingulate cortices, the medial prefrontal cortex, the cingulum bundle and the fornix. Exp Brain Res 113:509–519
Estes WK (1986) Memory for temporal information. In: Michon JA, Jackson JL (eds) Time, mind and behavior. Springer, New York, pp 151–168
Ferbinteanu J, Holsinger RM, McDonald RJ (1999) Lesions of the medial or lateral perforant path have different effects on hippocampal contributions to place learning and fear conditioning to context. Behav Brain Res 101:65–84
Foreman N, Save E, Thinus-Blanc C, Buhot MC (1992) Visually guided locomotion, distractibility, and the missing-stimulus effect in hooded rates with unilateral or bilateral lesions of parietal cortex. Behav Neurosci 106:529–538
Gilbert PE, Kesner RP (2002a) The amygdala but not the hippocampus is involved in pattern separation based on reward value. Neurobiol Learn Mem 77:338–353
Gilbert PE, Kesner RP (2002b) Role of the rodent hippocampus in paired-associate learning involving associations between a stimulus and a spatial location. Behav Neurosci 116:63–71
Gilbert PE, Kesner RP (2003a) Recognition memory for complex visual discrimination is influenced by stimulus interference in rodents with perirhinal cortex damage. Learn Mem 10:525–530
Gilbert PE, Kesner RP (2003b) Localization of function within the dorsal hippocampus: the role of the CA3 subregion in paired-associate learning. Behav Neurosci 117:1385–1394
Gilbert PE, Kesner RP (2006) The role of dorsal CA3 hippocampal subregion in spatial working memory and pattern separation. Behav Brain Res 169:142–149
Gilbert PE, Kesner RP, Lee I (2001) Dissociating hippocampal subregions: a double dissociation between dentate gyrus and CA1. Hippocampus 11:626–636
Gold E, Kesner RP (2005) The role of the CA3 subregion of the dorsal hippocampus in spatial pattern completion in the rat. Hippocampus 15:808–814
Hafting T, Fyhn M, Molden S, Moser MB, Moser EI (2005) Microstructure of a spatial map in the entorhinal cortex. Nature 436:801–806

Hargreaves EL, Rao G, Lee I, Knierim JJ (2005) Major dissociation between medial and lateral entorhinal input to dorsal hippocampus. Science 308:1792–1794
Harrison LD, Mair RG (1996) A comparison of the effects of frontal cortical and thalamic lesions on measures of spatial learning and memory in the rat. Behav Brain Res 75:195–206
Hasselmo ME, Wyble BP (1997) Free recall and recognition in a network model of the hippocampus: simulating effects of scopolamine on human memory function. Behav Brain Res 89:1–34
Ho JW-T, Narduzzo KE, Outram A, Tinsley CJ, Henley JM, Warburton EC, Brown MW (2011) Contributions of area Te2 to rat recognition memory. Learn Mem 18:493–501
Hunsaker MR, Mooy GG, Swift JS, Kesner RP (2007) Dissociations of the medial and lateral perforant path projections into dorsal DG, CA3, and CA1 for spatial and nonspatial (visual object) information processing. Behav Neurosci 121:742–750
Hunsaker MR, Lee B, Kesner RP (2008) Evaluating the temporal context of episodic memory: the role of CA3 and CA1. Behav Brain Res 188:310–315
Jackson PA, Kesner RP, Amann K (1998) Memory for duration: role of hippocampus and medial prefrontal cortex. Neurobiol Learn Mem 70:328–348
Jay TM, Witter MP (1991) Distribution of hippocampal CA1 and subicular efferents in the prefrontal cortex of the rat studied by means of anterograde transport of Phaseolus vulgaris-leucoagglutinin. J Comp Neurol 313:574–586
Kesner RP (1989) Retrospective and prospective coding of information: role of the medial prefrontal cortex. J Exper Brain Res 74:163–167
Kesner RP (1990) Memory for frequency in rats: role of the hippocampus and medial prefrontal cortex. Behav Neural Biol 53:402–410
Kesner RP (1998a) Neural mediation of memory for time: role of the hippocampus and medial prefrontal cortex. Psychon Bull Rev 5:585–596
Kesner RP (1998b) Neurobiological views of memory. In: Martinez JL, Kesner RP (eds) The neurobiology of learning and memory. Academic, San Diego, pp 361–416
Kesner RP (1999) Perirhinal cortex and hippocampus mediate parallel processing of object and spatial location information. Behav Brain Sci 22:455–479
Kesner RP (2000a) Subregional analysis of mnemonic functions of the prefrontal cortex in the rat. Psychobiology 28:219–228
Kesner RP (2000b) Behavioral analysis of the contribution of the hippocampus and parietal cortex to the processing of information: Interactions and dissociations. Hippocampus 10:483–490
Kesner RP (2007) Neurobiological views of memory. In: Kesner RP, Martinez JL (eds) The neurobiology of learning and memory, 2nd edn. Academic Press, San Diego, CA, pp 271–304
Kesner RP, Churchwell JC (2011) An analysis of rat prefrontal cortex in mediating executive function. Neurobiol Learn Mem 96:417–431
Kesner RP, Creem-Regehr SH (2013) Parietal contributions to spatial cognition. In: Nadel L, Waller D (eds) Handbook of spatial cognition. American Psychological Association, Washington, pp 35–64
Kesner RP, Gilbert P (2006) The role of the medial caudate nucleus, but not the hippocampus, in a matching-to sample task for a motor response. Eur J Neurosci 23:1888–1894
Kesner RP, Holbrook T (1987) Dissociation of item and order spatial memory in rats following medial prefrontal cortex lesions. Neuropsychologia 25:653–664
Kesner RP, Hopkins RO (2006) Mnemonic functions of the hippocampus: a comparison between animals and humans. Biol Psychol 73:3–18
Kesner RP, Warthen DK (2010) Implications of CA3 NMDA and opiate receptors for spatial pattern completion in rats. Hippocampus 20:550–557
Kesner RP, DiMattia BV, Crutcher KA (1987) Evidence for neocortical involvement in reference memory. Behav Neural Biol 47:40–53
Kesner RP, Farnsworth G, DiMattia BV (1989) Double-dissociation of egocentric and allocentric space following medial prefrontal and parietal cortex lesions in the rat. Behav Neurosci 103:956–961

Kesner RP, Fansworth G, Kametani H (1991) Role of parietal cortex and hippocampus in representing spatial information. Cereb Cortex 1:367–373
Kesner RP, Bolland B, Dakis M (1993) Memory for spatial locations, motor responses, and objects: triple dissociations among the hippocampus, caudate nucleus and extrastriate visual cortex. Exp Brain Res 93:462–470
Kesner RP, Hunt ME, Williams JM, Long JM (1996) Prefrontal cortex and working memory for spatial response, spatial location, and visual object information in the rat. Cereb Cortex 6: 311–318
Kesner RP, Lee I, Gilbert P (2004) A behavioral assessment of hippocampal function based on a subregional analysis. Reviews Neuroscience 15:333–351
Kesner RP, Hunsaker MR, Warthen MW (2008) The CA3 subregion of the hippocampus is critical for episodic memory processing by means of relational encoding in rats. Behav Neurosci 122:1217–1225
Kim J, Ragozzino ME (2005) The involvement of the orbitofrontal cortex in learning under changing task contingencies. Neurobiol Learn Mem 83:125–133
King VR, Corwin JV (1992) Comparison of hemi-inattention produced by unilateral lesions of the posterior parietal cortex or medial agranular prefrontal cortex in rats: neglect, extinction, and the role of stimulus distance. Behav Brain Res 54:117–131
Kohler C (1985) Intrinsic projections of the retrohippocampal region in the rat brain. I. The subicular complex. J Comp Neurol 236:504–522
Kolb B, Walkey J (1987) Behavioural and anatomical studies of the posterior parietal cortex in the rat. Behav Brain Res 23:127–145
Lee I, Kesner RP (2003) Time-dependent relationship between the dorsal hippocampus and the prefrontal cortex in spatial memory. J Neurosci 23:1517–1523
Levy WB (1996) A sequence predicting CA3 is a flexible associator that learns and uses context to solve hippocampal-like tasks. Hippocampus 6:579–590
Lipton PA, Alvarez P, Eichenbaum H (1999) Crossmodal associative memory representation in rodent orbitofrontal cortex. Neuron 22:349–359
Long JM, Kesner RP (1998) The effects of hippocampal and parietal cortex lesions on memory for egocentric distance and spatial location information in rats. Behav Neurosci 112:480–495
Long JM, Mellem JE, Kesner RP (1998) The effects of parietal cortex lesions on an object/spatial location paired-associate task in rats. Psychobiology 26:128–133
Maaswinkel H, Gispen WH, Spruijut BM (1996) Effects of an electrolytic lesion of the prelimbic area on anxiety-related and cognitive tasks in the rat. Behav Brain Res 79:51–59
Marr D (1971) Simple memory: a theory for archicortex. Philos Trans R Soc B: Biol Sci 262: 23–81
McAlonan K, Brown VJ (2003) Orbital prefrontal cortex mediates reversal learning and not attentional set shifting in the rat. Behav Brain Res 146:97–103
McNaughton BL, Morris RGM (1987) Hippocampal synaptic enhancement and information storage within a distributed memory system. Trends Neurosci 10:408–415
Muir JL, Everitt BJ, Robbins TW (1996) The cerebral cortex of the rat and visual attentional function: dissociable effects of mediofrontal, cingulate, anterior dorsolateral, and parietal cortex lesions on a five-choice serial reaction time ask. Cereb Cortex 6:470–481
Nadel L (1994) Multiple memory systems: what and why, an update. In Schacter DL, Tulving E (eds) Memory systems 1994. MIT Press, Cambridge, pp 39–63
Neave N, Lloyd S, Sahgal A, Aggleton JP (1994) Lack of effect of lesions in the anterior cingulate cortex and retrosplenial cortex on certain tests of spatial memory in the rat. Behav Brain Res 65:89–101
Neill WT, Mathis KM (1995) Transfer-inappropriate processing: negative priming and related phenomena. Psychol Learn Motiv 38:1–44
O'Reilly RC, McClelland JL (1994) Hippocampal conjunctive encoding, storage, and recall: avoiding a trade-off. Hippocampus 4:661–682
O'Keefe J, Nadel L (1978) The hippocampus as a cognitive map. Clarendon Press, Oxford

Olton DS (1983) Memory functions and the hippocampus. In: Seifert W (ed) Neurobiology of the hippocampus. Academic, New York

Otto T, Eichenbaum H (1992) Complementary roles of the orbital prefrontal cortex and the perirhinal-entorhinal cortices in an odor-guided delayed-nonmatching-to-sample task. Behav Neurosci 106:762–775

Passingham RE, Myers C, Rawlins N, Lightfoot V, Fearn S et al (1988) Premotor cortex in the rat. Behav Neurosci 102:101–109

Poucet B (1989) Object exploration, habituation, and response to a spatial change in rats following septal or medial frontal cortical damage. Behav Neurosci 103:1009–1016

Ragozzino ME, Kesner RP (1998) The effects of muscarinic cholinergic receptor blockade in the rat anterior cingulate and prelimbic/infralimbic cortices on spatial working memory. Neurobiol Learn Mem 69:241–257

Ragozzino ME, Kesner RP (1999) The role of the agranular insular cortex in working memory for food reward value and allocentric space in rats. Behav Brain Res 1:103–112

Ragozzino ME, Adams S, Kesner RP (1998) Differential involvement of the dorsal anterior cingulate and prelimbic-infralimbic areas of the rodent prefrontal cortex in spatial working memory. Behav Neurosci 112:293–303

Ragozzino ME, Wilcox C, Raso M, Kesner RP (1999a) Involvement of the medial prefrontal cortex subregions in strategy switching. Behav Neurosci 113:32–41

Ragozzino ME, Detrick S, Kesner RP (1999b) Involvement of the prelimbic-infralimbic areas of the rodent prefrontal cortex in behavioral flexibility for place and response learning. J Neurosci 19:4585–4594

Reep RL, Chandler HC, King V, Corwin JV (1994) Rat posterior parietal cortex: topography of cortico-cortical and thalamic connections. Exp Brain Res 100:67–84

Rogers JL, Kesner RP (2007) Hippocampal-parietal cortex interactions: evidence from a disconnection study in the rat. Behav Brain Res 179:19–27

Rolls ET (1996) A theory of hippocampal function in memory. Hippocampus 6:601–620

Rolls ET, Kesner RP (2006) A computational theory of hippocampal function, and empirical tests of the theory. Prog Neurobiol 79:1–48

Rolls ET, Treves A (1998) Neural networks and brain function. Oxford University Press, Oxford

Save E, Moghaddam M (1996) Effects of lesions of the associative parietal cortex on the acquisition and use of spatial memory in egocentric and allocentric navigation tasks in the rat. Behav Neurosci 110:74–85

Save E, Poucet B (2000) Involvement of the hippocampus and associative parietal cortex in the use of proximal and distal landmarks for navigation. Behav Brain Res 109:195–206

Save E, Buhot M-C, Foreman N, Thinus-Blanc C (1992) Exploratory activity and response to a spatial change in rats with hippocampal or posterior parietal cortical lesions. Behav Brain Res 47:113–127

Schacter DL (1987) Implicit memory: history and current status. J Exp Psychol Learn Mem Cogn 13:501–518

Schacter DL, Tulving E (eds) (1994) *Memory systems 1994*. MIT Press, Cambridge

Schoenbaum G, Chiba AA, Gallagher M (1999) Neural encoding in orbitofrotal cortex and basolateral amygdala during olfactory discrimination learning. J Neurosci 19:1876–1884

Seamans JK, Floresco SB, Phillips AG (1995) Functional differences between the prelimbic and anterior cingulate regions of the rat prefrontal cortex. Behav Neurosci 109:1063–1073

Shapiro ML, Olton DS (1994) Hippocampal function and interference. In: Schacter DL, Tulving E (eds) Memory systems 1994. MIT Press, Cambridge, pp 141–146

Shaw C, Aggleton JP (1993) The effects of fornix and medial prefrontal lesions on delayed non-matching-to-sample by rats. Behav Brain Res 54:91–102

Spear NF (1976) Retrieval of memories: a psychobiological approach In Estes WK (ed) Handbook of learning and cognitive processes (vol 4). Attention and memory. Erlbaum, Hillsdale

Squire LR (1994) Declarative and nondeclarative memory: multiple brain systems supporting learning and memory. In: Schacter DL, Tulving E (eds) Memory systems 1994. MIT Press, Cambridge, pp 203–231

Squire LR, Stark CE, Clark RE (2004) The medial temporal lobe. Annu Rev Neurosci 27:279–306
Tees RC (1999) The effects of posterior parietal cortex and posterior temporal cortical lesions on multimodal spatial and nonspatial competencies in rats. Behav Brain Res 106:55–73
Tulving E (1983) *Elements of episodic memory*. Clarendon, Oxford
Underwood BJ (1969) Attributes of memory. Psychol Rev 76:559–573
Uylings HBM, van Eden CG (1990) Qualitative and quantitative comparison of the prefrontal cortex in rat and in primates, including humans. Prog Brain Res 85:31–61
Van Groen T, Wyss JM (1990) The connections of presubiculum and parasubiculum in the rat. Brain Res 518:227–243
Vertes RP (2006) Interactions among the medial prefrontal cortex, hippocampus and midline thalamus in emotional and cognitive processing in the rat. Neuroscience 142:1–20
Weeden CSS, Hu NJ, Ho LUN, Kesner RP (2014) The role of the ventral dentate gyrus in olfactory pattern separation. Hippocampus 24:553–559
Whishaw IQ, Tomie J, Kolb B (1992) Ventrolateral prefrontal cortex lesions in rats impair the acquisition and retention of a tactile-olfactory configural task. Behav Neurosci 106:597–603
Winocur G (1991) Functional dissociation of the hippocampus and prefrontal cortex in learning and memory. Psychobiology 19:11–20
Winocur G, Eskes G (1998) Prefrontal cortex and caudate nucleus in conditional associative learning: dissociated effects of selective brain lesions in rats. Behav Neurosci 112:89–101
Wise SP, Murray EA, Gerfen CR (1996) The frontal cortex-basal ganglia system in primates. Crit Rev Neurobiol 10:317–356
Witter MP, Groenewegen HJ, Lopes da Silva FH, Lohman AH (1989) Functional organization of the extrinsic and intrinsic circuitry of the parahippocampal region. Prog Neurobiol 33:161–253

The manufacturer's authorised representative in the EU is Springer Nature Customer Service Centre GmbH, Europaplatz 3, 69115 Heidelberg, Germany. If you have any concerns regarding our products, please contact ProductSafety@springernature.com

Printed and bound by CPI Group (UK) Ltd, Croydon, CR0 4YY
25/09/2026
02222857-0013